I0051660

Metal-Organic Framework Nanocomposites

Metal-Organic Framework Nanocomposites

From Design to Application

Edited by
Anish Khan, Mohammad Jawaid,
Abdullah Mohammed Ahmed Asiri,
Wei Ni, and Mohammed Muzibur Rahman

CRC Press
Taylor & Francis Group
Boca Raton London New York

CRC Press is an imprint of the
Taylor & Francis Group, an **informa** business

First edition published 2021
by CRC Press
6000 Broken Sound Parkway NW, Suite 300, Boca Raton, FL 33487-2742

and by CRC Press
2 Park Square, Milton Park, Abingdon, Oxon, OX14 4RN

© 2021 Taylor & Francis Group, LLC

CRC Press is an imprint of Taylor & Francis Group, an Informa business

No claim to original U.S. Government works

Reasonable efforts have been made to publish reliable data and information, but the author and publisher cannot assume responsibility for the validity of all materials or the consequences of their use. The authors and publishers have attempted to trace the copyright holders of all material reproduced in this publication and apologize to copyright holders if permission to publish in this form has not been obtained. If any copyright material has not been acknowledged please write and let us know so we may rectify in any future reprint.

Except as permitted under U.S. Copyright Law, no part of this book may be reprinted, reproduced, transmitted, or utilized in any form by any electronic, mechanical, or other means, now known or hereafter invented, including photocopying, microfilming, and recording, or in any information storage or retrieval system, without written permission from the publishers.

For permission to photocopy or use material electronically from this work, access www.copyright.com or contact the Copyright Clearance Center, Inc. (CCC), 222 Rosewood Drive, Danvers, MA 01923, 978-750-8400. For works that are not available on CCC please contact mpkbookspermissions@tandf.co.uk

Trademark notice: Product or corporate names may be trademarks or registered trademarks, and are used only for identification and explanation without intent to infringe.

ISBN: 978-0-367-35783-2 (hbk)
ISBN: 978-0-429-34626-2 (ebk)

Typeset in Times
by Deanta Global Publishing Services, Chennai, India

This book is dedicated
to all the
COVID-19 warriors around the world

Contents

Preface

Metal-organic frameworks (MOFs) are nanoporous polymers, consisting of inorganic metal focuses connected by natural ligands. These entities have gotten extraordinary consideration because of their exceptional physical and chemical properties that make them pertinent in various fields, such as designing, medicine, and the environment. Since combination conditions affect the properties of these compounds incredibly, it is specifically important to choose an appropriate synthesis technique that produces the product with homogeneous morphology, dispersion, and high thermal stability with a good surface area suitable for its broad applications in different fields. This book covers different aspects of composite fabrication for energy storage and catalysts including the preparation, design, characterization techniques of MOF materials and their application. The special features in this book consist of illustrations and tables that summarize up-to-date information on research carried out on manufacturing, design, characterization, and applications of MOF composites. This book assembles the information and knowledge on MOF materials and emphasizes the concept of the latest technology in the field of manufacturing and design. This book benefits lecturers, students, researchers, and industrialists who are working in the field of material science, especially MOF composites, as a valuable reference handbook for teaching, learning, and research. This book is a new edition and differs from the previous edition in its focus on the usage of composites. It elaborates the design and manufacturing process of MOF materials in all aspects. The latest trends in the application of MOF composites are described to the readers. The replacement of conventional composite materials with MOF composites in the area of manufacturing and design to achieve sustainable practice is highlighted with real applications in this book, which are not covered in previous editions.

The focus of this book is about the current demand of the MOF composites that respond to the environment just like biological entities. The specialism in MOFs is understood by its outstanding repairing and sustainable properties. Unlike other composites, this composite is combined with resins, carbon materials, and nanoparticles to form a MOF composite. MOF composite materials currently stand best among the replacement materials to conventional engineering composite materials because of their outstanding features like their light weight, lower cost, environmentally friendly properties, and sustainability. The unique feature of this book is that it presents a unified knowledge of the human-friendly material based on characterization, design, manufacture, and applications.

We are highly thankful to all authors who contributed chapters and provided their valuable ideas and knowledge in this edited book. We attempt to gather all the scattered information of authors from diverse fields around the world (Malaysia, Jordan, USA, Turkey, India, Saudi Arabia, Bangladesh, Oman, and Sweden) in the areas of metal composites and finally complete this venture in a fruitful way. We greatly appreciate the contributors' commitment and support to compiling the ideas. We are also highly thankful to the CRC team for their generous cooperation at every stage of the book production.

Editors

Anish Khan is an assistant professor in the Chemistry Department and Centre of Excellence for Advanced Materials Research, Faculty of Science, King Abdulaziz University, Jeddah, Saudi Arabia. In 2010 he earned a PhD at Aligarh Muslim University, India. He has research experience of working in the field of synthetic polymers and organic-inorganic electrically conducting nanocomposites. In 2010–2011 he completed postdoctoral work at the School of Chemical Sciences, University Sains Malaysia (USM) in electroanalytical chemistry. He has written more than 150 research papers, 20 books, and 70 book chapters published by a referred international publisher. He has attended more than 20 international conferences/workshops and completed around 20 research projects.

Mohammad Jawaid is a High Flyer Fellow (professor) at the Biocomposite Technology Laboratory, Institute of Tropical Forestry and Forest Products (INTROP), Universiti Putra Malaysia, Malaysia. He is also a visiting professor at the Department of Chemical Engineering, College of Engineering, King Saud University, in Riyadh, Saudi Arabia. His areas of research interests include hybrid reinforced/filled polymer composites, advance materials (graphene/nanoclay/fire-retardant), lignocellulosic reinforced/filled polymer composites, modification and treatment of lignocellulosic fibers and solid wood, biopolymers, and biopolymers for packaging applications, nanocomposites and nanocellulose fibers, and polymer blends. He earned a PhD at Universiti Sains Malaysia and has published 34 books, 65 book chapters, more than 300 peer-reviewed international journal papers, and five published review papers under top 25 hot articles in ScienceDirect during 2016–2019.

Abdullah Mohammed Ahmed Asiri is a professor in the Chemistry Department, Faculty of Science, King Abdulaziz University. He earned a PhD in 1995 at the University of Wales College of Cardiff, UK, on tribochromic compounds and their applications. He is the chairman of the Chemistry Department, King Abdulaziz University, and the director of the Center of Excellence for Advanced Materials Research. He is a highly cited researcher according to Thomson Reuters. He is the director of the Education Affairs Unit, deanship of Community Services. He is a member of the advisory committee for advancing materials for the National

Technology Plan, King Abdulaziz City of Science and Technology, Riyadh, Saudi Arabia. He is interested in color chemistry, synthesis of novel photochromic and thermochromic systems, synthesis of novel colorants and coloration of textiles and plastics, molecular modeling, applications of organic materials into optics (such as OEDS, high-performance organic dyes and pigments), new and novel applications of organic photochromic compounds, organic synthesis of heterocyclic compounds as precursors for dyes, synthesis of polymers functionalized with organic dyes, preparation of some coating formulations for different applications, and photodynamics using organic dyes and pigments, virtual labs, and experimental simulations. He is a member of the editorial board of *Journal of Saudi Chemical Society*, *Journal of King Abdulaziz University*, *Pigment and Resin Technology Journal*, *Organic Chemistry Insights*, *Libertas Academica*, and *Recent Patents on Materials Science*. He holds a professional membership in many international and national societies and professional bodies.

Wei Ni earned a BE in polymer materials science and engineering at Zhengzhou University, Zhengzhou, China (2005) and a PhD in chemistry at the Institute of Chemistry, Chinese Academy of Sciences (ICCAS), Beijing, China (2011). After a research fellow experience at Nanyang Technological University (NTU), Singapore and the University of Oulu, Oulu, Finland, he became a research scientist at Panzhihua University, Panzhihua, China and a member of the Chinese Chemical Society (CCS) and the American Vacuum Society (AVS). He has published over 40 peer-reviewed SCI papers, two book chapters, and six authorized patents. His research interests include advanced nanomaterials for energy storage and environmental protection.

Mohammed Muzibur Rahman earned a BSc (1999) and an MSc (2001) at Shahjalal University of Science and Technology, Sylhet, Bangladesh. He earned a PhD (2007) at the Chonbuk National University, South Korea. From 2007 to 2011 he was a postdoctoral fellow and an assistant professor at pioneer research centers and universities in South Korea, Japan, and Saudi Arabia. Since 2011 he has been an associate professor at the Center of Excellence for Advanced Materials Research (CEAMR) and the Chemistry Department at King Abdulaziz University, Saudi Arabia. He has attended more than 245 international and domestic conferences, published several book chapters, and edited ten books. His research work includes photocatalysis, semiconductors, nanoparticles, carbon nanotubes, nanotechnology, electrocatalysis, sensors, ionic liquids, surface chemistry, electrochemistry, nanomaterials, etc.

Contributors

Hilal Acıdereli
Biochemistry Department
Dumlupınar University
Kütahya, Turkey

Afzal Ansari
Department of Applied Sciences and
    Humanities
Jamia Millia Islamia
New Delhi, India

S.R. Sundara Bharathi
Department of Mechanical Engineering
National Engineering College
Kovilpatti, India

Kemal Cellat
Biochemistry Department
Dumlupınar University
Kütahya, Turkey

Nhamo Chaukura
Department of Physical and Earth
    Sciences
Sol Plaatje University
Kimberley, South Africa

C. Dhanasekaran
Department of Mechanical Engineering
Vels Institute of Science, Technology &
    Advanced Studies
Chennai, India

Mpitloane J. Hato
Department of Chemistry
University of Limpopo
Polokwane, South Africa

Emmanuel I. Iwouha
Department of Chemistry
University of the Western Cape
Bellville, South Africa

Neslihan Karaman
Biochemistry Department
Dumlupınar University
Kütahya, Turkey

Rufaro B. Kawondera
Chemistry Department
Bindura University of Science
    Education
Bindura, Zimbabwe

Anish Khan
Department of Chemistry
King Abdulaziz University
Jeddah, Saudi Arabia

and

Center of Excellence for Advanced
    Materials Research
King Abdulaziz University
Jeddah, Saudi Arabia

R. Kumar
Department of Mechanical Engineering
Vels Institute of Science, Technology &
    Advanced Studies
Chennai, India

Tatenda C. Madzokere
Department of Metallurgy and
   Materials Engineering
Midlands State University
Gweru, Zimbabwe

and

Department of Metallurgy
University of Johannesburg
Johannesburg, South Africa

Katlego Makgopa
Department of Chemistry
Tshwane University of Technology
Pretoria, South Africa

Edwin Makhado
Department of Chemistry
University of Limpopo
Polokwane, South Africa

Sibongile M. Malunga
Chemistry Department
Bindura University of Science
   Education
Bindura, Zimbabwe

Thabiso C. Maponya
Department of Chemistry
University of Limpopo
Polokwane, South Africa

M.S. Maubane-Nkadimeng
School of Chemistry
University of Witwatersrand
Johannesburg, South Africa

Kwena D. Modibane
Department of Chemistry
University of Limpopo
Polokwane, South Africa

K.M. Molapo
Chemical Science Department
University of the Western Cape
Bellville, South Africa

Gobeng R. Monama
Department of Chemistry
University of Limpopo
Polokwane, South Africa

and

Department of Chemistry
University of the Western Cape
Bellville, South Africa

Tshaamano C. Morudu
Department of Chemistry
University of Limpopo
Polokwane, South Africa

Abdullah Arul Marcel Moshi
Department of Mechanical Engineering
National Engineering College
Kovilpatti, India

Norman Mudavanhu
Chemistry Department
Bindura University of Science
   Education
Bindura, Zimbabwe

Wisdom A Munzeiwa
Chemistry Department
Bindura University of Science
   Education
Bindura, Zimbabwe

M. Muthukrishnan
Department of Mechanical Engineering
Kalaignarkarunanidhi Institute of
   Technology
Coimbatore, India

Wei Ni
Vanadium and Titanium
   Resource Comprehensive
   Utilization Key Laboratory
   of Sichuan Province
Panzhihua University
Panzhihua, China

and

Material Corrosion and Protection
   Key Laboratory of Sichuan
   Province
Sichuan University of Science and
   Engineering
Zigong, China

and

Faculty of Technology
University of Oulu
Oulu, Finland

Nasani Rajendar
School of Chemistry
University of Hyderabad
Hyderabad, India

Phuti S. Ramaripa
Department of Chemistry
University of Limpopo
Polokwane, South Africa

M. Ramesh
Department of Mechanical
   Engineering
Kalaignarkarunanidhi Institute
   of Technology
Coimbatore, India

Kabelo E. Ramohlola
Department of Chemistry
University of Limpopo
Polokwane, South Africa

Arivumani Ravanan
Department of Mechanical
   Engineering
Kalaignarkarunanidhi Institute of
   Technology
Coimbatore, India

S.S. Saravanakumar
Department of Mechanical
   Engineering
Kamaraj College of Engineering
   and Technology
Virudhunagar, India

Tridib K. Sarma
Discipline of Chemistry
Indian Institute of Technology Indore
Indore, India

Fatih Şen
Biochemistry Department
Dumlupınar University
Kütahya, Turkey

P. Senthamaraikannan
Department of Mechanical
   Engineering
Kamaraj College of Engineering and
   Technology
Virudhunagar, India

Bhagwati Sharma
Materials Research Centre
Malaviya National Institute of
   Technology
Jaipur, India

Weqar Ahmad Siddiqi
Department of Applied Sciences and
   Humanities
Jamia Millia Islamia
New Delhi, India

Vasi Uddin Siddiqui
Department of Applied Sciences and
    Humanities
Jamia Millia Islamia
New Delhi, India

S. Sivaganesan
Department of Mechanical Engineering
Vels Institute of Science, Technology &
    Advanced Studies
Chennai, India

T.R. Somo
South African Institute of Advanced
    Chemistry
University of the Western Cape
Bellville, South Africa

M.D. Teffu
Department of Chemistry
University of Limpopo
Polokwane, South Africa

1 Significance of Metal-Organic Frameworks Consisting of Porous Materials

R. Kumar, Abdullah, Arul Marcel Moshi,
S.R. Sundara Bharathi, C. Dhanasekaran,
S. Sivaganesan, P. Senthamaraikannan,
S.S. Saravanakumar, and Anish Khan

CONTENTS

1.1 INTRODUCTION

Materials of porous nature are abundantly available in nature in a variety of forms. A few porous materials are mentioned in Figure 1.1. Metal-organic frameworks (MOFs) are a new class of hybrid porous solids, which are potentially a type of prominent porous adsorbent; and they can also exist in an empty guest-free state [1].

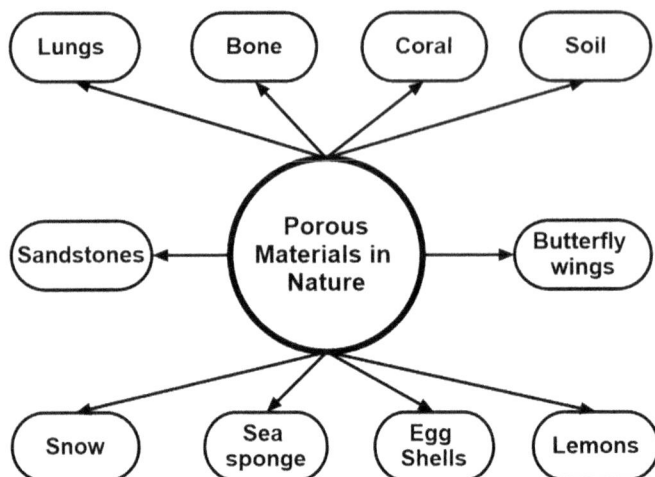

FIGURE 1.1   Materials with a porous nature [4].

MOFs are defined by Yaghi et al. as porous structures made with coordinative bonding between metal ions and organic linkers [2]. MOFs have grown to become the leading domain of solid-state chemistry [3,4]. This special case of crystalline materials presents a high degree of functional and structural tenability [5,6] which is not possible with other traditional porous materials like zeolites and activated carbons [3].

Even though the general porous materials have many valuable attributes, [7] techniques for controlling the individual crystal locations, and coatings with particularly designed pore sizes, their arrangement/distribution is not yet optimized [8]. Among all kinds of porous materials, MOFs are a special kind of ultra-porous material with an extraordinary accessible surface area because of the framework generated by the inorganic nodes and organic compounds [2,9]. These surface areas range between 1000 and 10,000 m^2/g, which exceed the values of other porous materials such as carbons, zeolites, and mesoporous-based oxides [10]. A few artificially made commonly used products with a porous nature are illustrated in Figure 1.2.

It is significant to note that MOFs are called by many names, such as porous coordination networks, porous coordination polymers, etc. The fast rate of growth in the synthesis, characterization, and analysis of MOFs could be noted in recent years. These kinds of materials are produced in such a way that they have permanent porosity [11]. The flexibility with MOFs is that their secondary building units (SBUs) and organic linkers can be varied, which has led to the formation of thousands of MOF compounds. Specifically, they have been extensively used in the energy domain, including fuel cell technology, super capacitors, and catalytic converters [12,13]. In order to utilize the positive features of both inorganic and organic porous compounds, porous hybrids (MOFs) are being generated which are stable, ordered, and have high surface areas.

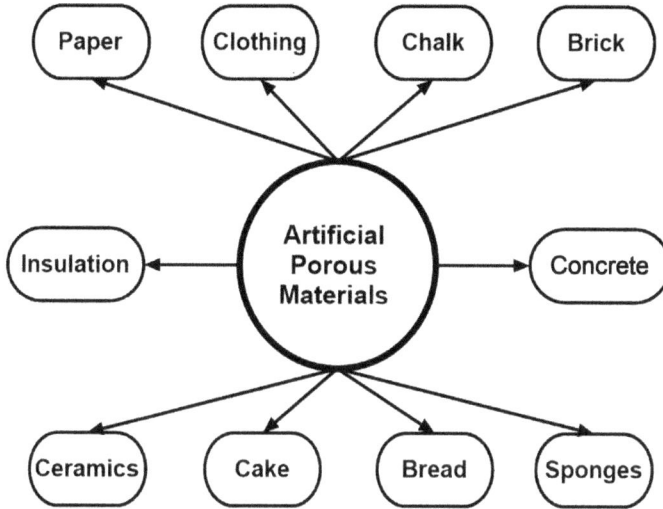

FIGURE 1.2 Artificially developed common porous materials [7].

1.1.1 Definition of Porosity

Porosity of any solid material can be realized with the presence of cavities, void space, and/or inter-channels. Materials consisting of a regular organic-inorganic hybrid framework and acting as a regular porous structure with pores of the size range 0.2×10^9 to 100×10^9 m are called nanoporous materials [13].

1.2 INFERENCES OBTAINED FROM THE WIDE RANGE OF RELEVANT RESEARCH ARTICLES

Various published research articles related to porosity for MOF materials have been referred to and the important elements are presented in this section.

1.2.1 Introduction to Porous MOFs

At present, MOF chemistry has grown well enough to the point where the chemical composition, structure of the compounds, specific functionality, and the nature of porosity of a metal-organic structure can be made for the desired application. This exclusive control over the assembly of compounds propels this area further into a new domain area for synthetic chemistry, in which further, more sophisticated materials may be approached. For example, materials can be visualized which have:

i. combined compartments which operate separately, but their function is integrated;
ii. ability to perform simultaneous operations; and
iii. dexterity to count, classify, and code data [14].

In recent years, researchers have carried out extensive works on crystalline extended structures [15,16]. Even though these structures are extended crystal structures and do not have large detached molecules like polymers, they are dubbed coordination "polymers" –MOFs [17],because these structures are constructed from long organic linkers which are surrounded by void space. MOFs are known to have the potential to be permanently porous like in the case of zeolites. The porosity of MOFs was investigated in the 1990s by forcibly sending gas molecules into the narrow openings at high pressure [18].

1.2.2　Zeolites—An Amorphous and Inorganic Porous Material

Zeolites are an ideal type of structure which belong to the group of purely inorganic materials, and which are a benchmark in the field of solid-state porous materials. Zeolites are readily rehydrated and dehydrated which makes them useful in various commercial areas [7]. Porous materials include a wide range of applications in industry, such as catalysis and absorption. Zeolites are the most perfect examples among the group of crystalline alumino silicate materials with interlinked pores of size 4 to 13 A [19,20]. In comparison with zeolites, activated carbons have high degrees of porosity and specific surface area. Activated carbon also belongs to amorphous porous materials, which rule a major area of the market of solid-state porous materials [21].

Inorganic porous frameworks exhibit a highly ordered structure (e.g. zeolites). Synthesis processes often require an organic or inorganic template with strong interactions between the template formed during the process and inorganic framework. As the outcome, elimination of the template can result in the collapse of the framework. Inorganic frameworks are also influenced by many factors such as lack of diversity. On the other hand, inorganic frameworks are being used in applications like catalysis and separation of gases [3].

1.2.3　Activated Carbon—An Organic Porous Material

Porous materials are utilized widely in gas storage, adsorption-based gas and vapor separation, selective catalysis, storage and delivery of drugs, and as templates in the synthesis of low geometric materials [22]. Conventionally, porous materials are of either inorganic or organic type. Possibly the most general organic type of porous material is activated carbon which is normally produced by decomposing carbon-rich materials at high temperature. Activated carbon has a high surface area and a good degree of adsorption capability, but it does not have an ordered structure. Though activated carbon has a lack of order, porous carbon materials are being used in many application areas including the separating and storing of gases, purifying water, and removing and recovering solvents [23].

1.2.4　Formation of Pores in MOFs

Pores are known to be the voids present within the porous materials while removing the guest molecules [24]. Even though MOFs are constructed by combining inorganic and organic compounds to have a large number of pores, frameworks will often merge with one another to improve the packing efficiency [25]. In such cases, the sizes of the pores are considerably reduced, but this would also be useful for a few applications. In fact, merged frameworks are being deliberately produced and this phenomenon has been found to be useful in improving the performance. Example: in the storage of H_2 [26]. Assessment of porous materials is currently focused on the adsorption of pure methane. Even though methane is the major constituent (95%), commercial natural gas also contains other impurities, including 3.2% of ethane, 0.2% of propane, and 0.5% of carbon dioxide [27]. Porous carbon materials, especially carbon nanotubes and activated carbons, are the most focused kinds of porous materials for storing methane [28].

1.2.5 TYPES OF PORES

The adhesion of the guest molecules and the surface of the adsorbent, as well as the relationship between the adsorbent's surface and guest molecules, plays needful roles in predicting the characteristics of the porous structures, which are strongly ruled by the shape and size of the pores. In the physical system, pores are categorized based on their sizes as listed and shown in Figure 1.3.

Liu et al. stated that microporous systems can be used in upgrading the energy density. Also, they have reported that mesoporous systems are used to improve the power density [29]. It is possible to have a flexible structure just by altering the inorganic or organic linkers [30].

Microporous materials are used in valuable applications like redox catalysts, in the petroleum industry, and in the synthesis of chemical items for different kinds of shape-selective transformation and detachment processes. They create the fundamentals of new environment-friendly technologies which involve cheaper and more efficient conditions for performing chemical reactions. Transition metals modified

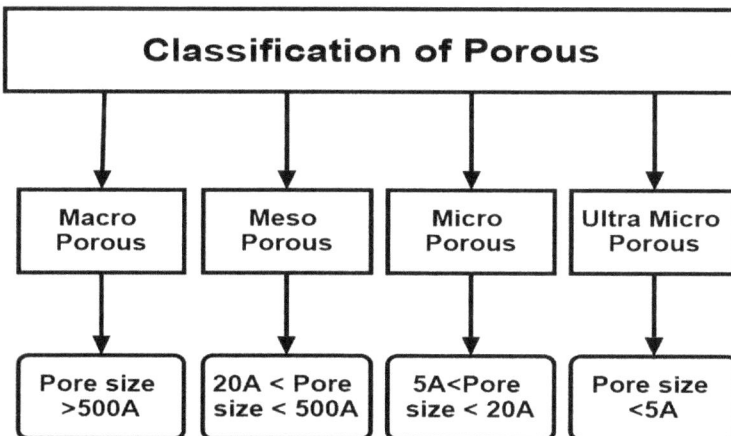

FIGURE 1.3 Types of pores based on pore size [29].

microporous molecular filters with alumino silicate and aluminophosphate frameworks accelerate a wide range of artificially effective oxidizing transformations with impurity-free oxidants like hydrogen peroxide under comparably light conditions, providing the advantage of recovering and recycling complex structures. MOFs are utilized in a significant number of applications in waste treatment activities including detachment of heavy metals and radio active species, ammonia, various kinds of phosphates, and harmful gases from soil, water, and air because they have unique structural and physicochemical characteristics. In earlier times, silica-aluminum-based zeolite microporous materials were mainly used. In recent times, several types of microporous materials are being produced with the aid of metal oxides, metal phosphates, and inorganic-organic hydride materials [31].

1.2.6 CHARACTERIZATION OF POROUS MOFs

As clearly explained, pore engineering is a powerful path to direct the structure and functionality of pores, which drastically promotes the development of MOFs for recognizing differential molecules. It is possible to engineer the pores of MOFs by tuning their sizes or channels, functional sites, and surface areas to achieve unique MOF materials to be utilized for specific gas separations [32].

This special characteristic makes MOFs unique regarding their structural properties, and which offers the desirable potential to use in different sectors. The comparably easiest production method of MOFs is one of the main reasons that they are a better choice for different applications. Their chemical properties, customized pore structure, and thermal stability make them preferable for sound application domains such as separation of specific gases [33], catalysis [34], and conduction of protons [35]. MOFs are usually characterized with the help of the X-ray diffraction (XRD) method, surface area analysis, electron microscopy (EM), thermogravimetric analysis for predicting thermal stability of various constituents, and Fourier transform infrared (FTIR) technique to characterize the molecules and atomic structures. Since MOFs hold the nature of both crystalline materials as well as highly porous materials, powder X-ray diffraction (PXRD) is usually used to characterize the adsorption measurements, phase purity, and crystalline nature to check for the porosity [36].

1.2.7 CHECKING FOR PERMANENT POROSITY

In order to ensure permanent porosity, it needs determination of reversible gas sorption isotherms at low temperatures and low-pressure conditions. Nevertheless, as we stated at that time [37], it has become commonplace to refer to the materials as "open framework" and "porous" even though such validation was missing. The ultimate proof for permanent porosity of MOFs can be obtained by estimating the carbon dioxide and nitrogen isotherms on layered zinc terephthalate MOF [37]. The overall order of porosity in a microporous molecular sieve depends on the template molecule, composition of inorganic materials, the condition of reaction, and the formation mechanism.

A significant advancement in the chemistry of MOFs happened in 1999 when the synthesis and determination of a single crystal structure using XRD, and low temperature-low pressure gas sorption characteristics were described for the first powerful porous MOF [38]. In order to prepare MOFs with a further higher surface area, which require an increased storage space per weight of the component, that phenomenon is termed as ultrahigh porosity. Organic linkers with large lengths offer a larger storage space and higher number of adsorption sites within a given component. Nevertheless, the more space within the framework makes it liable to make impregnating structures, in which two or more crystal frameworks get increased in size and mutually twirled together. By making MOFs with topology that obstructs impregnation, it can be effectively prevented from impregnation, which will require the further framework to be with a different topology [10,39].

1.2.8 ADVANTAGES OF MOF POROUS MATERIALS

The common advantages of MOF porous materials are presented in Figure 1.4. The foremost advantage of MOF porous materials is that the quantity of possible combinations of organic and inorganic parts to make the resulting expected structures is excellent [40]. Moreover, MOFs hold some unique properties like luminescence [41] and magnetism [42] in comparison with other porous materials.

MOFs have a few other advantages such as low-cost design, light weight, etc., which can probably be used to decrease the growing carbon dioxide level in the atmosphere due to the burning of fossil fuels [5].

A tank charged with a porous adsorbent permits storage of a gas at a very low pressure compared to a similar tank without adsorbent. Thus, tanks with high pressure and multi-stage compressors can be modified by providing a safe and economical gas storage method. Many gas storage analyses have been performed on porous adsorbents such as carbon nanotubes, activated carbon, and zeolites [31].

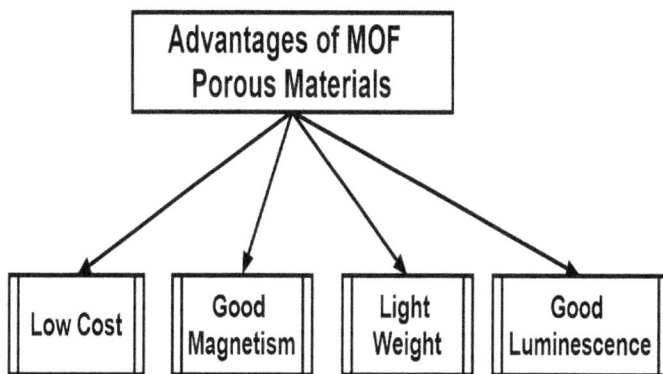

FIGURE 1.4   Common advantages of MOF porous materials [40].

1.2.9 Porous MOFs in Separation of Gases

As already discussed, MOF materials play vital roles in many application areas. Major application areas of MOF materials are tabulated in Table 1.1.

The insight of the first few porous MOFs with permanent porosity implemented by gas adsorption analyses significantly facilitated the development of these novel adsorbents for storing and separating gases [37,43]. For separating specific gases, the early researchers conducted tests based on single component adsorption and/or desorption isotherm prediction of pure gases. The accumulation of sorption information (from a number of MOFs) exhibited a positive potential for purification and separation of vapor or gas mixtures. However, normal separation of mixtures of gases with the help of MOFs was hardly understood until gas chromatography (a new evaluating method) was introduced into this field in 2005 [44]. In 2007, an experimental fixed bed breakthrough was also applied to the prediction of separation of MOFs [45].

Based on these technologies, MOFs have revealed real separation of gas mixtures from their intrinsic porous properties. In turn, the breakthrough experiment became a powerful tool for evaluating the separation of MOFs that can imitate the industrial process which cannot be implemented by a simple static single-component gas sorption analysis. Then, a number of significant and challenging separations such as capturing CO_2, separating CO_2 from nitrogen or methane, separating light hydrocarbons, separating isomers, separating noble gases, etc., have been accomplished by utilizing the unique MOFs as adsorbent materials [46,47].

1.2.10 Nano Porous MOFs

Currently, nano porous materials are being focused with the view of nano science and nanotechnology that is an evergreen multi-disciplinary domain of analysis, which attracted lots of effort in R&D around the world. Nano porous materials (as a subset of nano structured materials) possess unique surface, structural, and chemical

---

TABLE 1.1
General Application Areas of MOF Porous Materials [37]

Sl.No	Application Area	Description
1	Pollution control	Low-cost and light-weight MOF porous materials are used to reduce the CO_2 level in the atmosphere.
2	Gas storage and separation	CO_2 capture and CO_2 separation from CH_4 and N_2, light HC separation, isomers separation, noble gases separation, etc., are performed with the help of MOF porous materials.
3	Drug delivery	MOF porous materials are used to provide a therapeutic amount of drug to the proper site in the body to achieve prompt application and to maintain the desired drug concentration.
4	Catalysis	MOF porous materials are employed to catalyze transformations of existing petrochemical feed stocks.

properties which show their importance in various sectors such as ion exchange [48], separation [49], catalysis [50], gas storage [51–53], lithium ion batteries [54], biological molecular isolation [55], and purification fields because of their flexible frameworks, uniform pore size, controlled chemistry, and high internal surface area. In recent years, porous materials have also been used extensively in optical transparency, [56] photovoltaic solar cells [57], nano generators [58], nanotechnology [59], sensors [60], optoelectronic devices [61], biomedical imaging [62], and biomedical sciences [63]. However, they are capable of interacting with the atoms, ions, and molecules at their surfaces as well as throughout the portion of the materials [7].

Hence, nano porous materials also have scientific technological importance due to their flexibility to interact with atoms, ions, and molecules on their large interior surfaces and in their nanometer-size pore spaces. They provide new opportunities in many areas including inclusion chemistry and guest-host synthesis [64].

1.3 CONCLUSION

A detailed description of the porosity of MOFs is presented in this chapter. The following salient points have been discussed throughout the chapter:

- MOFs are obtained by combining inorganic and organic contents with the help of strong bonds.
- Porous materials are naturally occurring. In addition to that, materials with a porous nature are being intentionally artificially produced, such as brick, concrete, cake, sponge, etc.
- MOFs can be produced with comparable porosity to zeolites and activated carbon.
- Characterization methods used to characterize the porous MOF materials have been discussed in detail.
- MOFs possess a high degree of structural and functional tenability.
- Porous MOF materials have many advantages such as good magnetism, light weight, low cost, and good luminescence.
- Due to their porous nature, MOFs are used in applications such as gas separation, gas storage, catalysis, drug delivery, etc.

REFERENCES

1. C. Prestipino, L Regli, J.G. Vitillo, F Bonino, A. Damin, C. Lamberti, A. Zecchina, P.L. Solari, K.O. Kongshaug, S. Bordiga. Local structure of framework Cu(II) in HKUST-1 metallorganic framework: Spectroscopic characterization upon activation and interaction with adsorbates. *Chem. Mater.* 18 (2006) 1337–1346.
2. S.L. James. Metal organic frameworks. *Chem. Soc. Rev.* 32 (2003) 276–288.
3. G. Ferey. Hybrid porous solids: Past, present, future. *Chem. Soc. Rev.* 37 (2008) 191–214.
4. M. Eddaoudi, D.B. Moler, H.L. Li, B.L. Chen, T.M. Reineke, M. O'Keeffe, O.M. Yaghi. Modular chemistry: Secondary building units as a basis for the design of highly porous and robust metal-organic carboxylate frameworks. *Acc. Chem. Res.* 34 (2001) 319–330.

5. Z.J. Lin, J. Lu, M. Hong, R. Cao. Metal-organic frameworks based on flexible ligands (FL-MOFs): Structures and applications. *Chem. Soc. Rev.* 43 (2014) 5867–5895.
6. W. Lu, Z. Wei, Z.-Y. Gu, T.-F. Liu, J. Park, J. Park, J. Tian, M. Zhang, Q. Zhang, T. Gentle Iii, M. Bosch, H.-C. Zhou. Tuning the structure and function of metal-organic frameworks via linker design. *Chem. Soc. Rev.* 43 (2014) 5561–5593.
7. M.E. Davis. Ordered porous materials for emerging applications. *Nature* 417 (2002) 813–821.
8. P. Falcaro, D. Buso, A.J. Hill, C.M. Doherty. Patterning techniques for metal organic frameworks. *Adv. Mater.* 24 (2012) 3153–3168.
9. S.I. Noro, S. Kitagawa. The supramolecular chemistry of organic-inorganic hybrid materials. *Angew. Chem. Int. Ed.* 45 (2010) 235–269.
10. H. Furukawa, K.E. Cordova, M. O'Keeffe, O. Yaghi. The chemistry and applications of metal-organic frameworks. *Science* 341 (2013) 974–986.
11. O.M. Yaghi, M.O. Keeffe, N.W. Ockwig, H.K. Chae, M.Eddaoudi, J. Kim. Reticular synthesis and the design of new materials. *Nature* 423 (2003) 705–714.
12. U. Mueller, M.Schubert, F. Teich, H. Puetter, K.S.Arndt, J. Pastre. Metal-organic frameworks-prospective industrial applications. *J. Mater. Chem.* 16 (2006) 626–636.
13. M. Jacoby. Heading to market with MOFs. *Chem. Eng. News* 86 (2008) 13–16.
14. H. Deng, C.J. Doonan, H. Furukawa, R.B. Ferreira, J. Towne, C.B. Knobler, B. Wang, O.M. Yaghi. Multiple functional groups of varying ratios in metal-organic frameworks. *Science* 327 (2010) 846–850.
15. A.F. Wells. *Structural Inorganic Chemistry* (New York, NY: Oxford Univ. Press, 1984).
16. Y. Kinoshita, I. Matsubara, T. Higuchi, Y. Saito. The crystal structure of Bis(adiponitrilo) copper(I) nitrate. *Bull. Chem. Soc. Jpn.* 32 (1959) 1221–1226.
17. O.M. Yaghi, H. Li. Hydrothermal synthesis of a metal-organic framework containing large rectangular channels. *J. Am. Chem. Soc.* 117 (1995) 10401–10402.
18. M. Kondo, T. Yoshitomi, H. Matsuzaka, S. Kitagawa, K. Seki. Three-dimensional framework with channelling cavities for small molecules: [M2(4,4'-bpy)3(NO3)4]·xH2O} n(M = Co, Ni, Zn). *Angew. Chem. Int. Ed. Engl.* 36 (1997) 1725–1727.
19. Z.Y. Gu, C.X.Yang, N. Chang, X.P. Yan. Metal-organic frameworks for analytical chemistry: From sample collection to chromatographic separation. *Acc. Chem. Res.* 45 (2012) 734–745.
20. J.R. Li, J.Sculley, H.C. Zhou. Metal-organic frameworks for separations. *Chem. Rev.* 112 (2012) 869–932.
21. R.C. Bansal, M. Goyal. Activated carbon adsorption from solutions. In *Activated Carbon Adsorption* (CRC Press, 2005) 145–199.
22. J.R. Li, R.J. Kuppler, H.C. Zhou. Selective gas adsorption and separation in metal–organic frameworks. *Chem. Soc. Rev.* 38 (2009) 1477.
23. S.M. Manocha. Porous carbons. *Sadhana* 28 (2003) 335–348.
24. K.S.W. Sing, D.H. Everett, R.A.W. Haul, L. Moscou, R.A.Pierotti, J. Rouquerol, T. Siemieniewska. Physical and biophysical chemistry division commission on colloid and surface chemistry including catalysis. *Pure Appl. Chem.* 57 (1985) 603–619.
25. S.R. Batten, S.M. Neville, D.R. Turner. Coordination polymers: Design, analysisand application. R. Society of Chem., Cambridge, (2009) 1–424.
26. S. Ma, J.Eckert, P.M. Forster, J.W. Yoon, Y.K.Hwang, J.S. Chang, C.D. Collier, J.B.Parise, H.C. Zhou. Further investigation of the effect of framework catenation on hydrogen uptake in metal-organic frameworks. *J. Am. Chem. Soc.* 130 (2008) 15896–15902.
27. W.E. Liss, W.H. Thrasher, G.F. Steinmetz, P. Chowdiah, A. Attari. *Variability of Natural Gas Composition in Select Major Metropolitan Areas of the United States*, (1992) PB92-224617.

28. S.H. Yeon, S. Osswald, Y. Gogotsi, J.P. Singer, J.M. Simmons, J.E. Fischer, M.A. Lillo-Ro´denas, A.L. Solano. Enhanced methane storage of chemically and physically activated carbide-derived carbon. *J. Power Source.* 191 (2009) 560–567.

29. B. Liu, H. Shioyama, H. Jiang, X. Zhang, Q. Xu. Metal-organic framework (MOF) as a template for syntheses of nanoporous carbons as electrode materials for supercapacitor. *Carbon.* 48 (2010) 456–463.

30. I. Spanopoulos, I. Bratsos, C. Tampaxis, A. Kourtellaris, A. Tasiopoulos, G. Charalambopoulou, T.A. Steriotis, P.N. Trikalitis. Enhanced gas-sorption properties of a high surface area, ultramicroporous magnesium formate. *Cryst. Eng. Comm.* 17 (2015) 532–539.

31. S.A. Jenekhe, X.L. Chen. Self-assembly of ordered microporous materials from rod-coil block copolymers. *Science* 283 (1999) 372–375.

32. Y. Cui, B. Li, H. He, W. Zhou, B. Chen, G. Qian. Metal-organic frameworks as platforms for functional materials. *Acc. Chem. Res.* 49 (2016) 483–493.

33. A. Car, C. Stropnik, K.V. Peinemann. Hybrid membrane materials with different metal-organic frameworks (MOFs) for gas separation. *Desalination.* 200 (2006) 424–426.

34. F.X.L.I. Xamena, A. Abad, A. Corma, H. Garcia. MOFs as catalysts: Activity, reusability and shape-selectivity of a Pd-containing MOF. *J. Catal.* 250 (2007) 294–298.

35. K.S. Park, Z. Ni, A.P. Cote, J.Y. Choi, R. Huang, F.J.U. Romo, H.K. Chae, M. O'keeffe, O.M. Yaghi. Exceptional chemical and thermal stability of zeoliticimidazolate frameworks. *Proc. Natl. Acad. Sci.* 103 (2006) 10186–10191.

36. N.L. Rosi, J. Eckert, M. Eddaoudi, D.T. Vodak, J. Kim, M. O'Keefe, O.M. Yaghi. Hydrogen storage in microporous metal-organic frameworks. *Science* 300 (2003) 1127–1129.

37. H. Li, M. Eddaoudi, T.L. Groy, O.M. Yaghi. Establishing microporosity in open metal-organic frameworks: Gas sorption isotherms for Zn(BDC) (BDC = 1,4-benzenedicarboxylate). *J. Am. Chem. Soc.* 120 (1998) 8571–8572.

38. H. Li, M. Eddaoudi, M. O'Keeffe, O.M. Yaghi. Design and synthesis of an exceptionally stable and highly porous metal-organic framework. *Nature* 402 (1999) 276–279.

39. H.K. Chae, D.Y.S. Perez, J. Kim, Y.B. Go, M. Eddaoudi, A.J. Matzger, M.O. Keeffe, O.M. Yaghi. A route to high surface area, porosity and inclusion of large molecules in crystals. *Nature* 427 (2004) 523–527.

40. S. Ma. Gas adsorption applications of porous metal-organic frameworks. *Pure Appl. Chem.* 81 (2009) 2235–2251.

41. M.D. Allendorf, C.A. Bauer, R.K. Bhakta, R.J.T. Houk. Luminescent metal organic frameworks. *Chem. Soc. Rev.* 38 (2009) 1330–1352.

42. Y.F. Zeng, X. Hu, F.C. Liu, X.H. Bu. Azido-mediated systems showingdifferent magnetic behaviors. *Chem. Soc. Rev.* 38 (2009) 469–480.

43. S.S.Y. Chui, S.M.F. Lo, J.P.H. Charmant, A.G. Orpen, I.D. Williams. A chemically functionalizablenanoporous material[Cu$_3$(TMA)$_2$(H$_2$O)$_3$]. *Science* 283 (1999) 1148–1150.

44. B. Chen, C. Liang, J. Yang, D.S. Contreras, Y.L. Clancy, E.B. Lobkovsky, O.M. Yaghi, S. Dai. Angew. A microporous metal-organic framework for gas-chromatographic separation of alkanes. *Chem. Int. Ed.* 45 (2006) 1390–1393.

45. P.S. Bárcia, F. Zapata, J.A.C. Silva, A.E. Rodrigues, B. Chen. Kinetic separation of hexane isomers by fixed-bed adsorption with a microporous metal–organic framework. *J. Phys. Chem.* 111 (2007) 6101–6103.

46. Z. Bao, G. Chang, H. Xing, R. Krishna, Q. Ren, B. Chen. Potential0 of microporous metal-organic frameworks for separation of hydrocarbon mixtures. *Energy Environ. Sci.* 9 (2016) 3612–3641.

47. K. Adil, Y. Belmabkhout, R.S. Pillai, A. Cadiau, P.M. Bhatt, A.H. Assen, G. Maurin, M. Eddaoudi. Gas/vapour separation using ultra-microporous metal-organic frameworks: Insights into the structure/separation relationship. *Chem. Soc. Rev.* 46 (2017) 3402–3430.

48. A.K. Patra, A. Dutta, A. Bhaumik. Self-assembled mesoporousγ–Al2O3, spherical nanoparticles and their efficiency for the removal of arsenic from water. *J. Hazard. Mater.* 201–202 (2012) 170–177.

49. M. Hartmann. Ordered mesoporous materials for bioadsorption and biocatalysis. *Chem. Mater.* 17 (2005) 4577–4593.

50. A. Taguchi, F. Schiith. Ordered mesoporous materials in catalysis. *Micropor. Mesopor. Mater.* 77 (2005) 1–45.

51. D.J. Tranchemontagne, K.S. Park, H. Furukawa, J. Eckert, C.B. Knobler, O. M Yaghi. Hydrogen storage in new metal-organic frameworks. *J. Phy. Chem. C* 776 (2012) 13143–13151.

52. A. Dutta, M. Pramanik, A.K. Patra, M. Nandi, H.Uyama, A. Bhaumik. Hybrid porous tin (IV) phosphonate: An efficient catalyst for adipic acid synthesis and a very good adsorbent for CO_2 uptake. *Chem. Commun.* 48 (2012) 6738–6740.

53. M. Nandi, K. Okada, A. Dutta, A. Bhaumik, J. Maruyama, D. Derks, H. Uyama. Unprecedented CO_2 uptake over highly porous N-doped activated carbon monoliths prepared by physical activation. *Chem. Commun.* 48 (2012) 10283–10285.

54. H. Liu, Z. Bi, X.-G. Sun, R.R. Unocic, M.P. Paranthaman, S. Dai, G.M. Brown. Mesoporous TiO_2-B microspheres with superior rate performance for lithium ion batteries. *Adv. Mater.* 23 (2011) 3450–3454.

55. H. Wu, S. Zhang, J. Zhang, G. Liu, J. Shi, L. Zhang, X. Cui, M. Ruan, Q. He, W. Bu. A hollow-core, magnetic, and mesoporous double-shell nanostructure: In situ decomposition/reduction synthesis, bioimaging, and drug-delivery properties. *Adv. Fund. Mater.* 21 (2011) 1850–1862.

56. B.J. Scott, G. Wirnsberger, G.D. Stucky, mesoporous and mesostructured materials for optical applications. *Chem. Mater.* 13 (2001) 3140–3150.

57. S.E. Habas, H.A.S. Platt, M.F.A.M. Van Hest, D.S. Ginley. Low-cost inorganic solar cells: From ink to printed device. *Chem. Rev.* 110 (2010) 6571–6594.

58. A. Stein, B.J. Melde, R.C. Schroden. Hybrid inorganic-organic mesoporous silicates-nanoscopic reactors coming of age. *Adv. Mater.* 12 (2000) 1403–1419.

59. I.W. Hamley. Nanotechnology with soft materials. *Angew. Chem. Int. Ed. Engl.* 42 (2003) 1692–1712.

60. D. Mao, J. Yao, X. Lai, M. Yang, J. Du, D. Wang. Hierarchically mesoporous hematite microspheres and their enhanced formaldehyde-sensing properties. *Small* 7 (2011) 578–582.

61. R.C. Hayward, P.A. Henning, B.F. Chmelka, G.D. Stucky. The current role of mesostructures in composite materials and device fabrication. *Micropor. Mesopor. Mater.* 44–45 (2001) 619–624.

62. S. Wang. Ordered mesoporous materials for drug delivery. *Micropor. Mesopor. Mater.* 117 (2009) 1–9.

63. M. Vallet-Regi, F. Balas, D. Arcos. Mesoporous materials for drug delivery. *Angew. Chem. Int. Ed. Engl.* 46 (2007) 7548–7558.

64. Y. Wan, H. Yang, D. Zhao. Host-Gues chemistry in the synthesis of ordered nonsiliceousmesoporous materials. *Acc. Chem. Res.* 39 (2006) 423–432.

2 Metal-Organic Frameworks for Heavy Metal Removal from Water

Thabiso C. Maponya, Katlego Makgopa,
Kwena D. Modibane, and Mpitloane J. Hato

CONTENTS

2.1   INTRODUCTION

Members of the platinum group elements/metals (PGEs/PGMs) are economically valuable metals which are very popular for a variety of uses worldwide. This is due to various chemically and physically inherent properties that PGMs have, such as very high melting points, chemical inertness to most compounds even at elevated temperatures, corrosion resistance, and catalytic behavior [1,2]. These PGMs, including platinum (Pt), palladium (Pd), ruthenium (Ru), rhodium (Rh), osmium (Os), and iridium (Ir), occur together in nature in combination with minor gold (Au)) [2]. Amongst them, the most widely used in industries such as automotive, electronics, hydrogen fuel cells, jewelry, and oil refineries are Pt, Pd, and Ru [3]. PGMs are predominantly produced in South Africa from the Bushveld complex, followed by other countries such as Russia, Canada, Zimbabwe, and the United States of America [1–3]. However, the increasing demand for PGMs as a result of increasing population has put a strain on the available resources which poses concerns for the future [4–6]. Moreover, the possibility of environmental issues, such as water and land pollution, affecting the health of human beings and animals has been realized [4,7–9]. Hence, it is essential to recover PGMs from mining wastewater prior to it being released into the environment. The recovery of PGMs has led to the development of many separation techniques such as solvent extraction, chemical precipitation, ion exchange, photocatalytic degradation, and membrane technology (Figure 2.1) [8]. Nevertheless, some of the traditional technologies possess several drawbacks such as insufficient recovery, generation of harmful byproducts, high capital cost, and poor selectivity [10]. Adsorption technology has been identified as the beneficial technology owing to its ease of operation at low capital cost, ability to regenerate spent adsorbent, generation of fewer secondary products, fast kinetics, and flexibility to merge with other methods [6,11]. The success of the adsorption process requires the careful selection of a suitable adsorbent material.

Metal-organic frameworks (MOFs) are coordination polymers which have emerged as nanomaterials with interesting physical and chemical properties including tunable pore size and pore character, lower density, higher surface area, elevated thermal stability, enormous porosity, and abundant active functional sites [12,13]. These properties have led to the application of MOFs in various fields as catalysts, gas sensors, nonlinear optics, gas storage, and drug delivery materials [13–15]. Recently, MOFs have also found new application as adsorbent materials for the removal of pollutants such as metal ions, organics, and pathogens from wastewater owing to their enormous surface area [11,13,16,17]. They can serve as adsorption active sites and their incredible porosity allows dissemination of pollutants over the framework [18].

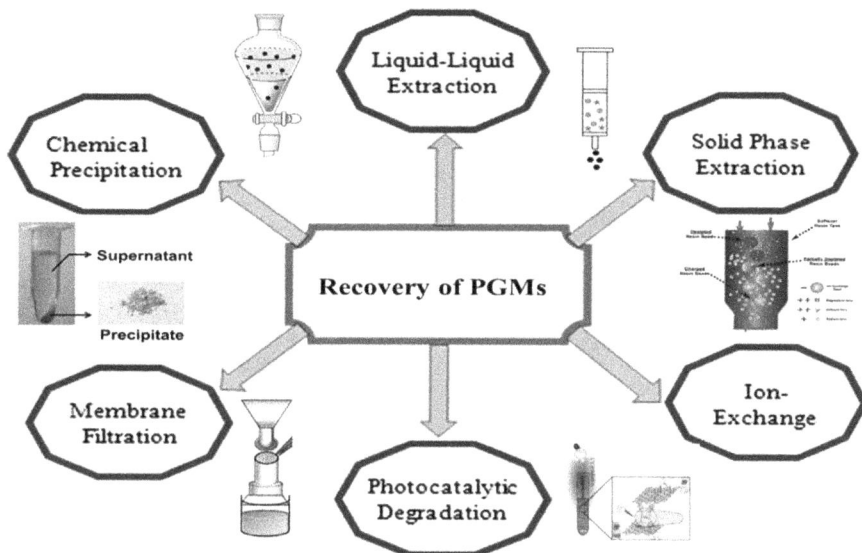

FIGURE 2.1 Methods used for platinum group metals recovery.

However, poor selectivity by MOFs towards PGMs in the presence of interfering ions is still a challenge [10]. Ion imprinting has appeared as a promising technique to fabricate adsorbent materials which are highly selective towards targeted metal ions [19,20]. In this chapter, MOFs, as new materials, are introduced and discussed in detail, from their various synthetic routes to the different composites and nano-composites that can be formed. This is followed by the discussion of several applications of MOFs that have been studied, including application in wastewater treatment, with more focus on adsorption of metal ions. Furthermore, we describe the challenges faced by MOFs and the possibility of addressing their limitations through the employment of the ion-imprinting method. Finally, we summarize and give a future perspective for realization of MOFs as ion-imprinted adsorbents.

2.2 METAL-ORGANIC FRAMEWORKS

2.2.1 BACKGROUND

The MOF first arose through the study of zeolites [21]. The term MOF was first introduced by Yaghi in 1995 for the newly synthesized copper-4,4′-bipryridal complex that exhibited extended metal-organic interactions [18,22]. Essentially, MOFs belong to the general family of coordination polymers. MOFs are more specific for one-, two-, or three-dimensional crystallized networks with porous properties when compared to the general coordination polymers [23,24]. Since the beginning of the millennium, there is an exponential growth of interest in the preparation of MOFs and publications. Figure 2.2a represents a Scopus database from 1998–2019 showing a significant increase in the number of publications on MOFs in general as

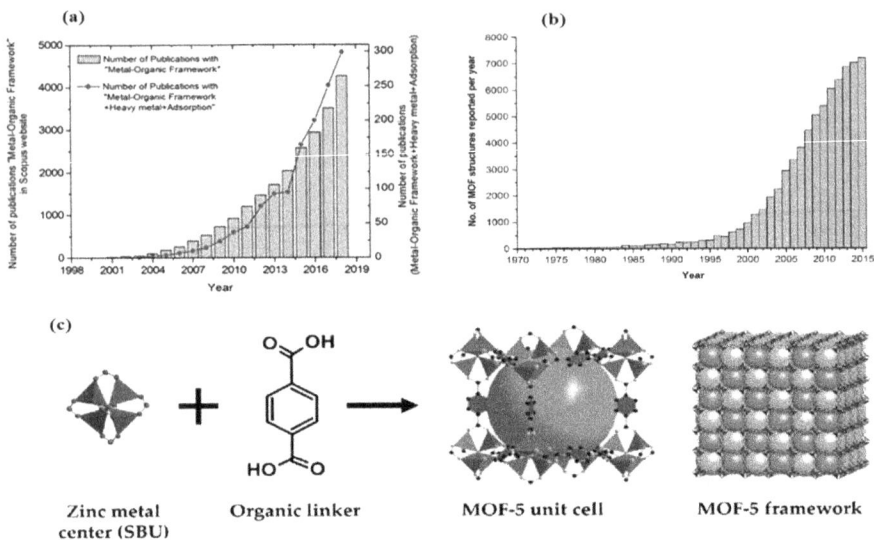

(a)

(b)

(c)

Zinc metal   Organic linker   MOF-5 unit cell   MOF-5 framework
center (SBU)

FIGURE 2.2 (a) Graph of number of MOFs published per year since 1998 to 2019 [16]. (b) The change in percentage of the MOFs in the overall Cambridge Structural Database (CSD) since 1970 to 2015 [25]. (c) MOF structure [27].

well as articles of MOFs for heavy metal adsorption [16]. It was seen in Figure 2.2b that over 4500 MOF structures have been published in the Cambridge Structural Database (CSD) from 1970 to 2015 [25]. These MOFs, as novel types of porous crystalline materials, have attracted increasing attention in water treatment applications due to their high surface area, permanent porosity, and controllable structures [12,18,26].

The structure of MOFs as shown in Figure 2.2c indicates that the metal nodes function as joining points and the organic linkers work as bridging ligands. The framework of the two components is held together by covalent bonds to form extended 3D infinite network structures [24,27]. The defined crystallinity of MOFs allows for identification of the exact positions of all atoms in the framework. By definition, a porous solid is one in which permanent channels or pores permeate the structure and have dimensions large enough to allow solvents or other guest molecules to diffuse into the structure. The porosity and structure are maintained even when guest particles/compounds within the channels are removed by heating or vacuum [28]. Furthermore, the MOF structures are highly tunable by varying the metal nodes or organic ligands which make it possible to obtain tailor-made MOF materials with the required structures and functionalities for specific application [29,30]. The porosity of known MOFs varies between 20 and 95%, where it is measured as the ratio of the accessible pore volume to the total volume of the solid [22]. This porosity relates to high internal surface areas, allowing for increased adsorption of guest molecules. As a result, MOFs have attracted considerable attention due to their high surface area and porosity, tunable structure, and functionality [21,31].

2.2.2 DIFFERENT STRUCTURES OF MOFs

The structural arrangement in MOFs extends from one- to two- and three- dimensional (1D, 2D, and 3D) networks, which are assembled from organic linkers and metal ion or cluster nodes [32]. For a 1D MOF geometry, the bonds between metal clusters and organic linkers form a 1D expansion throughout the polymer with the probable cavities being occupied by molecules of smaller size. 2D MOF polymers are composed of single type layers superimposed from either preceding edges or staggered type layers which are stacked by weak interactions existing between layers. Ligand modification constituting the layers are able to govern the way in which channel interiors stack and function [24]. The two possible ways in which 2D MOF structures can accommodate guest species are through the spaces found amongst the grids of layers and in between the layers. The structural frameworks of 3D MOFs have high porosity and stability as a result of the spreading of coordination bonds in three directions and is common in many MOFs [33,34], as shown in Figure 2.3a–l. The coordination that results in the above geometries (between metal ions and organic ligands) is facilitated by the non-covalent p-p stacking and hydrogen bonds. These interactions are responsible for converting the framework in MOFs to an infinitely dimensional network, as well as controlling their strength and direction. Moreover, the metal ion geometries have a great effect on the structures of MOF polymers [33].

Metal sites (referred to as unsaturated metal sites or accessible metal sites) in MOFs have tremendous influence on their adsorption properties. The metal ions, as the center connectors, are usually chosen from transition metals such as Cu, Zn, Mn, Co, etc. The metals in the MOF structure often serve as Lewis acids which can activate the coordinated organic ligands for succeeding organic transformation [33–36]. The partially positive charges of metal sites in MOF structures have been proved to increase adsorption capacity [36,37]. The commonly used transition metals give different geometries depending on their number of oxidation state [38,39]. The commonly observed geometries of each metal are noted below and also in Figure 2.3a–l:

a. Copper—electron configuration of zero-valent metal is $[Ar]3d^{10}4s_1$. Copper nodes are often in distorted octahedral and square-planar geometries. The distortion occurs mainly due to Jahn-Teller distortion commonly resulting in octahedral geometries having four short bonds and two longer bonds as the dz^2 orbital is filled whilst the dx^2-y^2 orbital is only partially filled.

b. Cobalt—electron configuration of the zero-valent metal is $[Ar]3d^74s^2$. There is a range of common coordination numbers for Co^{2+}. Geometries of cobalt(II) nodes are generally octahedral and tetrahedral.

c. Nickel—electron configuration of the zero-valent metal is $[Ar]3d^84s^1$. The most prevalent nickel geometries are octahedral and square-planar with consistent bond lengths.

d. Manganese—electron configuration of the zero-valent metal is $[Ar]3d^54s^2$. Manganese(II) compounds take on a range of coordination modes and geometries in MOFs, the most common being octahedral.

e. Zinc—electron configuration of the zero-valent metal is $[Ar]3d^{10}4s^2$. Zinc(II) compounds often take up octahedral or tetrahedral geometries.

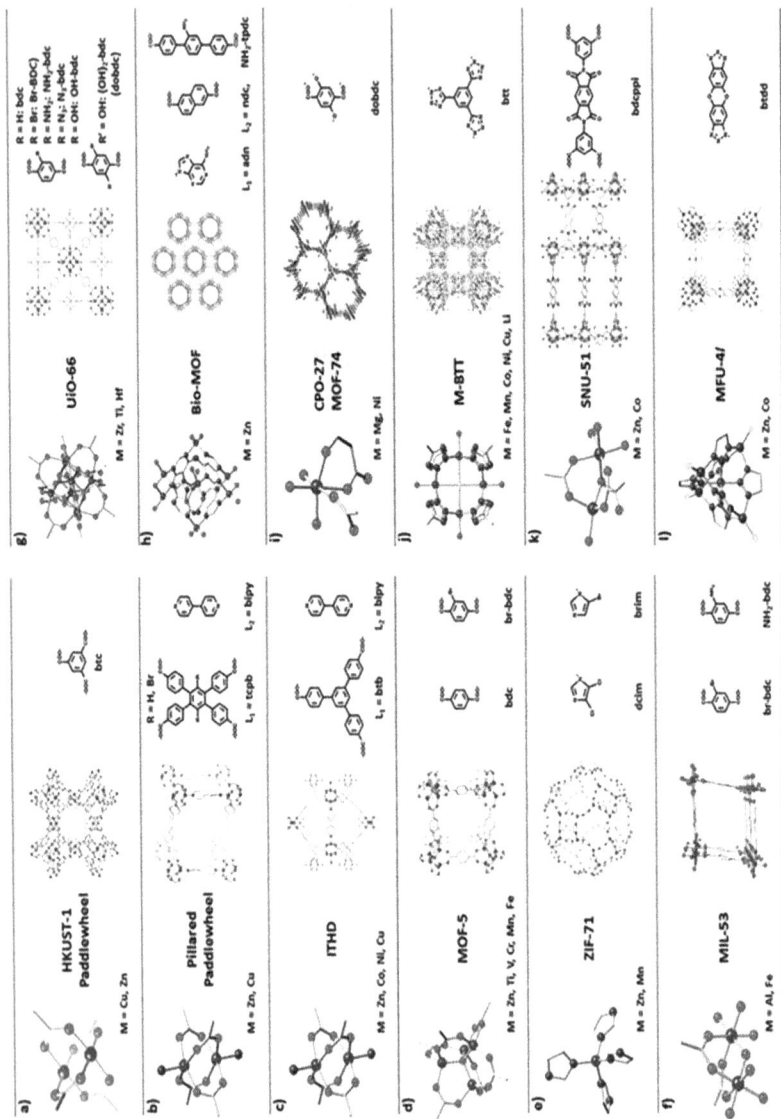

FIGURE 2.3 Lattice structures (middle) and corresponding SBUs (metal nodes (left), and organic linkers (right)) of some of the MOFs discussed in this review. (Atom definition: blue—metal, red—oxygen, purple—nitrogen, grey—carbon, green—chlorine.) [40].

The organic linkers used to connect the metal nodes in coordination polymers are usually multidentate organic molecules [33]. The linkers contain aromatic rings in frameworks, which help maintain the structural integrity of the complex and direct the geometry of the framework. The pore volume and surface area of MOFs can be organized by modifying organic ligands which act as spacers and create an open porous structure. The common organic linkers are shown in Figure 2.3 [40].

Secondary building units (SBUs) play an important role as they dictate the final geometry of MOFs. The structure and chemical properties of the SBUs and organic ligands lead to the prediction of the schemes and preparations of MOFs [41]. Literature reports have shown that under careful selection of reaction preconditions, multidentate linkers can aggregate and thus lock metal ions at definite locations to form SBUs [42]. The SBUs will subsequently join with rigid organic links to produce MOFs that exhibit high structural stability [24,41]. It was reported that the structure of the SBU is controlled by the metal-to-ligand ratio, solvent used, as well as source of anions used in balancing metal ion charges [37,43]. Since the structure of MOF polymers is governed by SBU geometries and organic linkers (shape and size), they are tunable to a certain degree by careful selection of SBUs and organic ligands having appropriate pore size, structure, and functionality for the precise application.

MOFs have gained interest in adsorption applications because they have easily modifiable surface pores which can lead to improved selectivity in adsorbing certain guest species with specific functional groups [11,44]. Hence, they have been extensively explored for the adsorption of numerous toxic materials which are organic and inorganic in nature. The ability of MOFs to remove pollutants from the environment is owed to their enormous porosity and incredible geometry of the pores. Furthermore, some of the MOF components that have been effectively employed to improve the adsorption interactions include coordinatively unsaturated metal sites, functionalized ligands, and added active species. These properties make MOF materials have better adsorption performance towards toxic species than most porous materials [11].

2.2.3 Physical Properties of MOFs

MOFs are prominent due to their capability of conveying various functionalities through appropriate selection of the metal ions and organic ligands. Various studies available in literature have already demonstrated a number of synthesis methods for tuning the chemistry, stability, particle size, and flexibility of MOF structures [45]. Moreover, these coordination polymers can be further improved by "post-synthetic modification" for further tuning the properties by swapping, altering, or removal of the ligand or metal ion from the structure. The mechanical properties of MOFs can be modified by introducing malleable ligands, modulating the strength of host-guest interactions, constructing multi-metallic frameworks, and manipulating the size of crystals.

The most important property in designing MOFs is stability. MOF polymers have to be stable in order for them to be characterized fully and be applied in various fields like adsorption, catalysis, and sensing [46]. The stabilities in the MOFs include chemical, thermal, hydrothermal, and mechanical. To improve the chemical stability of MOF polymers, metal clusters having higher valence, such as Cr^{3+}, Fe^{3+}, and Zr^{4+} are used, together with soft ligands including triazolates, imidazolates, and tetrazolates [34,36,47]. Furthermore, heterocyclic molecules containing nitrogen atoms can be utilized in the preparation of MOFs together with soft divalent metal clusters like Zn^{2+}, or Co^{2+} to increase the stability. The thermal stability of MOFs shows degradation when metal-ligand bonds break and the organic linkers get combusted [46]. This can however be improved by coordination of oxy-anion ligands to high valance metal ions [36]. On the other hand, the hydrothermal stability of MOFs refers to the stability of materials in the existence of moisture at higher temperatures, and it was also shown that it can be improved by introduction of hydrophobic functional groups into the MOF framework [48,49]. Mechanical stability in MOF materials is directly linked to their high porosity and they have an inversely proportional relationship. The mechanical stability is more elevated in MOFs that have solvent-filled pores than in those that have vacant pores [36,46]. Figure 2.4 depicts various MOF materials with improved Brunauer-Emmet-Teller (BET) results. It was shown that MOF-177 possessed the highest surface area in 2004 with BET surface area of 3780 m^2/g and porosity of 83% in 2007 [50]. Later in 2010, the surface area was doubled by MOF-200 and MOF-210 produced surface areas of 4530 and 6240 m^2/g, and porosities of 90 and 89%, respectively [51,52]. On the other hand, in 2007, MOF-5 material showed a surface area of 3800 m^2/g owing to active adsorption sites from zinc oxide SBU and the edges of the organic linker [52]. Furthermore, it has been reported that expanded tritopic linkers based on alkyne rather than phenylene units should increase the number of adsorption sites and enhance the surface area [51]. NU-110 MOFs in 2012 have demonstrated a high surface area of 7140 m^2/g [50], whereas

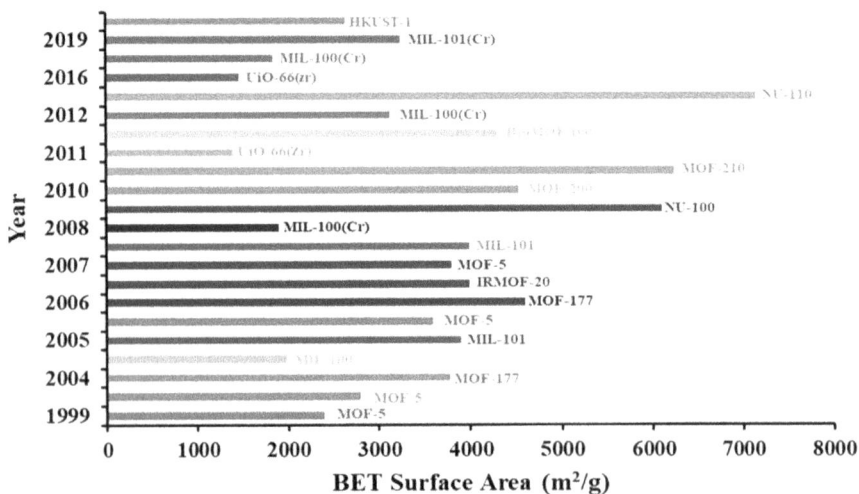

FIGURE 2.4 BET surface areas of MOFs and typical conventional materials were estimated from gas adsorption measurements.

UiO-66(Zr), MIL-100(Cr), MIL-101(Cr), and HKUST-1 showed a BET surface of 1473, 1842, 3250, and 2642 m²/g in 2013–2019, respectively [38,53,54]. For application in wastewater treatment a number of MOF polymers which are water-stable such as chromium-based MIL-101 series [38,53,54], zeolitic imidazolate frameworks (ZIFs) [55], zirconium-based carboxylates [56], aluminum-based carboxylates, and pyrazole-based MOFs have been reported [16,49]. These materials have improved stability which was achieved by increasing the coordination bonds strength of SBUs to organic ligands [49,37,57].

2.2.4 SYNTHESIS OF MOFs

There are numerous methods that are available for the synthesizing of MOFs. Many of these methods take place in liquid phase which involves the mixing of metal salt and ligand in a suitable solvent or preparing their solutions separately before mixing them. The important part in synthesizing MOFs with good properties is choosing a proper solvent by looking at features such as redox potential, reactivity, stability constant, and solubility [58]. Figure 2.5 illustrates a summary of several approaches for the preparation of MOFs. The main aspects and purpose of investigating different methods of MOF preparations are to determine the synthetic routes which can result in distinct inorganic building units without disintegration of the organic ligand. In this chapter, more focus is given to the selected preparation methods for synthesis of MOF structures and some are illustrated in Figures 2.5 and 2.6a–c.

2.2.4.1 Hydro/Solvothermal Method

The hydro/solvothermal process has been effectively implemented in preparing MOFs that exhibit nanoscale structural morphologies which are unattainable by

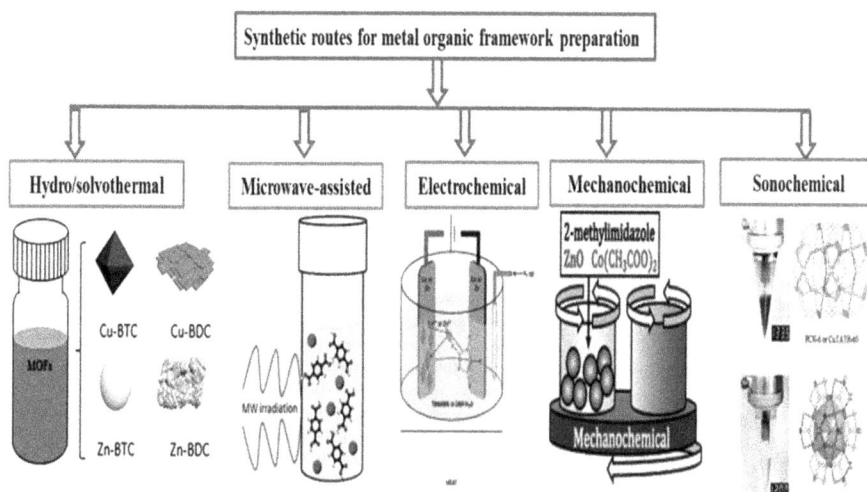

FIGURE 2.5 Overview of synthesis methods for preparation of MOFs.

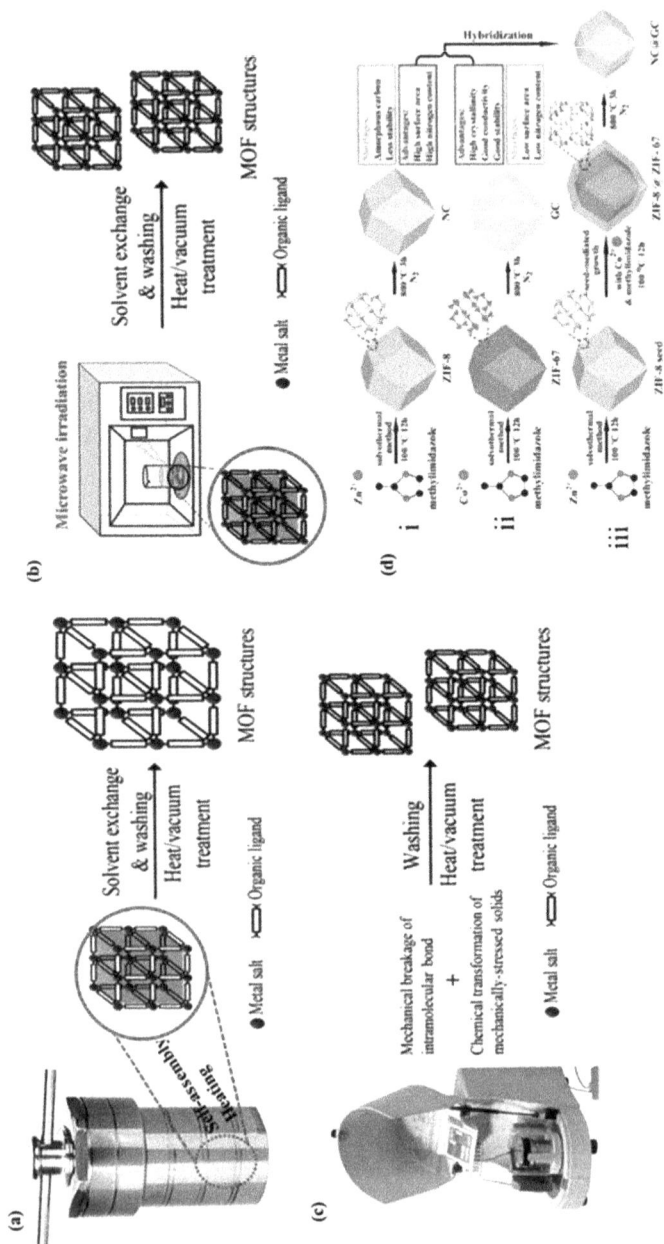

FIGURE 2.6 (a) Hydrothermal synthesis of MOF structures, (b) microwave-assisted solvothermal preparation of MOF structures. (c) mechano-chemical synthesis of MOF structures, and (d) schematic representation for the preparation of (i) MOFs (ZIF-8), (ii) ZIF-67 crystals and NC@GC, and core-shell ZIF-8@ZIF-67 crystals, and (iii) NC@GC via a conventional solvothermal method [61].

traditional techniques [33,59]. The important parameters in the preparation of MOFs are temperature, the choice of metal and organic linker, concentration of metal salt, and the solubility of the reactants [39]. The metal ions in this case are coordinated via functional groups of organic linkers to create a 3D porous paddle-wheel-like unit with a cubic structure. In addition, the use of traditional preparation schemes has often resulted in powdered MOFs with low densities. However, the low density property is not suitable for hydrogen storage as MOFs require high volume for hydrogen storage density with improved stability in humid and reactive environments, which are unavailable in slack powdered materials [60,61]. The hydrothermal method is usually carried out in a stainless steel autoclave as shown in Figure 2.6a. The method permits a precise regulation on the size and shape of the material to be synthesized, unlike conventional methods [36,62]. For example, Figure 2.6d shows schematic representation for the synthesis of ZIFs prepared by solvothermal techniques. It was seen that crystals slowly developed after heated solution of a hydrated metal salt, imidazole organic linker, and solvent [63]. This process is ideal for generating monocrystalline MOF materials.

2.2.4.2 Microwave/Ultrasonic Method

The microwave-synthesis technique has demonstrated to be an appealing route for fast preparation of nanostructured porous materials through hydrothermal conditions [35,64]. In addition to the rapid crystallization and high efficiency, this method offers potential advantages such as phase selectivity, narrow particle size distribution, and facile morphology control [33]. The microwave-assisted synthesis involves heating a substrate mixture in a suitable solvent with a microwave for an hour in order to yield nanosized crystals (Figure 2.6b). The quality of the crystals produced by the microwave-assisted process matches that produced by the ordinary solvothermal methods, however it is much faster [64].

2.2.4.3 Electrochemical Synthesis

The method of synthesizing MOFs electrochemically was initially described by BASF researchers in 2005 [33,36]. Electrochemical synthesis was developed in order to avoid the effects of anions from metal salts, such as nitrates, chlorides, and perchlorates. As a result, this method offered simplicity, high purity of the produced materials, and process controllability [33,36,65]. The process in electrochemical synthesis involves continuous introduction of metal ions instead of their salts into the anodic dissolution which moves them through the reaction medium to interact with the dissolved organic linkers at the cathode in the electrolyte [33,65,66]. To prevent metal from being deposited onto the cathode, protic solvents and compounds like acrylonitrile, acrylic, or maleic esters were used and H_2 gas was released during the process [23,61,65]. Furthermore, electrochemical synthesis offers an opportunity carry out a continuous process and attain a high solid content in industrial operation in comparison with normal batch reactions [65,67]. This method provides fast reaction rates and mild conditions which are essential in producing MOF materials on a large scale [36].

2.2.4.4 Mechanochemical Synthesis

The mechanochemical synthetic process is regarded as a solvent-free preparation procedure since it does not require any solvent for MOFs synthesis (Figure 2.6c). It proceeds through two steps, wherein the first step involves the mechanical disintegration of intramolecular bonds which is accompanied by chemical transformation [68]. The advantage of this solvent-free process is that reactions can proceed at room temperature without the use of an organic solvent. A qualitative yield of a small amount of product is obtainable in a short period of time ranging between 10 and 120 minutes. In several cases, the preference for the starting material was metal oxides rather than metal salts as they resulted in water as the only side product [69,70]. In recent years, mechanochemical synthesis has been effectively implemented for the fast preparation of MOFs through the use of liquid-assisted grinding (LAG). This process involves the addition of a small amount of solvent into a solid reaction mixture where it acts as a structure-directing agent. Furthermore, the method was developed to ion and liquid assisted grinding (ILAG) which demonstrated high efficiency for the selective manufacturing of pillared-layered MOFs [65,71]. Nonetheless, the process is restricted to explicit MOF types which cannot produce large amounts of product [72–74].

2.2.4.5 Sonochemical Synthesis

The sonochemical technique involves the use of high ultrasonic energy of about 20 kHz–10 MHz in a chemical reaction mixture. The ultrasound makes chemical/physical modifications through the cavitation method, where smaller bubbles are created and grow in a liquid state [43]. The bubbles collapse to form local hotspots having a short lifespan with temperature and pressure generating homogeneous nucleation centers and reducing the time for crystallization in comparison to traditional solvothermal processes [43,75].

2.2.4.6 Diffusion Method

The diffusion method is normally used in the preparation of materials to avoid the creation of polycrystalline powdered compounds by generating crystals that are suitable for single X-ray diffraction analysis [33,67]. The working principle behind this technique involves slowly bringing different species into contact and this can be achieved in different phases including gas, liquid, and gel diffusion [33,43]. The rate of reaction in all of these diffusions is influenced by one of the reactants. In solvent liquid diffusion, two layers with dissimilar densities form where one solvent contains the product and another solvent is for the precipitant. The two layers are further separated from each other by another solvent layer and the precipitant solvent gradually penetrates the separating layer and allows crystal formation at the interface [33,43]. Liquid phase diffusion involves the dissolution of the metal ions and organic linkers in immiscible solvents, whereas gas phase diffusion uses an organic ligand solvent that is volatile. In the gel diffusion method, MOF crystals form from mixing metal ion solution and organic ligands prior to dispersing them in a gel substance in order to prevent bulk material from precipitating. The gel substance helps to slow down the diffusion rate and the preparation of MOFs by the diffusion method, which is usually carried out in mild conditions and it is time-consuming [43].

2.2.4.7 Solvent Evaporation Technique

Solvent evaporation is another available technique for the synthesis of MOF coordination polymers. This synthetic route involves the formation of crystals through a gradually increasing concentration of mother liquor [33]. In the solvent evaporation method, the first step entails mixing precursors in a suitable solvent, followed by stirring the mixture to obtain a clear solution. The mixture is then exposed to a specific temperature in an inert environment for the solvent to evaporate and allow the crystals to grow [33,36]. The main advantage of this traditional technique of synthesizing MOFs is that it does not require any external energy supply and can be used at room temperature. However, it is time-consuming and poor solubility of reactants poses some drawbacks for the preparation of MOFs. The poor solubility of the reactants is then improved by mixing different solvents and the process is accelerated by using low-boiling-point solvents which can evaporate quicker [36,43].

2.2.4.8 Post Synthesis Method

Amongst other advantages of MOFs, it is their capability to incorporate complex functionalities into their framework, therefore generating a series of MOFs with diverse functionality while maintaining the same topology [76]. However, it remains a challenge to introduce functional groups onto the structure of an MOF during its synthesis. The above-mentioned drawback can be addressed through the post-synthesis modification (PSM) method, which is the chemical modification or functionalization of MOFs following their formation (Figure 2.7). The introduction of functional groups onto the structure of MOFs can be achieved by through noncovalent, coordinative, or covalent interactions [39,40,77]. Some of the simplest approaches employed in carrying out PSM are protonation and doping.

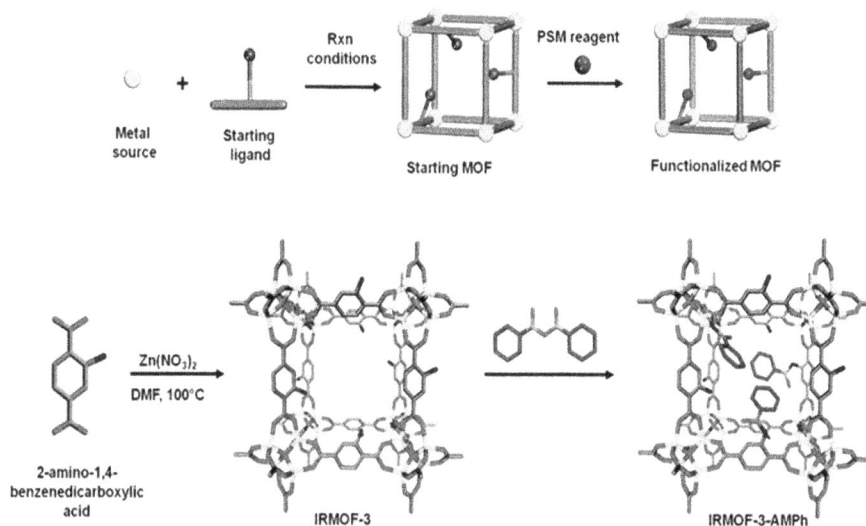

FIGURE 2.7   PSM of MOFs and functionalized ligands during MOFs synthesis [79].

Covalent interactions have been successfully used in PSM for amino function-alization of MOFs owing to the high reactivity of amino-functional groups. It must be noted that many PSM reaction methods make the functional groups increase in size and complexity in the pores, but there is the possibility of making the groups smaller and instantaneously unmask the protected functional group [47,78]. For dative PSM processes, there are two types that exist. Firstly, the one in which metal nodes bond coordinatively to the neutral sites of the linker. The second one occurs when metal ions coordinate to the organic linker by deprotonation-like transforma-tion of hydroxyl groups into alkoxides [47].

2.3   MOF NANOMATERIALS

The application of MOFs in various fields has led to the development of MOF nano-structures and MOF composites. Nanostructures of MOFs are formed when the size of the materials is reduced to nanoscale [80]. These type of materials have gained interest due to their properties which include luminescence, electrical and magnetic properties, enormous surface area, narrow pore volumes, and unique size-dependent optical behavior in comparison to traditional MOF polymers [65,80,81]. Furthermore, nanostructured MOFs are highly diverse in terms of composition, morphology, and characteristics, and have shown to be greatly dispersible and biocompatible. When comparing nanostructured MOFs with other nanostructured materials, their diver-sity in morphology includes nanocavities, hollow spheres, and polyhedrons [41]. The synthesis of these types of structures involves doping of inorganic nodes, which allows for modification of functional characteristics of MOF polymers while main-taining their coordination characteristics [65,81]. MOF materials are sometimes mixed with appropriate metals and graphene-based nanomaterials to obtain highly effective composite structures Sometimes, graphene-based nanomaterials are com-bined both with suitable metals and MOF structures. As a result, very effective composites are obtained. An example is a study conducted by Asadian and cowork-ers [82]. They were able to prepare nickel copper (NiCo) layered double hydroxide (LDH) nanosheets. They put the synthesized material on GCE and utilized it for detecting non-enzymatic glucose as depicted in Figure 2.8.

2.4   MOF COMPOSITES

A composite is a material that has multiple components with various phase domains, wherein one type of the domain is continuous [83]. Such materials are regularly employed in industrial systems, due to the combination of phase properties which are tunable. Hence, there are some studies on composites which have been conducted, such as improving gas sorption/separation and heavy metal adsorption [65,83,84]. Additionally, composites can be easily handled when comparing them to crystalline MOF polymers. In catalysis, the interest in composites lies in the combination of the catalytic activity of the dispersed phase and the stability of the continuous phase [65,83]. Formation of composites by modifying the surface of MOFs has been also investigated in biomedicine and found to improve the water dispersity and stability of

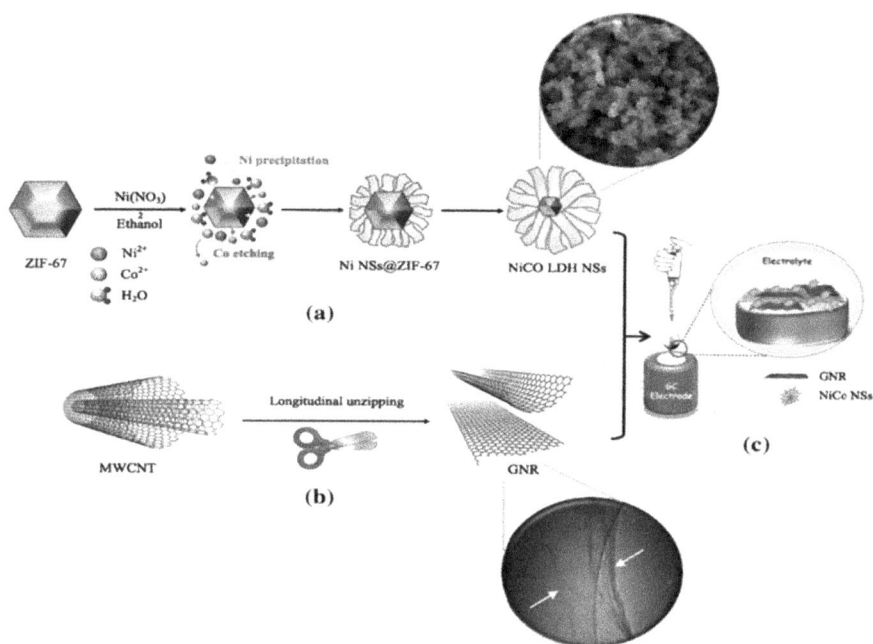

FIGURE 2.8   A carbon nanomaterial/MOF composite based sensor [82].

MOFs. This results in the enhancement of drug loading, reduction in plasma protein binding, avoided uptake by the reticuloendothelial system, etc. [85]. In addition to other applications, various MOF composites have been synthesized and studied, with many of them showing potential in enhancing efficiency for a specified application [84,86,87]. It was reported that incorporation of an ultrathin layer of MOF-derived nanocarbon on graphene oxide sheets resulted in the formation of a nanocarbon/ graphene oxide/nanocarbon sandwich-like structure with high specific surface area and excellent electronic conductivity (Figure 2.9) [88].

2.5   APPLICATIONS OF MOFS

The fascinating structural diversity of MOFs and their nanostructures and composites have earned these coordination polymers unlimited possibilities for a wide range of applications. MOFs can be tailored for a particular application through appropriate alteration of the structure. Literature reports have shown that properties of the system that have been measured correlate completely with chemical functions and matrices that have changed [85,89–91]. The surveys have shown that there are complete correlations between the measured properties of the systems and the changes occurring in their chemical functionality and matrix [92,93]. For this reason, a variety of design methodologies can be implemented in order to obtain the anticipated performances of the MOF materials. The following section provides an insight on some of the significant applications of MOF materials.

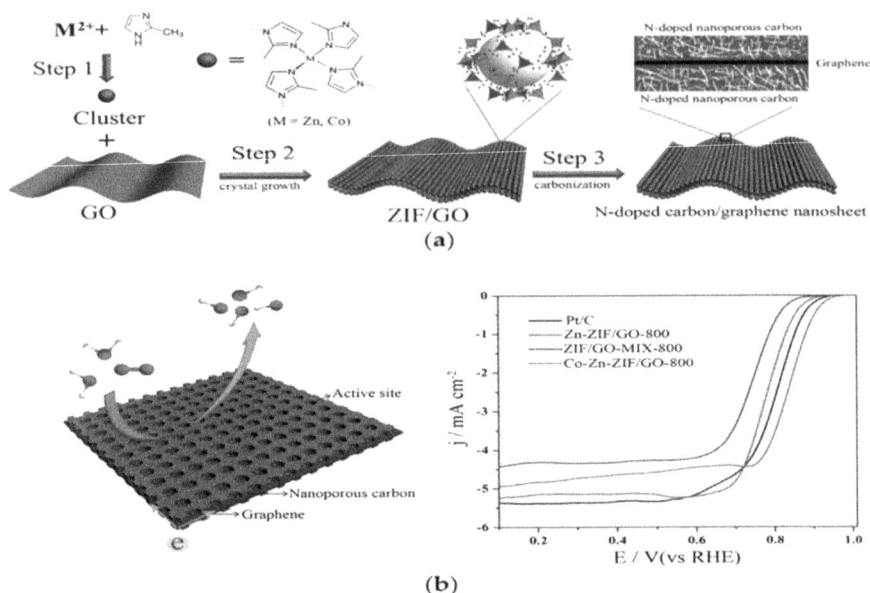

FIGURE 2.9 (a) MOF composite of ZIF/GO and N-doped nanoporous carbon/graphene/ nanoporous carbon sandwiches; and (b) their linear sweep voltammetry curves and oxygen reduction process and conducted in alkaline conditions [88].

2.5.1 BIOMEDICINE

The development of MOF composites and nanomaterials for drug delivery has been a growing research area in the field of medicine since the emergence of the use of pristine MOFs for drug delivery in the mid-2000s [94]. MOFs possess numerous ideal properties as drug carriers such as a high surface area which offers high drug loading ability, large pore size, and improved controlled release kinetics. Furthermore, the labile metal-ligand coordinated bonds and flexible functionality for post synthetic grafting of drug molecules offer the potential of the biodegradability of these materials [95]. MOFs' application as drug carriers is hindered by their insolubility in water which is addressed by surface modification and composite formation with components that are hydrophilic or coordinating with water-attracting ligands [94].

The utilization of MOF materials as drug carries can improve efficiency in drug delivery and reduce the side effects of the active pharmaceutical ingredient. The main objective in developing MOF drug carriers is to use metal ions and organic ligands that display insignificant toxicity in the human body system. Fewer toxic metals, which are widely used, include iron, calcium, copper, nickel, zinc, titanium, manganese, and their oral lethal dose is below 50 (LD50) within different metabolic body systems. In terms of organic ligands, the chosen ones are those prepared from natural compounds which have no effect on the body systems [33,94]. As an example, Haydar et al. [96] synthesized four different MOFs, namely Fe-100, Fe-MIL-101, Fe-MIL-53, and Ca-MOF using the hydrothermal method for the delivery of flurbiprofen. The Fe-MIL-100 and Fe-MIL-101 MOFs showed the loading percentages of 46 and 37%, respectively.

2.5.2 Sensors

The diversity in the structural designs and tunable surface chemistry of MOF properties has motivated their utilization in various sensing applications such as gas sensing, biosensing, electrochemical, and chemical sensing [81]. Generally, a sensor comprises two components, (i) a sensing unit which collects information and (ii) a transduction unit that is used for translating the collected information into either an electrical or optical signal [97]. MOF materials having magnetic or luminescence characteristics, as well as size/shape-selective adsorption, can offer potential for application in sensing devices. The preparation of luminescent networks can be achieved by using organic linkers and metal ions which are luminescent, organic, or by using the metal-to-ligand charge transfer [98]. Hence, the properties of MOFs that contribute significantly to their usage in developing an MOF luminescent sensing device include their structural characteristics, coordination environments, porosity, interactions with guest species, and non-covalent interactions. Lanthanides are promising candidates in preparing MOF-based sensors owing to their large coordination sphere (coordination number up to 12) as well as lacking strong coordination geometry [33,67,97].

MOFs may be used in electrochemical sensors which operate based on the oxidation and reduction reactions of the analytes. The measurement in electrochemistry takes place in an electrochemical system of two or three electrodes made of a working electrode, a counter electrode, or reference electrode. The quantitative measurement of species involved in the reaction can be achieved by measuring the current, electric potential, or other electrical signals. MOF composites have shown to be a good candidate for electrochemical gas sensing owing to their high surface area, pore volume, good absorbability, and high catalytic activity [97]. Mashao et al. [99] reported an electrochemical hydrogen gas sensor based on polyaniline/ZIF (Figure 2.10). They showed that electrochemical gas sensors based on MOFs have the capability to sense low concentrations (ppm level) of hydrogen gas without the requirement of heating.

2.5.3 Catalysis

In catalysis, the porosity and the thermal stability of MOF materials play a vital role. Moreover, the tunable nature of the size of the pores in MOFs allows them to be easily designed for the exact application as catalysts. The preferred catalysts are usually homogeneous catalysts owing to their higher efficiency and selectivity; however, their study is limited by the instability and limited recovery and separation of the spent catalyst. MOFs as solid materials offer potential in addressing the above-mentioned shortcomings since their framework can be altered to allow various catalytic reactions. An MOF possesses a heterogeneous catalytic characteristic due to its porous nature and structural properties such as high surface area, permanent

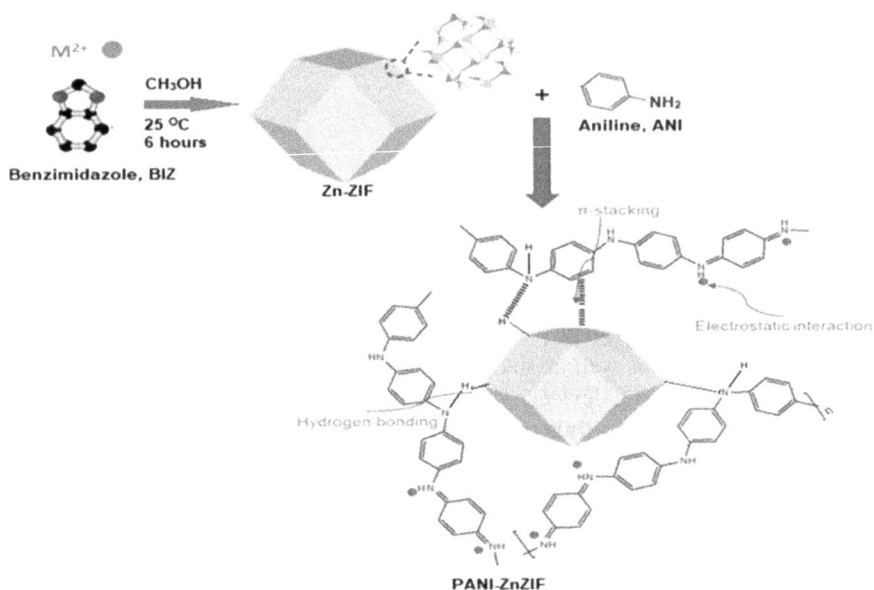

FIGURE 2.10 Schematic representation of polyaniline doped with ZIF for electrochemical hydrogen gas sensing [99].

porosity, and multifunctional ligands [100]. It was seen that the most vital characteristic responsible for its heterogeneous catalytic ability is lack of non-accessible volume. Literature shows that there are numerous of MOFs that have been used as heterogeneous catalysts for formation of organic oxidation, epoxidation, knovenagel condensation, aldol condensation, hydrogenation, Suzuki cross-coupling, or Friedel-Craft's alkylation reactions [101]. The advantage of employing an MOF catalyst in synthesis of organic compounds (Figure 2.11) is its ease in separation from the target product by centrifugation, and its reusability [102]. It was reported that rates of organic transformation depend on the surface area and pore size of the MOF since the catalytic reaction takes place on the outer surfaces of the MOF particles under UV radiation [102].

Figure 2.12 presents four strategies for the important role that MOFs play in catalysis [103]. The first approach is on the transition metal ions which can act as active sites, the requirement being vacant coordination sites on the metal centers. The second approach is based on the metal nanoparticles which are saturated into the pores of the MOF structure, however, they occupy the space in the pores that leads to a decrease in the surface area and pore volume which will reduce the catalytic activities. The third approach is on the organic linker which features the functional side-groups and possesses a possibility of post functionalization to enhance its catalytic activity. The last approach is on the preparation of mixed organic linker MOFs which are made from two isoreticular linkers that are disseminated in a mixed structure. It was seen that the challenges in the preparation of MOF catalysts are the introduction of unsaturated metal centers (UMC) and creation of void spaces into the

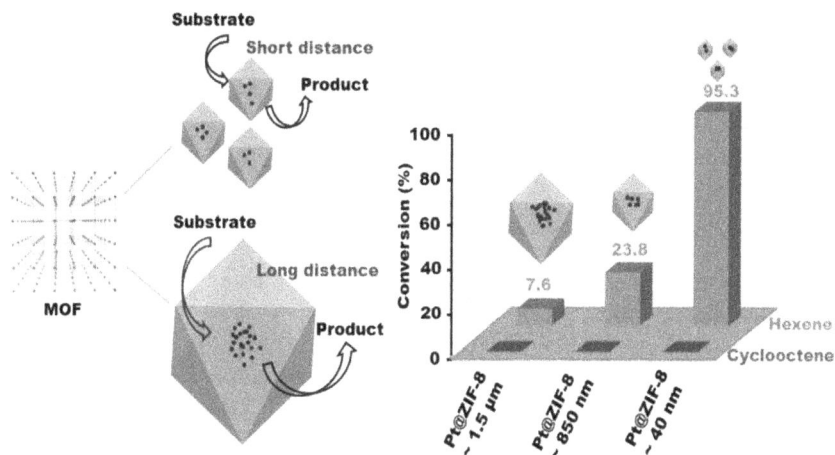

FIGURE 2.11 Formation of organic compounds using MOF catalyst and transport limitations in MOFs indicated by particle size effects on conversions [32].

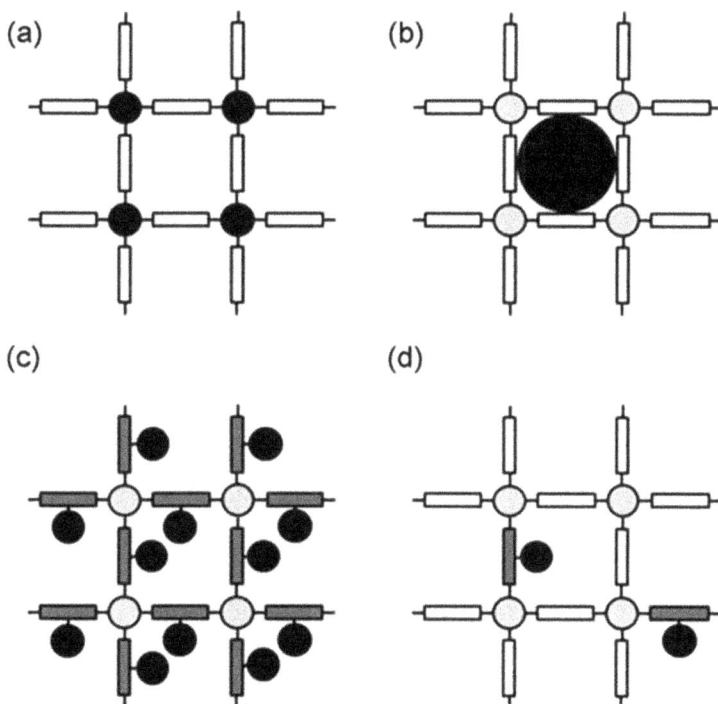

FIGURE 2.12 Different approaches for the use of MOFs in catalysis: (a) framework metals as active centers, (b) generation of metal nanoparticles inside the pores, (c) immobilization of active metal complexes via functional side-groups of the linker molecule, and (d) synthesis of mixed-linker MOFs (MIXMOFs). Black circles represent active metal centers and gray circles symbolize metal centers which are only used to build up the framework (no free coordination sites) [103].

framework structure. Current studies have shown that transition metal ions and multifunctional organic ligands are used as the building blocks and specific structure to obtain microporous MOFs [95,104]. In addition, the introduction of N-heterocyclic carbenes (NHC) into MOFs makes powerful heterogeneous catalysts and allows multiple catalytic active sites [33,36].

2.5.4 GAS STORAGE AND SEPARATION

It was documented that a porous adsorbent material has the ability to safely and economically store gas at lower pressure due to high surface area [33]. These kinds of porous materials are the candidates to replace the conventional storage methods of high pressure and multi-stage compressors. The literature survey shows that the reports on the utilization of activated carbon, carbon nanotubes, zeolites for gas storage, and interest in MOFs as gas storage materials is increasing owing to their tunable structural properties. MOF materials have been applied in hydrogen gas storage whereby hydrogen molecules are adsorbed on the surface of the pores of the MOF structure through a physisorption process [39,45]. There are number of MOF materials that exhibit outstanding performance for hydrogen adsorption under cryogenic conditions and high pressure up to 100 bar [85,105], However, their hydrogen store efficiency was shown to be low at ambient conditions [39,51,106]. Recently, extensive studies have been reported for the fabrication of MOF materials having high interaction energies for hydrogen gas at ambient conditions [51,67,103]. It was shown that the hydrogen uptake relates to the surface area of the material, therefore the enhancement of hydrogen adsorption may be achieved by an increase in surface area [105]. So far more than 150 microporous MOFs have been investigated for H_2 adsorption at different conditions. For instance, the hydrogen uptake was tested for the documented carboxylate-based MOFs such as MOF-5 and MOF-177 prepared from [Zn_4O] cluster with di- and tri-carboxylate ligands, respectively [51]. It was seen that MOF-5 possessed a BET surface area of 3800 m^2/g with 7.1 wt% of H_2 uptake at 40 bar and 77 K [51]. On the other hand, MOF-177 material with BET surface area of 4750 m^2/g showed 7.5 wt% of H_2 adsorption at 70 bar and 77 K [107]. Jihoon et al. [108] reported the preparation of Pt and carbon black (CB) impregnated MOF-5 composite, CB/Pt/MOF-5, possessing high hydrogen storage potential up to 0.62 wt% over pristine MOF-5 (0.44 wt%) at moderate temperatures and pressures. They further showed that the CB/Pt/MOF-5 composite has moisture-resistant capability as compared to pure MOF-5. Hu et al. [109] have synthesized Pt impregnated on MOF/graphene oxide (GO) composites to form Pt@HKUST-1/GO and Pt@ZIF-8/GO. They indicated that the composites possessed increment in hydrogen uptake with respect to their pure MOFs [28].

2.5.5 WATER PURIFICATION

The selection of a water treatment method is based on the initial quality of water, parameters established by regulations, and the intended usage of the water after purification. These methods are regarded as physical, chemical, or biological treatment based on the mechanism of removal of pollutants [59,110], as presented in Figure 2.13a.

The conventional treatment methods are inefficient for potable uses especially against raw water containing low concentrations of pollutants [49,111]. Treated water obtained from conventional methods can be reused for irrigation of crops or landscapes, refilling of aquifers, and non-potable urban uses. Nevertheless, traditional remediation methods do not produce water that is sufficient enough for reusability in industrial applications (such as cooling and boiler feed) and also drinking [112]. Hence, there is a need to develop an improved wastewater treatment technology as demonstrated in Figure 2.13.

2.5.5.1 Adsorption of Organic Pollutants

MOFs are promising porous adsorbent materials for the removal of organic contaminants from the environment due to their easy separation, high surface area, and high selectivity towards removal of contaminants. MIL-101 material as one

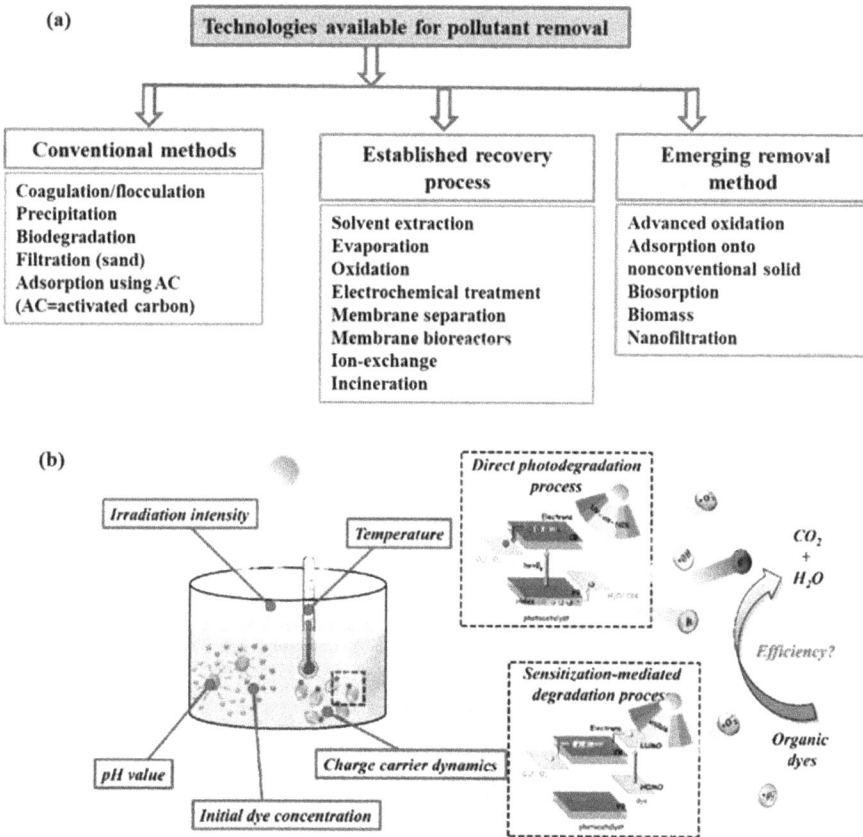

FIGURE 2.13 (a) Several technologies available for removing contaminants from wastewater. (b) Elimination of organic dyes from wastewater through photodegradation process [21].

of the MOF materials has a zeotype crystal structure with high resistance to air, water, and common solvents. It is a key property for an adsorbent for application in the pretreatment of aqueous-containing samples. Zhou et al. [113] prepared novel magnetic Fe_3O_4@MIL-100(Fe) nanoparticles for removal of organochlorine pesticides from tealeaves in a mechanochemical magnetic solid phase extraction (MCMSPE). The Fe_3O_4@MIL-100(Fe) material was prepared employing the step-by-step method, shown to be effective towards organochlorine pesticide recovery and also be reused with no significant changes in adsorption capacity after several cycles [113]. The findings showed that MOF materials are ideal recyclable adsorbents for removal of organic contaminants. Furthermore, metal oxides have been used to adsorb organic dyes which have become a hazardous contaminant from industries in the environment [30]. Wang et al. [114] investigated the utilization of magnetic Fe_3O_4/MIL-101(Cr) nanomaterial for removal of organic dyes such as acid red 1 (AR1) and orange G (OG). They reported that the prepared Fe_3O_4/MIL-101(Cr) nanomaterial possessed adsorption capacities of 142.9 and 200.0 mg/g for AR1 and OG, respectively. The results showed that MOF-based magnetic core-shell materials as good candidates as for dye removal from wastewater. Moreover, Yang et al. [115] reported a preparation of Fe_3O_4-PSS@ZIF-67 nanocomposite for selective adsorption of methyl orange (MO) from solution mixture of MO and methylene blue (MB) (Figure 2.14). They demonstrated that the adsorption capacity of the magnetic nanocomposites for MO was measured to be 738 mg/g with the separation rate of up to 92%.

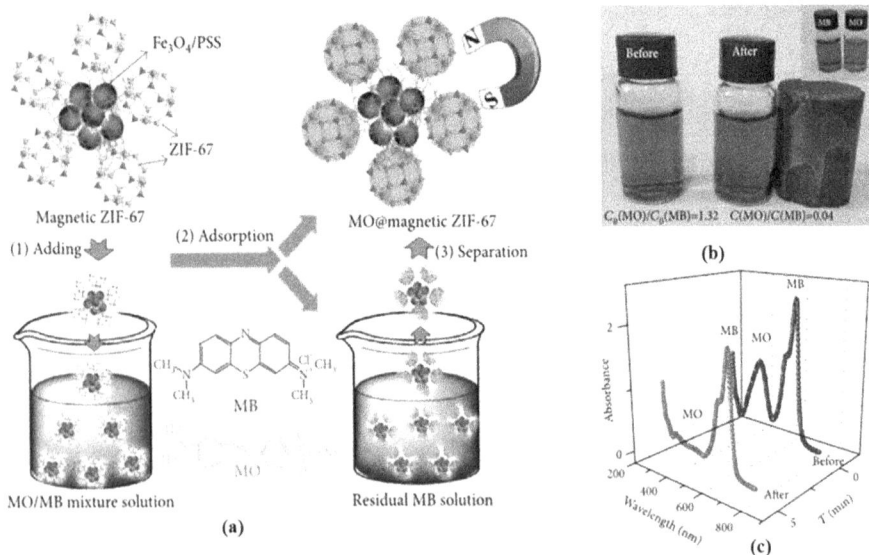

FIGURE 2.14 Scheme of selective adsorption of MO from the mixed MO/MB solution by Fe3O4-PSS@ZIF-67 magnetic composites; photographs (b) and UV-vis spectra (c) of the mixed MO/MB solution before and after magnetic separation [115].

2.5.5.2 Photodegradation of Organic Pollutants

It was well documented that heterogeneous photocatalysis (HP) (Figure 2.13b), is regarded as an emerging itinerary for water treatment for removal of organic dyes through photodegradation [21,116]. In HP, the photocatalyst is irradiated with UV light to separate charges followed by production of the reactive oxygen species (ROS) [70,117,118]. The material which is used in HP is known as a photocatalyst. It is a semiconductor, possessing the valence band (VB) and conduction band (CB). The VB is known as the highest occupied molecular orbital (HOMO) whereas the CB is the lowest unoccupied molecular orbital (LUMO) level. The distance between these levels is called band gap energy (E_g) [70]. For instance, the photocatalytic mechanism of pollutant removal is presented in Figure 2.13b. Once a semiconductor is irradiated with UV light, it absorbs light with energy which is equivalent to its bandgap energy ($\geq E_g$) [119]. The electrons (e^{-s}) in the material are promoted from the VB to the CB for photogeneration of charges and leave an electron hole, h^+ behind [70]. At excited state there are several pathways that photogenerated charges can take, such as recombination, releasing the excitation energy as heat, migration to the surface of the photocatalyst, or production of the ROS [117,119]. Lastly, OH is produced through the water oxidation process which is achieved by the h^+, whereas superoxide radical anions ($O_2\cdot^-$) are generated via adsorbed oxygen reduction mechanism. On the other hand, it was seen that the protonation process may take place to oxidize this $O_2\cdot^-$ to hydroperoxyl radicals ($HO_2\cdot^-$) [43,120]. It was observed that these oxidant species, together with direct oxidation by h^+, are capable of mineralizing the organic dye to CO_2 and H_2O [116].

2.5.5.2.1 Photocatalysts for Wastewater Treatment

Fast-growing interest in the area of photocatalysis for wastewater treatment has arisen in the manufacture of different photocatalysts such as metal oxide [21,121] and metal sulfides [122]. Table 2.1 presents some of the most investigated photocatalysts in water purification. Several investigations have concentrated on the use

TABLE 2.1

Some of the Most Investigated Photocatalysts in Water Purification

Photocatalyst	Purpose	Refs.
TiO$_2$	Photocatalytic degradation activity	[123]
	Photocatalysis for treating bacteria	[131]
	Photocatalyst in water treatment technology	[134]
ZnO	Photocatalytic degradation activity	[123]
	Photocatalyst in water treatment technology	[131]
WO$_3$	Photocatalytic degradation activity	[123]
CuS/ZnS	Exceptional visible-light driven photocatalytic activity	[124]
MoS$_2$/CdS	Enhanced visible-light photocatalytic activities	[130]
MOF	Photocatalytic degradation activity	[135]

of TiO_2 for water purification [123,124] owing to its high activity in photocatalysis. However, the application of TiO_2 powder has several setbacks such as poor porosity, low adsorption, and its difficulty in recovery [117]. In addition, it was seen that the photocatalytic activity of TiO_2 anatase, possesses a band gap energy of 3.2 eV ($\lambda \geq$ 387 nm) which is high and needs to be activated by UV radiation [21,123,125]. As a result, surface modification of TiO_2 by addition of carbon, graphene, or metal deposition, has been used in photocatalysis [123,125]. However, the preparation of a photocatalyst based on TiO_2 for practical wastewater treatment using visible and solar light is still a challenging boundary. Consequently, it is imperative to explore competent, robust, and cost-effective photocatalysts for replacement of the traditional ones. In the last two decades, a type of crystalline material called MOFs have received consideration in photocatalysis. MOFs offer a wide range of applications owing to their structural arrangement of coordination bonds between unsaturated metal core/node and multidentate organic linkers (catalytically active) [126]. Furthermore, their large surface area and well-ordered porous structures have significantly contributed towards their interest in numerous fields. In HP, the utilization of MOFs as photocatalysts is mainly based on three aspects: (1) encapsulating chromophores in the internal structure of the MOF, (2) promoting e^-/h^+ separation in the metal core, or (3) preparing MOFs using materials which have absorption bands in the visible region [122,119]. Moreover, some MOFs can serve as semiconductors (e.g. MOF-5 [127], NTU-9 [128], and UiO-66 [129]), in which the energy transfer takes place from the organic linker to the metal-oxo cluster [130]. Nonetheless, most MOF photocatalysts have a large band gap due to their poor conductivity caused by the insulator characteristic of the organic linker [21,124,131] that can simply harvest UV light, which immensely limits their further application [34]. There are several dissimilar approaches, including dye sensitization [132], decoration of linker or metal center [130,133], and combination with other semiconductors [134]. Hence, surface modification and functionalization of MOFs are required for their application as suitable photocatalytic materials.

2.5.5.2.2 Photocatalytic Degradation of Dyes using MOFs

The application of MOFs in water purification is subject to their photochemical reaction and stability. This means that MOFs must retain their structural and textural properties during photocatalytic activity. Several organic dyes such as MB, MO, and rhodamine blue (RhB) have been used as target pollutants and given in Table 2.2. The studies showed some of the approaches to establish photoactive materials which are founded from MOFs, as presented in Figure 2.15a [43,84,132]. Type I strategy uses the semiconductor dots characteristics of the metal cores in MOFs that behave like isolated nano-semiconductors which are sequestered by the organic ligands [122]. The type I MOFs are very effective compared to traditional semiconductors owing to their excessively porous nature. This favors the adsorption of contaminants that are near the semiconductor and photogenerated charges. Moreover, their high density of photoactive dots and the organic ligands which serve as antennae to absorb light, will accordingly enhance the photoresponse of the MOF as a catalyst [56,136]. In type II MOFs, dye-based organic linkers with photoresponse are used for absorption of light as well as transfer of photogenerated charges to the metal cores [118,137].

TABLE 2.2
Photocatalytic Degradation of MOF-Based Materials for Organic Dyes

MOF-Based Photocatalysts	Organic Dye	Irradiation Time (min)	PDE%	Refs.
MOF, $[Cu(4,4'-bipy)Cl]_n + H_2O_2$	MB	150	94	[119]
MOF, $[Co(4,4'-bipy)(HCOO)_2]_n + H_2O_2$	MB	150	55	[119]
Fe_2O_3/MIL-53(Fe)	MB	240	70	[143]
BiOBr/NH_2-MIL-125	RhB	100	100	[144]
Bi_2MoO_6/MIL-100	RhB	90	90	[145]
Ag_3PO_4/MIL-53(Fe)	RhB	90	100	[129]
g-C_3N_4/MIL-125	RhB	60	100	[146]
g-C_3N_4/MIL-100	RhB	240	100	[133]
g-C_3N_4/MIL-53(Al)	RhB	75	100	[147]
rGO/NH_2-MIL-125	MB	30	100	[148]
rGO/MIL-88(Fe)	RhB, MB	20	100	[136]
GO/MIL-101(Cr)	MG	60	92	[149]
Au@MIL-100(Fe)	MO	150	100	[150]
Pd@MIL-100(Fe)	MO	150	100	[150]
Pt@MIL-100(Fe)	MO	150	100	[150]
MIL-53(Fe)	Phenol	180	99	[137]
NH_2-MIL-53(Fe)	Phenol	180	92	[137]
Fe(BDC)(DMF)	Phenol	180	99	[137]
MIL-53(Fe)	RhB	50	98	[151]

Type III MOFs are regarded as the simplest route where MOFs act as a porous matrix in which the photoactive species are compressed within its structure [138]. However, the main drawback of these MOF materials in photodegradation is their stability in water. In 2007, Alvaro et al. [139] described the photocatalytic degradation of phenol in water by MOF-5. In addition, Hausdorf et al. [138], observed that the instability of MOF-5 depends on structural modification and water environment. The photodegradation of rhodamine 6G (R6G) by Fe-MOFs under visible light (550 nm) was studied by Laurier et al. [140]. They have observed that Fe-MOFs were better catalysts than the conventional TiO_2 and their structural properties were reasonably maintained after photocatalytic activities. Other MOFs for photodegradation of MB organic dye have demonstrated high photodegradation efficiency (PDE%). Typical examples in this chapter are presented in Figure 2.15b,c and Table 2.2. It was reported that Cd(II)-imidazole MOFs for photodegradation of the MB and MO under UV light generates the photogenerated charges that are vital for photocatalytic degradation of organic dyes [111]. Furthermore, Zhang et al. [141] examined other types of Cd(II)-imidazole MOFs for photodegradation of MO and revealed that the bandgap energy, efficiency in the transference, and separation of charges are the factors controlling the photogeneration of charges. On the other hand, Zn(II)-imidazolate MOF (ZIF-8)

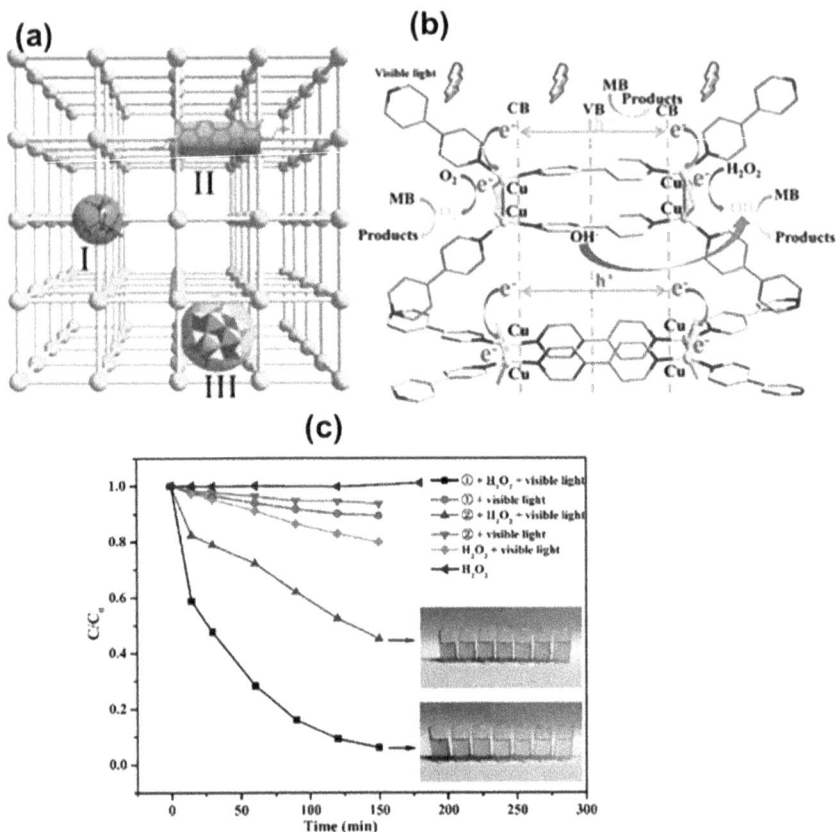

FIGURE 2.15 (a) Demonstration of the types of MOF developed for photocatalytic applications [70], (b) typical degradation mechanism of organic dyes using MOF as a photocatalyst, and (c) effect of irradiation time on MB using MOF under UV radiation [142].

also showed a high photocatalytic efficiency for the removal of MB UV radiation [121]. Du et al. [121] reported a synthesis and application of MIL-101 in the photodegradation of Remazol black B (RBB) dye. They have found that MIL-101 has high crystallinity, specific surface area (3360 m^2g^{-1}) and high stability in water and several organic solvents. The results showed that MIL-101 as a heterogeneous photocatalyst in the degradation reaction of RBB with 95% PDE after fourth cycle and the photocatalytic mechanism was through electron transfer from photoexcited organic ligands to metallic clusters in MIL-101 [121].

2.6   ADSORPTION OF HEAVY METAL IONS

Heavy metals are originally found in the natural environment from different types of rocks (i.e. igneous, metamorphic, and sedimentary). They interact with their surrounding environment through several processes including weathering, soil erosion, and soil formation and result in the accumulation of these metal ions in a higher toxic concentration [152]. The pollution of water by heavy metal ions such as copper (Cu), chromium (Cr), lead (Pb), cobalt (Co), nickel (Ni) PGMs, and mercury (Hg) expelled from industrial activities is of major concern owing to their numerous toxicological effects to human health and the environment [9,112]. The exposure to some of these heavy metal ions is accompanied by severe and irreversible effects even in lower concentrations [153]. The employment of adsorption technology as a feasible technique for wastewater treatment has been widely investigated owing to its simple designs and operation at a lower cost, production of less harmful secondary byproducts, and possible regeneration of the adsorbent. In the adsorption process, the adsorbent (usually porous solids) interacts with the adsorbate of suitable size and shape through physical (adsorptive) or chemical (reactive) adsorption [11]. The physical/adsorptive adsorption mechanism involves the entrapment of adsorbates into the pores of the adsorbents via van der Waals forces. In chemical/reactive adsorption, the mechanism of adsorbate-adsorbent interaction is through the formation of a chemical bond. The advantage of physical adsorption is the easy regeneration of the spent adsorbent using solvents exchange or by physical treatment like sonication, as compared to chemical adsorption which requires chemical treatments. Various mechanisms of adsorption that are possible between the adsorbate and adsorbent are represented in Figure 2.16 [37]. The effectiveness of the adsorption process is evaluated based on the uptake capacity by the adsorbents, specific selectivity, and of the rates mass transfer [112].

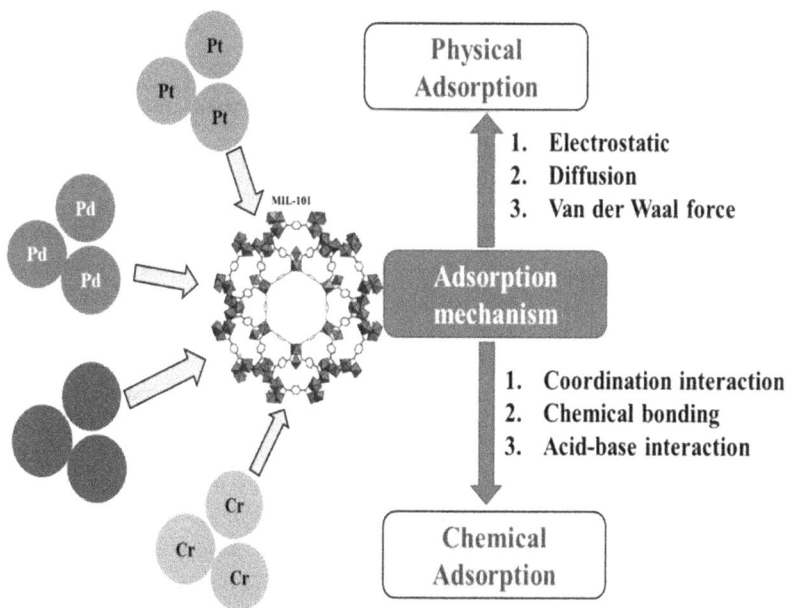

FIGURE 2.16 Various mechanisms of adsorption that are possible between the heavy metal and MIL-10l.

MOFs offer good aspects in the removal of heavy metal pollutants from wastewater due to their structural diversity, higher surface area, tenability of their pore size, and enormous porosity. Furthermore, MOFs have shown to have significant partitioning coefficients that indicate higher adsorption capacity for heavy metals regardless of the initial conditions [59,154]. In some cases, MOFs capture inorganic pollutants on their nodes via pseudo-ion-exchange processes whereby less strongly coordinated organic linkers are removed by the inbound contaminant [56]. As an example, Wu et al. [155] synthesized thiol-functionalized copper terephthalate $Cu_4O(BDC)$ MOF via PSM for the removal of Hg^{2+} ions. They obtained the maximum adsorption capacity of 405.6 mg/g at an equilibrium time of 90 min. In another study, Lu and coworkers [156] prepared a sandwich-structured MOF/graphene oxide (MIL-101(Fe)/GO) composite for the adsorption of Pb^{2+} ions from simulated wastewater. Though the composite showed the decrease in specific surface area from 1777 to 377 m^2/g, the Langmuir adsorption capacity was increased from 71.2 to 128.6 mg/g at an equilibrium adsorption time of 15 min. Lin et al. [157] synthesized three zirconium based MOFs UiO-66, UiO-66-NH$_2$, and UiO-66-NHCOCH$_3$ through the hydrothermal method. The MOF materials were tested for the selective removal of Pd(II) and Pt(VI) over Co(II) and Ni(II) competing ions. The functionalized MOFs showed an increase in the adsorption capacity towards Pd(II) and Pt(VI). However, the presence of competing ions was shown to have an effect on the selectivity of the uptake of PGMs by the MOF-based materials. Table 2.3 shows some of the reported MOF-based adsorbent for the removal of various heavy metal ions.

TABLE 2.3
Adsorption Capacity of MOFs for Heavy Metal Removal

MOF-Based Adsorbent	Targeted Metal Ion	*Q_{max} (mg/g)	Refs.
UiO-66 and UiO-66-NH$_2$	U(VI)	109.9 and 114.9	[158]
ED-MIL-101	Pb(II)	87.64	[15]
Fe$_3$O$_4$@AMCA-MIL53(Al)	U(VI) and Th(IV)	227.3 and 285.7	[159]
Cu-MOFs/Fe$_3$O$_4$	Pb(II)	219	[154]
Azine-Decorated Zn(II) MOF (TMU-4, TMU-5, and TMU-6)	Pb(II)	237, 251, and 224	[160]
Fe$_3$O$_4$–Pyridine)/Cu$_3$(BTC)	Pd(II)	105.1	[161]
MOF-802, UiO-66 and MOF-808	Pd(II)	25.8, 105.1, and 163.9	[91]
MIL-101-triglycine	Co(II)	232.6	[162]
MIL-101(Cr)-NH$_2$	Pd(II) and Pt(IV)	277.6 and 119.5	[12]
MIL-101-TEPA@CA Beads	Pb(II)	273.59	[163]

* = Langmuir Adsorption Capacity

2.6.1 PHOTODEGRADATION OF HEAVY METALS

The MIL-101, ZIF-8, UiO-66(Zr), and MIL-125(Ti) materials are the most studied MOFs in photocatalytic degradation [164]. Their photocatalytic mechanism relates to type II, MIL-53, and MIL-88B (all Fe-based type I MOFs) where the light is absorbed by the Fe-O cluster to photogenerate electrons and transferred from O^{2-} to Fe^{3+} [118,165]. In addition, the additional strategies were reported in literature showing the promotion of photocatalytic activity of MOFs by enlarging their visible light absorption, thus favoring their behavior under solar light [118]. One methodology comprises of functionalizing the organic linker or metal core to shift its photoresponse to a lower energy band gap. Shi et al. [165] prepared different amine-functionalized Fe-MOFs (NH_2-MIL-88B, NH_2-MIL-53, and NH_2-MIL-101) on the organic linker with high stability and activity for treatment of Cr(VI) via the photoreduction mechanism under visible light. They established that the presence of an amine group in an MOF structure increased the photodegradation efficiency (Figure 2.17a). Based on the observation, they projected a possible mechanism of photoreduction based on a dual excitation as presented in Figure 2.17b. The figure shows the promotion of electron transfer and reduction of charge recombination. Based on the reported literature about adsorption and photodegradation, MOF-based materials offer good potential as candidates for the removal of heavy metal ions from wastewater. However, selectivity for some of the reported composites is still a challenge.

2.6.2 ION-IMPRINTING TECHNIQUE

The ion-imprinting technique is a promising technology that has been investigated for the removal of targeted pollutants from wastewater [166]. This method

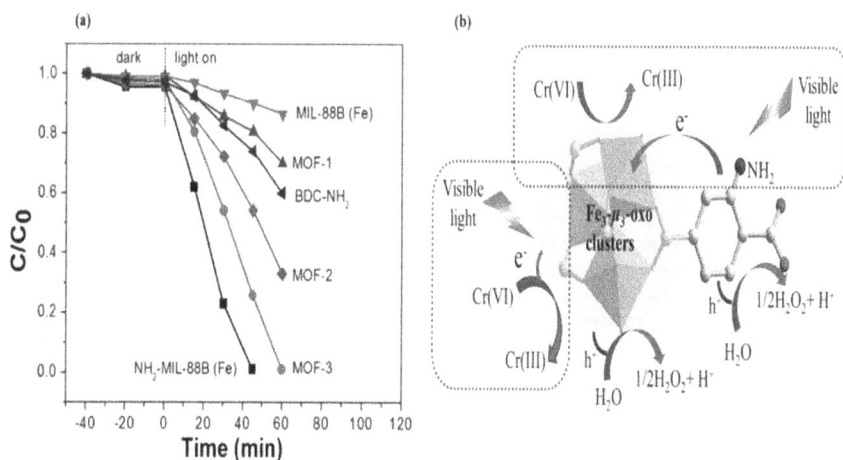

FIGURE 2.17 Photocatalytic reduction of Cr(VI) to Cr(III) using MOFs; (a) photocatalytic efficiency and (b) possible mechanism [165].

involves the synthesis of polymer materials that can selectively interact with the specific targeted metal ions in the presence of competing ions. The first step of an ion-imprinting polymer (IIP) entails the interaction of the targeted metal template with an appropriate functional monomer to form a complex. This is followed by the polymerization of the formed complex in the presence of an appropriate linker to form a 3D crosslinked structure [167,168]. Once the structure has formed, the metal ions template is removed from the framework by leaching to allow the selective rebinding of the targeted metal ions into the remaining vacant sites even in the presence of competing ions [169]. Different composite materials have been reported for the selective removal of heavy metal ions [10,169,170]. MOFs offer great potential as adsorbent material due to their diverse structures which can be easily tuned by ion imprinting to improve their adsorption functionality and selectivity for the removal of heavy metal ions from wastewater. In a study conducted by Yuan and coworkers in 2018 [20], an ion-imprinted glycine UiO-66-NH_2 MOF was successfully synthesized for the selective adsorption of Co(II) ions from wastewater. The ion-imprinted composite (Co(II)-IIP) showed an improved maximum adsorption capacity of 175 mg/g as compared to the non-imprinted composite (NIP) which had a maximum adsorption capacity of 125 mg/g (Figure 2.18a) [20]. Furthermore, the ion-imprinted UiO-66-NH_2 MOF had a higher surface area of 482.46 m^2/g than the non-imprinted UiO-66-NH_2 MOF which had a surface area of 471.65 m^2/g [20]. They came with adsorption mechanisms of Co(II) on Co(II)-IIP which are shown in Figure 2.18b. The cobalt ions in the figure act as a template agent in the preparation process of Co(II)-IIP. When the cobalt ions are removed by suitable solvent, the composite selectively adsorbs Co(II) from the aqueous solution because of the template recognition for cobalt ions. It was seen that the Schiff base nitrogen, carboxyl oxygen, and carbonyl oxygen in the framework are the main coordination atoms in the Co(II)-IIP [20].

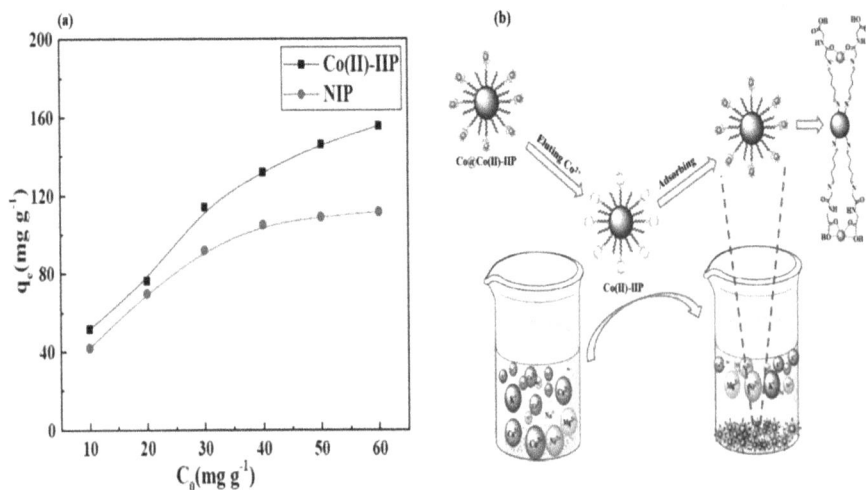

FIGURE 2.18 Effect of the cobalt(II) concentration on the sorption of cobalt (II) onto the Co(II)-IIP and NIP (t = 5 h; T = 298 K; V = 50 mL; m = 10 mg; pH = 8.4) [20].

Their discovery of ion-imprinted polymers based on MOFs provides new opportunities for the functionalization of MOFs in recovery of PGMs from wastewater.

2.7   SUMMARY AND FUTURE PERSPECTIVES

The presence of high concentrations of toxic heavy metal ions in the available water resources continues to be a global challenge. The priority on the removal and recovery of some of the precious metals is a growing research topic in the field of wastewater treatment. Amongst the number of technologies that have been investigated, the adsorption process has shown to be effective in removing various heavy metal ions from wastewater with the appropriate choice of the adsorbent material. MOF materials have gained popularity for various applications including in heavy metal ions removal. In this review, MOF materials were briefly discussed starting from their interesting structural arrangement, followed by their various synthetic routes then finally to their various applications. MOFs have been exploited for various applications due to their highly porous nature, higher surface area and tunable pore sizes which affect their functionality. Numerous data have been collected for comparison of MOF-based materials with other porous materials for each specific application. For application in wastewater treatment, MOF-based materials have demonstrated some significant improvements in both the adsorption and degradation of pollutants. This review summarized some of the work done on the adsorption of heavy metal ions (including PGMs) by MOFs which showed an increase in the adsorption capacity. However, there were some MOF-based composites which lacked selectivity towards the adsorption of heavy metals of targeted contaminants. To overcome this challenge, the ion-imprinting technique was employed for the selective removal of specific pollutants from wastewater and the capacity improved further. In literature, there are only a few reports on ion-imprinted MOFs for the adsorption of heavy metal. The process of ion imprinting is implemented during the synthesis of the adsorbent materials to generate footprints of the targeted pollutants on the surface of the adsorbent. Considering the tunable nature of MOFs and the ion-imprinting technique, there is a future perspective for the development of a highly porous MOF composite with higher removal uptake and high selectivity towards targeted pollutants.

ACKNOWLEDGMENTS

The authors immensely acknowledge the financial support from the National Research Foundation (NRF) (Grant Nos. 117727, 117984 and 118113), Tshwane University of Technology and University of Limpopo Research Grants (R202, R232 and R355), South Africa.

REFERENCES

1. Glaister BJ, Mudd GM, The environmental costs of platinum—PGM mining and sustainability: Is the glass half-full or half-empty? *Miner Eng* 2010;23: 438–50.
2. Mpinga CN, Eksteen JJ, Aldrich, C, Dyer L, Direct leach approaches to platinum group metal (PGM) ores and concentrates: A review, *Miner Eng* 2015;78: 93–13.

3. Mudd GM, Jowitt SM, Werner TT, Science of the total environment global platinum group element resources, reserves and mining—A critical assessment, *Sci Total Environ* 2018;622–623: 614–25.

4. Maponya TC, Ramohlola KE, Kera NH, Modibane KD, Maity A, Katata-Seru LM, et al., Influence of magnetic nanoparticles on modified polypyrrole/m-phenylediamine for adsorption of Cr(VI) from aqueous solution, *Polymers* 2020;12: 679–95.

5. Suoranta T, Zugazua O, Niemelä M, Perämäki P, Hydrometallurgy recovery of palladium, platinum, rhodium and ruthenium from catalyst materials using microwave-assisted leaching and cloud point extraction, *Hydrometallurgy* 2015;54: 56–2.

6. Can M, Bulut E, Özacar M, Reduction of palladium onto pyrogallol-derived nano-resin and its mechanism, *Chem Eng J* 2015;275: 322–30.

7. Sharma P, Bhardwaj D, Tomar R, Tomar R, Recovery of Pd (II) and Ru (III) from aqueous waste using inorganic ion-exchanger, *J Radioanal Nucl* 2007;274: 281–86.

8. Wei W, Lin S, Reddy DHK, Bediako JK, Yun Y, Poly(styrenesulfonic acid) -impregnated alginate capsule for the selective sorption of Pd(II) from a Pt(IV) -Pd (II) binary solution, *J Hazard Mater* 2016;18:79–9.

9. Utembe W, Matatiele P, Gulumian M, Hazards identified and the need for health risk assessment in the South African mining industry, *Hum Exp Toxicol* 2015;34: 1212–21.

10. Mao J, Lin S, Juan X, Hui X, Zhou T, Yun Y, Ion-imprinted chitosan fiber for recovery of Pd (II): Obtaining high selectivity through selective adsorption and two-step desorption, *Environ Res* 2020;182: 108995.

11. Khan NA, Hasan Z, Jhung SH, Adsorptive removal of hazardous materials using metal-organic frameworks (MOFs): A review, *J Hazard Mater* 2013;244–245: 444–56.

12. Lim C, Lin S, Yun Y, Highly efficient and acid-resistant metal-organic frameworks of MIL-101(Cr)-NH$_2$ for Pd(II) and Pt(IV) recovery from acidic solutions: Adsorption experiments, spectroscopic analyses, and theoretical computations, *J Hazard Mater* 2020;387: 121689.

13. Luo X, Shen T, Ding L, Zhong W, Luo J, Novel thymine-functionalized MIL-101 prepared by post-synthesis and enhanced removal of Hg^{2+} from water, *J Hazard Mater* 2016;306: 13–22.

14. Qian K, Deng Q, Fang G, Wang J, Pan M, Wang S, Biosensors and bioelectronics metal-organic frameworks supported surface-imprinted nanoparticles for the sensitive detection of metolcarb, *Biosens Bioelectron* 2016;79: 359–63.

15. Luo X, Ding L, Luo J, Adsorptive removal of Pb(II) ions from aqueous samples with amino-functionalization of metal-organic frameworks MIL-101(Cr), *J Chem Eng* 2015;60: 1732–43.

16. Ramanayaka S, Vithanage M, Sarmah A, An T, Kim KH, Ok YS, Performance of metal-organic frameworks for the adsorptive removal of potentially toxic elements in a water system: A critical review, *RSC Adv* 2019;9: 4359–76.

17. Zhao G, Qin N, Pan A, Wu X, Peng C, Ke F, Iqbal M, Ramachandraiah K, Zhu J, Magnetic nanoparticles@metal-organic framework composites as sustainable environment adsorbents, *J Nanomater* 2019; 2019: 1–11.

18. Kobielska PA, Howarth AJ, Farha OK, Nayak S, Metal-organic frameworks for heavy metal removal from water, *Coord Chem Rev* 2018;358: 92–7.

19. Yu H, Shao P, Fang L, Pei J, Ding L, Pavlostathis SG, Palladium ion-imprinted polymers with PHEMA polymer brushes: Role of grafting polymerization degree in anti-interference, *Chem Eng J* 2019;359: 176–85.

20. Yuan G, Tu H, Liu J, Zhao C, Liao J, Yang Y, Yang J, Liu N, A novel ion-imprinted polymer induced by the glycylglycine modified metal-organic framework for the selective removal of Co(II) from aqueous solutions, *Chem Eng J* 2018;333: 280–88.

21. Chiu YH, Chang TFM, Chen CY, Sone M, Hsu YJ, Mechanistic insights into photodegradation of organic dyes using heterostructure photocatalysts, *Catalysts* 2019;9: 430–61.
22. Meng Z, Lu R, Rao D, Kan E, Catenated metal-organic frameworks: Promising hydrogen purification materials and high hydrogen storage medium with further lithium doping, *Int J Hydrogen Energy* 2017;38; 9811–18.
23. Farha OK, Hupp T, Activation of metal-organic framework materials, *Acc Chem Res* 2010;43: 1166–75.
24. Jiang H, Makal TA, Zhou H, Interpenetration control in metal-organic frameworks for functional applications, *Coord Chem Rev* 2013;257: 2232–49.
25. Rungtaweevoranit B, Diercks CS, Kalmutzki MJ, Yaghi OM, Spiers memorial lecture: Progress and prospects of reticular chemistry, *Faraday Discuss* 2017;201: 9–45.
26. Petit C, Present and future of MOF research in the field of adsorption and molecular separation, *Curr Opin Chem Eng* 2018;20: 132–42.
27. Perez EV, Karunaweera C, Musselman IH, Balkus KJ, Ferraris JP, Origins and evolution of inorganic-based and MOF-based mixed-matrix membranes for gas separations, *Processes* 2016;4: 32–9.
28. Yun W, Perera SP, Burrows AD, Microporous and mesoporous materials manufacturing of metal-organic framework monoliths and their application in CO_2 adsorption, *Microporous Mesoporous Mater* 2015;214: 149–55.
29. Lee SJ, Yoon JW, Seo YK, Kim MB, Lee SK, Lee UH, Hwang YK, Bae YS, Chang JS, Microporous and mesoporous materials effect of purification conditions on gas storage and separations in a chromium-based metal-organic framework MIL-101, *Microporous Mesoporous Mater* 2014;93: 160–65.
30. Huang Z, Kee H, Micro-solid-phase extraction of organochlorine pesticides using porous metal-organic framework MIL-101 as sorbent, *J Chromatogr A* 2015;1401: 9–6.
31. Wu Z, Huang X, Zheng H, Wang P, Hai G, Dong W, Wang G, Environmental aromatic heterocycle-grafted NH_2-MIL-125(Ti) via conjugated linker with enhanced photocatalytic activity for selective oxidation of alcohols under visible light, *Appl Catal B Environ* 2018;224: 479–87.
32. Yang D, Gates BC, Catalysis by metal-organic frameworks: Perspective and suggestions for future research, *ACS Catal* 2019;9: 1779–98.
33. Kumar K, Maddila S, Babu S, Jonnalagadda SB, A review on contemporary metal-organic framework materials, *Inorganica Chim Acta* 2016;446: 61–4.
34. Zhang S, Yang Q, Liu X, Qu X, Wei Q, Xie G, High-energy metal-organic frameworks (HE-MOFs): Synthesis, structure and energetic performance, *Coord Chem Rev* 2016;307: 292–12.
35. Lee YR, Kim J, Ahn WS, Synthesis of metal-organic frameworks: A mini review, *Korean J ChemEng* 2013;30: 1667–80.
36. Kurian VRRM, Synthesis and catalytic applications of metal-organic frameworks: A review on recent literature, *Int Nano Lett* 2019;9: 17–9.
37. Elsaidi SK, Mohamed MH, Banerjee D, Thallapally PK, Flexibility in metal-organic frameworks: A fundamental understanding, *Coord Chem Rev* 2018;358: 125–52.
38. Kim SI, Yoon TU, Kim MB, Lee SJ, Hwang YK, Chang JS, Kim HJ, Lee HN, Lee UH, Bae YS, Metal-organic frameworks with high working capacities and cyclic hydrothermal stabilities for fresh water production, *Chem Eng J* 2016;286: 467–75.
39. Alhamami M, Doan H, Cheng C, A review on breathing behaviors of metal-organic frameworks (MOFs) for gas adsorption, *Materials* 2014;7: 3198–50.
40. Deria P, Mondloch JE, Karagiaridi O, Bury W, Hupp JT, Farha OK, Beyond post-synthesis modification: Evolution of metal-organic frameworks via building block replacement, *Chem Soc Rev* 2014;43: 5896–12.

41. Cook TR, Zheng Y, Stang PJ, Metal-organic frameworks and self-assembled supramolecular coordination complexes: Comparing and contrasting the design, synthesis, and functionality of metal-organic materials, *Chem Rev* 2020;113: 734–77.
42. Langseth E, Swang O, Arstad B, Lind A, Cavka JH, Jensen TL, TKristensen TE, Moxnes J, Unneberg E, Heyn RH, Synthesis and characterization of Al@MOF materials, *Mater Chem Phys* 2019;226: 220–25.
43. Maponya TC, Hato MJ, Modibane KD, Makgopa K. Photocatalysts in advanced oxidation processes for wastewater Treatment. In Fosso-Kankeu E, Pandey S, Sinha Ray S, editors. *Metal-Organic Frameworks as Possible Candidates for Photocatalytic Degradation of Dyes in Wastewater*. John Wiley & Sons Publishers; 2020. pp. 65–92.
44. Hashemi B, Zohrabi P, Raza N, Kim K, Metal-organic frameworks as advanced sorbents for the extraction and determination of pollutants from environmental, biological, and food media, *Trends Anal Chem* 2017;97: 65–2.
45. Chen B, Xiang S, Qian G, Metal-organic frameworks with functional pores for recognition of small molecules, *Acc Chem Res* 2010;43: 1115–24.
46. Ren J, Langmi HW, North BC, Mathe M, Review on processing of metal-organic framework (MOF) materials towards system integration for hydrogen storage, *Int J Energy Res* 2014;39: 607–20.
47. Kirchon A, Feng L, Drake HF, Joseph EA, Zhou HC, From fundamentals to applications: A toolbox for robust and multifunctional MOF Materials, *Chem Soc Rev* 2018;47: 8611–38.
48. Bosch M, Zhang M, Zhou H, Increasing the stability of metal-organic frameworks, *Adv Chem* 2014;2014: 1–8.
49. Dhaka S, Kumar R, Deep A, Kurade MB, Ji S, Jeon B, Metal-organic frameworks (MOFs) for the removal of emerging contaminants from aquatic environments, *Coord Chem Rev* 2019;380: 330–52.
50. Li Y, Yang RT, Gas adsorption and storage in metal-organic framework MOF-177, *Langmuir* 2007;23: 12937–44.
51. Thomas KM, Hydrogen adsorption and storage on porous materials, *Catal Today* 2007;120: 389–98.
52. Furukawa H, Cordova KE, O'Keeffe M, Yaghi OM, The chemistry and applications of metal-organic frameworks, *Science* 2013;341: 1230444–55.
53. Wang S, Bromberg L, Schreuder-gibson H, Hatton TA, Organophophorous ester degradation by chromium(III) terephthalate metal-organic framework (MIL-101) chelated to N,N-dimethylaminopyridine and related aminopyridines, *ACS Appl Mater* 2013;5: 1269–78.
54. Wang Z, Ge L, Li M, Lin R, Wang H, Zhu Z, Orientated growth of copper-based MOF for acetylene storage, *Chem Eng J* 2019;357: 320–27.
55. Zhang L, Wang J, Ren X, Zhang W, Zhang T, Liu X, et al., Internally extended growth of core-shell NH_2-MIL-101(Al)@ZIF-8 nanoflowers for the simultaneous detection and removal of Cu(II), *J Mater Chem A* 2018;6: 21029–38.
56. Drout RJ, Robison L, Chen Z, Islamoglu T, Farha OK, Zirconium metal-organic frameworks for organic pollutant adsorption, *Trends Chem* 2019;3: 304–17.
57. Jiao L, Yen J, Seow R, Skinner WS, Wang ZU, Jiang H, Metal-organic frameworks: Structures and functional applications, *Mater Today* 2018;27: 43–8.
58. Rowsell JLC, Yaghi OM, Metal-organic frameworks: A new class of porous materials, *Microporous Mesoporous Mater* 2004;73: 3–14.
59. Joseph L, Jun BM, Jang M, Park CM, Muñoz-Senmache JC, Hernández-Maldonado AJ, et al., Removal of contaminants of emerging concern by metal-organic framework nanoadsorbents: A review, *Chem Eng J* 2019;369: 928–46.

60. Song Y, Yan B, Chen Z, Hydrothermal synthesis, crystal structure and luminescence of four novel metal-organic frameworks, *J Solid State Chem* 2006;179: 4037–46.

61. Lee Y, Kim J, Ahn W, Synthesis of metal-organic frameworks: A mini review, *Korean J Chem Eng* 2013;30: 1667–80.

62. Monama GR, Hato MJ, Ramohlola KE, Maponya TC, Mdluli SB, Molapo KM, et al., Hierarchiral 4-tetranitro copper (II) phthalocyanine based metal organic framework hybrid composite with improved electrocatalytic efficiency towards hydrogen evolution reaction, *Res Phys* 2019;15: 102564–75.

63. Zhao Y, Song Z, Li X, Sun Q, Cheng N, Lawes S, Metal-organic frameworks for energy storage and conversion, *Energy Storage Mater* 2016;2: 35–2.

64. Choi JY, Kim J, Jhung SH, Kim H, Chang J, Chae HK, Microwave synthesis of a porous metal-organic framework , zinc terephthalate MOF-5, *Bull Korean Chem Soc* 2006;27: 1523–24.

65. Stock N, Biswas S, Synthesis of metal-organic frameworks (MOFs): Routes to various MOF topologies, morphologies, and composites, *Chem Rev* 2012;112: 933–69.

66. Ren J, Dyosiba X, Musyoka NM, HW Langmi, Mathe M, Review on the current practices and efforts towards pilot-scale production of metal-organic frameworks (MOFs), *Coord Chem Rev* 2017;352: 187–19.

67. Dey C, Kundu T, Biswal BP, Mallick A, Banerjee R, Crystalline metal-organic frameworks (MOFs): Synthesis, structure and function, *Acta Crystallogr B* 2014;70: 3–10.

68. Chen Y, Zhang R, Jiao L, Jiang H, Metal-organic framework-derived porous materials for catalysis, *Coord Chem Rev* 2018; 362: 1–23.

69. Volkova EI, Vakhrushev AV, Suyetin M, Improved design of metal-organic frameworks for efficient hydrogen storage at ambient temperature: A multiscale theoretical investigation, *Int J Hydrogen Energy* 2014; 39: 8347–50.

70. Zeng L, Guo X, He C, Duan C, Metal-organic frameworks: Versatile materials for heterogeneous photocatalysis, *ACS Catal* 2016; 6: 7935–47.

71. Tranchemontagne DJ, Hunt JR, Yaghi OM, Room temperature synthesis of metal-organic frameworks: MOF-5 , MOF-74 , *Tetrahedron* 2008;64: 8553–57.

72. Ahmed I, Jhung SH, Composites of metal-organic frameworks: Preparation and application in adsorption, *Biochem Pharmacol* 2014;17:136–46.

73. Burnett BJ, Barron PM, Hu C, Choe W, Stepwise synthesis of metal-organic frameworks: Replacement of structural organic linkers, *J Am Chem Soc* 2011;133: 9984–87.

74. Azad FN, Ghaedi M, Dashtian K, Hajati S, Pezeshkpour V, Ultrasonically assisted hydrothermal synthesis of activated carbon-HKUST-1-MOF hybrid for efficient simultaneous ultrasound-assisted removal of ternary organic dyes and antibacterial investigation: Taguchi optimization, *Ultrason Sonochem* 2016;31: 383–93.

75. Safaei M, Mehdi M, Ebrahimpoor N, A review on metal-organic frameworks: Synthesis and applications, *Trends Anal Chem* 2019;118: 401–25.

76. Candu N, Tudorache M, Florea M, Ilyes E, Vasiliu F, Mercioniu I, et al., Postsynthetic modification of a metal-organic framework (MOF) structure for enantioselective catalytic epoxidation, *Chempluschem* 2013;78: 443–50.

77. Burrows AD, Jiang D, Keenan LL, Burrows AD, Edler KJ, Synthesis and post-synthetic modification of MIL-101(Cr)-NH$_2$ via a tandem diazotisation process, *ChemComm* 2012;48: 99–102.

78. Lei J, Qian R, Ling P, Cui L, Ju H, Design and sensing applications of metal-organic framework composites, *Trends Anal Chem* 2014;58: 71–8.

79. Kim J, Kim DO, Kim DW, Sagong K, Synthesis of MOF having hydroxyl functional side groups and optimization of activation process for the maximization of its BET surface area, *J Solid State Chem* 2013;197: 261–65.

80. Wang L, Zheng M, Xie Z, Nanoscale metal-organic frameworks for drug delivery: A conventional platform with new promise, *J Mater Chem B* 2018;6: 707–17.

81. Hui M, Loon K, Zheng G, Synthesis and applications of MOF-derived porous nanostructures, *Green Energy Environ* 2017;2: 218–45.

82. Asadian E, Shahrokhian S, Iraji A, Highly sensitive nonenzymetic glucose sensing platform based on MOF-derived NiCo LDH nanosheets/graphene nanoribbons composite, *J Electroanal Chem* 2018;808: 114–23.

83. Li S, Huo F, Metal-organic framework composites: From fundamentals to applications, *Nanoscale* 2015;7: 7482–501.

84. Liu Y, Liu Z, Huang D, Cheng M, Zeng G, Lai C, Metal or metal-containing nanoparticle@MOF nanocomposites as a promising type of photocatalyst, *Coord Chem Rev* 2019;388: 63–8.

85. Beg S, Rahman M, Jain A, Saini S, Midoux P, Pichon C, et al., Nanoporous metal-organic frameworks as hybrid polymer-metal composites for drug delivery and biomedical applications, *Drug Discov Today* 2017;22: 625–37.

86. Deria P, Mondloch JE, Karagiaridi O, Bury W, Hupp JT, Farha OK, Beyond post-synthesis modification: Evolution of metal-organic frameworks via building block replacement, *Chem Soc Rev* 2014;43: 41–4.

87. Yin Z, Wan S, Yang J, Kurmoo M, Zeng M, Recent advances in post-synthetic modification of metal-organic frameworks: New types and tandem reactions, *Coord Chem Rev* 2019;378: 500–12.

88. Wei J, Hu Y, Liang Y, Kong B, Zhang J, Song J, et al., Nitrogen-doped nanoporous carbon/graphene nano-sandwiches: Synthesis and application for efficient oxygen reduction, *Adv Funct Mater* 2015;25: 768–77.

89. Chen Y, Mu X, Lester E, Wu T, High efficiency synthesis of HKUST-1 under mild conditions with high BET surface area and CO_2 uptake capacity, *Prog Nat Sci Mater Int* 2018;28: 584–89.

90. Deng K, Hou Z, Li X, Li C, Zhang Y, Deng X, et al., Aptamer-mediated up-conversion core/MOF shell nanocomposites for targeted drug delivery and cell imaging, *Sci Rep* 2015;5: 7851–57.

91. Lin S, Zhao Y, Bediako JK, Cho C, Sarkar AK, Structure-controlled recovery of palladium (II) from acidic aqueous solution using metal-organic frameworks of MOF-802 , UiO-66 and MOF-808, *Chem Eng J* 2019;362: 280–86.

92. Ahmed I, Jhung SH, Remarkable adsorptive removal of nitrogen-containing compounds from a model fuel by a graphene oxide/MIL-101 composite through a combined effect of improved porosity and hydrogen bonding, *J Hazard Mater* 2016;314: 318–25.

93. Kareem HM, Alrubaye RTA, Synthesis and characterization of metal-organic frameworks for gas storage, *IOP Conf Ser Mater Sci Eng* 2019;518: 062013–21.

94. Chowdhury MA, Metal-Organic-Frameworks for biomedical applications in drug delivery, and as MRI contrast agents, *J Biomed Mater Res A* 2017;105: 1184–94.

95. Katoch A, Goyal N, Gautam S, Applications and advances in coordination cages: Metal-Organic Frameworks, *Vacuum* 2019;167: 287–300.

96. Al MH, Abid HR, Sunderland B, Wang S, Metal-organic frameworks as a drug delivery system for flurbiprofen, *Drug Des Dev Ther* 2017;11: 2685–95.

97. Fang X, Zong B, Mao S, Metal-organic framework-based sensors for environmental contaminant sensing, *Nano-Micro Lett* 2018;10: 1–19.

98. Qiu S, Zhu G, Molecular engineering for synthesizing novel structures of metal-organic frameworks with multifunctional properties, *Coordin Chem Rev* 2009;253: 2891–11.

99. Mashao G, Ramohlola KE, Mdluli SB, Monama GR, Hato MJ, Makgopa K, et al., Zinc-based zeolitic benzimidazolate framework/polyaniline nanocomposite for electrochemical sensing of hydrogen gas, *Mater Chem Phys* 2019;230: 287–98.

100. Zhao M, Huang Y, Peng Y, Huang Z, Ma Q, Two-dimensional metal-organic framework nanosheets: Synthesis and applications, *Chem Soc Rev* 2018;47: 6267–95.
101. Xu W, Bahadur K, Ju Q, Fang Z, Huang W, Heterogeneous catalysts based on mesoporous metal-organic frameworks, *Coord Chem Rev* 2018;373: 199–32.
102. Chughtai AH, Ahmad N, Younus HA, Laypkov A, Verpoort F, Metal-organic frameworks: Versatile heterogeneous catalysts for efficient catalytic organic transformations, *Chem Soc Rev* 2015;44: 6804–49.
103. Kleist W, Maciejewski M, Baiker A, MOF-5 based mixed-linker metal-organic frameworks: Synthesis, thermal stability and catalytic application, *Thermochim Acta* 2010;499: 71–8.
104. Wang B, Liu W, Zhang W, Liu J, Nanoparticles@nanoscale metal-organic framework composites as highly efficient heterogeneous catalysts for size- and shape-selective reactions, *Nano Res* 2017;10: 3826–35.
105. Xue D, Wang Q, Bai J, Amide-functionalized metal-organic frameworks: Syntheses, structures and improved gas storage and separation properties, *Coord Chem Rev* 2019;378: 2–16.
106. Li H, Li L, Lin R, Zhou W, Zhang Z, Xiang S, Porous metal-organic frameworks for gas storage and separation: Status and challenges, *Energy Chem* 2019;1: 100006–44.
107. Zn O, Kaye SS, Dailly A, Yaghi OM, Long JR, Impact of preparation and handling on the hydrogen storage properties of $Zn_4O(1,4-benzenedicarboxylate)_3$ (MOF-5), *J Am Chem Soc* 2007;129: 14176–77.
108. Kim J, Yeo S, Jeon JD, Kwak SY, Enhancement of hydrogen storage capacity and hydrostability of metal-organic frameworks (MOFs) with surface-loaded platinum nanoparticles and carbon black, *Microporous Mesoporous Mater* 2015;202: 8–15.
109. Zhou H, Zhang J, Zhang J, Yan X, Shen X, High-capacity room-temperature hydrogen storage of zeolitic imidazolate framework/graphene oxide promoted by platinum metal catalyst, *Int J Hydrogen Energy* 2015;40: 12275–85.
110. Kumar P, Bansal V, Kim K, Kwon EE, Metal-organic frameworks (MOFs) as futuristic options for wastewater treatment, *J Ind Eng Chem* 2018;62: 130–45.
111. Pi Y, Li X, Xia Q, Wu J, Li Y, Xiao J, et al., Adsorptive and photocatalytic removal of persistent organic pollutants (POPs) in water by metal-organic frameworks (MOFs), *Chem Eng J* 2018;337: 351–71.
112. Peng Y, Huang H, Zhang Y, Kang C, Chen S, Song L, et al., A versatile MOF-based trap for heavy metal ion capture and dispersion, *Nat Commun* 2018;9: 1–9.
113. Zhou Y, Zhu J, Yang J, Lv Y, Zhu Y, Bi W, et al., Magnetic nanoparticles speed up mechanochemical solid phase extraction with enhanced enrichment capability for organochlorines in plants, *Anal Chim Acta* 2019;1066: 49–7.
114. Wang T, Zhao P, Lu N, Chen H, Zhang C, Hou X, Facile fabrication of Fe_3O_4/MIL-101(Cr) for effective removal of acid red 1 and orange G from aqueous solution, *Chem Eng J* 2016;295: 403–13.
115. Yang Q, Ren S, Zhao Q, Lu R, Hang C, Chen Z, et al., Selective separation of methyl orange from water using magnetic ZIF-67 composites, *Chem Eng J* 2017;333: 49–7.
116. Wang C, Zhang Y, Li J, Wang P, Photocatalytic CO_2 reduction in metal-organic frameworks: A mini review, *J Mol Struct* 2015;1083: 127–36.
117. Ngoepe NM, Hato MJ, Modibane KD, Hintsho-Mbita NC, Photocatalysts in advanced oxidation processes for wastewater treatment. In Fosso-Kankeu E, Pandey S, Sinha Ray S, editors. *Biogenic Synthesis of Metal Oxide Nanoparticle Semiconductors for Wastewater Treatment.* John Wiley & Sons Publishers; 2020. pp. 3–31.
118. Zhao H, Xia Q, Xing H, Chen D, Wang H, Construction of pillared-layer MOF as efficient visible-light photocatalysts for aqueous Cr(VI) reduction and dye degradation, *ACS Sustain Chem Eng* 2017;5: 4449–56.

119. Zhao S, Wang G, Poelman D, Metal organic frameworks based materials for heterogeneous photocatalysis, *Molecules* 2018;23: 2947–69.
120. Ahmed S, Rasul MG, Martens WN, Advances in heterogeneous photocatalytic degradation of phenols and dyes in wastewater: A review, *Water Air Soil Pollut* 2011; 215: 3–29.
121. Du PD, Thanh HTM, To TC, Thang HS, Tinh MX, Tuyen TN, et al., Metal-organic framework MIL-101: Synthesis and photocatalytic degradation of remazol Black B dye, *J Nanomater* 2019;2019: 1–15.
122. Yi X, Wang F, Du X, Fu H, Wang C, Highly efficient photocatalytic Cr(VI) reduction and organic pollutants degradation of two new bifunctional 2D Cd/Co-based MOFs, *Polyhedron* 2018;152: 216–24.
123. Kumar SG, Rao KSRK, Comparison of modification strategies towards enhanced charge carrier separation and photocatalytic degradation activity of metal oxide semiconductors (TiO$_2$, WO$_3$ and ZnO), *Appl Surf Sci* 2017;391: 124–48.
124. Mondal C, Singh A, Sahoo R, Sasmal AK, Negishi Y, Pal T, Preformed ZnS nanoflower prompted evolution of CuS/ZnS p-n heterojunctions for exceptional visible-light driven photocatalytic activity, *New J Chem* 2015;39: 5628–35.
125. Islam MJ, Kim HK, Reddy DA, Kim Y, Ma R, Baek H, et al., Hierarchical BiOI nanostructures supported on a metal organic framework as efficient photocatalysts for degradation of organic pollutants in water, *Dalton Trans* 2017;46: 6013–23.
126. Kojtari A, Ji H-F, Metal organic framework micro/nanopillars of Cu(BTC)3H$_2$O and Zn(ADC)DMSO, *Nanomaterials* 2015;5: 565–76.
127. Rodríguez NA, Parra R, Grela MA, Structural characterization, optical properties and photocatalytic activity of MOF-5 and its hydrolysis products: Implications on their excitation mechanism, *RSC Adv* 2015;5: 73112–18.
128. Kaur R, Rana A, Singh RK, Chhabra VA, Kim K, Deep A, Efficient photocatalytic and photovoltaic applications with nanocomposites between CdTe QDs and an NTU-9 MOF, *RSC Adv* 2017;7: 29015–24.
129. Sha Z, Chan HSO, Wu J, Ag$_2$CO$_3$/UiO-66(Zr) composite with enhanced visible-light promoted photocatalytic activity for dye degradation, *J Hazard Mater* 2015;299: 132–40.
130. Wang C, Lin H, Xu Z, Cheng H, Zhang C, One-step hydrothermal synthesis of flowerlike MoS$_2$/CdS heterostructures for enhanced visible-light photocatalytic activities, *RSC Adv* 2015;5: 15621–26.
131. Lee KM, Lai CW, Ngai KS, Juan JC, Recent developments of zinc oxide based photocatalyst in water treatment technology: A review, *Water Res* 2016;88: 428–48.
132. Wang CC, Li JR, Lv XL, Zhang YQ, Guo G, Photocatalytic organic pollutants degradation in metal-organic frameworks, *Energy Environ Sci* 2014;7: 2831–67.
133. Hong J, Chen C, Bedoya FE, Kelsall GH, O'Hare D, Petit C, Carbon nitride nanosheet/metal-organic framework nanocomposites with synergistic photocatalytic activities, *Catal Sci Technol* 2016;6: 5042–51.
134. Fagan R, McCormack DE, Dionysiou DD, Pillai SC, A review of solar and visible light active TiO$_2$ photocatalysis for treating bacteria, cyanotoxins and contaminants of emerging concern, *Mater Sci Semicond Process* 2016;42: 2–14.
135. Gao Y, Li S, Li Y, Yao L, Zhang H, Accelerated photocatalytic degradation of organic pollutant over metal-organic framework MIL-53(Fe) under visible LED light mediated by persulfate, *Appl Catal B, Environ* 2017;202: 165–74.
136. Wu Y, Luo H, Wang H, Synthesis of iron(iii)-based metal-organic framework/graphene oxide composites with increased photocatalytic performance for dye degradation, *RSC Adv* 2014;4: 40435–38.

137. Sun Q, Liu M, Li K, Zuo Y, Han Y, Wang J, et al., Facile synthesis of Fe-containing metal-organic frameworks as highly efficient catalysts for degradation of phenol at neutral pH and ambient temperature, *CrystEngComm* 2015;17: 7160–68.

138. Hausdorf S, Mossig R, Mertens FORL, Proton and water activity-controlled structure formation in zinc carboxylate-based metal organic frameworks, *J Phys Chem A* 2008;112: 7567–76.

139. Alvaro M, Carbonell E, Ferrer B, Llabrøs FX, Semiconductor behavior of a metal-organic framework (MOF), *Chem Eur J* 2007;13: 5106–12.

140. Laurier KGM, Vermoortele F, Ameloot R, De Vos DE, Hofkens J, Roe MBJ, Iron(III)-based metal-organic frameworks as visible light photocatalysts, *J Am Chem Soc* 2013;135: 14488–91.

141. Zhang C, Ma D, Zhang X, Ma J, Liu L, Xu X, Preparation, structure and photocatalysis of metal-organic frameworks derived from aromatic carboxylate and imidazole-based ligands, *J Coord Chem* 2016;69: 985–95.

142. Zhang M, Wang L, Zeng T, Shang Q, Zhou H, Pan Z, et al., Two pure MOF-photocatalysts readily prepared for the degradation of methylene blue dye under visible light, *Dalton Trans* 2018;47: 4251–58.

143. Feng X, Chen H, Jiang F, In-situ ethylenediamine-assisted synthesis of a magnetic iron-based metal-organic framework MIL-53(Fe) for visible light photocatalysis, *J Colloid Interface Sci* 2017;494: 32–7.

144. Zhu SR, Liu PF, Wu MK, Zhao WN, Li GC, Tao K, et al., Enhanced photocatalytic performance of BiOBr/NH$_2$-MIL-125(Ti) composite for dye degradation under visible light, *Dalt Trans* 2016;45: 17521–29.

145. Yang J, Niu X, An S, Chen W, Wang J, Liu W, Facile synthesis of Bi$_2$MoO$_6$-MIL-100(Fe) metal-organic framework composites with enhanced photocatalytic performance, *RSC Adv* 2017;7: 2943–52.

146. Wang H, Yuan X, Wu Y, Zeng G, Chen X, Leng L, et al., Synthesis and applications of novel graphitic carbon nitride/metal-organic frameworks mesoporous photocatalyst for dyes removal, *Appl Catal B Environ* 2015;174–175: 445–54.

147. Guo D, Wen R, Liu M, Guo H, Chen J, Weng W, Facile fabrication of g-C$_3$N$_4$/MIL-53(Al) composite with enhanced photocatalytic activities under visible-light irradiation, *Appl Organomet Chem* 2015;29: 690–97.

148. Huang L, Liu B, Synthesis of a novel and stable reduced graphene oxide/MOF hybrid nanocomposite and photocatalytic performance for the degradation of dyes, *RSC Adv* 2016;6: 17873–79.

149. Fazaeli R, Aliyan H, Banavandi RS, Sunlight assisted photodecolorization of malachite green catalyzed by MIL-101/graphene oxide composites, *Russ J Appl Chem* 2015;88: 169–77.

150. Liang R, Jing F, Shen L, Qin N, Wu L, M@MIL-100(Fe) (M = Au, Pd, Pt) nanocomposites fabricated by a facile photodeposition process: Efficient visible-light photocatalysts for redox reactions in water, *Nano Res* 2015;8: 3237–49.

151. Ai L, Zhang C, Li L, Jiang J, Iron terephthalate metal-organic framework: Revealing the effective activation of hydrogen peroxide for the degradation of organic dye under visible light irradiation, *Appl Catal B Environ* 2014;148–149: 191–200.

152. Abney CW, Gilhula JC, Lu K, Lin W, Metal-organic framework templated inorganic sorbents for rapid and efficient extraction of heavy metals, *Adv Mater* 2014;26: 7993–97.

153. Salarian M, Ghanbarpour A, A metal-organic framework sustained by a nanosized Ag12 cuboctahedral node for solid-phase extraction of ultra-traces of lead (II) ions, *Microchim Acta* 2014;181: 999–1007.

154. Shi Z, Xu C, Guan H, Li L, Fan L, Wang Y, et al., Magnetic metal organic frameworks (MOFs) composite for removal of lead and malachite green in wastewater, *Colloids Surf A Physicochem Eng Asp* 2018;539: 382–90.

155. Wu Y, Xu G, Liu W, Yang J, Wei F, Li L, et al., Postsynthetic modification of copper tere-phthalate metal-organic frameworks and their new application in preparation of samples containing heavy metal ions, *Microporous Mesoporous Mater* 2015;210: 110–15.

156. Lu M, Li L, Shen S, Chen D, Han W, Highly efficient removal of Pb^{2+} by a sandwich structure of metal-organic framework/GO composite with enhanced stability, *New J Chem* 2019;43: 1032–37.

157. Lin S, Bediako JK, Cho CW, Song MH, Zhao Y, Kim JA, et al., Selective adsorption of Pd(II) over interfering metal ions (Co(II), Ni(II), Pt(IV)) from acidic aqueous phase by metal-organic frameworks, *Chem Eng J* 2018;345: 337–44.

158. Luo BC, Yuan LY, Chai ZF, Shi WQ, Tang Q, U(VI) capture from aqueous solution by highly porous and stable MOFs: UiO-66 and its amine derivative, *J Radioanal Nucl Chem* 2016;307: 269–76.

159. Alqadami AA, Naushad M, Alothman ZA, Ghfar AA, Novel metal-organic framework (MOF) based composite material for the sequestration of U(VI) and Th(IV) metal ions from aqueous environment, *ACS Appl Mater Interfaces* 2017;9: 36026–37.

160. Tahmasebi E, Masoomi MY, Yamini Y, Morsali A, Application of mechanosynthe-sized azine-decorated zinc(II) metal-organic frameworks for highly efficient removal and extraction of some heavy-metal ions from aqueous samples: A comparative study, *Inorg Chem* 2015;54: 425–33.

161. Bagheri A, Taghizadeh M, Behbahani M, Akbar A, Synthesis and characterization of magnetic metal-organic framework (MOF) as a novel sorbent, and its optimization by experimental design methodology for determination of palladium in environmental samples, *Talanta* 2012;99: 132–39.

162. Yuan G, Tu H, Li M, Liu J, Zhao C, Liao J, et al., Glycine derivative-functionalized metal-organic framework (MOF) materials for Co (II) removal from aqueous solution, *Appl Surf Sci* 2019;466: 903–10.

163. Wang N, Yang LY, Wang YG, Ouyang XK, Fabrication of composite beads based on calcium alginate and tetraethylenepentamine-functionalized MIL-101 for adsorption of Pb(II) from aqueous solutions, *Polymers* 2018;10: 750–64.

164. Wang C, Du X, Li J, Guo X, Wang P, Zhang J, Photocatalytic Cr(VI) reduction in metal-organic frameworks: A mini-review, *Applied Catal B, Environ* 2016;193: 198–16.

165. Shi L, Wang T, Zhang H, Chang K, Meng X, Liu H, An amine-functionalized iron (III) metal-organic framework as efficient visible-light photocatalyst for Cr (VI) Reduction, *Adv Sci* 2015;2: 1500006–13.

166. Yu H, Shao P, Fang L, Pei J, Ding L, Pavlostathis SG, Palladium ion-imprinted poly-mers with PHEMA polymer brushes: Role of grafting polymerization degree in anti-interference, *Chem Eng J* 2019;359: 176–85.

167. Guo Z, Florea A, Jiang M, Mei Y, Zhang W, Jaffrezic-renault N, Molecularly Imprinted polymer/metal organic framework based chemical sensors, *Coatings* 2016;6: 42–50.

168. Lin S, Wei W, Wu X, Zhou T, Mao J, Yun Y, Selective recovery of Pd (II) from extremely acidic solution using ion-imprinted chitosan fiber: Adsorption performance and mechanisms, *J Hazard Mater* 2015;299: 10–7.

169. Tao HC, Gu YH, Liu W, Huang SB, Cheng L, Zhang LJ, et al., Preparation of palladium (II) Ion-imprinted polymeric nanospheres and its removal of palladium (II) from aque-ous solution, *Nanoscale Res Lett* 2017;12: 583–91.

170. Mafu LD, Mamba BB, Msagati TAM, Synthesis and characterization of ion imprinted polymeric adsorbents for the selective recognition and removal of arsenic and selenium in wastewater samples, *J Saudi Chem Soc* 2016;20: 594–605.

3 Metal-Organic Framework Nanocomposites for Adsorptive Applications

*Nhamo Chaukura, Wisdom A Munzeiwa,
Rufaro B Kawondera, Tatenda C Madzokere,
Norman Mudavanhu, and Sibongile M Malunga*

CONTENTS

3.1 INTRODUCTION

Metal-organic frameworks (MOFs) are crystalline porous compounds with facile synthesis, functional tunability, and diversity, ordered porosity, and high and adsorption capacity [1–3]. They are constructed from inorganic metals or clusters and polytopic organic bridging ligands which assemble into open networks of diverse geometries [3]. The MOFs are therefore inorganic-organic hybrid materials. Through a judicious choice of ligands, the intrinsic properties such as pore size, geometry, and functional groups can be tailored for specific applications [2].

Using rational design, and structural and functional group modification, MOFs have been integrated into nanocomposites to produce MOF nanocomposites (MOFNs) with enhanced properties. Moreover, their hybrid nature bequeaths the frameworks with flexibility where characteristically linker-based motions can give rise to dynamic guest-responsive performances [4]. Because MOFNs combine the properties of both MOFs and nanoparticles, their adsorptive properties are enhanced, and thus they have shown great promise in the removal of many pollutants such as heavy metals, dyes, persistent organic pollutants (POPs), endocrine-disrupting compounds (EDCs), rare earth elements (REEs), and a whole host of emerging pollutants [3]. These applications largely emanate from the capability of MOFNs to act as excellent host matrices for the immobilization of various fugitive molecules in the environment. For example, relative to traditional adsorbents, many MOFNs have also shown superior performance in the adsorptive removal of pollutants from water [5].

Relative to entirely inorganic porous materials, MOFNs offer diverse chemical composition, enabling structural and functional modification through the selection of precursors. Current research is progressively more targeted at the development and fabrication of application-specific MOFNs including powders, thin films, membranes, and granules. Combining MOFs with ceramics, biopolymers, nanoparticles, and proteins leads to the development of a new generation of functional materials with superior properties [4]. Consequently, MOFNs have been used in removing various pollutants from contaminated environmental matrices. Some of the synthesis methods are relatively simple. For instance, the encapsulation of magnetite nanoparticles into zeolitic imidazolate framework-67 (ZIF-67) by the hydrothermal method using cobalt nitrate [6]. A separate study synthesized a new MOF/GO composite based on [Ni_2(BDC-NH_2)$_2$(DABCO)].xDMF.yH_2O MOF and graphene oxide by heating in an autoclave at 383 K for 24 h [7]. Other MOFNs which have been synthesized include Fe_3O_4/RGO nanocomposites [8], MOF/GO composites [7], and an amino-functionalization magnetic multi-MOF [9]. These have been used for a range of applications including removal of dyes (e.g., [10,11], CO_2 capture [7], oxidation of organics [12], heavy metals and radionuclides [13–15], and the adsorptive removal of organic pollutants [16,17].

In this chapter, the properties desirable for MOFNs to be good adsorbents, and applications for the adsorption of various pollutants are discussed. Research gaps and future research directions are subsequently presented.

3.2 CHARACTERISTICS OF MOFNs FOR ADSORPTION

The resultant behavior of nanostructured materials is chiefly dependent on features such as morphology, size, and structure. Tailoring of such features is achieved through various synthetic methods (bottom-to-top or top-to-bottom approaches) leading to novel magnetic, electronic, optical, mechanical, and chemical properties [18]. For instance, size imparts surface area characteristics of nanostructured materials responsible for the development of their unique properties at the nanoscale regime. It has been established that changes in surface area lead to changes in the reaction time of a substance [19]. Nano-based MOFs are currently drawing a lot of interest from researchers because of tunable properties such as structure, porosity, surface architecture, and final composition [20]. MOFs can therefore be predictably tailored or nanoengineered so as to assume desirable adsorptive characteristics for various applications.

3.2.1 POROSITY, SURFACE AREA, AND CRYSTALLINITY

MOFs are fabricated from metal ions/clusters and organic linker molecules or bridging-ligands the orientation of whose architecture and donor groups determine the structure. Target design can be achieved by adjusting the linker geometry, type, and length, and this has a bearing on the size, shape, and internal surface properties [21]. Consequently, MOFNs have exhibited a superior specific surface area compared to other porous carbon-based materials and inorganic polymers. For instance, a remarkable MOF-NU-110E with an extended acetylene hexadentate linker has been shown to exhibit a specific surface area of 7140 m^2/g which is higher compared to zeolites and carbon nanotubes [22]. MOFs generally exhibit permanent porosity and show pore windows in the microporous, mesoporous, and microporous range, which is vital for selective separation and adsorption [17]. The ultrahigh porosity and enormous internal surface areas of mesoporous MOFNs through ligand extension opens possibilities for host-guest interactions.

3.2.2 FUNCTIONAL GROUPS AND SURFACE CHARGE

The surface charge and functionalities of MOFN frameworks can be modified to suit certain applications using various strategies. The organic linkers can be pre-synthesized with specific functional groups for targeted applications. Additionally, the MOFN frameworks can be engineered by a post-synthetic modification approach, opening up possibilities for a range of applications [23]. Acidic functional groups (e.g., –COOH and –SO_3H), and basic groups (e.g., –NH_2), which are important in heavy metal adsorptive removal, can be introduced [9,24,25]. Depending on the pH, these groups can be ionized and be involved in electrostatic interactions especially for the removal of cationic and anionic pollutant species. The presence of alkyl chains in the organic ligand increases the hydrophobicity, while phenyl groups increase π-π interactions [26]. The use of positive metal clusters and neutral linkers results in cationic MOFNs where the charge is concentrated at the corners of the framework. The combination of cationic ligands and nodes results in uniform charge distribution around the framework [27].

3.2.3 MORPHOLOGY AND SIZE

The composition, structure, and morphology of MOFNs influence their properties and subsequent application. In MOFN engineering, the size and morphology have been accomplished by various methodologies such as changing the synthetic method varying reaction conditions, and metal:linker ratios [28]. However, regulating the shape and size of MOFNs is still to be realized. Common morphologies that have been observed include nanorods, spherical, hexagonal, rectangular prism nanocubes, and nanosheets [14,29–31]. MOFNs have been shown to exhibit superior physical properties compared microscale MOFs. Of interest, microwave-assisted, sonochemical and miniemulsion synthetic methods have been used to selectively modulate MOF shape and size at nanoscale [30]. Generally, nanomaterials present a greater surface area, hence MOFNs are a good candidate for adsorptive applications. Although a lot of research has established the superior adsorption capabilities of MOFNs, nevertheless, structure-property relationships are not fully considered.

3.2.4 DEGRADABILITY AND HYDROSTABILITY

Current research is targeting the development of hydro-, thermally-, and chemically-stable MOFNs specifically for applications such as catalysis, gas storage, and water remediation. In aqueous media MOFNs tend to leach out the metal ions, hence a good combination of metal-ligand bonds, and ligand shielding ability enhances water resistance [32]. Furthermore, increasing the oxidation state of the metal and using hard donor ligands usually leads to an improvement of the hydrothermal stability [33]. This increase in Lewis acidity results in strong metal-donor coordination, hence making the MOFNs less prone to hydrolysis. The chemical and hydrothermal stability can be considerably improved by varying the porosity and pore surface hydrophobicity through altering functional groups and the length of organic ligands [21]. These approaches tentatively reduce the water absorption capability, minimizing ligand displacement and hydrolysis, resulting in an enhanced adsorption capacity.

3.2.5 OTHER PROPERTIES

In conjunction with structural properties, MOFNs exhibit other properties such as magnetism, optical, electrical, and fluorescence, which have opened many possibilities for these materials [34]. These properties can be conferred to these materials through careful selection of metal ions and ligands, coordination manipulation, and inclusion of guest molecules and nanoparticles within the pores. Magnetic properties can be specifically induced by using paramagnetic transition elements and open-shell organic ligands, or through the inclusion of magnetic nanoparticles such as iron oxide within the pores [35]. Long organic ligands tend to give weak magnetic interaction, hence short ligands are preferred, but at the expense of porosity. The use of π-electron rich aromatic ligands and lanthanides ions results in the inclusion of chromophoric molecules and generates MOFNs with good optical and fluorescent properties [34]. These properties can be manipulated in different ways to enhance the adsorption capacities of MOFNs.

3.3 APPLICATIONS FOR ADSORPTION

MOFs are commonly prepared in powder form which is restrictive to many practical applications including adsorption [36]. In this regard, powder is deposited onto supports to make them more useful for adsorption [37]. Shaping produces objects such as extrudates, beads, and pellets with enough mechanical and amorphous properties while maintaining the porosity [36,38]. The choice of the method of shaping is dependent on the application. Thus, tailored applications can be achieved by appropriately altering the structure [39]. While applications impose the shape of the object, the operating conditions will constrain the choice of binders and heat treatments [38]. For example, silk fiber was used as the substrate for coating MOF-76(Tb) powder, which selectively adsorbs Cu^{2+} [40].

Adsorptive removal of pollutants is either through batch or column techniques. The batch technique provides a simple way to investigate the experimental factors influencing the adsorption process, while column studies generate data to help decide the quantity of adsorbent required for adsorption [41]. Because in columns the adsorbent is constantly in contact with new feed solution, it has superior removal efficiency relative to batch processes [41]. Using batch experiments, an amino-functionalization magnetic multi-MOF material (Fe_3O_4/MIL-101($Al_{0.9}Fe_{0.1}$)/NH_2) showed a maximum adsorption capacity of 355.8 mg/g for methyl orange removal [9], while novel Fe_3O_4@C MOFNs used in columns was effective in removing chlorophenols from environmental samples [16]. Another study used a specialized continuous contactor loaded with graphitic carbon nitride (g-CN)/ATO MOFN for the removal of a mixture of volatile organic compounds from aqueous solution [42]. Magnetic composite microspheres containing a SiO_2-protected Fe_3O_4 core and a TiO_2 shell exhibited high adsorption capacity for fungicides [43].

3.3.1 REMOVAL OF HEAVY METALS

Many MOFNs possess desirable features for the adsorption of heavy metal pollutants. These include functional groups carrying electron-rich hetero-atoms such as O, N, or S, which can datively bond with the heavy metals. A number of MOFNs have been modified for specific heavy metal removal. Table 3.1 summarizes some of the heavy metals and the corresponding adsorption capacities of the MOFNs.

TABLE 3.1

Adsorption of HM over MOFs from Aquatic Systems

Metal Species	MOFN-Based Materials	Q_{max} (mg/g)	Reference
As(V)	Fe/Mg-MIL-88B	303.6	[44]
As(III)	Fe_3O_4@ZIF-8	100	[45]
Hg(II)	Fe_3O_4@Cu_3(btc)$_2$ magnetic microspheres	348.43	[46]
Cu(II)	ZIF-67/MIM/CA	18.9	[45]
Cr(VI)	GO-CS@MOF [Zn(BDC)(DMF)]	144.92	[47]
Pb(II)	MOF/GO	108	[48]
Cd(II)	PAN/chitosan/UiO-66-NH$_2$	415.6	[49]

3.3.2 REMOVAL OF ORGANICS

Synthetic dyes released from various industrial activities are highly toxic and carcinogenic, recalcitrant to removal, and thus pose a serious threat to ecological and public health [50,51]. The dyes commonly used in the textile industry are azo, basic, acid, reactive, direct, disperse, mordant, sulfur, and vat dyes, and their chemistry determines the choice of the MOFN that can be used for their removal [50]. Several studies have reported the use of MOFNs in removing dye in the aquatic environment. Rhodamine B, for instance, has been efficiently removed (Q_{max} = 84.5 mg/g) from aqueous solution by a Ni@C MOFN through batch experiments [52]. A separate study used MOFNs for the removal of methyl orange (MO), methylene blue (MB), and malachite green (MG) from aqueous systems [51]. Another study reported the maximum adsorption capacities (Q_{max}) for the removal of quinolone yellow on MOF-5/AC and MG on ZIF-8/CNT as 18.62 and 3300 mg/g, respectively. MB experiments with MOF/GO hybrids reported Q_{max} of 155 and 1231 mg/g for HKUST-1/GO/Fe_3O_4 and MIL-100(Fe)/GO, respectively. The high Q_{max} values in these studies suggest that these MOFNs are potentially useful for dye remediation [53].

3.3.3 REMOVAL OF EMERGING POLLUTANTS

Emerging pollutants (EPs) are non-traditional pollutants which include pesticides, dyes, antibiotics, POPs, REEs, and hormones [54]. These pollutants were traditionally not encountered, and/or not monitored, due to technological challenges, and are mainly of anthropogenic origin, such as the discharge of treated wastewater from wastewater treatment plants [51]. EPs have been removed to various extents using MOFNs (Table 3.2). The mechanisms of removal of EPs from water and wastewater include electrostatic interactions, acid-base interactions, and complexation. Due to their large surface area, permanent porosity, large adsorption capacities, and adjustable functional groups, MOFNs can be used to remove EPs from wastewater [14].

TABLE 3.2
Removal of Emerging Pollutants Using MOFNs

Pollutant	Nature of MOFN	Removal (%)	References
Chlorophenol	Fe₃O₄@C	92–99% (water), 87.6–97.8% (soil)	[55]
Nitro-polycyclic aromatic hydrocarbons	Indium (III) sulfide @MIL-125(Ti)	Recoveries = 71.3%–112.2%	[56]
Cyclohexene	Fe₃O₄@Au/MOF	Max 58% conversion with 90% selectivity	[57]
Cis-diol containing luteolin	Fe₃O₄@PDA @BA-MOFs	91.23%	[58]
Dibenzothiophene	Cu-BTC and MOF/ grapheme hybrid nanocomposite	92.2%, (MOF/graphe), 70% (Cu/MOF)	[59]
ibuprofen, diclofenac, naproxen, nalidixic acid	Fe₃O₄/Cu₃(BTC)₂	94–102%	[60]
Triadimenol, hexaconazole, diniconazole, myclobutanil, tebuconazole	Fe₃O₄@SiO₂ @MOF/TiO₂	90.2–104.2%	[43]
Rhodamine B	MIL–53–Fe MOF/ Fe₂O₃/Biochar	71%	[61]
Methylene blue (MB) and for Rhodamine B(Rhb)	H₆P₂W₁₈O₆₂/MOF-5	97 % (MO), (Rhb) 68 %	[62]
Tetracyline	MIL-68(Al)/GO	Qₑ = 228	[63]

3.3.3.1 Pharmaceutical and Personal Care Products

Pharmaceutical products are prescription, over the counter, veterinary, therapeutic drugs used to prevent or treat human and animal diseases, while personal care products are used mainly to improve the quality of life [64]. Pharmaceutical and personal care products (PPCPs) are used in daily life, and they include fragrances, UV filters, antimicrobials, illicit drugs, and surfactants. Although PPCPs have been removed from wastewater by solar/free chlorine systems, conventional wastewater treatment systems are unable to remove the PPCPs [65]. Hence the need for a method that can effectively remove PPCPs from aquatic systems.

The use of MOFNs in removing PPCPs is still emerging. However, the potential of MOFNs can be inferred from MOFs, which have been widely studied (e.g., [51,66]. The MOF, MIL-101 $(Cr_3O(F/OH)(H_2O)_2[C_6H_4(CO_2)_2])$, which has large pore sizes and huge porosity, is widely studied [67]. Various derivatives of MIL-101 have been used to remove a range of pollutants from aqueous systems including furosemide, sulfasalazine, clofibric acid, and naproxen [14]. Diclofenac, a common painkiller, has been removed using UiO-66(Zr) and its functionalized derivative (UiO-66 with SO_3H/NH_2) [51]. Zr-MOFs have also been used to remove antibiotics from aqueous solutions [68]. Veterinary drugs such as *p*-arsanilic acid, roxarsone, and phenylarsonic acid, have also be removed from wastewater using MIL-101(Cr) and MIL-101(Fe) [51].

3.3.3.2 Endocrine-Disrupting Compounds

Endocrine-disrupting compounds (EDCs) are exogeneous substances that alter the functions of the endocrine system. They may interfere with the body's endocrine system and produce adverse developmental, reproductive, neurological, and immune effects in both humans and animals [69]. EDCs include pharmaceuticals, dioxins, polychlorinated biphenyls, and pesticides, and are found in everyday products such as plastic bottles, metal food cans, detergents, and flame retardants [70]. The majority of EDCs are found in municipal wastewater and are recalcitrant to biodegradation [71]. They can, however, be removed by absorption, chemical degradation, biological degradation, transformation, volatization, and adsorption. Materials such as activated carbon tin pillared montmorillonite and high silica zeolites have been used to remove EDCs from wastewater [72]. The use of MOFNs in the removal of EDCs has been widely investigated. For example, polytetraflouroethylene modified with ZIF-8 was reported to remove 40% more progesterone compared to an unmodified membrane [73]. In another study, MOFNs have been used to remove EDCs such as bisphenol A from wastewater [66].

3.3.3.3 Antibiotic-Resistant Genes

The release of antibiotics into the environment induces the development of antibiotic-resistant genes (ARGs), which reduce the therapeutic potential of antibiotics against human and animal pathogens [74]. Antibiotic-resistant bacteria enter into the environment from human and animal sources. These bacteria are able to spread their genes into water-indigenous microbiomes, which also contain ARGs. Recently, MOFNs have also been used to remove ARGs from wastewater with good performance [66].

3.3.3.4 Persistent Organic Pollutants

POPs are toxic chemicals that originate from the production, use, and disposal of certain organic chemicals. They can be intentionally released into the environment in the case of pesticides and polychlorinated biphenyls, or unintentionally introduced as in the case of dioxins and furans, which are released by combustion of organic chemicals or as byproducts of industrial processes [75]. Because they are volatile and obstinate, bioaccumulating in living tissue, POPs pose major environmental and public health risks [76]. Moreover, POPs are recalcitrant to biodegradation, and are thus not effectively removed by water treatment processes [76]. The Stockholm Convention of May 2001 was established to protect the environment and public health from the risks associated with POPs [77].

Due to its simple design and low cost, adsorption is effective for the removal of POPs from water. A host of materials, including Zn-based MOFNs have been used to remove organophosphorus pesticides from water and fruit juice samples [2]. Fe_3O_4-MWCNT/ZIF-8 composites have been used for the removal of mixture of organophosphate pesticides, although low adsorption capacities were reported due to competitive adsorption [78].

3.3.3.5   Rare Earth Elements

Rapid industrialization during the past few decades has involved the extensive use of these REEs, and this has resulted in extensive pollution [79]. Several methods which include chemical precipitation, ion exchange extraction, coagulation, flocculation, solid-liquid phase and liquid-liquid extraction, and biosorption have been used to remove REEs from aqueous systems [80]. Adsorption is an effective mechanism of REEs because of its simplicity, efficiency, and wide range of availability of sorbents. Materials such as biomass, zeolites, activated carbon, and silicas have been used to remove REEs from water and wastewater [79].

Due to their versatility, MOFNs can be tuned so they can be used for many applications. For instance, MOFNs have been magnetized to efficiently remove REEs from brine [81]. A separate study used Cr-MIL-101 to remove La^{3+}, Ce^{3+}, Nd^{3+}, Sm^{3+}, and Gd^{3+} from aqueous solution [82]. Overall, compared to the traditional adsorbents, MOF-based materials have superior adsorption capacities and higher efficiencies for REEs removal.

3.4   MECHANISMS OF ADSORPTION

Adsorption mechanisms are governed by the nature of the targeted pollutant, and the chemistry of the MOFNs. For instance, the mechanisms for the adsorption of heavy metal ions primarily involve acid-base and electrostatic interactions, coordinative bonding, diffusion, and van der Waals forces [15]. For the adsorption of organics on MOFNs, the mechanisms include chemical bonding, π-π, electrostatic, and acid-base interactions (Table 3.3) [51]. In acid-base interactions, soft acids react strongly with the soft bases, and conversely the hard acids react strongly with the hard bases.

TABLE 3.3
Mechanism for the Adsorption of Pesticides in Aquatic Systems by MOFNs

Pesticide	MOFN	Removal Capacity (mg/g)	Adsorption Mechanism	References
Methylchlorophenoxy-propionic Acid	Cotton@ UiO-66(Zr)	12.8	-	[83]
2,4-dichlorophenoxyacetic acid	Carbon derived from IL@ZIF-67	448	hydrophobic and π-π interactions	[84]
Diuran	Carbon derived from IL @ZIF-67	284	hydrophobic and π-π interactions	[84]
Ethion	Cu-BTC@Cotton	182	Coordination linkage and H-bonding	[85]
Glyphosate	UiO-67(Zr)/GO	482	Chemisorption	[86]
2,4-dichlorophenoxyacetic acid	Ce-BTC	95.78	π-π stacking and electrostatic interactions	[29]

Coordinative interactions usually occur between heavy metals and MOFNs modified with functional groups such as $-NH_2$, $-COOH$, and $-SH$ [15]. In some adsorption processes, new chemical bonds are formed between the heavy metal pollutants and MOFNs, which is another major adsorption mechanism. Electrostatic interactions occur when charged MOFNs interact with oppositely charged pollutants and are influenced by the surface charge on the MOFNs. The surface charge in turn varies with the pH of the aqueous system, and with surface functional groups, which can be protonated or deprotonated [15]. The other adsorption mechanisms applicable to all adsorbates when surface area or porosity are taken into consideration, are van der Waals forces and pore entrapment [15].

3.5 MOFN-BASED MEMBRANES

With continuing research on novel MOFNs, recent studies are focusing on the growth and processing of MOFs into application-specific configurations such as thin films and membranes [3]. MOFN thin films are mostly composed of and supported by polymeric substrates. A common thin film MOFN membrane used in water treatment is the polyamide film that is made up of MOFNs supported on a plastic sheet [87]. Generally, membrane-based separation technologies are widely applied in the petrochemical, medical, energy, water and wastewater treatment, desalination, environmental, and food industries [88]. Membrane technologies involving MOFNs are used in the extraction and determination of solutes from biological, environmental, and food media [16]. This technique involves passing the solute through a MOFN membrane and preventing the interference of large molecules and particles to facilitate the separation of analytes from the matrix. This method is rapid, simpler, and cost effective with concurrent analyte extraction and a clean-up process. Following the enrichment process, target analytes can be eluted using small volumes of solvent [16].

Engineered forms of MOFNs such as membranes and thin films are being upscaled from laboratory to industrial products for separation applications such as solvent nanofiltration and desalination, pervaporation, and gas separations [89]. For example, an ultrahigh MOF loading MOF-polyethylene hybrid membrane was successfully synthesized for high flux separations [90]. The membrane exhibited high organic dyes rejections (99%), excellent chiral compounds, and protein separation performance. Antifouling and long-term stability properties are some of the robust attributes of MOFN-based membranes. According to Shekhah et al. [91] the first MOF-based membrane consisting of a MOF-5 membrane on alumina support was tailored by using a seeding approach in order to achieve good gas separation characteristics. Owing to their tunable properties, applications of MOF-based membranes in liquid separation are rapidly evolving. For example, in pervaporative separation of alcohols. Pervaporative separation of alcohols from alcohol-water mixtures was achieved by a continuous ZIF-71 membrane generated via a secondary growth route [40]. These classical examples highlight the effectiveness of conventional MOFs in separation processes.

Novel properties of nanomaterials present an opportunity to exploit nanomaterial properties to enhance the performance of MOF-based sorbents in the domain of separation/membrane technology. Membrane-based MOFs can be fabricated by various techniques such as solvent evaporation, slip casting, and electrospinning. Electrospinning is a simple and low-cost technique that affords the generation of nanofibers with a high porosity and huge surface area to volume ratio [92]. These exceptional features of electrospun nanofibers are important in imparting improved performance of MOFNs. A considerable number of conventional materials and nanoparticles have been incorporated into nanofibers. For example, porous silica electrospun nanofibers containing catalytic silver nanoparticles have been generated using the electrospinning technique [93]. In another study, nylon 6 and poly(styrene-co-divinylbenzene) electrospun nanofibers were fabricated and used for the adsorption of 1-hydroxypyrene with recoveries of around 72% [94].

There are, however, very few reports on electrospun MOFN fibers. For example, a hybrid membrane, ESF@MOF, composed of electrospun-silk-nanofiber and ZIF-8 or ZIF-67, achieved 95 and 99% removal for rhodamine B and MG dyes, respectively [37]. In another study, a MOF-electrospun nanofiber mat from a blend of polytactic acid and ZIF-8 was evaluated for capturing particulate matter [95]. The ZIF-8 improved the particulate matter capture capacity of the membranes due to its high porosity and surface area. Notably, addition of the ZIF-8 led to a marked decrease of nanofiber diameter and a rough surface morphology. The decrease in fiber diameter suggests a higher specific surface area, which is conducive for particulate matter capture. It was also concluded that rougher fiber surfaces may have a positive effect on the performance of the membrane. ZIF8 has also been used for adsorption of Congo red dye in the form of a nanocomposite elctrospun ZIF-8@PVA nanofibers. The resultant MOFN mat was flexible and structurally stable and exhibited excellent adsorption performance. This can be upscaled to industrial scale wastewater treatment targeting organic pollutant removal. An electrospun nanofiber composite mat made of MOFNs and polyacrylonitrile (PAN) nanofibers with a BET specific surface area of 1217 m^2/gm was fabricated for CO_2 adsorption [96]. MOFNs were integrated by electrospraying them onto an electrospun PAN nanofiber mat. Strong interactions between CO_2 and the open metal sites of MOFNs with higher binding energy were observed [96]. This was an indicator of good sorption capabilities of the MOFNs/PAN membrane.

MOFNs possess tunable void volumes such as cavities, pores, and channels suitable for the adsorption of toxic inorganic gases such as NH_3 H_2S, NO_x, CO, and SO_x. For example, H_2S has been successfully removed using MOFNs with high capacity and selectivity. MOFNs, such as HKUST-21, MIL-47(V), and MIL-53(Cr, Al), have been used for the physisorption of H_2S gas [97]. Novel magnetic properties of nanomaterials have been successfully used for formulating efficient hybrid MOFNs for the removal of hazardous substances such as heavy metals and organic dyes from environmental matrices. MOFNs have the advantages of recyclability, good dispersion, and excellent selectivity [98]. Abdi et al. [1] fabricated a nano-superparamagnetic MOFN (ZIF-@SiO_2@$MnFe_2O_4$) with a large surface area of 830 m^2/g for the selective removal of organic dyes from multicomponent systems. The MOFNs retained 90% of the original adsorption capacity and magnetic separation efficiency after the three successive runs.

Apart from use in adsorption of pollutants from environmental matrices, MOFNs present excellent opportunities in nanomedicine for loading "cargo" such as chemotherapy drugs, DNA or RNA, and enzyme carriers [20]. In this regard, the first MOFNs developed for photodynamic therapy applications exhibited an enhanced activity as the oxygen generation capacity was doubled compared to free porphyrin [35]. Other less-researched areas include the integration with solid phase extraction strategies in the microextraction of biological and natural essential compounds, such as flavonoids, from natural materials [58].

3.6   RECOMMENDATIONS AND FUTURE OUTLOOK

With the rapid industrialization accompanied by the discharge of EPs into the environment, future research needs to focus on:

(1) Designing and fabricating new MOFNs with specificity towards certain EPs.
(2) Due to their desirable characteristics which are important for adsorption, it is interesting to note that MOFNs have the potential to be a new generation of antidotes to hazardous chemical poisoning in humans and animals.
(3) To safeguard human, animal, and aquatic life and the environment, it is important for researchers to perform rigorous nanotoxicity and risk assessment studies of emerging MOFNs before full-scale commercialization. To this effect, researchers could generate Material Safety Data Sheets with potential hazards (health, fire, reactivity, and environmental), and enough information on how to work safely with the product.
(4) Governments, especially in developing countries, are encouraged to develop sector-specific regulatory frameworks and legislation which bind manufacturers, importers, and users of MOFN products to ensure safety from the manufacturing line to the market, and disposal.

3.7   CONCLUSION

The application of MOFNs as adsorbents is dependent on their intrinsic properties such as high surface area, architecture, tunable porosity, functionalities, and high crystallinity. The porosity and high specific surface area enable high loading of guest molecules. The metal nodes, organic ligands, and topological configurations of MOFNs are varied. These characteristics can thus be tailored to suit desired applications. Various functional groups can be homogeneously distributed both on the surface and in the pores of MOFNs, which facilitates interaction with molecules to enhance adsorption. The excellent crystallinity confers distinct networks and defined structural information, which is important in investigating the adsorptive mechanisms of pollutants. These exceptional properties of MOFNs allow them to function as supports for the integration of other moieties and materials to form composites for the removal of complex and emerging pollutants. The resulting composites can combine the key features of both components, where MOFNs and the other moieties are synergistic. As a result,

MOFNs have been used to remove a variety of pollutants such as dyes, pharmaceuticals, heavy metals, and REEs from different environmental matrices. With the growth of synthesis technology involving new modification possibilities, the range of MOFNs is likely to increase, and new sorption applications are expected to emerge.

REFERENCES

1. Abdi, J., Mahmoodi, N.M., Vossoughi, M., Alemzadeh, I. Synthesis of magnetic metal-organic framework nanocomposite (ZIF-8@SiO2@MnFe2O4) as a novel adsorbent for selective dye removal from multicomponent systems. *Microporous and Mesoporous Materials* 2019;273:177–88.
2. Amiri, A., Tayebee, R., Abdar, A., Sani, F.N. Synthesis of a zinc-based metal-organic framework with histamine as an organic linker for the dispersive solid-phase extraction of organophosphorus pesticides in water and fruit juice samples. *Journal of Chromatography A* 2019;1597:39–45.
3. An, H., Li, M., Gao, J., Zhang, Z., Ma, S., Chen, Y. Incorporation of biomolecules in Metal-Organic Frameworks for advanced applications. *Coordination Chemistry Reviews* 2019; 384:90–106.
4. Aguilera-Sigalat, J., Bradshaw, D. Synthesis and applications of metal-organic framework-quantum dot (QD@MOF) composites. *Coordination Chemistry Reviews, Part 2* 2016;30:267–91.
5. Guo, X., Kang, C., Huang, H., Chang, Y., Zhong, C. Exploration of functional MOFs for efficient removal of fluoroquinolone antibiotics from water. *Microporous and Mesoporous Materials* 2019;286:84–91.
6. Archana, K., Pillai, N.G., Rhee, K.Y., Asif, A. Super paramagnetic ZIF-67 metal-organic framework nanocomposite. *Composites Part B: Engineering* 2019;158:384–9.
7. Asgharnejad, L., Abbasi, A., Shakeri, A. Ni-based metal-organic framework/GO nanocomposites as selective adsorbent for CO2 over N2. *Microporous and Mesoporous Materials* 2018;262:227–34.
8. Babu, C.M., Vinodh, R., Selvamani, A., Kumar, K.P., Parveen, A.S., Thirukumaran, P., Srinivasan, V.V., Balasubramaniam, R., Ramkumar, V. Organic functionalized Fe3O4/RGO nanocomposites for CO2 adsorption. *Journal of Environmental Chemical Engineering* 2017;5:2440–7.
9. Bao, S, Li, K., Ning, P., Peng, J., Jin, X., Tang, L. Synthesis of amino-functionalization magnetic multi-metal organic framework (Fe3O4/MIL-101(Al0.9Fe0.1)/NH2) for efficient removal of methyl orange from aqueous solution. *Journal of the Taiwan Institute of Chemical Engineers* 2018;87:64–72.
10. Zhao, S, Chen, D., Wei, F., Chen, N., Liang, Z., Luo, Y. Removal of Congo red dye from aqueous solution with nickel-based metal-organic framework/graphene oxide composites prepared by ultrasonic wave-assisted ball milling. *Ultrasonics Sonochemistry* 2017;39:845–52.
11. Chakraborty, A., Acharya, H. Facile synthesis of MgAl-layered double hydroxide supported metal-organic framework nanocomposite for adsorptive removal of methyl orange dye. *Colloid and Interface Science Communications* 2018;24:35–9.
12. Chen, X., Chen, X., Cai, S., Yu, E., Chen, J., Hongpeng, J. MnOx/Cr2O3 composites prepared by pyrolysis of Cr-MOF precursors containing in situ assembly of MnOx as high stable catalyst for toluene oxidation. *Applied Surface Science* 2019;475:312–24.
13. Feng, M., Zhang, P., Zhou, H. C., Sharma, V. K. Water-stable metal-organic frameworks for aqueous removal of heavy metals and radionuclides: A review. *Chemosphere* 2018;209:783–800.

14. Gao, G., Nie, L., Yang, S., Jin, P., Chen, R., Ding, D., Wang, X.C., Wang, W., Wu, K., Zhang, Q. Well-defined strategy for development of adsorbent using metal-organic frameworks (MOF) template for high performance removal of hexavalent chromium. *Applied Surface Science* 2018;457:1208–17.

15. Wen, J., Fang, Y., Zeng, G. Progress and prospect of adsorptive removal of heavy metal ions from aqueous solution using metal-organic frameworks: A review of studies from the last decade. *Chemosphere* 2018;201:627–43.

16. Hashemi, B., Zohrabi, P., Raza, N., Kim, K.H. Metal-organic frameworks as advanced sorbents for the extraction and determination of pollutants from environmental, biological, and food media. *Trends in Analytical Chemistry* 2017;97:65–82.

17. Zhang, B., Luo, Y., Kanyuck, K., Saenz, N., Reed, K., Zavalij, P., Mowery, J., Bauchan, G. Facile and template-free solvothermal synthesis of mesoporous/macroporous metal-organic framework nanosheets. *RSC Advances* 2018;8:33059–64.

18. Murty, B.S., Shankar, P., Raj, B., Rath, B.B., Murday, J. Unique properties of nanomaterials. In *Textbook of Nanoscience and Nanotechnology*. Berlin, Heidelberg: Springer; 2013. pp. 29–65.

19. Sanjay, S.S., Pandey, A.C. A brief manifestation of nanotechnology. In Shukla, Ashutosh Kumar, editor. *EMR/ESR/EPR Spectroscopy for Characterization of Nanomaterials*. New Delhi, India: Springer; 2017.

20. Wang, L., Zheng, M., Xie, Z. Nanoscale metal-organic frameworks for drug delivery: A conventional platform with new promise. *Journal of Materials Chemistry B* 2017;6:707–17.

21. Lu, W., Wei, Z., Gu, Z., Liu, T., Park, J., Park, J., Tian, J., Zhang, M., Zhang, Q., Gentle III, T., Bosch, M., Zhou, H. Tuning the structure and function of metal-organic frameworks via linker design. *Chemical Society Reviews* 2014;43:5561–93.

22. Farha, O.K., Eryazici, I., Jeong, N.C., Hauser, B.G., Christopher E. Wilmer, C.E., Sarjeant, A.A., Snurr, R.Q., Nguyen, S.B., Yazaydın, A.O., Hupp, J.T. Metal-organic framework materials with ultrahigh surface areas: Is the sky the limit? *Journal of American Chemical Society* 2012;134:15016–21.

23. Yin, Z., Wan, S., Yang, J., Kurmoo, M., Zeng, M. Recent advances in post-synthetic modification of metal-organic frameworks: New types and tandem reactions. *Coordination Chemistry Reviews* 2019;378:500–12.

24. Belmabkhout, Y., Mouttaki, H., Eubank, J.F., Guillerm, V., Eddaoudi, M. Effect of pendant isophthalic acid moieties on the adsorption properties of light hydrocarbons in HKUST-1-like tbo-MOFs: Application to methane purification and storage. *RSC Advances* 2014;4:63855–9.

25. Andriamitantsoa, R.S., Wang, J., Dong, W., Gao, H., Wang, G. SO3H-functionalized metal organic frameworks: An efficient heterogeneous catalyst for the synthesis of quinoxaline and derivatives. *RSC Advances* 2016;6:35135–43.

26. Li, N., Xu, J., Feng, R., Hu, T., Bu, X. Governing metal-organic frameworks towards high stability. *Chemical Communications* 2016;52:8501–13.

27. Zhou, M., Ju, Z., Yuan, D. A new metal-organic framework constructed from cationic nodes and cationic linkers for highly efficient anion exchange. *Chemical Communications* 2018;54:2998–3001.

28. Lee, H.J., We, J., Kim, J.O., Kim, D., Cha, W., Lee, E., Sohn, J., Oh, M. Morphological and structural evolutions of metal-organic framework particles from amorphous spheres to crystalline hexagonal rods *Angewandte Chemie International Edition* 2015;54:10564–8.

29. Elhussein, A.A.A., Sahin, S., Bayazit, S.S. Preparation of CeO$_2$ nanofibers derived from Ce-BTC metal organic frameworks and its application on pesticide adsorption. *Journal of Molecular Liquids* 2019;225:10–7.

30. Burgaz, E., Erciyes, A., Andac, M., Andac, O. Synthesis and characterization of nano-sized metal-organic framework-5 (MOF-5) by using consecutive combination of ultra-sound and microwave irradiation methods. *Inorganica Chimica Acta* 2019;485:118–24.
31. Kaur, R., Kaur, A., Umar, A., Anderson, W.A., Kansal, S.K. Metal-organic framework (MOF) porous octahedral nanocrystals of Cu-BTC: Synthesis, properties and enhanced adsorption properties. *Materials Research Bulletin* 2019;109:124–33.
32. Low, J.J., Benin, A.I., Jakubczak, P., Abrahamian, J.F., Faheem, S.A., Willis, R.R. Virtual high throughput screening confirmed experimentally: Porous coordination polymer hydration. *Journal of the American Chemical Society* 2009;131:15834–42.
33. Canivet, J., Fateeva, A., Guo, Y., Coasne, B., Farrusseng, D. Water adsorption in MOFs: Fundamentals and applications. *Chemical Society Reviews* 2014;43:5594–617.
34. Tanase, S., Mittelmeijer-Hazeleger, M.C., Rothenberg, G., Mathonière, C., Jubera, V., Smits, J.M.M., de Gelder, R. A facile building-block synthesis of multifunctional lanthanide MOFs. *Journal of Materials Chemistry A* 2011;21:15544–51.
35. Liu, Y., Liu, Z., Huang, D., Cheng, M., Zeng, G., Lai, C., Zhang, C., Zhou, C., Wang, W., Jiang, D., Wang, H., Shao, B. Metal or metal-containing nanoparticle@MOF nanocomposites as a promising type of photocatalyst. *Coordination Chemistry Reviews* 2019:388:63–78.
36. Abbasi, Z., Cseri, L., Zhang, X., Ladewig, B.P., Wang, H. Metal-organic frameworks (MOFs) and MOF-derived porous carbon materials for sustainable adsorptive waste-water treatment In Gyorgy Szekely, AL, editor. *Sustainable Nanoscale Engineering* Elsevier; 2020. pp. 163–94.
37. Li, J., Yuan, X., Wu, Y., Ma, X., Li, F., Zhang, B., Wang, Y., Lei, Z., Zhang, Z. From powder to cloth: Facile fabrication of dense MOF-76(Tb) coating onto natural silk fiber for feasible detection of copper ions. *Chemical Engineering Journal* 2018;350:637–44.
38. Bazer-Bachi, D., Assié, L., Lecocq, V., Harbuzaru, B., Falk, V. Towards industrial use of metal-organic framework: Impact of shaping on the MOF properties. *Powder Technology* 2014;255:52–9.
39. Gangu, K.K., Maddila, S., Mukkamala, S.B., Jonnalagadda, S.B. A review on contemporary Metal-Organic Framework materials. *Inorganica Chimica Acta* 2016;446:61–74.
40. Li, G., Si, Z., Cai, D., Wang, Z., Qin, P., Tan, T. The in-situ synthesis of a high-flux ZIF-8/polydimethylsiloxane mixed matrix membrane for n-butanol pervaporation. *Separation and Purification Technology* 2019;236:116263.
41. Arora, C., Soni, S., Sahu, S., Mittal, J., Kumar, P., Bajpai, P.K. Iron based metal-organic framework for efficient removal of methylene blue dye from industrial waste. *Journal of Molecular Liquids* 2019;284:343–52.
42. Ojha, D.P., Song, J.H., Kim, H.J. Facile synthesis of graphitic carbon-nitride supported antimony-doped tin oxide nanocomposite and its application for the adsorption of volatile organic compounds. *Journal of Environmental Sciences* 2019;79:35–42.
43. Su, H., Lin, Y., Wang, Z., Wong, Y. L.E., Chen, X., Chan, T.W D. Magnetic metal-organic framework-titanium dioxide nanocomposite as adsorbent in the magnetic solid-phase extraction of fungicides from environmental water samples. *Journal of Chromatography A* 2016;1466:21–8.
44. Gu, Y., Xie, D., Wang, Y., Qin, W., Zhang, H., Wang, G., Zhang, Y., Zhao, H. Facile fabrication of composition-tunable Fe/Mg bimetal-organic frameworks for exceptional arsenate removal. *Chemical Engineering Journal* 2019;357:579–88.
45. Hou, X., et al. High adsorption pearl-necklace-like composite membrane based on metal-organic framework for heavy metal ion removal. *Particle & Particle Systems Characterization* 2018;35:1–8.
46. Ke, F, Jiang, J., Li, Y., Liang, J., Wan, X., Ko, S. Highly selective removal of Hg2+ and Pb2+ by thiol-functionalized Fe3O4@metal-organic framework core-shell magnetic microspheres. *Applied Surface Science* 2017;4103:266–74.

47. Samuel, M.S., Subramaniyan, V., Bhattacharya, J., Parthiban, C., Chand, S., Singh, N.D.P. A GO-CS@MOF Zn(BDC)(DMF) material for the adsorption of chromium(VI) ions from aqueous solution. *Composites Part B: Engineering* 2018;152:116–25.
48. Jun, B., Kim, S., Kim, Y., Her, N., Heo, J., Han, J., Jang, M., Park, C.M., Yoon, Y. Comprehensive evaluation on removal of lead by graphene oxide and metal-organic framework. *Chemosphere* 2019;231:82–92.
49. Jamshidifard, S., Koushkbaghi, S., Hosseini, S., Rezaei, S., Karamipour, A., Jafarirad, A., Irani, M. Incorporation of UiO-66-NH2 MOF into the PAN/chitosan nanofibers for adsorption and membrane filtration of Pb(II), Cd(II) and Cr(VI) ions from aqueous solutions. *Journal of Hazardous Materials* 2019;368:10–20.
50. Sivashankar, R., Sathya, A.B., Vasantharaj, K., Sivasubramanian, V. Environmental Nanotechnology, Monitoring & Management Magnetic composite an environmental super adsorbent for dye sequestration - A review. *Environmental Nanotechnology, Monitoring & Management* 2014;1–2:36–49.
51. Dhaka, S., Kumar, R., Deep, A., Kurade, M.B., Ji, S W., Jeon, B.H. Metal-organic frameworks (MOFs) for the removal of emerging contaminants from aquatic environments. *Coordination Chemistry Reviews* 2019;380:330–52.
52. Song, Y., Qiang, T., Ye, M., Ma, Q., Fang, Z. Metal-organic framework derived magnetically separable 3-dimensional hierarchical Ni@C nanocomposites: Synthesis and adsorption properties. *Applied Surface Science* 2015;359:834–40.
53. Zhu, L., Meng, L., Shi, J., Li, J., Zhang, X., Feng, M. Metal-organic frameworks/carbon-based materials for environmental remediation : A state-of-the-art mini-review. *Journal of Environmental Management* 2019;232:964–77.
54. Gwenzi, W., Chaukura, N., Noubactep, C., Mukome, F.N.D. Biochar based water treatment as a potential low-cost and sustainable technology for clean water provision. *Journal of Environmental Management* 2017;197:732–49.
55. Hao, L., Liu, W., Wang, C., Wu, Q., Wang, Z. Novel porous Fe3O4@C nanocomposite from magnetic metal-phenolic networks for the extraction of chlorophenols from environmental samples. *Talanta* 2019;194:673–9.
56. Jia, Y., Zhao, Y., Zhao, M., Wang, Z., Chen, X., Wang, M. Core-shell indium (III) sulfide@metal-organic framework nanocomposite as an adsorbent for the dispersive solid-phase extraction of nitro-polycyclic aromatic hydrocarbons. *Journal of Chromatography A* 2018;1551:21–8.
57. Kohantorabi, M., Gholami, M.R. Cyclohexene oxidation catalyzed by flower-like core-shell Fe3O4@Au/metal -organic frameworks nanocomposite. *Materials Chemistry and Physics* 2018;213:472–81.
58. Liu, S., Ma, Y., Gao, L., Pan, J. pH-responsive magnetic metal-organic framework nanocomposite: A smart porous adsorbent for highly specific enrichment of cis-diol containing luteolin. *Chemical Engineering Journal* 2018;341:198–207.
59. Matlooba, A.M., El-Hafiza, D.R.A, Saada, L., Mikhaila, S., Guirguis, D. Metal-organic framework-graphene nanocomposite for high adsorption removal of DBT hazardous material in liquid fuel. *Journal of Hazardous Materials* 2019;373:447–58.
60. Mirzajana, R., Kardani, F., Ramezani, Z. Preparation and characterization of magnetic metal-organic framework nanocomposite as solid-phase microextraction fibers coupled with highperformance liquid chromatography for determination of non-steroidal anti-inflammatory drugs in biological fluids and tablet formulation samples. *Microchemical Journal* 2019;144:270–84.
61. Navarathna, C.M., Dewage, N.B., Karunanayake, A.G., Farmer, E.L., Perez, F., Hassan, E.B., Mlsna, T.E., Pittman Jr, C.U. Rhodamine B adsorptive removal and photocatalytic degradation on MIL-53-Fe MOF/magnetic magnetite/biochar composites. *Journal of Inorganic and Organometallic Polymers and Materials* 2020;30:214–229.

62. Liu, X., Gong, W., Luo, J., Zou, C., Yang, Y., Yang, S. Selective adsorption of cationic dyes from aqueous solution by polyoxometalate-based metal-organic framework composite. *Applied Surface Science* 2016;362:517–24.

63. Yu, L., Cao, W., Wu, S., Yang, C., Cheng, J. Removal of tetracycline from aqueous solution by MOF/graphite oxide pellets: Preparation, characteristic, adsorption performance and mechanism. *Ecotoxicology and Environmental Safety* 2018;164: 289–96.

64. Suarez, S., Lema, J.M., Omil, F. Pre-treatment of hospital wastewater by coagulation-flocculation and flotation. *Bioresource Technology* 2009;100:2138–46.

65. Hua, Z., Guo, K., Kong, X., Lin, S., Wu, Z., Wang, L., Huang, H., Fang, J. PPCP degradation and DBP formation in the solar/free chlorine system: Effects of pH and dissolved oxygen. *Water Research* 2019;150:77–85.

66. Joseph, L., Jun, B., Jang, M., Park, C.M., Muñoz-Senmached, J.C., Hernández-Maldonadod, A.J., Heydene, A., Yuf, M., Yoon, Y. Removal of contaminants of emerging concern by metal-organic framework nanoadsorbents: A review. *Chemical Engineering Journal* 2019;369:928–46.

67. Seo, Y.S., Khan, N.A., Jhung, S.H. Adsorptive removal of methylchlorophenoxypropionic acid from water with a metal-organic framework. *Chemical Engineering Journal* 2015;270:22–7.

68. Wang, B., Lv, X., Feng, D., Xie, L., Zhang, J., Li, M., Xie, Y., Li, J., Zhou, H. Highly stable Zr (V)- based metal-organic frameworks for the detection and removal of antibiotics and organic explosives in water. *Journal of the American Chemical Society* 2016;138:6204–16.

69. Annamalai, J., Namasivayam, V. Endocrine disrupting chemicals in the atmosphere: Their effects on humans and wildlife. *Environment International* 2015;201:78–97.

70. van Zijl, M.C., Aneck-Hahn, N.H., Swart, P., Hayward, S., Genthe, B., De Jager, C. Estrogenic activity, chemical levels and health risk assessment of municipal distribution point water from Pretoria and Cape Town, South Africa. *Chemosphere* 2017;186:305–13.

71. Byrne, C., Subramanian, G., Pillai, S.C. Recent advances in photocatalysis for environmental applications. *Journal of Environmental Chemical Engineering* 2018;6: 3531–55.

72. Fukahori, S., Fujiwara, T., Ito, R., Funamizu, N. *pH dependent adsorption of sulfa drugs on high silica zeolites. Desalination* 2011;275:237–42.

73. Ragab, D., Gomaa, H.G., Sabouni, R., Salem, M., Ren, M., Zhu, J. Micropollutants removal from water using microfiltration membrane modified with ZIF-8 metal-organic framework (MOFs). *Chemical Engineering Journal* 2016;300:273–9.

74. Rizzo, L., Manaia, C., Merlin, C., Schwartz, T., Dagot, C., Ploy, M.C., Michael, I., Fatta-Kassinos, D. Urban wastewater treatment plants as hotspots for antibiotic resistant bacteria and genes spread into the environment, a review. *Science of The Total Environment* 2013;447:345–60.

75. Mwakalapa, E.B., Simukoko, C.K., Mmochi, A.J., Mdegela, R.H., Berg, V., Müller, M.H.B., Lyche, J.V., Polder, A. Heavy metals in farmed and wild milkfish (*Chanos chanos*) and wild mullet (*Mugil cephalus*) along the coasts of Tanzania and associated health risk for humans and fish. *Chemosphere* 2019;224:176–86.

76. Musterat, C., Teodosin, C. Removal of persistant organic pollutants from textile wastwawater by membrane processes. *Environmental Engineering and Management Journal* 2007;6:175–87.

77. Loha, K.M., Lamoree, M., Weiss, J.M., de Boer, J. Import, disposal, and health impacts of pesticides in the East Africa Rift (EAR) zone: A review on management and policy analysis. *Crop Protection* 2018;112:322–31.

78. Liu, G., Li, L., Huang, X., Zheng, S., Xu, X., Liu, Z., Zhang, Y., Wang, J., Lin, H., Xu, D. Adsorption and removal of organophosphorus pesticides from environmental water and soil samples by using magnetic multi-walled carbon nanotubes @ organic framework ZIF-8. *Journal of Materials Science* 2018;53:10772–83.

79. Gwenzi, W., Mangori, L., Danha, C., Chaukura, N., Dunjana, N. Sources, behaviour and environmental risks of high-technology rare earth elements as emerging contaminants. *Science of The Total Environment* 2018;636:299–313.

80. Yesiller, S.U., Eroğlu, A.E., Shahwan, T. Removal of aqueous rare earth elements REEs using nano-based materials. *Journal of Industrial and Engineering Chemistry* 2013;19:898–907.

81. Elsaidi, S.K., Sinnwell, M.A., Devaraj, A., Droubay, T.C., Nie, Z., Murugesan, V., McGrail, B.P., Thallapally, P.K. Extraction of rare earth elements using magnetite@ MOF composites. *Journal of Materials Chemistry A* 2018;6:18438–43.

82. Lee, YaA. Selective adsorption of rare earth elements over functionalized Cr-MIL-101. *ACS Applied Materials & Interfaces* 2018;10:23918–27.

83. Schelling, M., Kim, M., Otal, E., Hinestroza, J. Decoration of cotton fibers with a water-stable metal-organic framework (UiO-66) for the decomposition and enhanced adsorption of micropollutants in water. *Bioengineering* 2018;5:1–11.

84. Sarker, M., Ahmed, I., Jhung, S.H. Adsorptive removal of herbicides from water over nitrogen-doped carbon obtained from ionic liquid@ ZIF-8. *Chemical Engineering Journal* 2017;323:203–11.

85. Abdelhameed, R.M., Abdel-Gawad, H., Elshahat, M., Emam, H.E. Cu-BTC@ cotton composite: Design and removal of ethion insecticide from water. *RSC Advances* 2016;6:42324–33.

86. Yang, Q., Wang, J., Zhang, W., Liu, F., Yue, X., Liu, Y., Yang, M., Li, Z., Wang, J. Interface engineering of metal -organic framework on graphene oxide with enhanced adsorption capacity for organophosphorus pesticide. *Chemical Engineering Journal* 2017;313:19–26.

87. Kadhom, M., Deng, B. Metal-organic frameworks (MOFs) in water filtration membranes for desalination and other applications. *Applied Materials Today* 2018;11:219–30.

88. Fard, A.K., McKay, G., Buekenhoudt, A., Sulaiti, H.A., Motmans, F., Khraisheh, M. Atieh, M. Inorganic Membranes: Preparation and application for water treatment and desalination. *Materials* 2018;11:1–47.

89. Denny, M.S., Moreton, J.C., Benz, L., Cohen, S.M. Metal-organic frameworks for membrane-based separations. *Nature Reviews Materials* 2016;1:1–17.

90. Wang, H., Zhao, S., Liu, Y., Yao, R., Wang, X., Cao, Y., Ma, D., Zou, M., Cao, A., Feng, X., Wang, B. Membrane adsorbers with ultrahigh metal-organic framework loading for high flux separations. *Nature Communications* 2019;10:1–9.

91. Shekhah, O., Chernikova, V., Belmabkhout, Y., Eddaoudi, M. Metal-Organic Framework Membranes: From fabrication to gas separation. *Crystals* 2018;8:1–55.

92. Tijing, L.D., Woo, Y.C., Yao, M., Ren, J., Shon, H.K., Drioli, E., Giorno, L., Fontananova, E. Electrospinning for membrane fabrication: Strategies and applications. In Drioli, E., Giorno, L., and Fontananova, E., editors. *Comprehensive Membrane Science and Engineering*. 2nd ed. Oxford, UK: Elsevier; 2017. pp. 418–44.

93. Chigome, S., Darko, G., Torto, N. Electrospun nanofibers as sorbent material for solid phase extraction. *Analyst* 2011;136:2879–89.

94. Ifegwu, O.C., Anyakora, C., Chigome, S., Torto, N. Electrospun nanofiber sorbents for the pre-concentration of urinary 1-hydroxypyrene. *Journal of Analytical Science and Technology* 2015;6:1–12.

95. Wang, Y., Dai, X., Li, X., Wang, X. The PM2.5 capture of poly (lactic acid)/nano MOFs eletrospinning membrane with hydrophilic surface. *Material Research Express* 2018;5:1–6.

96. Wahiduzzaman, A.K., Stone, J., Harp, S., Khan Mujibur, K. Synthesis and electro-spraying of nanoscale MOF (Metal Organic Framework) for high performance CO_2 adsorption membrane nanoscale. *Research Letters* 2017;12.
97. Wen, M., Li, G., Liu, H., Chen, J., An, T., Yamashita, H. Metal-organic framework-based nanomaterials for adsorption and photocatalytic degradation of gaseous pollutants: Recent progress and challenges. *Environmental Science: Nano* 2019;6:1006–25.
98. Zhao, G., Qin, N., Pan, A., Wu, X., Peng, C., Ke, F., Iqbal, M., Ramachandraiah, K., Zhu, J. Magnetic Nanoparticles@Metal-Organic Framework composites as sustainable environment adsorbents. *Journal of Nanomaterials* 2019;2019:1–11.

4 Metal-Organic Framework-Derived Carbon-Coated Nanocomposites for Electrochemical Capacitors

Wei Ni and Anish Khan

CONTENTS

4.1 INTRODUCTION

With the rising concerns of environmental pollution, climate change, and the forthcoming energy crisis, the development of clean and sustainable energy instead of fossil fuels has attracted intensive attention. It is of great urgency and importance to develop efficient energy storage and conversion technologies and materials aimed at the utilization of various clean energy sources [1–18]. Electrochemical capacitors (ECs), also known as supercapacitors (SCs), are a new type of energy storage device between conventional capacitors and rechargeable batteries. They are based on the electrochemical processes of either electrical double-layer capacitance (EDLC) or pseudocapacitance (involving ultrafast surface redox reactions), i.e., maintaining the advantages of EDLC-level high power density combined with battery-level high energy density (viz. short charging time and long life cycle). The development of new electrode materials with high specific capacity for excellent supercapacitive energy storage and conversion is highly desirable [19–28]. Notably, many important break-throughs for a new generation of ECs have been reported in recent years, related to the theoretical understanding, material synthesis, and electrode/device design, e.g., hybrid supercapacitors (HSCs), microsupercapacitors (MSCs), and flexible superca-pacitors (FSCs) to bridge the performance gap between batteries, conventional capac-itors, and wearable electronics [16,21,29–31]. Metal-organic frameworks (MOFs), as a novel class of intrinsically porous inorganic-organic hybrid crystalline materials with well-defined structures, are acting as a superior candidate in recent years based on their high surface area, controllable pore size and structures/topologies, and excellent electrochemical performances for next-generation electrochemical energy-storage systems [1,15,20,32–49]. Theoretically, MOFs can be ideal supercapacitor electrode materials due to their unique structures. The direct utilization of pristine MOFs as a versatile supercapacitor-active material has preliminarily proved to be very promis-ing owing to the tremendous pseudocapacitive redox centers and a porous skeleton with facilitated fast-ion diffusion [38,43,50–55]. However, the intrinsically insulating nature and poor chemical stability of most reported MOFs are the main obstacles for their extended electrochemical applications, especially for high-power ECs [56,57]; and the MOF-derived carbon-coated transition-metal oxide or chalcogenide nano-composites with the additional merits of enhanced conductivity and structural stabil-ity have endowed the MOFs with promising practical applications [25,33,55,57–63].

Based on the intense research on MOF-derived nanostructures with great advantages in promising ECs, significant advances in the recent development of MOF-derived high-conductivity materials (e.g., carbon-coated metal oxides/chal-cogenides) for next-generation high-power clean energy applications are selectively reviewed, with special emphasis on the applications of MOFs as versatile platforms for emerging advanced ECs. Firstly, a fundamental understanding of the mechanism is focused on the relationship between the structure, composition of MOF-derived electrode composite materials, and their electrochemical performances, as well as performance-optimizing strategies. Secondly, some emerging electrode materials of carbon coated or supported MOF-derived hybrids/composites and their promising applications in ECs, hybrid supercapacitors, and flexible/wearable supercapacitors are summarized. Finally, the future perspectives and key technical challenges are presented for further research in this thriving field.

4.2 MOF-DERIVED NANOARCHITECTURES: STRUCTURE, PROPERTIES, AND STRATEGIES FOR ECS

MOFs are a class of crystalline materials formed via self-assembly of inorganic metal ions (or clusters) and organic ligands connected by coordination interactions

[20,34,35,64], whose capacitance comes from the redox reaction of center metal ions and the electrochemical double-layer (EDL) capacitance on the large internal surface area [29,65]. The extremely high porosity and openness, structural tailorability, and chemical versatility, as well as low density of MOFs, give superior capacity for storage and robust cycling of ions, which give them a key role in high-performance electrochemical energy storage [34,36,50,66–71]. However, direct adoption of conventional pristine MOFs as electrode materials often shows impoverished electrochemical performance owing to their low electrical conductivity and poor chemical stability [56,57]. MOF-derived nanostructures or nanocomposites, including carbonaceous materials and transition-metal oxides/chalcogenides (TMOs/TMCs), are gaining momentum in the electrochemical conversion and storage fields [20,32,33,45,60,63,64,72–77]. Via the MOF-derived synthetic approaches, various novel compositional and structural features (indicated by high porosity and large surface area and the resulting large electrochemically active surface, facile charge/mass transport, and efficient accommodation of strain) may be easily created in well-controlled micro-/nanometer scales, providing enhanced and/or unique performance not expected in their conventional counterparts. Generally, two typical strategies for controlling the chemical composition of MOF-derived materials are adopted, i.e., tuning the composition of MOF precursors and/or manipulating the conversion process, which may simultaneously happen in many cases [25].

Pyrolysis of MOFs in air usually leads to corresponding metal oxides and nanocomposites comprised of metal-containing components, and carbon shells/skeletons can be produced by pyrolysis of MOFs in nonoxidative/inert atmospheres without further etching, e.g., acid leaching (viz. these distinct derivatives may be transformed into both asymmetric electrodes based on a "one for two" (also called two-for-one) strategy) [25,60,77–81]. The inorganic components may vary, and could be metals, metal oxides/chalcogenides, or even metal carbides/phosphates, depending on the type of metal in the specific MOFs and corresponding synthetic conditions (e.g., reacting with secondary reactants) [25,60,82]. And the derived composite/hybrid materials typically exist with metal-containing nanoparticles or even atoms/clusters combined with carbonaceous components owing to the homogeneous and localized reactions between the metal nodes and organic linkers in the MOF precursors [25,59,83].

On one hand, the presence of carbon matrix/skeleton not only enhances the conductivities but would also effectively suppress or avoid the excessive growth and agglomeration of metal-containing moieties as well as the diminishment of surface area during a high-temperature synthesis process, thus for a well-organized porous structures, e.g., carbon-coated/supported porous metal compound-based composites [25, 84]. On the other hand, the porous carbons (including the MOFs derived) widely used as electrode materials/components of ECs, featuring extremely high surface area, controlled pore structure, excellent chemical stability, and relatively low cost, concurrently contribute to the capacitance of the composite electrodes [58,77,85–89]. With the assistance of carbon coating, the MOF-derived nanocomposites obtained exhibit superior cyclability and a greatly enhanced rate performance when employed as electrode materials for ECs.

To achieve excellent performance in all aspects of power density and energy density, the high conductivity and large electrolyte-accessible surface area are two critical prerequisites for the active materials used in ECs [58]. The electrochemical properties of MOF-derived materials may be morphology/size dependent [90,91], composition/doping/valence related [92–96], carbon coated or supported [84,97–105], and affected by carbonization temperature [106]. Moreover, owing to the high catalytic activity of transition-metal clusters or nanoparticles, the carbon coating may often evolve into carbon nanotubes (CNTs), thus for C/CNT-incorporated nanocomposites [107,108]. Of course, besides the carbon as superior conductive component, many others such as metals (e.g., silver nanoclusters [109]), conducting polymers or other conductive inorganic compounds also work, and some of them even offer additional pseudocapacitance for enhanced energy density. Overall, the ones with higher specific surface areas, more active sites, and higher conductivity usually exhibit superior electrochemical performances.

According to the equation $E = 1/2 \times CV^2$, construction of asymmetric supercapacitors (ASCs, or hybrid capacitors, HCs) is an efficient way to improve the energy density by increasing either capacitance (C) or cell potential (V) [97,110]. For the HCs, one of the key limitations for high performance is the kinetic imbalance between the anode and cathode, which results in inferior rate capacity and cycle life. Developing appropriate nanoscale structures that endow anode materials with a rapid pseudocapacitive reaction mechanism has proved to be an effective resolution. Since the capacity of the HCs is based on a classic equation ($1/C_{HC} = 1/C_{anode} + 1/C_{cathode}$), the improvement of capacitive electrodes (mostly activated carbons), especially in organic electrolytes, is also vital to avoid diluting the capacity of battery-type anodes and the configured full cells [111]. Moreover, the addition of redox components, e.g., $K_4Fe(CN)_6$, to the electrolytes could further enhance the capacitance of the device [112].

4.3 MOF-DERIVED METAL OXIDE/CARBON COMPOSITES

Transition-metal oxides (TMOs) demonstrate great potential in energy storage and conversion owing to their reversible surface redox reactions. In particular, the first-row transition metal (e.g., Mn, Fe, Co, and Ni) oxides with mutable valence states have been intensively investigated as promising electrode materials for ECs [20]. MOF-derived highly porous TMOs with large accessible internal surface areas will facilitate the diffusion of ions, which is critical for the electrochemical activity of the electrodes [58].

4.3.1 MOF-Derived TMOs

TMOs are typical derivatives with superior electrochemical and structural stability for ECs, and their related carbon-coating composites mainly include the following types of composites such as MnO_x/CSs nanocomposites (CSs: ultrathin carbon sheets) [113], GO/CuO hybrid (GO: graphene oxide) [114], Co_3O_4/ZnO/rGO (rGO: reduced graphene oxide) [115], MOF-derived hollow cage $Ni_xCo_{3-x}O_4$/graphene

[116], CuO_x@mC@PANI@rGO hybrid (mesoporous carbon: mC, polyaniline: PANI) [117], and MOF-derived composite aerogels (rGO/Fe_2O_3, rGO/NiO/Ni) [118].

Among a variety of electrode materials, graphene-based composites are one of the most promising candidates for ECs [42,118,119]. 3D graphene architectures could further generate synergistic effects (e.g., increased electrical conductivity, reduced volume change, and improved structural stability) to achieve enhanced performances [116,120,121]. As a typical example, Cao and coworkers developed a facile and general method for the large-scale preparation of various 3D GO/MOF composite hydrogels with controllable composition and the subsequent MOF-derived composite aerogels (e.g., rGO/Fe_2O_3 and rGO/NiO/Ni) through freeze-drying and calcination processes (Figure 4.1a–d). When used as a supercapacitor electrode, the rGO/Fe_2O_3 composite shows high specific capacitances and a good rate capability (869 and 290 F g^{-1} at 1 and 20 A g^{-1}, respectively), as well as a long cycle life without obvious capacitance decay over 5000 cycles (Figure 4.1e–h) [118]. It has the potential for the flexible all-solid-state supercapacitor device with excellent mechanical flexibility. And the optimized transition-metal mixed oxides (e.g., MOF-derived hollow cage $Ni_xCo_{3-x}O_4$) synergy with 3D graphene can achieve even higher specific capacitance (2871 and ca. 2100 F g^{-1} at 1 and 10 A g^{-1}, respectively) and thus higher specific energy density and power density (50.2 and 23.3 W h kg^{-1} at 750 and 3750 W kg^{-1}, respectively) [116].

4.3.2 MOF-DERIVED CARBON

Owing to their controllable morphologies, tunable porous structures, diverse compositions, and facile fabrication, MOFs are an ideal class of precursor material to develop high-performance carbon-based materials for energy applications [122]. Compared with discrete carbon-supported MOF-derived nanomaterials, carbon-coated nanocomposites with easily accessible external conductive surfaces will be more competent [123]. Typical examples include MOF-derived nanoporous carbon (MOF-NPC)/MnO_2 hybrids [97], MOF-NPC/RuO_2 composite [124], Na^+-MnO_2/MOF-NPC hybrid [125], porous N-doped carbon/CNTs [126], and ZnO@C@$NiCo_2O_4$ core–shell structures [127].

For example, Liu et al. constructed a hybrid material of T-Nb_2O_5 quantum dots (QDs) embedded in a MOF (ZIF-8)-derived N-doped carbon (NQD-NC) with uniform rhombic dodecahedral morphology (Figure 4.2a and b), and studied its advanced hybrid supercapacitors (HSCs) application. The use of a highly dispersed QDs structure shows great advances in energy storage devices to buffer the volume change of active materials, and the MOFs have offered a new platform for novel nanoporous carbon materials including high-content N-doped carbons with a high surface area, large pore volume, and superior electronic characteristics for diverse electrochemical applications. Owing to these merits, the Li-HSCs based on the NQD-NC hybrid exhibit enhanced electrochemical performances, including excellent cycle stability (82% capacity retention at 5 A g^{-1} after 3000 cycles in a voltage window of 0.5–4.0 V; 85% retention after 4500 cycles at 0.5–3.0 V), high energy density (76.9 W h kg^{-1}), and high power density (11.3 kW kg^{-1}) (Figure 4.2c and d) [100].

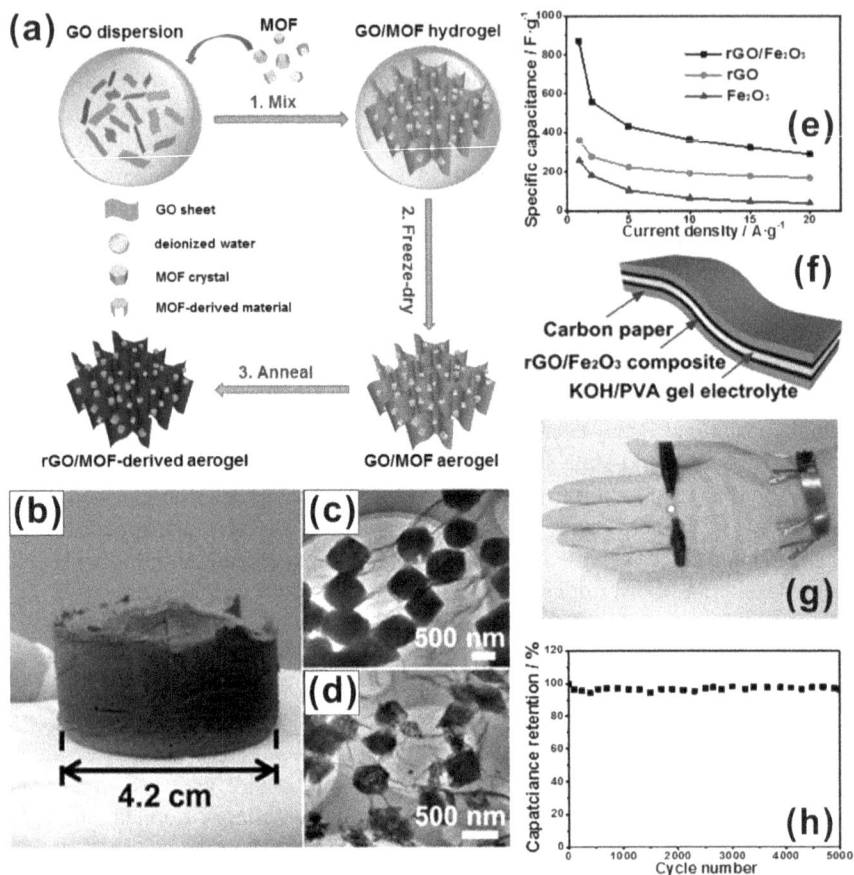

FIGURE 4.1 (a) Scheme of preparation processes for GO/MOF and rGO/MOF-derived composite aerogels. (b and c) Photograph and TEM image of GO/Fe-MOF aerogel, and (d) TEM image of the derived rGO/Fe$_2$O$_3$ composite aerogel. Electrochemical performance of (e) the prepared rGO/Fe$_2$O$_3$, rGO, and Fe$_2$O$_3$ electrodes. Electrochemical performance of the all-solid-state flexible supercapacitor device fabricated with rGO/Fe$_2$O$_3$ composite. (f and g) Schematic illustration and photograph of the fabricated device. (h) The GCD cycling performance at 50.4 mA cm^{-3} for 5000 cycles. Reprinted with permission from Ref. [118]. Copyright 2017 American Chemical Society.

4.3.3 MOF-Derived Composites

The TMO/TMC composites can be directly achieved and these typical samples include Fe$_x$O$_y$/porous carbon and Cr$_2$O$_3$/porous carbon [128], Fe$_3$O$_4$/Fe nanoparticles decorated in carbon shell [106], Cl/O-doped C/CoO nanoparticles [93], Co$_3$O$_4$@C nanocomposite [81], nanoporous carbon/Co$_3$O$_4$ [129], mesoporous Co$_3$O$_4$@carbon composites [130], NiO@ graphite carbon nanocomposites [131], dilute NiO/CNF composite [132], NiO/Ni encapsulated in N-doped carbon nanotubes (NiO/

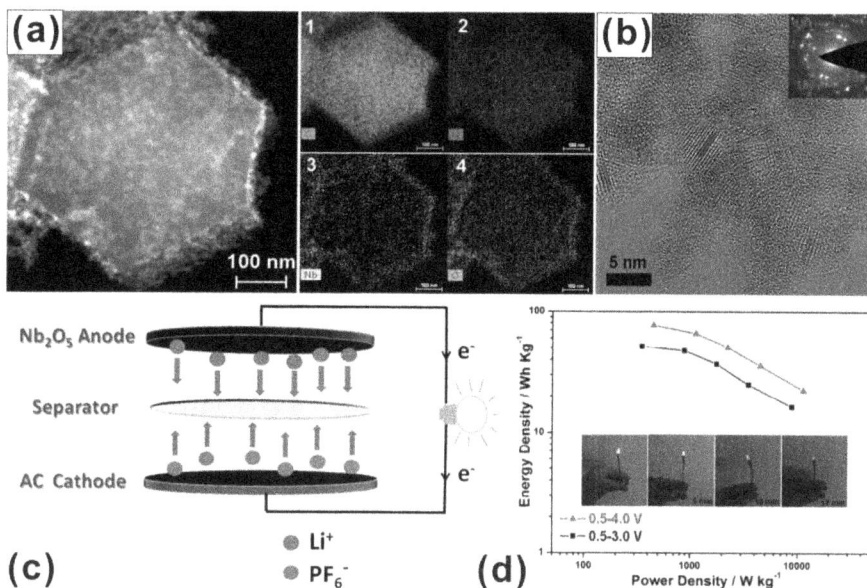

FIGURE 4.2 (a) EDX-elemental mapping images (1–4: C, N, Nb, O) and (b) HRTEM image of Nb_2O_5 quantum dots embedded in MOF-derived N-doped porous carbon (NQD–NC). (c) Schematic of the discharge mechanism of the AC//NQD–NC HSCs. (d) Ragone plots (energy density vs. power density) of AC//NQD–NC HSCs (inset: lighting up a blue LED for 12 min). Reprinted with permission from Ref. [100]. Copyright 2016 The Royal Society of Chemistry.

Ni/NCNTs) [133], MnO_x/N-doped carbon/MnO_2 [134], Co/MnO@graphite carbon composites [26], carbon nanosheets-supported MnO_2 nanosheet arrays (MnO_2/CNS) (Figure 4.3) [99], ZnO/porous carbon composites [135], ZnO QDs/carbon/CNTs [126], rodlike CeO_2/carbon nanocomposite [112], $ZnMn_2O_4$/carbon nanorods [84], $MnCr_2O_4$/Mn_3O_4/C and $CoCr_2O_4$/C nanoparticles [101], nickel/porous carbon composite [136], hierarchical porous Ni@C nanospheres [137], nanoporous cobalt-rich carbons [138], Co/C composites [139], Ni-Co@carbon hybrid [140], Ni/Co metal nanoparticles incorporated in biomass carbon matrix [141], 2D CoNi nanoparticles@S,N-doped carbon composites (CoNi@SNC) [122], and hierarchical Ni/P/N/C composites [142]; and for typical examples, see Table 4.1. As an example, dilute NiO particles decorated carbon nanofiber (NiO/CNF) can be in situ fabricated by direct pyrolysis of assembled 1D Ni, Zn-containing MOF nanofibers (Figure 4.4). Through successful modulation of conductivity and porosity of final composites, the NiO/CNF composites display well-defined capacitive features and good rate capability as well as cycling stability up to 5000 times (only 10% loss), and it is exceptional that the 0.43 wt% NiO contributed to over 35% of the total capacitance (234 F g^{-1}) [132].

Besides the appropriate nanostructures, the metal carbon-based materials, especially these multicomponent carbon-based composites with further enhanced conductivity owing to the direct metals or indirect catalytic graphitization of amorphous

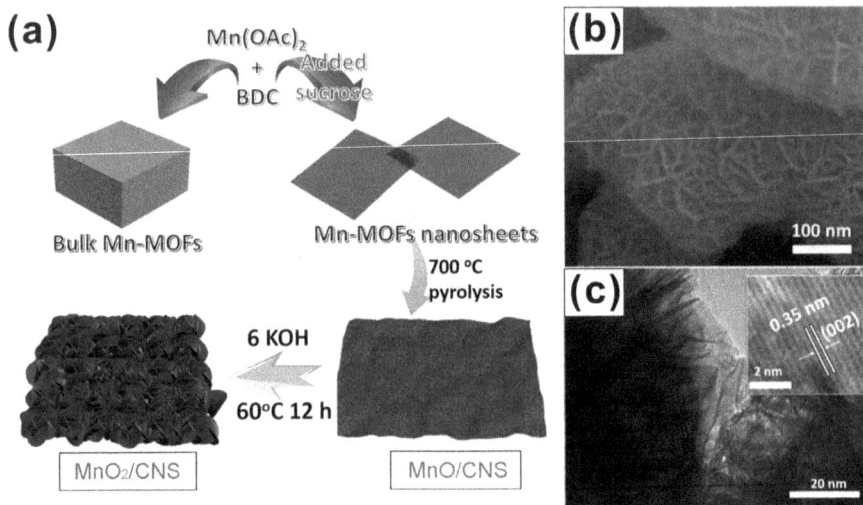

FIGURE 4.3 (a) Schematic illustration for the fabrication of the vertically aligned MnO_2 nanosheets strongly coupled with carbon nanosheets (MnO_2/CNS). (b and c) The corresponding SEM and TEM (inset: HRTEM) images of MnO_2/CNS. Reprinted with permission from Ref. [99]. Copyright 2018 Springer Nature.

carbon at relatively low temperature, are also a key to superior performances (e.g., the rate capability and power density, and these electroactive metals also contribute to the higher specific capacitance and energy density of composite electrodes) [26,7 9,81,107,108,122,136,143,144]. For example, Kim et al. designed a novel nanoarchitecture using selected Co^{2+}-excess bimetallic hybrid Co/Zn ZIFs (HZs) for nanoporous functional composites (NPFCs), i.e., those containing N-doped nanoporous carbon with a rich graphitic carbon nanotube (CNT) content on particle surfaces and embedded with Co and Co_3O_4 nanoparticles (NPs, 15 nm) through controlled carbonization and subsequent oxidation (Figure 4.5). The as-synthesized composites utilize both electric double-layer capacitance and pseudocapacitance within a single nanoporous composite particle, and the coexisting micro-/mesoporous structures exhibit superior electrochemical storage performance (545 F g^{-1} in a three-electrode system, 320% enhanced capacitance compared to that of pristine nanoporous carbon, and the mechanism for further improving cycle life is needed) [108]. Xu et al. developed a facile method to synthesize bimetallic Co/MnO@graphite carbon (GC) nanocomposites by an in situ conversion of Co/Mn-MOFs, and different carbon-coated composites can be obtained by precisely controlled calcination temperature. The as-prepared Co/MnO@GC hybrids, especially that carbonized at 700°C, preserve the nanocubic morphology of its precursor and show a superior specific capacitance of 2275 F g^{-1} at a current density of 4 A g^{-1}. When fabricated into an aqueous asymmetric supercapacitor device (aASC device) with active carbon as negative electrode, the device shows a high cell voltage of 1.7 V and a considerably high specific capacitance of 246 F g^{-1} at 2 A g^{-1} (Figure 4.6) [26]. Overall, the strategic

TABLE 4.1

Typical MOF-Derived TMOs/Carbon Hybrid Materials for ECs (VR, Voltage Range; CR, Capacitance Retention; CD, Current Density; SR, Scan Rate; ED/PD, Energy Density/Power Density)

Electrode Materials [MOF Precursors]	VR	Cycles, CR, CD or SR	ED@PD or Capacitance (CD) [W h kg⁻¹/W kg⁻¹] or [F g⁻¹ (A g⁻¹)]	Electrolyte	Year/Ref.
ASC (TiO$_2$/C//3D nanoporous carbon) [MIL-125 (Ti), ZIF-8]	1–4 V	10 k, >90%, 1.0 A g⁻¹	142.7@0.25, 61.8@25	1 M NaClO$_4$ EC/PC (1:1 v/v) + 5 wt% FEC	2018/[111]
ASC (MOF-NPC/MnO$_2$//MOF-NPC) [ZIF-8]	0–2.2 V	3 k, 87.3%, 5 A g⁻¹	76.0@2.2 49.6@22	1 M Na$_2$SO$_4$ aqueous	2017/[97]
SSC (MOF-NPC/RuO$_2$) [ZIF-8]	0–1.2 V	3 k, 80.9%, 3 A g⁻¹	23.4@0.6, 19.3@12	1 M H$_2$SO$_4$ aqueous	2018/[124]
MnO$_x$-CSs nanocomposites [Mn-MOF]	0–1.8 V	2 k, 94.4%, 4 A g⁻¹	27.5@0.225 12.0@5.4	1 M Na$_2$SO$_4$ aqueous	2017/[113]
Co/Co$_3$O$_4$ NPs@MOF-NFC/CNTs (HZ-NPFC) [Co/Zn ZIF]	−0.8 to 0.4 V (vs. Ag/AgCl)	2 k, 72%; 2 A g⁻¹	545/180 (2/10)	6 M KOH aqueous	2017/[108]
Ordered porous Mn$_3$O$_4$@N-doped carbon/graphene (MCG) [Mn-MOF]	−0.1 to 0.9 V (vs. SCE)	2 k, 98.1%; 5 A g⁻¹	456/246 (1/20)	1 M Na$_2$SO$_4$ aqueous	2017/[98]
Nickel/ porous carbon composite [Ni-MOF]	−1.0 to 0 V (vs. Hg/HgO)	5 k, 98.5%, 5 A g⁻¹	81 (0.2)	6 M KOH aqueous	2018/[136]
Nanoporous cobalt-rich carbons [Co-ZIF]	−1.0 to 0.33 V (vs. SCE)	6 k, 80%, 5 A g⁻¹	393 (0.5)	6 M KOH aqueous	2019/[138]
SSC	0–1.4 V	20 k, 95.4%, 5 A g⁻¹	61.2@0.7		
Ni-Co@carbon hybrid [nickelocene@ZIF-67]	−0.2 to 0.3 V (vs. Ag/AgCl)	1 k, 94.3%, 10 A g⁻¹	236/212/180/120 (1/25/10)	6 M KOH aqueous	2018/[140]

(Continued)

TABLE 4.1 (CONTINUED)

Typical MOF-Derived TMOs/Carbon Hybrid Materials for ECs (VR, Voltage Range; CR, Capacitance Retention; CD, Current Density; SR, Scan Rate; ED/PD, Energy Density/Power Density)

Electrode Materials [MOF Precursors]	VR	Cycles, CR, CD or SR	ED@PD or Capacitance (CD) [W h Kg^{-1}/W Kg^{-1}] or [F g^{-1} (A g^{-1})]	Electrolyte	Year/ Ref.
Ni/Co NPs@N-C [Ni/Co-MOF]	0–0.45 V (vs. Ag/AgCl)	–	2471 (1.0)	2 M KOH aqueous	2018/ [141]
ASC (Ni/Co NPs@N-C//N-C)	0–1.55 V	5 k, 76.6%, 10 A g^{-1}	31.8@6.2, 28.1@7.75	6 M KOH aqueous	2017/ [122]
2D CoNi nanoparticles@S,N-doped carbon composites (CoNi@SNC) [S,N-containing Co/Ni MOFs]	0–0.5 V (vs. Hg/HgO)	3 k, 95.1%, 10 A g^{-1}	1970/1897/1730/1543/1282 (1/2/5/10/20)	6 M KOH aqueous	
ASC (CoNi@SNC//AC)	0–1.6 V	4 k, 90.6%, 10 A g^{-1}	55.7@0.8		
Hierarchical Ni/P/N/C composites	0–0.3 V (vs. Hg/HgO)	5 k, 90%, 10 A g^{-1}	2888/1497 (1/20)	6 M KOH aqueous	2019/ [142]
Na$^+$-MnO$_2$/MOF-NPC hybrid [ZIF-67]	0–0.8 V (vs. SCE)	5 k, 91%, 1 A g^{-1}	217 (10 mV s^{-1})	1 M Na$_2$SO$_4$ aqueous	2018/ [125]
ASC (Na$^+$-MnO$_2$/NPC//AC)	0–2.1 V	3 k, 82%, 2 A g^{-1}	15.3 W h Kg^{-1}, 5.15 W Kg^{-1}		
NiO@graphite carbon nanocomposites [Ni-MOF]	0–0.45 V (vs. Ag/AgCl)	1.5 k, 74%, 2 A g^{-1}	323/280/200/133 (1/2/5/10)	6 M KOH aqueous	2019/ [131]
ZnO-porous carbon composites [MOF-74(Zn)]	–1.0–0 V (vs. SCE)	1 k, 97.8%, 1 A g^{-1}	198 (0.6)	1 M H$_2$SO$_4$ aqueous	2019/ [135]

(Continued)

TABLE 4.1 (CONTINUED)

Typical MOF-Derived TMOs/Carbon Hybrid Materials for ECs (VR, Voltage Range; CR, Capacitance Retention; CD, Current Density; SR, Scan Rate; ED/PD, Energy Density/Power Density)

Electrode Materials [MOF Precursors]	VR	Cycles, CR, CD or SR	ED@PD or Capacitance (CD) [W h Kg⁻¹/W Kg⁻¹] or [F g⁻¹] (A g⁻¹)]	Electrolyte	Year/Ref.
Nanostructured Fe_3O_4/Fe/C hybrid xerogel [MOX-Fe]	−0.25–0.4 V (vs. SCE)	5 k, ~100% (initial sharp increase due to activation)	600/500 (1/8)	6 M KOH aqueous	2016/ [180]
Nanoporous carbon (NPC) xerogel [MOX-Al]	−0.7–0.1 V (vs. SCE)	1 k, ~100%	272/258/207 (2/5/250 mV s⁻¹)	1 M KOH aqueous	
ASC (Fe_3O_4/Fe/C//NPC)	0–1.6 V	10 k, >80%	17.5@0.39	6 M KOH aqueous	
Fe_3O_4/Fe/C composite [MOX-Fe]	−1.1–0.1 V (vs. Ag/AgCl)	1 k, 94%, 3 A g⁻¹	246/127/105 (1/2/3)	6 M KOH aqueous	2018/ [179]
Co/CoO/MnO@C composites [CoMn-MOF-74]	−0.85–0.15 V (vs. SCE)	1 k, 85%, 20 mV s⁻¹	894/800/456/272 (0.5/1/2/4)	6 M KOH aqueous	2016/ [143]
Co/MnO@GC nanocomposites [Co/Mn-MOF]	0–0.46 V (vs. SCE)	—	2275/1620/1261/1108 (4/10/20/30)	1 M KOH aqueous	2018/ [26]
ASC (Co/MnO@GC//AC)	0–1.7 V	5 k, 84.2%, 10 A g⁻¹	320/246/144/105 (1/2/10/20)		
Fe_3O_4-doped porous carbon nanorods supported 3D kenaf stem-derived macroporous carbon (3D-KSPC/Fe_3O_4-DCN) [MIL-88A]	−0.8–0.2 V (vs. SCE)	—	373/285/212 (0.25/1/2)	2 M KOH aqueous	2016/ [146]

(Continued)

TABLE 4.1 (CONTINUED)

Typical MOF-Derived TMOs/Carbon Hybrid Materials for ECs (VR, Voltage Range; CR, Capacitance Retention; CD, Current Density; SR, Scan Rate; ED/PD, Energy Density/Power Density)

Electrode Materials [MOF Precursors]	VR	Cycles, CR, CD or SR	ED@PD or Capacitance (CD) [W h Kg⁻¹/W Kg⁻¹] or [F g⁻¹ (A g⁻¹)]	Electrolyte	Year/Ref.
Cl/O-doped C/CoO nanoparticles [Co-MOF]	0–0.49 V (vs. Hg/HgO)	–	1052/837 (0.5/10)	2 M KOH aqueous	2018/[93]
Cl/O-doped C nanoparticles [Co-MOF]	–1.1–0 V (vs. Hg/HgO)	–	207/164/81 (0.5/1/10)		
ASC	0–1.4 V	10 k, 61%, 1.5 A g⁻¹	25.0@3.5, 17.4@7.0		
Cr₂O₃/porous carbon [MIL–101(Cr)]	0–0.45 V (vs. SCE)		420/177 (2/100)	6 M KOH aqueous	2019/[128]
Fe$_x$O$_y$/porous carbon [MIL–101(Fe)]	–1.0–0 V (vs. SCE)		114 (2)		
ASC	0–1.6 V	3 k, 93%, 3.0 A g⁻¹	9.6@0.8, 2.1@80		
Fe₃O₄/Fe nanoparticles decorated in carbon shell [Fe-MOF]	0–0.8 V (vs. SCE)	1 k, 64%, 5 A g⁻¹	972 (1)	6 M KOH aqueous	2018/[106]
Mesoporous Co₃O₄@C composites [ZSA-1]	0–0.48 V (vs. Hg/HgO)	0.5 k, 88.9%, 2 A g⁻¹	205/183/173/151 (0.2/1/2/5)	6 M KOH aqueous	2016/[130]
Nanoporous carbon/Co₃O₄ composite (NPC–Co₃O₄) [ZIF-67]	0–1.0 V (vs. Ag/AgCl)	10 k, ~94%, 2.5 A g⁻¹	885/625/472/210 (2.5/5/8/20)	2 M KOH aqueous	2018/[129]

(Continued)

TABLE 4.1 (CONTINUED)

Typical MOF-Derived TMOs/Carbon Hybrid Materials for ECs (VR, Voltage Range; CR, Capacitance Retention; CD, Current Density; SR, Scan Rate; ED/PD, Energy Density/Power Density)

Electrode Materials [MOF Precursors]	VR	Cycles, CR, CD or SR	ED@PD or Capacitance (CD) [W h Kg⁻¹/W Kg⁻¹] or [F g⁻¹ (A g⁻¹)]	Electrolyte	Year/Ref.
Co_3O_4@C nanocomposite [Co-MOF]	0–0.3 V (vs. Ag/AgCl)	—	261/171/50 (1/2/10)	6 M KOH aqueous	2017/ [81]
Co@C nanocomposite [Co-MOF]	−0.3–0.2 V (vs. Ag/AgCl)	—	90/81 (1/2)		
ASC (Co@C//Co_3O_4@C)	0–1.5 V	1 k, ~100%, 10 A g⁻¹	8.8@0.375, 2.8@3		
Carbon nanosheets-supported MnO_2 nanosheet arrays (MnO_2/CNS) [Mn-MOF nanosheet-derived MnO/CNS]	0–1.0 V (vs. SCE)	5 k, 96.1%, 5 A g⁻¹	339/210 (0.5/20)	1 M Na_2SO_4 aqueous	2018/ [99]
ASC (MnO_2/CNS//UT-CNS)	0–1.8 V	—	30.1@0.46, 8.1@9.0	PVA/Na_2SO_4	
Co_3O_4/ZnO/rGO [Co/Zn-ZIF]	−0.1–0.45 V (vs. Ag/AgCl)	—	204 (1.0)	6 M KOH aqueous	2017/ [115]
ASC (Co_3O_4/ZnO/rGO/rGO)	0–1.6 V	2 k, 87%, 3 A g⁻¹	12.4@0.84, 7.6@8.5		
GO/CuO hybrid [Cu-MOF]	−0.4–1.4 V (vs. Ag/AgCl)	1 k, 93%, 3 A g⁻¹	580 (1.0)	0.5 M H_2SO_4 aqueous	2018/ [114]
$ZnMn_2O_4$/carbon nanorods [Zn/Mn-MOFs]	0–1.2 V (vs. SCE)	2 k, 98.1%, 10 A g⁻¹	606 (2 mV s⁻¹) 589/278 (1/20)	1 M Na_2SO_4 aqueous	2018/ [84]
Rodlike CeO_2/carbon nanocomposite [Ce-MOF]	0–0.4 V (vs. SCE)	5 k, 97.7%, 10 A g⁻¹	1102/621/418 (2/10/20)	2 M KOH + 0.1 M $K_4Fe(CN)_6$	2018/ [112]

(Continued)

TABLE 4.1 (CONTINUED)

Typical MOF-Derived TMOs/Carbon Hybrid Materials for ECs (VR, Voltage Range; CR, Capacitance Retention; CD, Current Density; SR, Scan Rate; ED/PD, Energy Density/Power Density)

Electrode Materials [MOF Precursors]	VR	Cycles, CR, CD or SR	ED@PD or Capacitance (CD) [W h Kg⁻¹/W Kg⁻¹] or [F g⁻¹ (A g⁻¹)]	Electrolyte	Year/Ref.
ZnO QDs-decorated CNF [ZIF-8]	$-1.0–0$ V (vs. Ag/AgCl)	5 k, 85%, 1 A g⁻¹	346/273 (0.5/8)	1 M Na₂SO₄ aqueous	2017/[145]
Dilute NiO/CNF composite [Ni/Zn-MOF]	$-1.0–0$ V (vs. Hg/HgO)	5 k, 90%, 5 A g⁻¹	33.4@0.125 29.0@1 241/226/214/209 (0.25/0.5/1/2)	6 M KOH aqueous	2017/[132]
NiO/Ni encapsulated in N-doped carbon nanotubes (NiO/Ni/NCNTs) [Ni-MOF]	0–0.4 V (vs. SCE)	10 k, 63%, 10 A g⁻¹	778/525 (1/10)	3 M KOH aqueous	2019/[133]
ASC (NiO/Ni/NCNTs//porous carbon)	0–1.6 V	40 k, 90%, 1 A g⁻¹	33.9@0.8, 21.1@8		
MnOₓ/N-doped carbon/MnO₂ composites [Mn-MOF]	0–0.8 V (vs. SCE)	5 k, 89%, 100 mV s⁻¹	281/207/192 (0.125/1.25/2.5)	0.5 M Na₂SO₄ aqueous	2019/[134]
ASC (MnOₓ/NC/MnO₂//AC)	0–2.0 V	2 k, 86%, 2 A g⁻¹	16.8@0.5		
CuOₓ@mC@PANI@rGO hybrid	0–0.75 V (vs. SCE)	2.5 k, ~70%, 2 A g⁻¹	569/535/494/408 (0.5/1/2/5)	1 M H₂SO₄ aqueous	2018/[117]
Carbonized NiCo₂O₄ (cNiCo₂O₄) anode [Ni/Co-MOF]	1.0–4.2 V	9 k, ~90%, 4 A g⁻¹	137@0.2 26.4@40	1 M LiPF₆ (EC/DEC, 1:1 v/v)	2019/[199]
Vertically aligned carbon nanoflakes (VACNFs) cathode					

(Continued)

TABLE 4.1 (CONTINUED)

Typical MOF-Derived TMOs/Carbon Hybrid Materials for ECs (VR, Voltage Range; CR, Capacitance Retention; CD, Current Density; SR, Scan Rate; ED/PD, Energy Density/Power Density)

Electrode Materials [MOF Precursors]	VR	Cycles, CR, CD or SR	ED@PD or Capacitance (CD) [W h Kg^{-1}/W Kg^{-1}] or [F g^{-1} (A g^{-1})]	Electrolyte	Year/ Ref.
ZnO QDs/carbon/CNTs [ZIF-8]	−0.3–0.7 V (vs. Ag/AgCl)	5 k, 91.8%, 20 A g^{-1}	185/179/160/152 (0.5/1/10/20)	1 M Na$_2$SO$_4$ aqueous	2016/ [126]
Porous N-doped carbon/CNTs [ZIF-8]	−1.0–0 V (vs. Ag/AgCl)	–		1 M Na$_2$SO$_4$ aqueous	
All-solid-state ASC	0–1.7 V	3 k, 99%, 5 A g^{-1}	250/166/102 (1/10/20) 23.6@0.85, 10.5@16.9	PVA-NaNO$_3$ gel polymer	
ASC (NQD-NC//AC)	0.5–3.0 V,	4.5 k, ~85%, 5 A g^{-1}	51.4@0.35, 16.3@8.75,	1 M LiPF$_6$ (EC/ DMC, 1:1 v/v)	2016/ [100]
NQD-NC anode (Nb$_2$O$_5$ QDs/ MOF-derived NC) [ZIF-8]	0.5–4.0 V	3 k, ~82%, 5 A g^{-1}	76.9@0.45, 22.4@11.3		
rGO/Fe$_2$O$_3$ composite aerogel [Fe-MOF]	−1.2 to −0.4 V (three-electrode)	5 k, 105%, 20 A g^{-1}	869/431/363/290 (1/5/1020)	6 M KOH aqueous	2017/ [118]
SSC	0–1.0 V	5 k, 95.3%, 50.4 mA cm^{-3}	35/2 μW h cm^{-3} @ 3/72 mW cm^{-3} (250 mF cm^{-3} at 6.4 mA cm^{-3})	PVA-KOH gel polymer	
Hollow cage Ni$_x$Co$_{3−x}$O$_4$/graphene (GNiCo) [Ni-doped ZIF-67]	0–0.45 V (vs. Ag/AgCl)	5 k, 88.1%	2871/2500/2100 (1/5/10)	2 M KOH aqueous	2017/ [116]
ASC (GNiCo//graphene hydrogel)	0–1.5 V	5 k, 81.1%	50.2@0.75, 23.3@3.75		

Note: Abbr.: *2D*: two-dimensional, *AC*: activated carbon, *ASC*: asymmetric supercapacitor, *CNF*: carbon nanofiber, *CNTs*: carbon nanotubes, *CSs*: carbon sheets, *DEC*: diethyl carbonate, *DMC*: dimethyl carbonate, *EC*: ethylene carbonate, *FEC*: fluoroethylene carbonate, *GC*: graphite carbon, *GO*: graphene oxide, *HZ-NPFC*: hybrid ZIFs derived nanoporous functional composites, *mC*: mesoporous carbon, *MOF-NPC*: MOF-derived nanoporous carbon, *MOX*: MOF xerogel, *N-C* or *NC*: N-doped carbon, *NPs*: nanoparticles, *PANI*: polyaniline, *PC*: propylene carbonate, *QDs*: quantum dots, *SCE*: saturated calomel electrode, *SSC*: symmetric super-capacitor, *UT-CNS*: ultrathin carbon nanosheets.

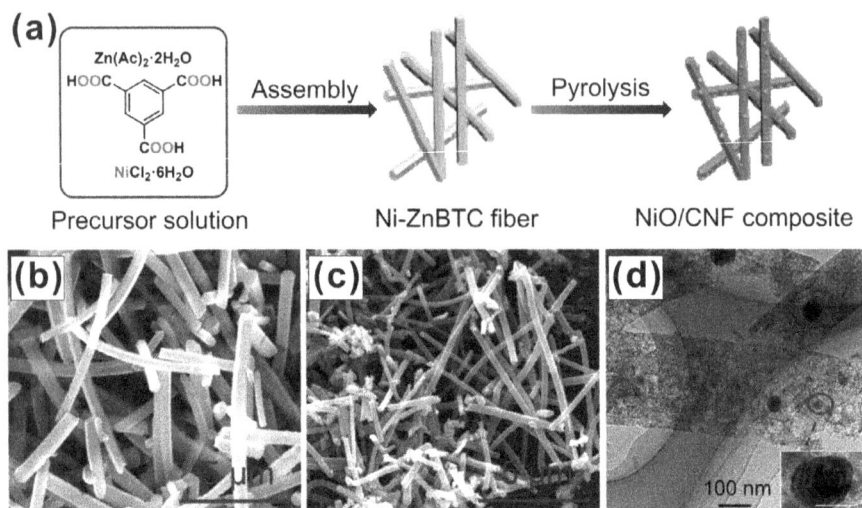

FIGURE 4.4 (a) Schematic description of the synthesis of NiO/CNF composite. (b) SEM image of Ni-ZnBTC MOF fibers, and (c and d) SEM and TEM images of the derived NiO/CNF (inset: high-magnification image for a single particle, scale = 50 nm). Reprinted with permission from Ref. [132]. Copyright 2016 Elsevier B.V.

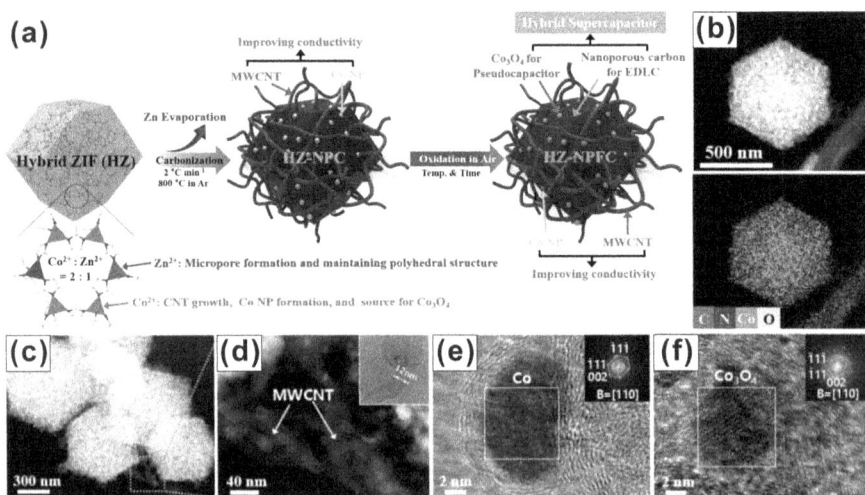

FIGURE 4.5 (a) Representation of the nanoarchitecture design for the functional MOF-derived nanoporous carbon (HZ-NPFC) from the Co^{2+}/Zn^{2+} hybrid ZIF obtained by the use of reductive carbonization and oxidation optimization. (b–d) HAADF-STEM images and EDS mapping of HZ-NPFC nanoparticle (inset of d: HRTEM image of MWCNTs on HZ-NPFC). (e and f) HRTEM images of Co and Co_3O_4 nanoparticles on HZ-NPFC, respectively (insets show the corresponding fast Fourier transform (FFT) patterns). Reprinted with permission from Ref. [108]. Copyright 2017 The Royal Society of Chemistry.

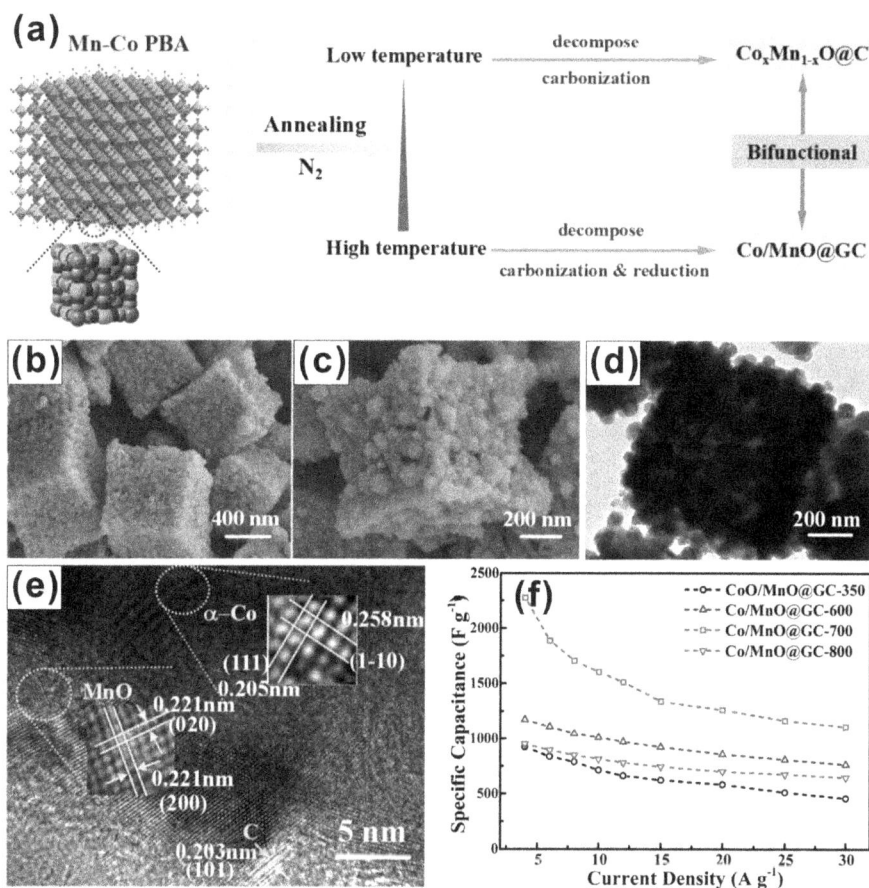

FIGURE 4.6 (a) Schematic illustration of the formation process of $Co_xMn_{1-x}O@GC$ and Co/$MnO@GC$ nanohybrids. (b–e) SEM, TEM and HRTEM images of Co/MnO@GC-700. (f) Specific capacitances as a function of current density for different electrodes. Reprinted with permission from Ref. [26]. Copyright 2018 The Royal Society of Chemistry.

in situ composite nanoarchitecture design (i.e., MOF-derived metal and metal oxide-embedded carbon nanocomposites) offers a new opportunity for future applications in high-performance energy storage systems.

The electrospinning and carbonization method is another easy route to increase the conductivity of MOF-derived materials [145]. By introducing the structurally advantageous quantum-sized metal oxides, Jang and coworkers fabricated hierarchical fibrous structures composed of uniformly dispersed ZnO quantum dots (~10 nm), amorphous carbon, and carbon nanofibers (ZnO QDs@C/CNF) via a single carbonization process of electrospun ZIF-8/PVA nanofibers (Figure 4.7). The as-designed electrode material for supercapacitors exhibits outstanding electrochemical performances with high capacitance (346 F g^{-1} at 0.5 A g^{-1}), reliable rate capability (79%

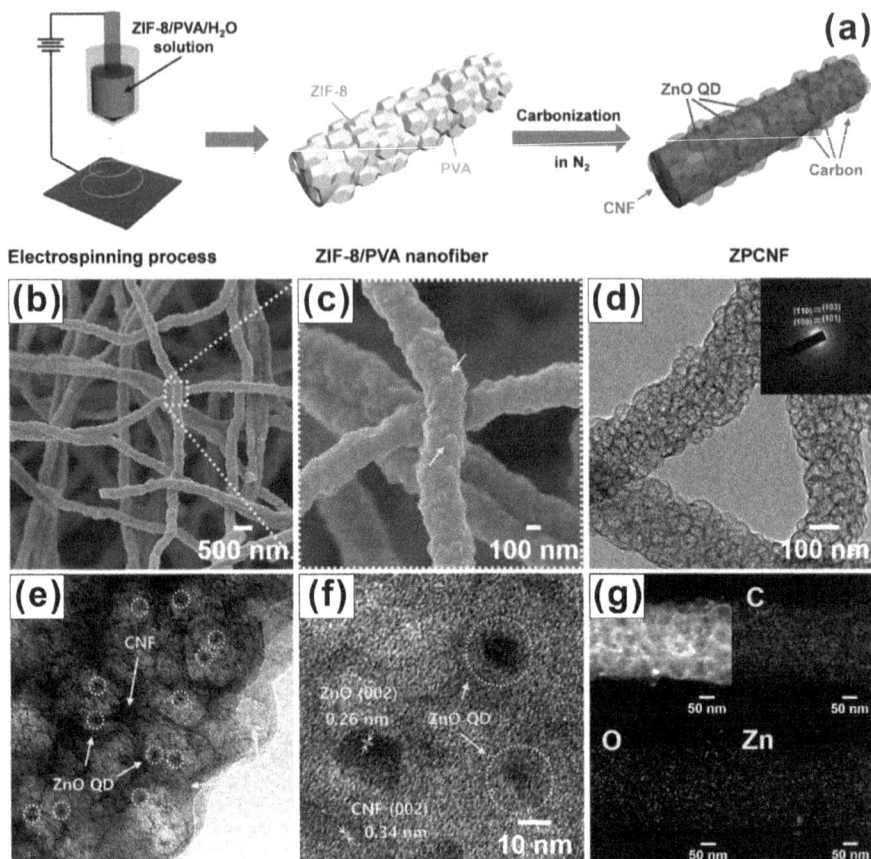

FIGURE 4.7 (a) Schematic illustration of the preparation process of ZnO QDs-decorated CNF (ZPCNF). (b and c) SEM images, (d and e) TEM images (inset: SAED pattern), (f) HRTEM image, and (g) STEM image with the corresponding elemental mapping of ZPCNF. Reprinted with permission from Ref. [145]. Copyright 2017 The Royal Society of Chemistry.

capacitance retention at 8 A g^{-1}), and long cycle life (85% of its maximum capacitance after 5000 charge/discharge cycles) [145].

A 3D porous carbon/metal oxide structure using MOFs as precursor would be a highly effective approach for high-performance supercapacitors, owing to its hierarchically interconnected features [98]. Zhao et al. designed an ordered porous Mn$_3$O$_4$@N-doped carbon/graphene (MCG) composite derived from Mn-MOFs with the assistance of monodispersed polymer colloid template followed by a facile carbonization process. For the MCG composite, ultrasmall Mn$_3$O$_4$ nanoparticles (average diameter 7 nm) are uniformly anchored and the carbon is formed in situ. The binary carbon components synergistically contribute to a continuous and intimate conductive framework for the pseudocapacitive components besides the advantages of enhanced active surface, mass transport, and structural stability, resulting a superior combination of EDLC capacitance and pseudocapacitance. Thus, the MCG, as a supercapacitor electrode, displays excellent electrochemical performances in a neutral aqueous 1 M Na_2SO_4 electrolyte with a maximum specific capacitance of 456 F g^{-1} at 1 A g^{-1} and a high rate capability of 246 F g^{-1} at 20 A g^{-1}, as well as a good cycling stability with 98.1% capacitance retention over 2000 cycles at 5 A g^{-1} (Figure 4.8) [98]. Alternatively, the MOF-derived composites may also be supported on other 3D macroporous carbon frameworks with large specific surface area and hierarchical pores and high conductivity for high-performance supercapacitors, e.g., Fe_3O_4-doped porous carbon nanorods (Fe_3O_4-DCN, derived from MIL-88A) supported by 3D kenaf stem-derived macroporous carbon (KSPC) (Figure 4.9) [146].

In addition, the MOF-derived metal-carbon composites could be utilized as either negative electrodes (anodes) [81,138,147] or positive electrodes (cathodes) [122,138,141,142]. For example, Tong et al. designed a kind of 2D Co/Ni MOFs nanosheets (molar ratio of Co/Ni of 1:1) synthesized at room temperature followed

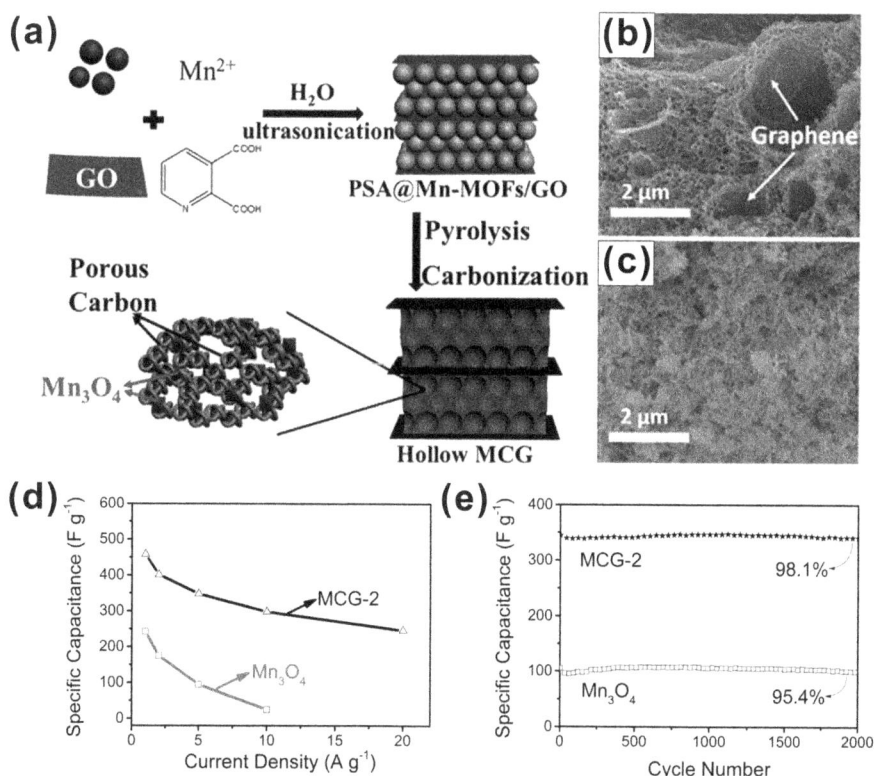

FIGURE 4.8 (a) Schematic illustration of the fabrication of ordered porous Mn_3O_4@N-doped carbon/graphene (MCG) hybrids. (b and c) SEM images of MCG and the derived Mn_3O_4. Electrochemical performance of MCG and Mn_3O_4: (d) calculated specific capacitance at different current densities; (e) cycling performance at a current density of 5 A g^{-1}. Reprinted with permission from Ref. [98]. Copyright 2016 Springer.

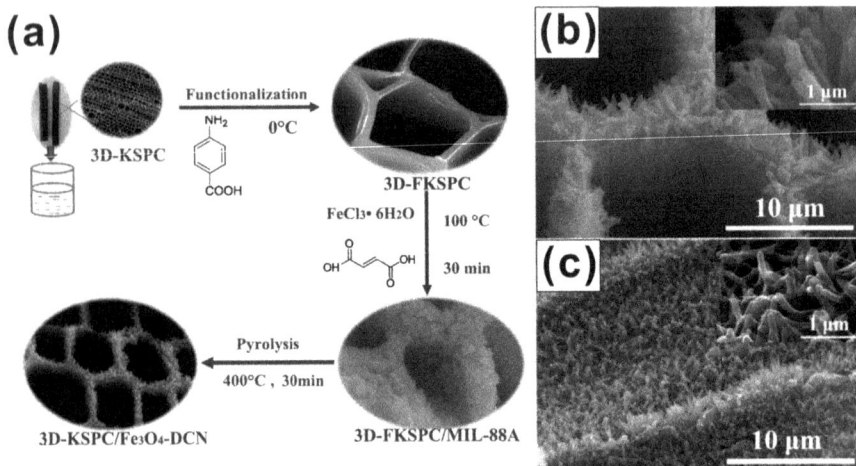

FIGURE 4.9 (a) Schematic illustration of the formation process for the hybrid of Fe_3O_4-doped porous carbon nanorods supported by 3D kenaf stem-derived porous carbon (3D-KSPC/Fe_3O_4-DCN) nanocomposites. SEM images of the obtained 3D-KSPC/Fe_3O_4-DCN: (b) top view and (c) side view. Reprinted with permission from Ref. [146]. Copyright 2016 American Chemical Society.

by further pyrolysis at 550°C in N_2 atmosphere to obtain 2D S,N-doped carbon nanosheets incorporated with CoNi alloy nanoparticles (CoNi@SNC) with high surface area, porous structure, and good conductivity [122]. Compared to the 2D Co/Ni MOFs nanosheets which can be directly used as electrode materials for supercapacitors (delivering a specific capacitance of 312 F g^{-1} at 1 A g^{-1}), the derived CoNi@SNC exhibits a vastly superior specific capacitance (1970, 1897, and 1730 F g^{-1} at 1, 2, and 5 A g^{-1}, respectively) with long cycling life (95.1% capacitance retention at 10 A g^{-1} over 3000 cycles), and excellent rate capability. Further, the constructed asymmetric supercapacitor (ASC) device, with CoNi@SNC as positive electrode and active carbon as negative electrode, exhibits a high energy density of 55.7 W h kg^{-1} at a power density of 0.8 Kw kg^{-1} and good cycle stability (capacitance retention of 90.6% over 4000 cycles). With regard to the charge-discharge mechanism, it reveals that the electrochemical activation-generated CoNi oxides/oxyhydroxides on the surface of CoNi alloy nanoparticles in alkaline electrolyte during electrochemical processes are the electrochemical active species of the CoNi@SNC-constructed supercapacitors. Overall, the high performance of the CoNi@SNC-constructed SCs can be attributed to the synergistic merits, i.e., high surface area for sufficient exposure of electrochemical active sites, porous structure for promoted redox-related mass transport, and the combination of CoNi alloy nanoparticles with graphitic carbon for improved electron transfer. To further derive the anticipated electrode materials, two-fold interpenetrating MOFs with a microporous structure and multicomponents (e.g. Ni, P, N, and O) could be uniformly incorporated into the final framework (i.e., the carbon materials) and render an excellent synergistic effect to improve the electrochemical energy storage performance. The maximum specific capacitance for the hierarchical Ni/P/N/C composite electrode could reach 2888 F g^{-1} at 1 A g^{-1}, superior to that of other hierarchical composites and established a new benchmark in the related field [142]. These findings may benefit the design and development of carbon electrode materials incorporating transition-metal nanoparticles for high-performance supercapacitors.

4.4 MOF-DERIVED METAL SULFIDES/CARBON COMPOSITES

Transition-metal chalcogenides (TMCs), especially sulfides (TMSs), have drawn considerable attention owing to their high theoretical capacitance, natural abundance, and low cost; also, compared with their oxide counterparts (TMOs), the improved electrical conductivity of sulfides can facilitate fast electron transfer or provide strong electronic coupling between different components [22,110,148–149]. However, these sulfides generally suffer from poor structural stability and low electronic conductivity, which compromise their practical capacity, high rate, and cycle performance. Effective ways to overcome these problems include designing nanostructured materials, constructing hierarchical/hollow structures or incorporation of conductive components (mainly carbons) to shorten the diffusion distance of ions, provide adequate contact area between electrode and electrolyte, restrict the aggregation and dissolution of active materials, alleviate the volume change, increase the active reaction sites, and facilitate the fast transport of electrons [58,61,67,149]. These carbon-encapsulated or supported TMSs include porous Cu_2S/C [150], Cu_7S_4/C nanocomposites [151], NiS_2/C [152], NiS nanorods/rGO [153], hierarchical Ni/Ni_3S_2 decorated carbon nanofibers (Ni/Ni_3S_2/CNFs) [154], Co_9S_8/C [155], 2D Co_xS_y/C hybrids [156,157], CoS_2@C/CNTs (Figure 4.10a and b) [107], hollow Co_9S_8@rGO [158, 159], hollow C/NiCo-LDH/Co_9S_8 with complexity [57], 3D hierarchical honeycomb-like Co_9S_8@C [160], bimetallic MOFs derived N-doped carbon coated $Cu_{0.5}Co_{2.5}S_4$ hollow spheres [161], rod-like NiS_2/CoS_2/NC (NC, N-doped carbon) composite [162], porous Ni–Co sulfides (Ni–Co–S) [163], hollow Ni–Co–S/NC composites [149], Ni-Co-S@graphene [164], cross-linking nanoporous $Zn_{0.76}Co_{0.24}$S@C-ZIF-$Zn_{0.76}Co_{0.24}$S core–shell nanosheet arrays [165]; and for typical examples, see Table 4.2. Besides the hydrothermal vulcanization [149,151,159,161,163,166], in situ gas H_2S, or sublimate sulfur-based sulfurization of transition-metal-containing composites is a facile process for the carbon-coated TMSs nanocomposites [107,152–153,162,167–168].

Co_9S_8 has attracted intensive attention as an anode material for ECs due to its unique structural and rich electrochemical features (e.g., high theoretical capacitance) as well as environmental compatibility, low cost, and high abundance [155,158–159,169]. Although suffering from the drawbacks of low conductivity and instability, the incorporation of carbon has guaranteed the high specific capacitance, rate performance, and stability. For example, the MOF (DUT-58) derived Co_9O_8 nanoparticles embedded in N/S codoped porous graphitic carbon (Co_9S_8/NS-C) exhibited a high specific capacitance (734 F g^{-1} at 1 A g^{-1}), excellent rate capability (653 F g^{-1} or 89% retention at 10 A g^{-1}) and superior cycling stability (capacitance retention of 99.8% after 140,000 cycles at a 10 A g^{-1}) [155]. The long-term stability is thrilling and will pave the way for practical application.

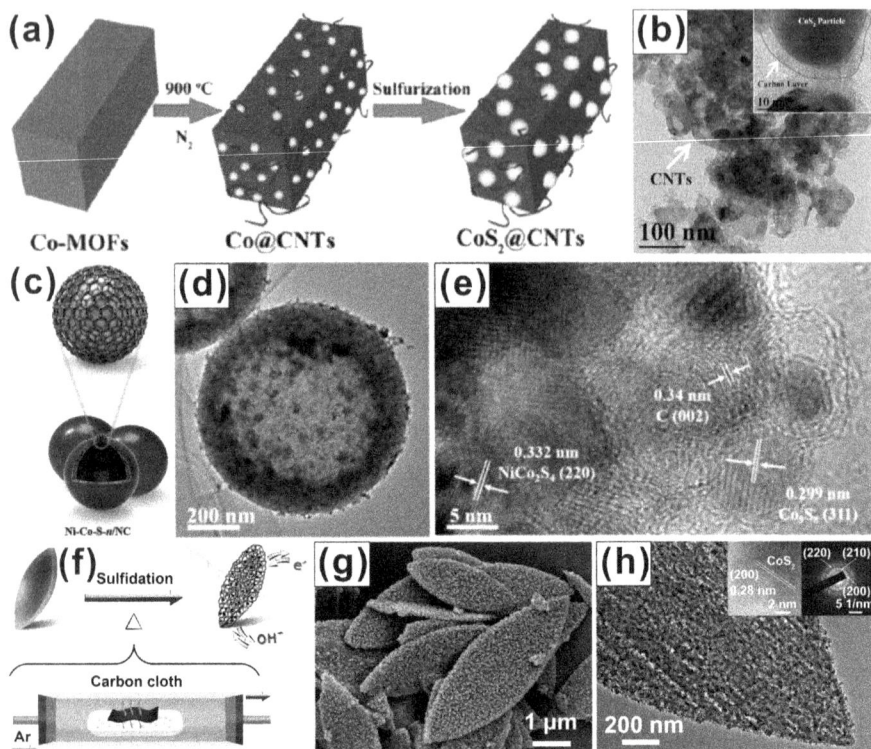

FIGURE 4.10 (a) Schematic illustration for the synthesis of CoS$_2$@CNT, and (b) the corresponding TEM images. Reprinted with permission from Ref. [107]. Copyright 2017 American Chemical Society. (c) Illustration of the hollow bimetallic nickel cobalt sulfide and nitrogen-doped carbon (Ni-Co-S/NC) composites, and (d and e) the corresponding TEM and HRTEM images. Reprinted with permission from Ref. [149]. Copyright 2019 American Chemical Society. (f) Schematic representation of the fabrication of leaf-like CoS$_2$ nanoparticles in N-doped carbon (CoSNC) nanocomposites by depositing ZIF-L-Co on the carbon cloth through simultaneous sulfidation and carbonization processes, and (g and h) SEM and TEM image (insets: HRTEM image and SAED pattern) of leaf-like CoSNC nanocomposites. Reprinted with permission from Ref. [167]. Copyright 2018 The Royal Society of Chemistry.

2D or layered materials, including the newly developed 2D MOFs, have attracted tremendous research interest in recent years owing to their unique physical and chemical properties [156]. Zhang and coworkers, for the first time, facilely developed a kind of 2D MOF (porphyrin paddlewheel framework-3, PPF-3) with thickness of ca. 12–43 nm via surfactant-assisted synthetic method, and followed by simultaneous sulfidation and carbonization the ultrathin 2D nanocomposite (thickness 24.5 ± 6.4 nm) of CoS$_{1.097}$ nanoparticles embedded in N-doped carbon matrix (CoSNC) are obtained (Figure 4.11a–h). When used as anode material for ECs, the 2D CoSNC nanocomposite demonstrates a specific capacitance of 360 F g^{-1} at 1.5 A g^{-1} and a high rate capability (74.3% and 56.8% retention at 15 and 30 A g^{-1}, respectively)

TABLE 4.2

Typical MOF-Derived TMSs/Carbon Hybrid Materials for ECs (VR: Voltage Range, CR: Capacitance Retention, CD: Current Density, SR: Scan Rate, ED/PD:Energy Density/Power Density)

Electrode Materials [MOF Precursors]	VR	Cycles, CR, CD or SR	ED@PD or Capacitance (CD) [W h Kg⁻¹/W Kg⁻¹] or [F g⁻¹ (A g⁻¹)]	Electrolyte	Year/Ref.
Co_9S_8@S,N-doped carbon (Co_9S_8@SNC) [Co-MOF]	−0.1–0.4 V (vs. Hg/HgO)	2 k, 98%, 5 A g⁻¹	429/368/336 (1/10/50)	6 M KOH aqueous	2017/ [169]
2D nanocomposite of $CoS_{1.097}$ NPs embedded in N-doped carbon matrix (CoSNC) [PPF-3]	0–0.6 V (vs. Ag/AgCl)	2 k, ~90%, 12 A g⁻¹	360/268/205 (1.5/15/30)	2 M KOH aqueous	2016/ [156]
Sandwich-like Co_xS@C/rGO [ZIF-67]	0–0.5 V (vs. SCE)	4 k, 99.7%, 1 A g⁻¹	240/175 (5/20)	1 M KOH	2017/ [157]
ASC (Co_xS@C/rGO//C/rGO)	0–1.8 V	2 k, ~100%, 2 A g⁻¹ (1 M KOH aqueous)	30@1.74	quasi-solid-state PVA-PAA/KOH	
N-doped carbon coated $Cu_{0.5}Co_{2.5}S_4$ hollow spheres [Cu/Co-MOF]	0–0.5 V (vs. Hg/HgO)	2/10 k, ~100/81.5%, 20 A g⁻¹	1228/1070/864/784 (1/2/10/20)	6 M KOH aqueous	2019/ [161]
Hollow Co_9S_8/rGO [ZIF-67]	0–0.45 V (vs. SCE)	9 k, 92.0%, 4 A g⁻¹	576/448 (2/10)	1 M KOH aqueous	2018/ [158]
Hollow polyhedral C/NiCo-LDH/Co_9S_8 (C/LDH/S) [ZIF-67]	0–0.45 V (vs. SCE)	3 k, 95.4%, 12 A g⁻¹	1653/1288/1025 (4/10/20)	1 M KOH aqueous	2017/ [57]
ASC (C/LDH/S//CNTs)	0–1.2 V	10 k, 90.9%, 8 A g⁻¹	39@2.4, 30.1@7.4		

(Continued)

TABLE 4.2 (CONTINUED)

Typical MOF-Derived TMSs/Carbon Hybrid Materials for ECs (VR: Voltage Range, CR: Capacitance Retention, CD: Current Density, SR: Scan Rate, ED/PD:Energy Density/Power Density)

Electrode Materials [MOF Precursors]	VR	Cycles, CR, CD or SR	ED@PD or Capacitance (CD) [W h Kg⁻¹/W Kg⁻¹] or [F g⁻¹ (A g⁻¹)]	Electrolyte	Year/Ref.
Honeycomb-like Co_9S_8@C [Co-MOF]	0–0.45 V (vs. SCE)	10 k, 98.8%, 5 A g⁻¹	1887/1343 (1/10)	2 M KOH aqueous	2018/[160]
ASC (Co_9S_8@C//AC)	0–1.6 V	10 k, 86%, 5 A g⁻¹	58@1, 38@17.2		
Porous $Cu_{1.96}S$–C octahedral [HKUST-1]	0–0.9 V (vs. Ag/AgCl)	3 k, ~80%, 50 mV s⁻¹	200 (0.5)	1 M H_2SO_4 aqueous	2019/[150]
Co_9O_8 NPs embedded in N/S codoped porous graphitic carbon (Co_9S_8/NS–C) [DUT-58]	0–0.5 V (vs. Hg/HgO)	140 k, 99.8%, 10 A g⁻¹	734/653 (1/10)	6 M KOH aqueous	2017/[155]
ASC (Co_9S_8/NS–C//AC)	0–1.2 V	2 k, 99.5%, 10 A g⁻¹	14.9@0.68, 6.63@6.8		
CoS_2@C/CNTs [Co-MOF]	0–0.4 V (vs. SCE)	5 k, 82.9%, 1 A g⁻¹	825 (0.5)	2 M KOH aqueous	2017/[107]
NiS nanorods/rGO (R-NiS/rGO) [Ni-MOF-74]	0–0.5 V (vs. Ag/AgCl)	20 k, 89%, 20 A g⁻¹	744/600 C g⁻¹ (1/50)	2 M KOH aqueous	2018/[153]
ASC (R-NiS/rGO//C/NG-A)	0–1.6 V	10 k, 93.2%, 20 A g⁻¹	93@0.96, 54@46		
Rod-like NiS_2/CoS_2/NC [Ni/Co-MOF]	0–0.5 V (vs. SCE)	5 k, 75%, 5 A g⁻¹	1325/1247/860 (1/2/10)	6 M KOH aqueous	2019/[162]
ASC (NiS_2/CoS_2/NC//AC)	0–1.6 V	20 k, 85.7%, 5 A g⁻¹	54@0.80, 19.6@16	3 M KOH aqueous	

(*Continued*)

TABLE 4.2 (CONTINUED)

Typical MOF-Derived TMSs/Carbon Hybrid Materials for ECs (VR: Voltage Range, CR: Capacitance Retention, CD: Current Density, SR: Scan Rate, ED/PD:Energy Density/Power Density)

Electrode Materials [MOF Precursors]	VR	Cycles, CR, CD or SR	ED@PD or Capacitance (CD) [W h Kg⁻¹/W Kg⁻¹] or [F g⁻¹ (A g⁻¹)]	Electrolyte	Year/ Ref.
Porous Ni-Co sulfides (Ni-Co-S) [Ni/Co-MOF]	0–0.43 V (vs. Hg/HgO)	3 k, 93.7%, 10 A g⁻¹	1378/1232 (1/10)	3 M KOH aqueous	2018/ [163]
ASC (Ni-Co-S//AC)	0–1.7 V	3 k, 83.0%, 6 A g⁻¹	36.9@1.07		
Hollow Ni–Co–S-0.5/NC submicrospheres [Ni/Co-MOF]	0–0.55 V (vs. Hg/HgO)	2 k, ~100%, 6 A g⁻¹	544/424/366 C g⁻¹ (1/10/20)	3 M KOH aqueous	2019/ [149]
ASC (Ni–Co–S-0.5/NC//AC)	0–1.6 V	3 k, 31.5%, 4 A g⁻¹	39.6@0.81		
Porous Ni–Co–S@graphene [nickelocene/Co-MOF]	0–0.3 V (vs. Ag/AgCl)	1 k, 87.4%, 17 A g⁻¹	1463/750 (1/10)	6 M KOH aqueous	2019/ [164]
ASC (Ni–Co–S@G//AC)	0–1.3 V	1 k, 93.3%, 17 A g⁻¹	51@0.65, 21@7.1	PVA/KOH gel electrolyte	
FeS₂/C nanoparticles [Fe-MOF] ASC (FeS₂/C//AC)	0–3.2 V	2.5 k, ~100%	63@0.152, 9@3.24	1 M LiTFSI/diglyme	2018/ [168]

Note: Abbr: *CNT:* carbon nanotubes, *LDH:* layered double hydroxide, *NG-A:* N-doped graphene aerogel, *rGO:* reduced graphene oxide, *ZIF:* zeolitic imidazole framework.

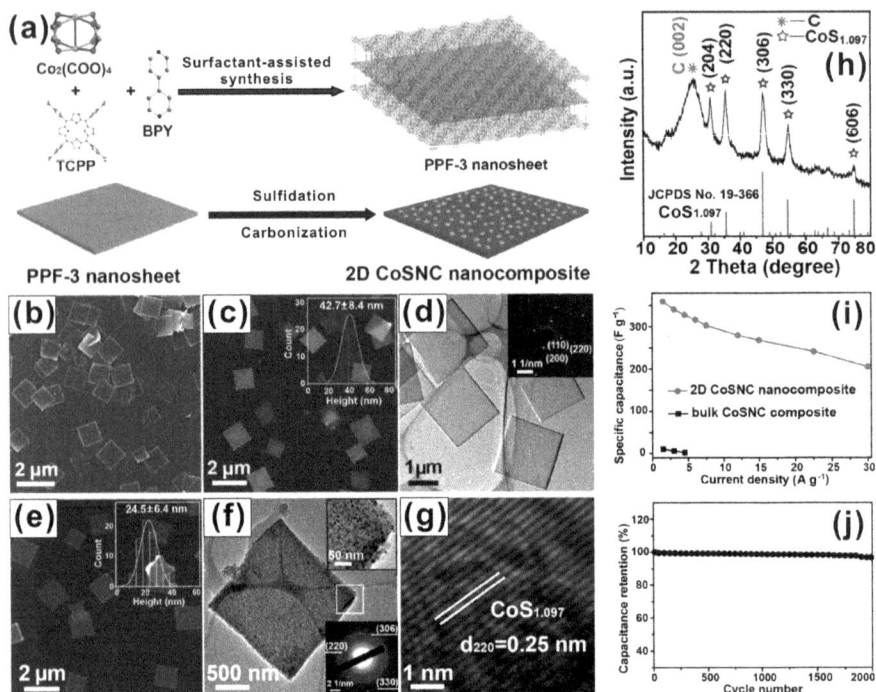

FIGURE 4.11 (a) Schematic illustration of the synthesis process of 2D porphyrin paddle-wheel framework-3 (PPF-3) nanosheets and the following synthesis process of 2D CoSNC nanocomposites. (b–d) SEM, AFM, and TEM images of PPF-3 nanosheets (inset of c and d: the corresponding statistical analysis of the thickness and SAED pattern). (e–h) AFM, TEM, HRTEM images, and XRD pattern of 2D CoSNC nanocomposites (inset of e and f: the corresponding statistical analysis of the thickness, zoomed-in TEM image and SAED pattern). (i) Specific capacitance of 2D CoSNC nanocomposite and bulk CoSNC composite as a function of current density. (j) Cycling stability of 2D CoSNC nanocomposite electrode measured at 12 A g^{-1} in 2 M KOH. Reprinted with permission from Ref. [156]. Copyright 2016 American Chemical Society.

(Figure 4.11i and j) [156]. Moreover, a dual-structural carbon (e.g., simultaneous N-doped porous carbon shells and graphene substrates) may further enhance the cycling stability [157].

Hollow structures with rich surface chemical active sites and robust structural stability are ideal candidate for high-performance energy storage devices [25,57,170–173]. As for hybrid MOF derivatives incorporated with carbon frameworks [57,149,161], Ho and coworkers developed an in situ self-templated pseudomorphic transformation strategy by which MOFs are evolved into hybrid materials with high complexity, i.e., hollow rhombic dodecahedral C/NiCo-LDH/Co$_9$S$_8$ (C/LDH/S) via the stepwise treatment including sulfurization of Ni/Co-layered double hydroxides (NiCo-LDHs) (Figure 4.12a–d). The incorporation of metal sulfide species at the LDH intergalleries also provides enhanced interfacing and stability of

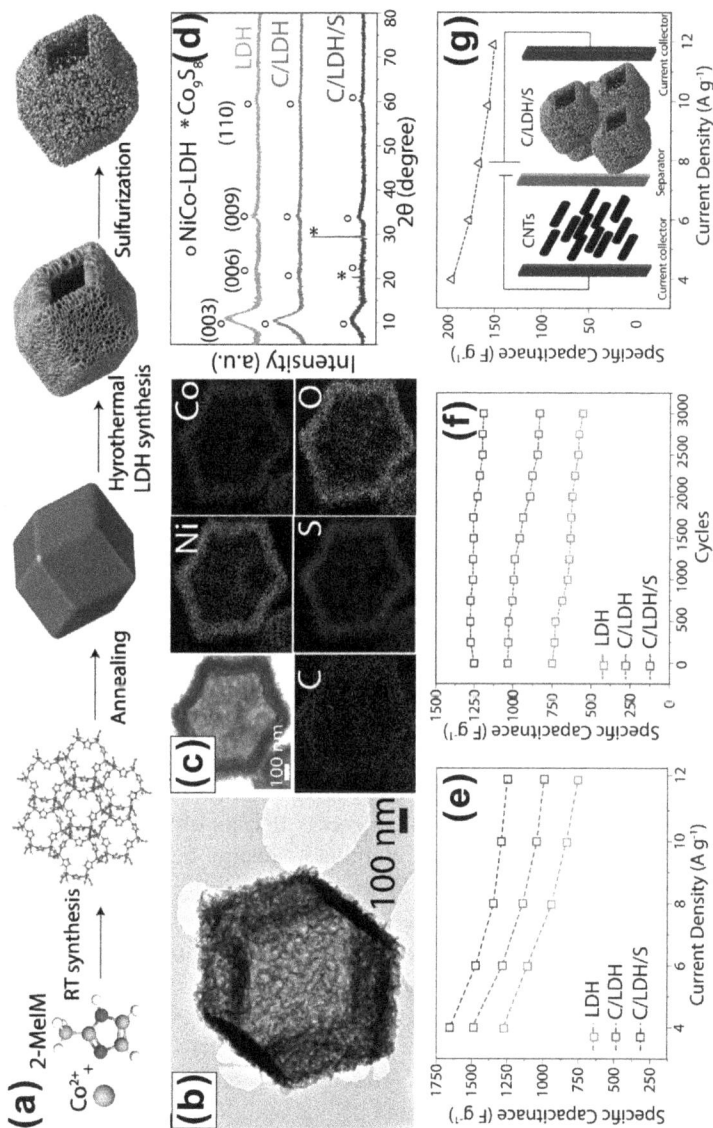

FIGURE 4.12 (a) Schematic illustration for the synthesis process of hollow NiCo-LDH/Co_9S_8 (C/LDH/S) hybrid. (b and c) TEM image and elemental mapping of C/LDH/S. (d) The representative XRD patterns for the as-prepared LDH, C/LDH, and C/LDH/S; and the corresponding (e) specific capacitances as a function of current density and (f) cycling stability tests over 3000 cycles. (g) Calculated specific capacitance values for C/LDH/S//CNTs asymmetric supercapacitor cell (inset: schematic illustration of the cell). Reprinted with permission from Ref. [57]. Copyright 2017 WILEY-VCH.

the hybrids amongst various collective merits, thus leading to a significant improvement in their supercapacitive energy storage performances (i.e., superior specific capacitances of 1653, 1288, and 1025 F g^{-1} at current densities of 4, 10, and 20 A g^{-1}, respectively) as well as excellent cycling stability (95.4% retention over 3000 cycles for C/LDH/S). And for the ASC device C/LDH/S//CNTs, it exhibits a high rate capability and superior stability (90.9% capacitance retention over 10,000 cycles at 8 A g^{-1}) and delivers a high energy density (39 W h kg^{-1} at a power density of 2.4 kW kg^{-1} and 30.1 W h kg^{-1} at a power density of 7.4 kW kg^{-1}) (Figure 4.12e–g) [57]. Yi et al. designed a kind of MOF-derived hybrid hollow submicrospheres of N-doped carbon-encapsulated bimetallic Ni-Co-S nanoparticles (Figure 4.10c–e). Through easily controlling of the Ni/Co molar ratio, it is found that the hollow Ni-Co-S-0.5/ NC (i.e., n = 0.5) composite exhibited optimal electrochemical performance, e.g., a high specific capacity of 543.9 C g^{-1} at 1 A g^{-1} and a capacity retention of 67.3% with increased current density up to 20 A g^{-1}, as well as a competent candidate for ASC with a high energy density of 39.6 and 27.1 W h kg^{-1} at a power density of 808 and 7910 W kg^{-1}, respectively [149].

Hierarchically porous electrodes comprised of electrochemically active materials and conductive additives (typically 2D graphene) usually display synergistic effects originating from the structure and the interaction between the constituent phases [153,164]. For example, Qu et al. designed and fabricated a hierarchically porous hybrid electrode made of NiS nanorods decorated on rGO (R-NiS/rGO) derived from Ni-MOF-74/rGO templates/precursors (Figure 4.13a and b). Owing to the abundant (101) and (110) surfaces on the edges with the strong affinity for OH^- in KOH electrolytes revealed by microanalyses and density functional theory (DFT) calculations, the hybrid electrode with enhanced electronic conductivity and highly exposed active surfaces favorable for fast redox reactions in a basic electrolyte (Figure 4.13c and d) showed a superior capacitance of capacity of 744 C g^{-1} at 1 A g^{-1} and 600 C g^{-1} at 50 A g^{-1}, respectively, and retained over 89% of the initial capacity after 20,000 cycles. When coupled with a N-doped graphene aerogel (C/NG-A) negative electrode, the hybrid supercapacitor (R-NiS/rGO//C/NG-A) can achieve an ultrahigh energy density and power densities (an energy density of 93 W h kg^{-1} at a power density of 962 W kg^{-1}, while still maintaining an energy density of 54 W h kg^{-1} at an elevated working power of 46,034 W kg^{-1}) [153]. Among the hierarchical structures, a honeycombed porous structure is a unique and effective strategy beyond complex hollow structures that lead to extraordinary electrochemical energy storage performance [160]. For example, Sun et al. designed a 3D honeycombed porous hybrid structure of cobalt sulfide dispersed in a carbon matrix via facile pyrolysis of Co-MOFs. Benefiting from the loose adsorption, plentiful surface area and high conductivity of the honeycomb-like Co_9S_8@C structure for enhanced Faradaic processes and redox reactions, the composite could deliver a high capacitance (1887 F g^{-1} at 1 A g^{-1}), rate capability (1343 F g^{-1} up to 10 A g^{-1}), and robust cycle stability (98.8% retention over 10,000 cycles at 5 A g^{-1}), thus promising for the application of high-performance hybrid supercapacitors (energy density of 58 and 38 W h kg^{-1} at 1.0 and 17.2 kW kg^{-1}, respectively, and capacitance retention of ~86% over 10,000 cycles at 5 A g^{-1}) (Figure 4.14) [160]. It is noteworthy that this kind of 3D hierarchical

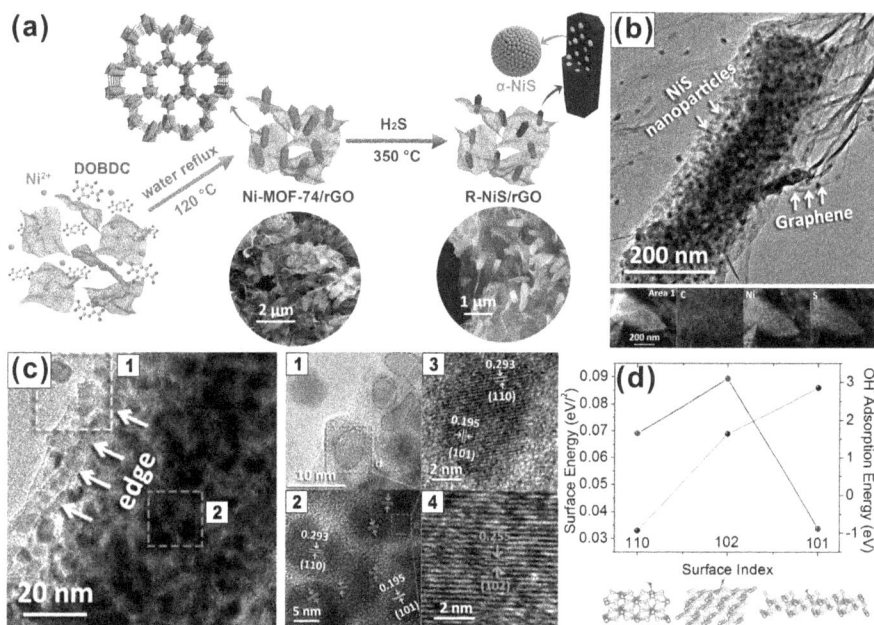

FIGURE 4.13 (a) Schematic illustration of the synthesis procedure of the R-NiS/rGO from MOF precursors, and (b) TEM image and EDS elemental mappings (C, Ni, and S) of the R-NiS/rGO. (c) HRTEM and the zoomed-in images of the R-NiS/rGO hybrid nanostructure. (d) Surface energy and OH$^-$ adsorption energy of different slabs of α-NiS from DFT calculations. Reprinted with permission from Ref. [153]. Copyright 2018 The Royal Society of Chemistry.

hybrid structure usually demonstrates superior specific capacitance or long-term cycling stability than those of solid structures [155] or low-dimensional structures [156]. These remarkable electrochemical performances synergistically contributed by the 3D continuous thin carbon frameworks and the coated TMSs are amazing and show promising application for fast energy storage systems.

4.5 BINDER-FREE, FREESTANDING, AND FLEXIBLE DEVICES

Although significant advances in carbon/metal-oxide composite electrodes have been made, binder-free or self-standing electrodes based on these materials with seamless interconnecting merits are quite attractive [4,6,174–176]; and for typical examples, see Table 4.3. Conductive substrate supported TMOs with hierarchically porous structures are considered as promising electrodes for efficient electrochemical energy storage and conversion [177–178]. For example, an integrated hybrid electrode of porous Co_3O_4/C nanowire arrays (NAs) on nickel foam was designed by thermally annealing a supported Co-MOF (Figure 4.15a–g) [177]. The as-synthesized Co_3O_4/C NAs electrode demonstrates a high areal capacitance of 1.32 F cm^{-2} (corresponding to a specific capacitance of 776.5 F g^{-1}) at a current density of

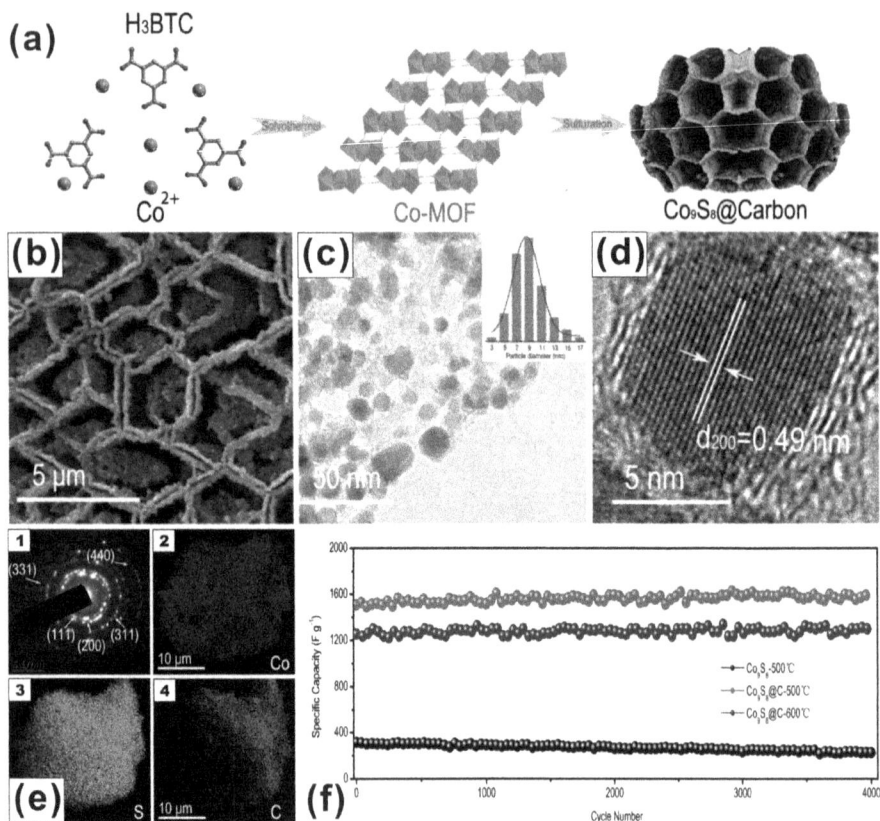

FIGURE 4.14 (a) Schematic illustration of the synthesis strategy for honeycomb Co_9S_8@C composites. (b–d) FESEM, TEM (inset: particle-size distribution), HRTEM images, and (e) SAED patterns and the element mapping images of Co_9S_8@C-500. (f) Electrochemical cycling performance of Co_9S_8@C-500, compared to Co_9S_8-500 and Co_9S_8@C-600 at a current density of 5 A g^{-1}. Reprinted with permission from Ref. [160]. Copyright 2018 WILEY-VCH.

1 mA cm^{-2}, much superior to (i.e., 4 times larger than) that of bare Co_3O_4 NAs electrode (Figure 4.15h). The configured symmetric supercapacitor based on Co_3O_4/C NAs electrodes exhibits an improved durability (21.7% capacitance decay over 5000 cycles). The excellent electrical conductivity of the interconnected carbon frameworks and ion diffusion within the hierarchical pores intrinsically and synergistically contributed to the promoted pseudocapacitive performance, and the synthesis strategy herein opens an avenue to the design of high-performance electrodes for energy storage and conversion. Similarly, a novel in situ method for porous 2D-layered carbon–metal oxide (2D-CMO) composite electrode via anodic electrodepostion of Ni/ Co-MOFs on nickel foam followed by pyrolysis and activation was exploited [174]. Besides the layered structures, an ultrahigh mass loading of 13.4 mg cm^{-2}, high capacitance but superior rate performance (2098 mF cm^{-2} at a current density of

TABLE 4.3

Typical MOF-Derived Binder-Free or Flexible TMOs/TMSs—Carbon Hybrid Electrodes for ECs (VR: Voltage Range, CR: Capacitance Retention, CD: Current Density, SR: Scan Rate, ED/PD: Energy Density/Power Density)

Electrode Materials [MOF Precursors]	VR	Cycles, CR, CD or SR	ED@PD or Capacitance (CD) [W h Kg⁻¹/W Kg⁻¹] or [F g⁻¹ (A g⁻¹)]	Electrolyte	Year/Ref.
SSC (freestanding flexible Mn@ZnO/CNF)	$0–1.0$ V	10 k, 92%, 1 A g^{-1} 10 k, 84%, 5 A g^{-1}	72.1@0.5, 33.3@5, 501/230 (1/10)	6 M KOH aqueous	2019/[103]
Porous Co_3O_4/C nanowire arrays (NAs) supported by Ni foam [Co-MOF]	$-0.1–0.4$ V (vs. Ag/AgCl)	2 k, 96%, 100 mV s^{-1}	1.32 F cm^{-2}/777 F g^{-1} (1 mA cm^{-2}), 1.08 F cm^{-2} (20 mA cm^{-2})	3 M KOH aqueous	2016/[177]
SSC	$0–1.0$ V	5 k, 78.3%	0.14 mW h cm^{-3}@8.54 W h kg^{-1}	PVA–KOH gel electrolyte	
Porous 2D-layered carbon–metal oxide composite (2D-CMO) (on Ni foam) [Ni/Co-MOFs]	$0–0.55$ V (vs. Hg/HgO)	(4.2 k, 92%, 20 mV s^{-1})	157/145 F g^{-1} or 2098/1945 mF cm^{-2} (1/20 mA cm^{-2})	4 M KOH aqueous	2017/[174]
ASC (2D-CMO//AC)	$0–1.3$ V	10 k, ~92%; 20 mV s^{-1}	36.4@0.09, 16.3@4.5		
$Zn_{0.76}Co_{0.24}S$ @C-ZIF-$Zn_{0.76}Co_{0.24}S$ core-shell nanosheet arrays (on Ni foam) [ZIF-ZnCo]	$0–0.55$ V (vs. SCE)	5 k, 92%, 10 A g^{-1}	1202 C g^{-1} (1)	3 M KOH aqueous	2019/[165]
ASC (with AC)	$0–1.6$ V	2 k, 97%, 10 A g^{-1}	79.2@0.625, 41.5@6.25		
ZnO@C@$NiCo_2O_4$ core-shell structures (on CC)	$0–0.56$ V (vs. SCE)	4 k, 76%, 10 mA cm^{-2}	3.18 F cm^{-2} (6 mA cm^{-2}) 2650 (5)	2 M KOH aqueous	2016/[127]

(Continued)

TABLE 4.3 (CONTINUED)

Typical MOF-Derived Binder-Free or Flexible TMOs/TMSs—Carbon Hybrid Electrodes for ECs (VR: Voltage Range, CR: Capacitance Retention, CD: Current Density, SR: Scan Rate, ED/PD: Energy Density/Power Density)

Electrode Materials [MOF Precursors]	VR	Cycles, CR, CD or SR	ED@PD or Capacitance (CD) [W h Kg⁻¹/W Kg⁻¹ or [F g⁻¹ (A g⁻¹)]	Electrolyte	Year/Ref.
Leaf-like CoSNC nanocomposites (on CC) (CoS$_2$ NPs in NC) [ZIF-L-Co]	0–0.5 V (vs. Ag/AgCl)	10 k, 91.0%, 10 A g^{-1}	383/277/226 (1/10/20)	2 M KOH aqueous	2018/[167]
N-doped hollow carbon spheres/Co$_9$S$_8$ embedded in graphene (GH@NC@Co$_9$S$_8$) [ZIF-67]	0–0.4 V (vs. Ag/AgCl)	–	842/552 F cm^{-3} (1/10) 540/354 (1/10)	6 M KOH aqueous	2019/[159]
ASC (GH@NC@Co$_9$S$_8$//GH@NC)	0–1.6 V	8 k, 95.8%	28.7 W h L^{-1}, @972 W L^{-1}		
S-α-Fe$_2$O$_3$@C/OCTNF [MIL-88-Fe]	–1.0–0 V (vs. Ag/AgCl)	4 k, 97.6%, 3.75 A g^{-1}/ 3 mA cm^{-2}	1538/1117/991 (2.5/10/20), 1232/776 mF cm^{-2} (2/20 mA cm^{-2})	1 M Na$_2$SO$_4$ aqueous	2018/[147]
FASCs (S-α-Fe$_2$O$_3$/OCNTF//Na-MnO$_2$ NSs/ CNTF)	0–2.2 V	10 k, ~87%	135/92 µW h cm^{-2}@2.2/22 mW cm^{-2}	Na$_2$SO$_4$–CMC gel electrolyte	

Note: Abbr. *CC*: carbon cloth, *CMC*: carboxymethyl cellulose sodium, *FASC*: fiber-shaped asymmetric supercapacitor, *(O)CNTF*: (oxidized) carbon nanotube fiber, *SSC*: symmetric supercapacitor.

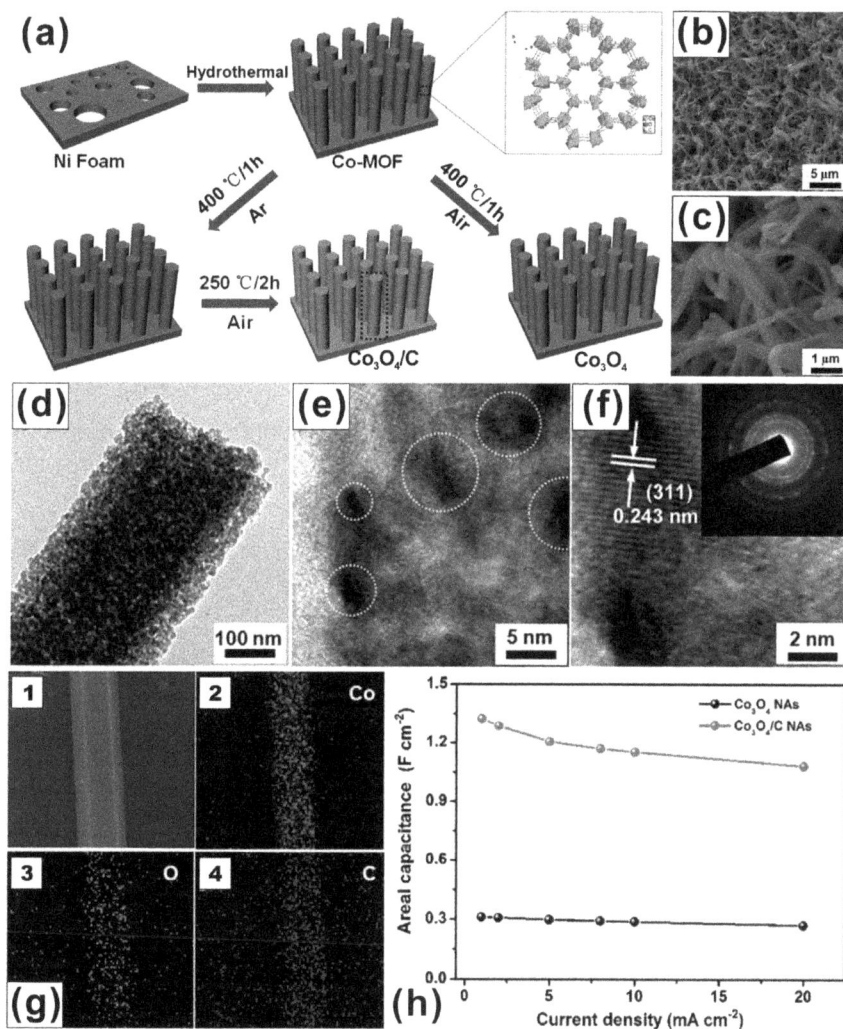

FIGURE 4.15 (a) Schematic illustration of the synthesis of hierarchically porous hybrid Co_3O_4/C nanowire arrays (NA) electrodes. (b and c) SEM images of Co_3O_4/C NAs at different magnifications. (d–f) TEM and HRTEM images (inset: SAED pattern) of the Co_3O_4/C NAs. (g) SEM image of a single Co_3O_4/C nanowire and the corresponding elemental mapping images (Co, O, and C). (h) Plots of the current density against areal capacitances of hybrid Co_3O_4/C NA and the pristine Co_3O_4 electrodes. Reprinted with permission from Ref. [177]. Copyright 2016 The Royal Society of Chemistry.

1 mA cm^{-2}, and 93% retention from 1 to 20 mA cm^{-2}) of the as-prepared electrode is amazing, making it a promising candidate for practical supercapacitor application. As a proof of concept, this anodic electrodesposition of 2D layered MOF may be a versatile method for the preparation of binder-free carbon/metal oxide composite electrodes for electrochemical application. Niu et al. fabricated a novel macroporous

nanostructure utilizing N-doped hollow carbon spheres and ZIF-67 derived Co_9S_8 embedded in graphene (GH@NC@Co_9S_8) with enhanced surface area and favorable transport environment for electrons and ions (Figure 4.16a). Furthermore, the Co_9S_8 polyhedra grown in situ derived from ZIF-67 with amorphous phase and hollow 3D structure can improve the reactivity and specific capacity of the composite. The as-prepared binder-free electrode for supercapacitors showed a high volumetric capacitance of 842.4 F cm^{-3} at a current density of 1 A g^{-1}. When assembled for an ASC fabricated with GH@NC@Co_9S_8 (positive electrode) and GH@NC (negative electrode), the ASC showed a superior volumetric energy density of 28.7 W h L^{-1} at a high power density of 972 W L^{-1} as well as a stable long cycle life (95.8% capacity retention over 8000 cycles), revealing great practical value in electrochemical energy storage (Figure 4.16b and c) [159]. And solid-state or gel-polymer electrolytes are usually adopted for these devices [126,177].

Despite the fact that MOFs are well investigated, the utilization of extended metal-organic gels (MOGs) derived from metal and organic linkers for an assembled

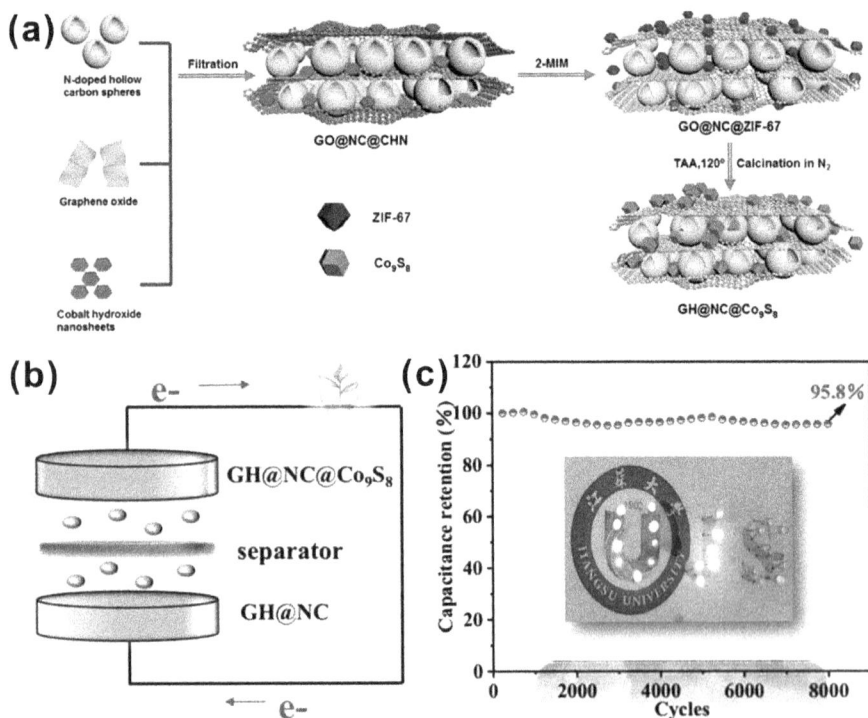

FIGURE 4.16 (a) Schematic illustration of the fabrication process of the macroporous thin film of N-doped hollow carbon spheres and ZIF-67 derived Co_9S_8 embedded in graphene (GH@NC@Co_9S_8), and (b) the configured ASC device composed of GH@NC@Co_9S_8 as positive electrode and GH@NC as negative electrode. (c) Cyclic performance of the GH@NC@Co_9S_8//GH@NC ASC (inset: photograph of parallel LEDs lit by two ASC devices in series). Reprinted with permission from Ref. [159]. Copyright 2019 The Royal Society of Chemistry.

structure such as metal-organic xerogels (MOXs) is emerging [179,180]. For example, Mahmood et al. designed a novel strategy to prepare MOX-derived nanoporous carbon (NPC) xerogels with or without decorated Fe_3O_4/Fe nanoparticles by the selection of central metals and the optimization of calcination temperature as highly active electrode materials for ASCs. The nanostructured Fe_3O_4/Fe/C hybrid demonstrated a high specific capacitance of 600 F g^{-1} at a current density of 1 A g^{-1} and excellent rate capabilities (capacitance retention up to 500 F g^{-1} at 8 A g^{-1}) (Figure 4.17a–c). When configured with the hierarchically NPC of high surface area and excellent electrochemical performance, the aqueous ASC (aASC) displayed a high energy density of 17.5 W h kg^{-1} at a power density of 389 W kg^{-1} (Figure 4.17d–f) [180]. Although the performances are assessed in a coin cell, the high energy density,

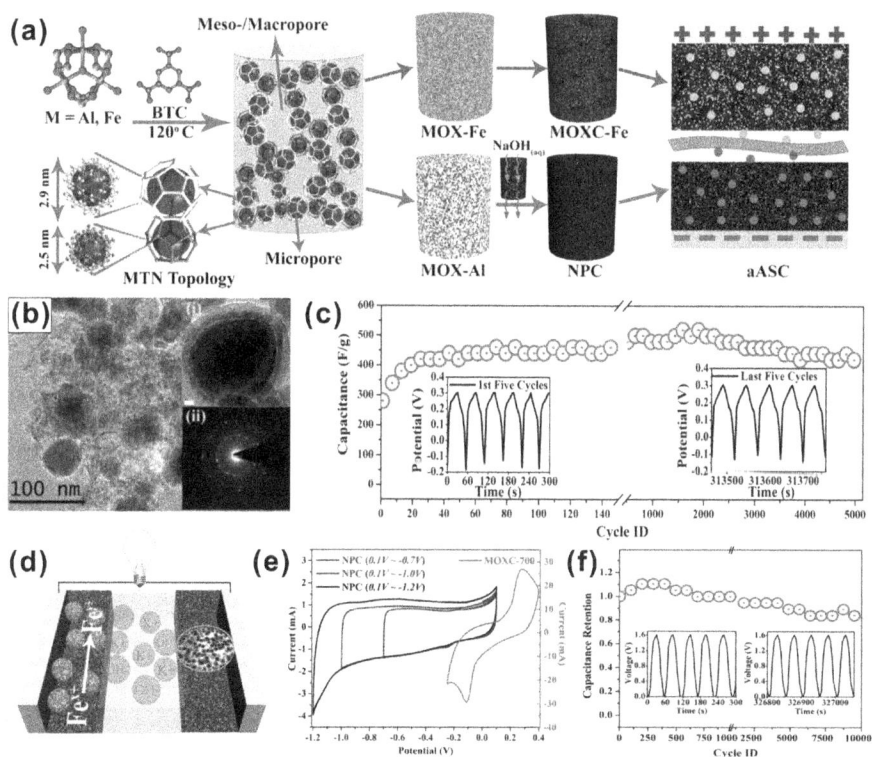

FIGURE 4.17   (a) Schematic representation for the derivation of MOF xerogels (MOXC-Fe) as positive and nanoporous carbon (NPC) as negative electrode materials for configuration of an aqueous asymmetric SC (aASC). (b) TEM image of MOX-Fe-700 (inset: HRTEM analysis, scale = 5 nm, and SAED diffraction patterns). (c) Cyclic performance of the MOXC-700 for 5,000 cycles at current density of 8 A g^{-1}. View and electrochemical performance of MOXC-700//NPC asymmetric cell: (d) aASC assembly, (e) comparative CV curves of MOXC-700 and NPC, (f) capacitance retention for assembled MOXC-700//NPC aASC over 10,000 cycles (inset: GCD curves showing symmetric behavior after initial activation). Reprinted with permission from Ref. [180]. Copyright 2016 American Chemical Society.

excellent capacity retention, low cost, and an environment benign of the developed materials in an aqueous system show great promise for the practical massive utilization of these energy storage devices.

Carbon cloth (CC), also called carbon fiber (CF), is an ideal substrate (serving both as the backbone and electron "superhighway" for charge storage and delivery) for diverse flexible energy-storage devices including ECs; among which, MOF-derived materials are playing a vital role in rational composite electrode design and performance promotion [92,127,181–187], e.g., hollow $NiCo_2O_4$ nanowall arrays [182], carbon nanowall arrays (CNWAs) [110], core-shell $ZnO@C@NiCo_2O_4$ nanorod sheet arrays (NRSAs) [127], leaf-like CoS_2/N-doped carbon (CoSNC) nanocomposites (Figure 4.10f–h) [167], and conductive polymer coated PPy/Cu_9S_8@C-CC nanocomposite electrode [166]. The synergistic effects between CC and nanostructures in the nanocomposites can not only effectively shorten the ion/electron transmission path and provide more exposed faradaic redox sites, but also enhance the electrochemical kinetics and electrode structural stability [167], thus for an enhanced electrochemical performance including high specific capacitance, good rate capacity, and cycling stability. For example, Duan and coworkers designed and synthesized a kind of unique core–shell nanostructure, i.e. MOF-derived $ZnO@C@NiCo_2O_4$ nanorod sheet arrays (NRSAs) on CC substrate via the hydrothermal and electrodeposition method, for high-performance supercapacitors. Benefiting from the synergy of carbon shell derived from MOFs (ZIF-8), the core-shell structure and the highly conductive matrix, the flexible hybrid electrode exhibited a significantly enhanced electrochemical performance (e.g., a high areal capacitance of 3.18 F cm^{-2} at 6 mA cm^{-2} and 2650 F g^{-1} at 5 A g^{-1}, good rate capability and cycling stability), compared to that of $ZnO@NiCo_2O_4$ NRSAs without a carbon coating (Figure 4.18) [127]. These facile and cost-effective syntheses of robust hybrid electrodes hold great promise for high-performance supercapacitor applications in the future.

Flexible and lightweight supercapacitors have recently attracted more and more attention for flexible, portable or wearable electronic devices [103,147,175,188–190]. Yoon and coworkers fabricated such electrodes comprising carbon nanofibers (CNFs) decorated with ZIF-8-derived Mn-doped zinc oxide (Mn@ZnO) nanoparticles (Figure 4.19a–d). The synergy between Mn@ZnO (active sites for Faradaic reactions beyond EDLC capacitance) and the highly electrically conductive carbon nanofiber (providing fast electron-transfer pathways) improves the performance of the supercapacitor electrode (Figure 4.19e). Thus, the Mn@ZnO/CNF electrodes demonstrate a high specific capacitance of 501 F g^{-1} and retain >92% of their initial capacitance even after 10,000 cycles. The optimized Mn@ZnO/CNF electrodes deliver impressive energy densities (72.1 W h kg^{-1} and 33.3 W h kg^{-1} at power densities of 500 W kg^{-1} and 5000 W kg^{-1}, respectively) (Figure 4.19f and g) [103]. And this electrospinning strategy for robust nanostructured composite electrode material sheds light on designing long-lifetime high-rate energy storage devices. For a more cost-effective choice, iron oxide (Fe_2O_3) may be adopted. Yao and coworkers designed a facile method to directly grow MIL-88-Fe MOF-derived spindle-like α-Fe_2O_3@C on highly hydrophilic oxidized carbon nanotube fiber (S-α-Fe_2O_3@C/OCNTF), which demonstrated a high areal capacitance of 1232 mF cm^{-2} at a current density

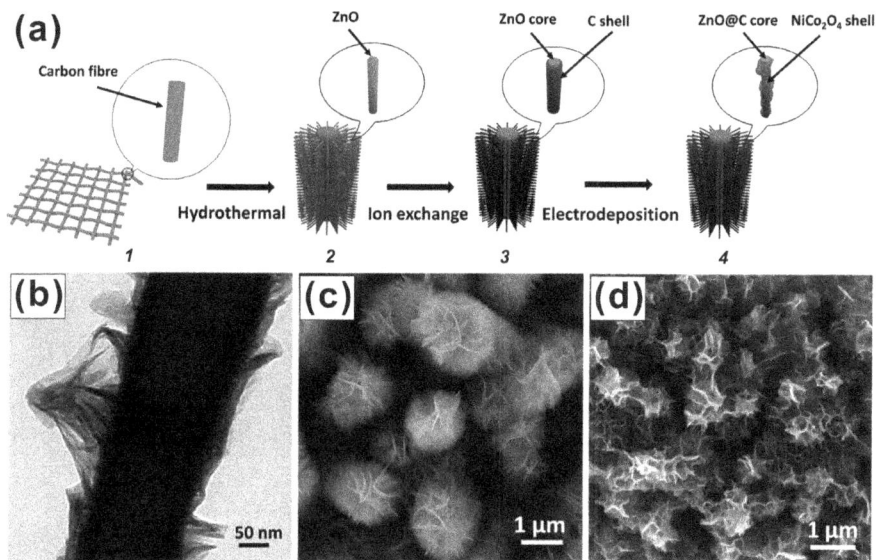

FIGURE 4.18 (a) Schematic illustration of the fabrication of $ZnO@C@NiCo_2O_4$ nanorod sheet arrays (NRSAs): (*1*) carbon cloth substrate, (*2*) ZnO NRs, (*3*) ZnO@C NRs, (*4*) $ZnO@C@NiCo_2O_4$ NRSAs. (b) TEM image of a $ZnO@C@NiCo_2O_4$ core–shell nanorod sheet. Morphological analysis of (c) $ZnO@C@NiCo_2O_4$ NRSAs after 4000 cycles and (d) $ZnO@NiCo_2O_4$ NRSAs after 3000 cycles. Reprinted with permission from Ref. [127]. Copyright 2016 The Royal Society of Chemistry.

of 2 mA cm^{-2} and considerable rate capability with 63% retention up to 20 mA cm^{-2} (Figure 4.20a and b). When matched with the Na-doped MnO_2 nanosheets on CNTF (Na-MnO_2 NSs/CNTF) as the cathode, the flexible solid-state twisted fiber-shaped asymmetric supercapacitors (FASCs) with maximum operating voltage of 2.2 V exhibited a high specific capacitance of 201 mF cm^{-2} and a superior energy density of 135 µW h cm^{-2} (Figure 4.20c–h). With the merits of high energy and power densities, excellent flexibility and neutral gel-type electrolytes, MIL-88-Fe MOF-derived S-α-Fe_2O_3@C will be a promising anode for next-generation wearable asymmetric supercapacitors [147]. However, if a slower scan rate or lower charge/discharge rate is applied (or incorporated with a battery-type cathode such as an MOF-derived NiZnCoP nanosheet array on CNTF), the battery-type features will be dominant, towards fiber-shaped aqueous rechargeable batteries (FARBs) [191]. Furthermore, these FASCs may be integrated with self-charging power systems such as flexible triboelectric nanogenerators to harvest and store energy simultaneously for flexible multifunctional electronic devices [192].

4.6 HYBRID CAPACITORS

Hybrid electrochemical capacitors (HECs), also known as HCs, hybrid supercapacitors (HSCs) or ASCs, can potentially combine the merits of high-energy density

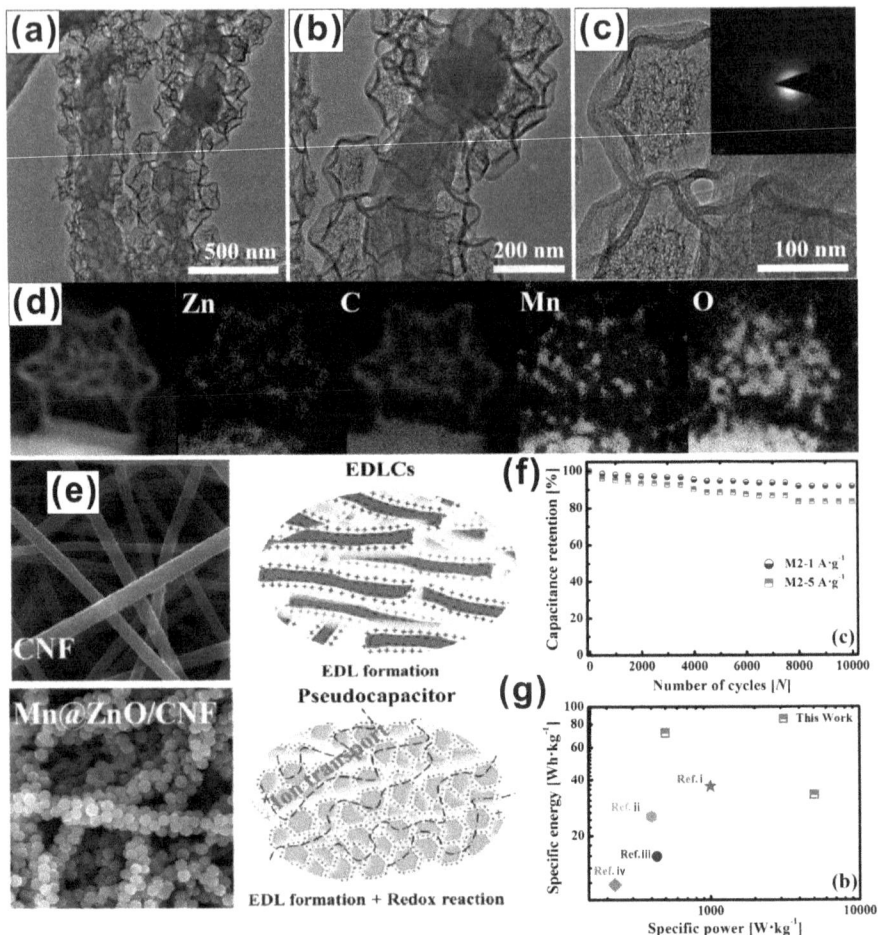

FIGURE 4.19 (a–c) TEM images (inset: the corresponding SAED pattern) and (d) EDS-based elemental maps (Zn, C, Mn, and O) in the carbonized Mn@ZnO/CNF composite (M2). (e) Morphology and mechanism depicting EDL and pseudocapacitance. (f) Cycling performance of Mn@ZnO/CNF (M2) at various current densities. (g) Comparison of energy and power densities with previous reports (Ref. i–iv, see the original sources). Reprinted with permission from Ref. [103]. Copyright 2019 Elsevier B.V.

of batteries and high-power output plus long cycle life of capacitors in one device, i.e., bridging the gap between conventional rechargeable-ion batteries and ECs [27,59,100,111]. The key point of configuring a high-performance HEC is to couple appropriate battery-type and capacitor-type electrode materials with well-matched capacity and kinetics behavior simultaneously, and an alkali metal salt-containing organic electrolyte will extend the voltage window and thus for a significantly enhanced energy density [111]. Since some of the HECs have been mentioned above, some other typical proofs-of-concept are depicted as follows.

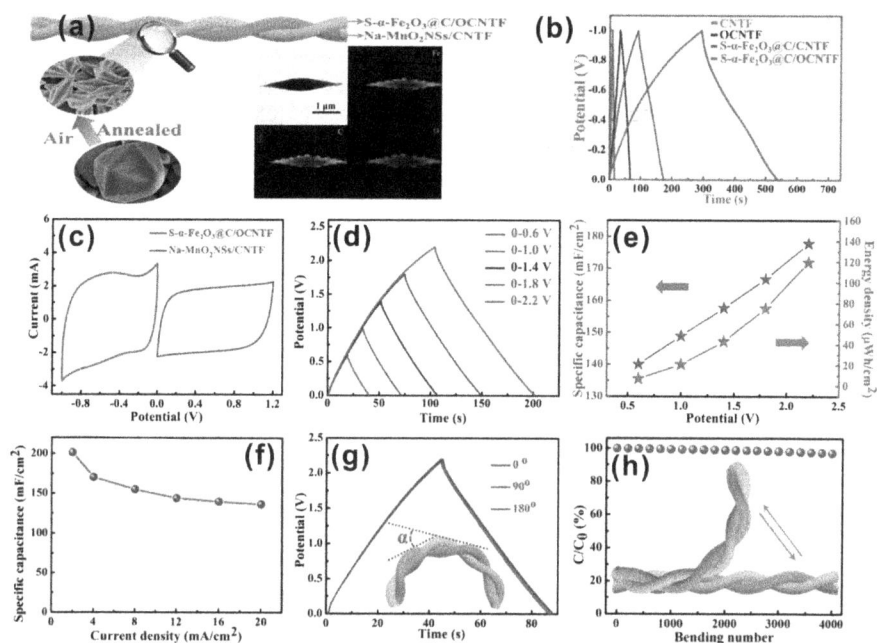

FIGURE 4.20 (a) Schematic illustration of a fiber-shaped asymmetric supercapacitor (FASC) device (including a TEM image and EDX mappings of the different elements of Fe, C, and O from an individual S-α-Fe$_2$O$_3$@C). (b) Comparison of GCD curves of pristine CNTF, OCNTF, S-α-Fe$_2$O$_3$@C/CTNF, and S-α-Fe$_2$O$_3$@C/OCTNF electrodes obtained at a current density of 5 A g^{-1} (4 mA cm^{-2}). (c) Comparative CV curves of S-α-Fe$_2$O$_3$@C/OCNTF and Na-MnO$_2$ NSs/CNTF at a scan rate of 25 mV s^{-1}. (d) GCD curves of the as-fabricated FASC device collected over different voltages from 0.6 to 2.2 V at a current density of 4 mA cm^{-2}. (e) Areal specific capacitance and energy density calculated based on GCD curves obtained at 4 mA cm^{-2}. (f) Areal-specific capacitances calculated from GCD curves as a function of the current density. (g) CV curves of the as-prepared FASC device measured at a current density of 4 mA cm^{-2} under different bending angles. (h) Normalized capacitances of the FASC with a bending angle of 90° for 4000 cycles. Reprinted with permission from Ref. [147]. Copyright 2018 American Chemical Society.

Aqueous asymmetric supercapacitors (aASCs) have attracted increasing attention recently owing to several advantages such as low cost, nonflammability, high ionic conductivity, acceptable voltage window (ca. 0–1.6 V), and facile assembly under ambient conditions; amongst them, layered double hydroxide (LDH)/carbon composites, beyond those of conventional hydroxides, have drawn increasing attention for use as high-performance electrodes owing to their high theoretical capacitance, large specific surface area and sufficient interspacing [180,193–197]. For example, Bai et al. designed a sandwiched Ni–Co LDH hierarchical hollow nanocages/graphene derived from ZIF-67 as sacrifice template [194]. The optimized composite electrode exhibited a maximum specific capacitance of 1265 F g^{-1}, high rate capability (50% capacitance retention at ten times higher current density), and good cycling life (92.9% capacitance retention over 2000 cycles). The combination of battery-type Ni-Co LDH hollow nanocages/graphene composite and EDLC-type active carbon allows for the excellent electrochemical performance of an aASC device, delivering a maximum specific capacitance of 171 F g^{-1} in a potential window of 0–1.7 V, as well as a high energy density (68.0 W h kg^{-1}) and excellent power output (4759 W kg^{-1}). And for a MOF-derived 2D assembled Ni-Mn-C ternary composite with hierarchical micro-/nanostructures comprised of $Ni(OH)_2$ and MnO_2 nanosheets decorating on a carbon matrix ($Ni(OH)_2$-MnO_2/C), owing to its high surface area, good electrical conductivity, and the synergistic effect among components, it demonstrated enhanced electrochemical performances, especially high-rate capabilities (574 F g^{-1} at 40 A g^{-1}) and cycling stability (capacitance retention of ~87% at 2 A g^{-1} over 10,000 cycles). Also, the $Ni(OH)_2$-MnO_2/C composite-based all-solid-state supercapacitor demonstrated high power density, good cycling stability, and excellent flexibility [198]. Furthermore, the incorporation of 2D graphene could significantly enhance the structural stability of the electrodes and thus deliver a better rate capability and a superior long-term cycle performance (e.g., merely 19% capacity loss from 1 to 20 A g^{-1}, 95% capacity retention over 10,000 cycles at 20 A g^{-1}) [193].

Compared to aASCs, the non-aqueous hybrid electrochemical capacitor (NHEC) with organic (e.g. LiTFSI/ether based) electrolytes with high operating voltage is highly desirable for its superior energy density as well as high power density [168]. Amongst NHECs, Li-ion capacitors (LICs) have recently emerged as a new promising type of electrochemical energy storage devices that fill the gap between Li-ion batteries (LIBs) and supercapacitors (SCs) by providing balanced high energy and power densities [199–201]. Benefiting from the highly porous capacitive carbon cathode, LIB-like intercalation/conversion anode and Li salt-containing organic electrolytes, the cell voltage of LIC could be operated around 4 V, which is significantly higher than that of commercial SCs (ca. 2.7 V) [199]. Zhu and coworkers developed an integrated LIC with MOF-derived bimetallic oxide (carbonized $NiCo_2O_4$) conversion anode and the binder-free vertically aligned carbon nanoflakes (VACNFs) cathode (Figure 4.21a–d). The carbonized $NiCo_2O_4$ nanocomposites with high surface areas and high pseudocapacitance from conversion reaction demonstrate large Li-ion storage capability, good charge/discharge rate and remarkable cyclability as an anode material, while the VACNFs cathode displays several favorable characteristics such as excellent intrinsic conductivity, high specific surface area and more electrochemically active sites for enhanced ion accessibility in the organic electrolyte. Owing to the advantages of the well-designed nanostructural electrode materials, the assembled LICs exhibited a superior energy density of up to 136.9 W h kg^{-1} (at 200 W kg^{-1}) and an ultrahigh power density of 40 kW kg^{-1} (i.e., charge/discharge within 4 s at 26.4 W h kg^{-1}), as well as a good cycle stability (~90% capacitance retention over 9000 cycles at a current density of 4 A g^{-1}) within the voltage range of 1.0–4.2 V (Figure 4.21e) [199]. Kim and coworkers exploited a cost-effective Fe-MOF precursor derived pyrite (FeS_2/C) nanoparticles for Faradaic electrode of NHEC using ether-based electrolyte (1 M LiTFSI/diglyme) [168]. The as-fabricated NHEC (FeS_2/C//AC) showed an enhanced voltage window of 0–3.2 V and could deliver energy densities in the range of 63–69 W h kg^{-1} at power densities

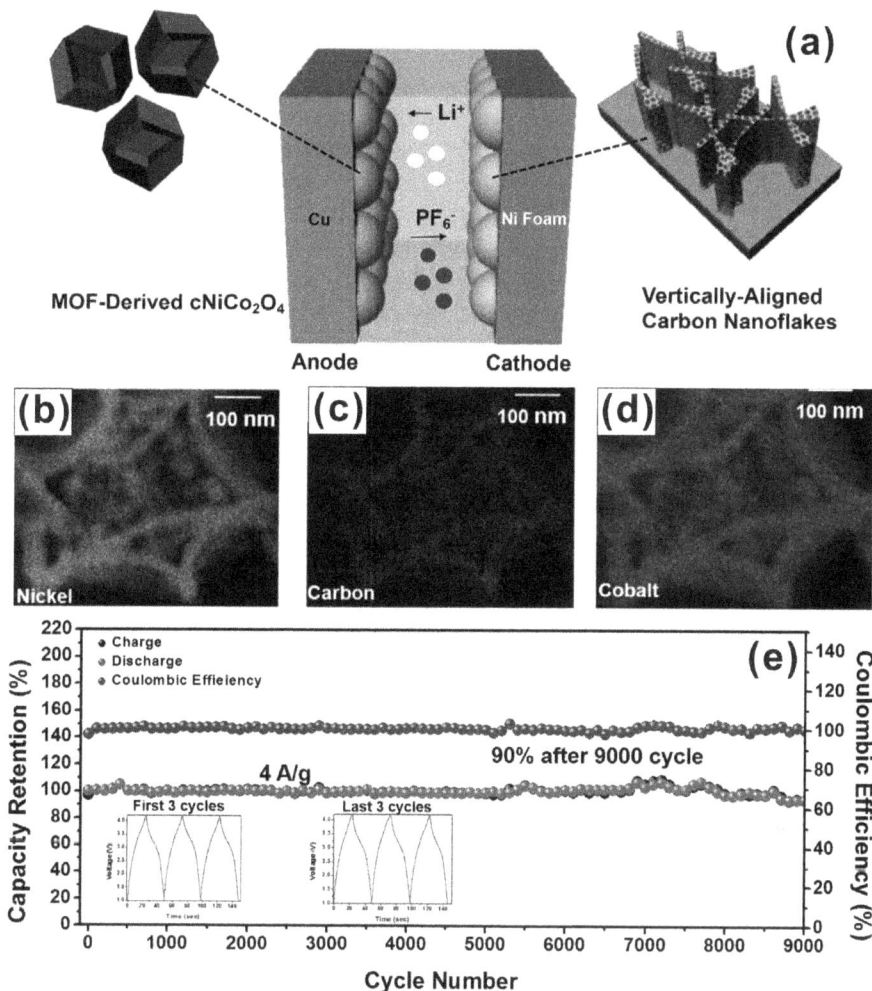

FIGURE 4.21 (a) Schematic illustration for the design of the Li-ion capacitor (with carbonized NiCo$_2$O$_4$ anode and vertically aligned carbon nanoflakes cathode). (b–d) Elemental mapping images of the carbonized NiCo$_2$O$_4$ nanocomposite. (e) Cycling performance of the LIC with current density of 4 A g^{-1} (insets: GCD curves for the first and last three cycles). Reprinted with permission from Ref. [199]. Copyright 2019 Elsevier B.V.

of 152–3240 W kg^{-1}, as well as remarkable cycling stability (over 2500 cycles at high power densities).

Sodium-ion hybrid capacitors (SIHCs) as a potential candidate have gained increasing attention, especially for the application in large-scale energy storage and smart grids, due to the high earth abundance and low cost of sodium, compared to lithium now widely used in LIB and LICs [111]. Li et al. designed and fabricated a novel SIHC via coupling a nanocubic TiO$_2$/C nanocomposite anode with a 3D nanoporous polyhedral carbon cathode, which are both derived by in situ pyrolysis of MOFs precursors (i.e. MIL-125 (Ti) and ZIF-8, respectively) (Figure 4.22a and b). The robust structure and extrinsic pseudocapacitance of TiO_2/C nanocomposite contribute to the superior cyclic stability and rate capability, while the hierarchical nanoporous polyhedral carbon possesses excellent capacity and rate performance as expected. Benefiting from the synergistic merits, the as-assembled organic electrolyte based SIHC can achieve a high energy density of 142.7 W h kg^{-1} (at a power density of 250 W kg^{-1}) and a ultrahigh power output of 25 kW kg^{-1} (remaining an energy density of 61.8 W h kg^{-1}) in the voltage window of 1–4 V. Moreover, the excellent life span of 10,000 cycles with over 90% capacity retention can be easily achieved along with a good linear and symmetric GCD curves (i.e., good reversibility, high Coulombic efficiency and capacitive-dominated mechanism), making it a competitive candidate in high energy and power-required electricity storage applications (Figure 4.22c–f) [111].

4.7 CONCLUSIONS AND OUTLOOK

In summary, MOFs and their derived materials have attracted huge research interest as emerging candidates for superior energy storage especially for ECs. The design and synthesis of conductive MOF-derived hybrids/composites are a promising and low-cost strategy to produce novel materials bearing a set of combined properties superior to those of the individual components (typically synergistic effect); and it is noteworthy that carbon-coated MOF-derived composites are promising electrode materials for energy storage applications in recent years. In future, much work should focus on overcoming some major challenges such as higher energy density, longer cycling performance, and higher mass loading of the active materials. However, given that MOF-derived materials possess well-established structure and property tunability, it will provide a great prospect for developing MOF derivatives as an ideal class of electrode materials for ECs. Several ongoing research hotspots and future trends of MOF-derived composite materials for high-performance ECs are listed as follows:

(i) Despite the fact that MOFs allow facile transformations to various functional materials in a well-controlled manner compared with traditional inorganic precursors, the structural design and the controlling of morphology, particle size, pore properties, and surface area need to be further investigated for optimized performances of MOF derivatives. The nanostructuring is substantially an efficient strategy, together with the greatly enhanced specific surface area and pore properties, it will provide more electrochemically active sites for higher and faster energy storage. Doping with heteroatoms can also tailor the physicochemical structures and improve the electrochemical performance thereof. The dissolution of the MOF-derived electrodes, especially the TMSs, which may severely limit the cycling stability of the electrode, should be further investigated and

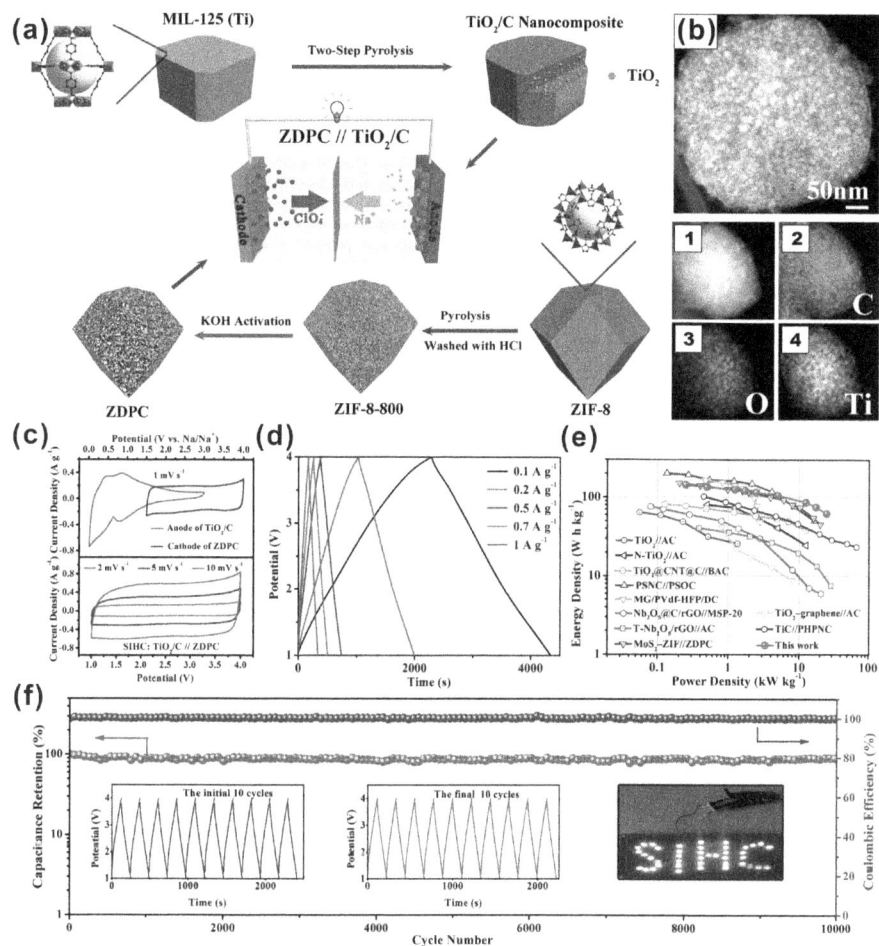

FIGURE 4.22 (a) Schematic illustration showing the construction process of a sodium-ion hybrid capacitor (SIHC) using MIL-125 (Ti) derived TiO_2/C nanocomposite as battery-type anode and ZIF-8 derived nanoporous carbon (ZDPC) as capacitor-type cathode. (b) HAADF-STEM and (1) STEM images of TiO_2/C-700 sample, and (2–4) the corresponding element mappings (C, O, and Ti). Electrochemical performance of TiO_2/C//ZDPC SIHC device: (c) CV curves of TiO_2/C-700 and ZDPC in a sodium half-cell at scan rate of 1 mV s^{-1} (top), and CV curves of TiO_2/C//ZDPC SIHC at various scan rates ranging from 2 to 10 mV s^{-1} (bottom); (d) GCD profiles of TiO_2/C//ZDPC SIHC at different current densities of 0.1–1 A g^{-1}; (e) Ragone plots of SIHC in this work compared with reported representative SIHCs; (f) Long-term cycle life and Coulombic efficiency of TiO_2/C//ZDPC SIHC under 1 A g^{-1} and a large voltage window of 1–4 V (insets: GCD profiles of the initial ten cycles and the final ten cycles and the photograph showing green LED arrays lightened by the SIHC device). Reprinted with permission from Ref. [111]. Copyright 2018 WILEY-VCH.

suppressed. Besides the enhancement of electrical conductivity and ion insertion/adsorption via tailoring the structure or pore size, the compatibility between these derivatives of MOFs and the electrolyte (including aqueous, organic, gel-type, and solid-state) should also be considered, thus the optimization of composition/structure and electrolytes are beneficial as well. The introduction of redox additives into the electrolyte will also give an additional pseudocapacitance contribution to the ECs.

(ii) Although the self-templating feature of a number of MOF-derived processes simplifies the synthetic procedures, general synthetic strategies and conversion approaches with enhanced controllability and scalability are in urgent need to meet the ultimate goal of real application. In view of the large volume loss for some MOFs during the pyrolysis/conversion process, these MOFs are especially suitable as precursors for the fabrication of hollow or frame-like structures with enhanced structural stability and rate performance owing to the buffering effect and capacious open channels. However, the volumetric capacitance or energy density may deteriorate, thus it should be balanced and comprehensively considered for an optimal solution.

(iii) As an efficient strategy to improve the conductivity of derivatives of MOFs, the integration of highly conductive components, especially carbons, will not only embody the advantages of all constituents but also overcome the drawbacks of individual components, i.e., achieve the goal of combining both the high energy density and high power density features. Carbon nanomaterials, especially carbon coating, graphene, and 3D carbon networks, will provide versatile platforms for integrating derivatives of MOFs for acting as most effective electrodes of ECs. Owing to the synergistic effect between the assembled tiny primary particles and the carbon coating, which shortens the diffusion length of ions, they possess an enhanced conductivity and tolerate the volume changes; MOF-derived composite materials exhibit good electrochemical properties as active electrode materials for ECs.

(iv) Further understanding of the formation/conversion mechanisms of MOF-derived materials and the relationship between the composition/structure and their ECs properties is still lacking, theoretical calculation or modeling may play vital roles in future studies. The selection of coordinated metals and optimization of carbonization temperature are critical to obtain active electrode materials with designed nanostructures and compositions, which still lack complete understanding and need further exploration.

(v) Hybrid capacitors or asymmetric supercapacitors are optional solutions to achieve higher energy and power density simultaneously. By designing the electrolytes (especially the non-aqueous ones) for optimum electrochemical performances, the high-voltage operation and the higher energy/power densities thereof can be obtained without compromising long-term cycling stability or lifetime. Despite some critical factors for optimizing pseudocapacitive derivatives of MOFs, it is quite promising for these high-rate capability electrodes to couple with the capacitive (e.g. carbon) negative electrode when used in hybrid (asymmetric) devices for practical applications.

(vi) Binder-free and flexible solid-state supercapacitors are stylish frontrunners in energy storage device technology and have attracted intensive attention owing to recent significant breakthroughs in modern wearable electronics. In view of the powder form of most MOF-derived materials, constructing 3D bulk architectures such as flexible or freestanding electrodes has gained increasing interest, where the micro-/nanometer scale subunits organized macroscopic structure will benefit their use in electrochemical devices. Although some encouraging progress has been made on flexible supercapacitors, their high energy density, mechanical robustness, bending-tolerant property, and performances under various severe service conditions are still limited; excellent mechanical properties of electrodes are required to guarantee the stable energy output in wearable devices; and several other intractable problems, especially for wearable applications (e.g., the potential risks of catching fire or corrosion, short cycling life, long charging time, and the difficulties in meeting the rigid requirement for various forming factors) are also concerned.

ACKNOWLEDGMENTS

This work was supported by National Natural Science Foundation of China (Grant No. 51403193), Opening Project of Sichuan University of Science and Engineering, Material Corrosion and Protection Key Laboratory of Sichuan Province (2019CL19), Vanadium and Titanium Resource Comprehensive Utilization Key Laboratory of Sichuan Province (2019FTSZ01), Municipal Sci-Tech Program of Panzhihua (2019ZD-G-3), and Doctoral innovation fund of Panzhihua University (20190106).

ABBREVIATIONS

1D/2D/3D	one-/two-/three-dimensional
AC	activated carbon
(a)ASC	(aqueous) asymmetric supercapacitor
CC	carbon cloth
CNF	carbon nanofiber
CNT	carbon nanotube
(O)CNTF	(oxidized) carbon nanotube fiber
ECs	electrochemical capacitors
EDLC	electrical double-layer capacitance
GH	graphene
GO	graphene oxide
GCD	galvanostatic charge-discharge
HCs	hybrid capacitors
HECs	hybrid electrochemical capacitors
HSCs	hybrid supercapcitors

LDH	layered double hydroxides
LIBs	lithium-ion batteries
LICs	Li-ion capacitors
LiTFSI	lithium bis(trifluoromethanesulfonyl)imide
MOFs	metal-organic frameworks
MOXs	metal-organic xerogels
NAs	nanowire arrays
N-C or NC	N-doped carbon
NHEC	non-aqueous hybrid electrochemical capacitor
NPC	nanoporous carbon
QDs	quantum dots
rGO	reduced graphene oxide
SCs	supercapacitors
SIHCs	sodium-ion hybrid capacitors
TMCs	transition-metal chalcogenides
TMOs	transition-metal oxides
TMSs	transition-metal sulfides

REFERENCES

1. Li S-L, Xu Q. Metal-organic frameworks as platforms for clean energy. *Energy Environ Sci* 2013;6(6):1656–83.
2. Wei Z, Wang L, Zhuo M, Ni W, Wang H, Ma J. Layered tin sulfide and selenide anode materials for Li- and Na-ion batteries. *J Mater Chem A* 2018;6(26):12185–214.
3. Huang J, Wei Z, Liao J, Ni W, Wang C, Ma J. Molybdenum and tungsten chalcogenides for lithium/sodium-ion batteries: Beyond MoS$_2$. *J Energy Chem* 2019;33:100–24.
4. Wu M, Xu B, Zhang Y, Qi S, Ni W, Hu J, et al. Perspectives in emerging bismuth electrochemistry. *Chem Eng J* 2020;381:122558.
5. Wu M, Ni W, Hu J, Ma J. NASICON-structured NaTi$_2$(PO$_4$)$_3$ for sustainable energy storage. *Nano-Micro Lett* 2019;11(1):44.
6. Ni W, Shi L. Layer-structured carbonaceous materials for advanced Li-ion and Na-ion batteries: Beyond graphene. *J Vac Sci Technol A* 2019;37(4):040803.
7. Huang L, Guan Q, Cheng J, Li C, Ni W, Wang Z, et al. Free-standing N-doped carbon nanofibers/carbon nanotubes hybrid film for flexible, robust half and full lithium-ion batteries. *Chem Eng J* 2018;334:682–90.
8. Ni W, Cheng J, Li X, Guan Q, Qu G, Wang Z, et al. Multiscale sulfur particles confined in honeycomb-like graphene with the assistance of bio-based adhesive for ultrathin and robust free-standing electrode of Li-S batteries with improved performance. *RSC Adv* 2016;6(11):9320–7.
9. Huang L, Cheng J, Li X, Yuan D, Ni W, Qu G, et al. Sulfur quantum dots wrapped by conductive polymer shell with internal void spaces for high-performance lithium-sulfur batteries. *J Mater Chem A* 2015;3(7):4049–57.
10. Ni W, Cheng J, Shi L, Li X, Wang B, Guan Q, et al. Integration of Sn/C yolk-shell nanostructures into free-standing conductive networks as hierarchical composite 3D electrodes and the Li-ion insertion/extraction properties in a gel-type lithium-ion battery thereof. *J Mater Chem A* 2014;2(45):19122–30.
11. Ni W, Wang Y, Xu R. Formation of Sn@C yolk–shell nanospheres and core-sheath nanowires for highly reversible lithium storage. *Part Part Syst Charact* 2013;30(10):873–80.

12. Ni W, Liang F, Liu J, Qu X, Zhang C, Li J, et al. Polymer nanotubes toward gelating organic chemicals. *Chem Commun* 2011;47(16):4727–9.
13. Ni W, Xu Q. Advances with supercritical fluid technologies for synthesis and processing of ordered materials. *Chin Sci Bull* 2009;54(6):707–16.
14. Xu Q, Ni W. Nanomaterials preparation in the supercritical fluid system. *Prog Chem* 2007;19(9):1419–27.
15. Hou C-C, Xu Q. Metal-organic frameworks for energy. *Adv Energy Mater* 2019;9(23):1801307.
16. Dubal DP, Ayyad O, Ruiz V, Gomez-Romero P. Hybrid energy storage: The merging of battery and supercapacitor chemistries. *Chem Soc Rev* 2015;44(7):1777–90.
17. Chu S, Majumdar A. Opportunities and challenges for a sustainable energy future. *Nature* 2012;488(7411):294–303.
18. Chu S, Cui Y, Liu N. The path towards sustainable energy. *Nat Mater* 2016;16:16–22.
19. Simon P, Gogotsi Y, Dunn B. Where do batteries end and supercapacitors begin? *Science* 2014;343(6176):1210–1.
20. Zheng S, Li X, Yan B, Hu Q, Xu Y, Xiao X, et al. Transition-metal (Fe, Co, Ni) based metal-organic frameworks for electrochemical energy storage. *Adv Energy Mater* 2017;7(18):1602733.
21. Wang F, Wu X, Yuan X, Liu Z, Zhang Y, Fu L, et al. Latest advances in supercapacitors: From new electrode materials to novel device designs. *Chem Soc Rev* 2017;46(22):6816–54.
22. Ni W, Wang B, Cheng J, Li X, Guan Q, Gu G, et al. Hierarchical foam of exposed ultrathin nickel nanosheets supported on chainlike Ni-nanowires and the derivative chalcogenide for enhanced pseudocapacitance. *Nanoscale* 2014;6(5):2618–23.
23. Ni W, Cheng J, Li X, Gu G, Huang L, Guan Q, et al. Polymeric cathode materials of electroactive conducting poly (triphenylamine) with optimized structures for potential organic pseudo-capacitors with higher cut-off voltage and energy density. *RSC Adv* 2015;5(12):9221–7.
24. Raza W, Ali F, Raza N, Luo Y, Kim K-H, Yang J, et al. Recent advancements in supercapacitor technology. *Nano Energy* 2018;52:441–73.
25. Wu HB, Lou XW. Metal-organic frameworks and their derived materials for electrochemical energy storage and conversion: Promises and challenges. *Sci Adv* 2017;3(12):eaap9252.
26. Xu J, Zhang H, Xu P, Wang R, Tong Y, Lu Q, et al. In situ construction of hierarchical Co/MnO@graphite carbon composites for highly supercapacitive and OER electrocatalytic performances. *Nanoscale* 2018;10(28):13702–12.
27. Sun J, Wu C, Sun X, Hu H, Zhi C, Hou L, et al. Recent progresses in high-energy-density all pseudocapacitive-electrode-materials-based asymmetric supercapacitors. *J Mater Chem A* 2017;5(20):9443–64.
28. Panda PK, Grigoriev A, Mishra YK, Ahuja R. Progress in supercapacitors: Roles of two dimensional nanotubular materials. *Nanoscale Adv* 2020;2:70–108.
29. Jiao Y, Pei J, Chen D, Yan C, Hu Y, Zhang Q, et al. Mixed-metallic MOF based electrode materials for high performance hybrid supercapacitors. *J Mater Chem A* 2017;5(3):1094–102.
30. Jiao Y, Pei J, Yan C, Chen D, Hu Y, Chen G. Layered nickel metal-organic framework for high performance alkaline battery-supercapacitor hybrid devices. *J Mater Chem A* 2016;4(34):13344–51.
31. Wu Z, Li L, Yan J-m, Zhang X-b. Materials design and system construction for conventional and new-concept supercapacitors. *Adv Sci* 2017;4(6):1600382.
32. Morozan A, Jaouen F. Metal organic frameworks for electrochemical applications. *Energy Environ Sci* 2012;5(11):9269–90.

33. Cao X, Tan C, Sindoro M, Zhang H. Hybrid micro-/nanostructures derived from metal-organic frameworks: Preparation and applications in energy storage and conversion. *Chem Soc Rev* 2017;46(10):2660–77.

34. Wang H, Zhu Q-L, Zou R, Xu Q. Metal-organic frameworks for energy applications. *Chem* 2017;2(1):52–80.

35. Furukawa H, Cordova KE, O'Keeffe M, Yaghi OM. The chemistry and applications of metal-organic frameworks. *Science* 2013;341(6149):1230444.

36. Zhao W, Peng J, Wang W, Liu S, Zhao Q, Huang W. Ultrathin two-dimensional metal-organic framework nanosheets for functional electronic devices. *Coord Chem Rev* 2018;377:44–63.

37. Zhao Y, Song Z, Li X, Sun Q, Cheng N, Lawes S, et al. Metal organic frameworks for energy storage and conversion. *Energy Storage Mater* 2016;2:35–62.

38. Wang L, Han Y, Feng X, Zhou J, Qi P, Wang B. Metal–organic frameworks for energy storage: Batteries and supercapacitors. *Coord Chem Rev* 2016;307:361–81.

39. Xu Y, Li Q, Xue H, Pang H. Metal-organic frameworks for direct electrochemical applications. *Coord Chem Rev* 2018;376:292–318.

40. Ke F-S, Wu Y-S, Deng H. Metal-organic frameworks for lithium ion batteries and supercapacitors. *J Solid State Chem* 2015;223:109–21.

41. Xu G, Nie P, Dou H, Ding B, Li L, Zhang X. Exploring metal organic frameworks for energy storage in batteries and supercapacitors. *Mater Today* 2017;20(4):191–209.

42. Zheng Y, Zheng S, Xue H, Pang H. Metal-organic frameworks/graphene-based materials: Preparations and applications. *Adv Funct Mater* 2018;28(47):1804950.

43. Liang Z, Qu C, Guo W, Zou R, Xu Q. Pristine metal-organic frameworks and their composites for energy storage and conversion. *Adv Mater* 2018;30(37):1702891.

44. Zhou J, Wang B. Emerging crystalline porous materials as a multifunctional platform for electrochemical energy storage. *Chem Soc Rev* 2017;46(22):6927–45.

45. Xie Z, Xu W, Cui X, Wang Y. Recent progress in metal-organic frameworks and their derived nanostructures for energy and environmental applications. *ChemSusChem* 2017;10(8):1645–63.

46. Valizadeh B, Nguyen TN, Stylianou KC. Shape engineering of metal-organic frameworks. *Polyhedron* 2018;145:1–15.

47. Zhao Y, Liu J, Horn M, Motta N, Hu M, Li Y. Recent advancements in metal-organic framework-based electrodes for supercapacitors. *Sci China Mater* 2018;61(2):159–84.

48. Xiao X, Zou L, Pang H, Xu Q. Synthesis of micro/nanoscaled metal-organic frameworks and their direct electrochemical applications. *Chem Soc Rev* 2020;49(1):301–31.

49. Kuyuldar S, Genna DT, Burda C. On the potential for nanoscale metal-organic frameworks for energy applications. *J Mater Chem A* 2019;7(38):21545–76.

50. Ni W, Shi L. Metal-organic-framework composites as proficient cathodes for supercapacitor applications. *Mater Res Found* 2019;58:177–238.

51. Cheng J, Chen S, Chen D, Dong L, Wang J, Zhang T, et al. Editable asymmetric all-solid-state supercapacitors based on high-strength, flexible, and programmable 2D-metal-organic framework/reduced graphene oxide self-assembled papers. *J Mater Chem A* 2018;6(41):20254–66.

52. Choi KM, Jeong HM, Park JH, Zhang Y-B, Kang JK, Yaghi OM. Supercapacitors of nanocrystalline metal–organic frameworks. *ACS Nano* 2014;8(7):7451–7.

53. Zheng S, Xue H, Pang H. Supercapacitors based on metal coordination materials. *Coord Chem Rev* 2018;373:2–21.

54. Sundriyal S, Kaur H, Bhardwaj SK, Mishra S, Kim K-H, Deep A. Metal-organic frameworks and their composites as efficient electrodes for supercapacitor applications. *Coord Chem Rev* 2018;369:15–38.

55. Du W, Bai Y-L, Xu J, Zhao H, Zhang L, Li X, et al. Advanced metal-organic frameworks (MOFs) and their derived electrode materials for supercapacitors. *J Power Sources* 2018;402:281–95.

56. Xue Y, Zheng S, Xue H, Pang H. Metal-organic framework composites and their electrochemical applications. *J Mater Chem A* 2019;7(13):7301–27.

57. Yilmaz G, Yam KM, Zhang C, Fan HJ, Ho GW. In situ transformation of MOFs into layered double hydroxide embedded metal sulfides for improved electrocatalytic and supercapacitive performance. *Adv Mater* 2017;29(26):1606814.

58. Xie X-C, Huang K-J, Wu X. Metal-organic framework derived hollow materials for electrochemical energy storage. *J Mater Chem A* 2018;6(16):6754–71.

59. Wu F, Zhang S, Xi B, Feng Z, Sun D, Ma X, et al. Unusual formation of CoO@C "dandelions" derived from 2D Kagóme MOLs for efficient lithium storage. *Adv Energy Mater* 2018;8(13):1703242.

60. Salunkhe RR, Kaneti YV, Yamauchi Y. Metal–organic framework-derived nanoporous metal oxides toward supercapacitor applications: Progress and prospects. *ACS Nano* 2017;11(6):5293–308.

61. Cai Z-X, Wang Z-L, Kim J, Yamauchi Y. Hollow functional materials derived from metal-organic frameworks: Synthetic strategies, conversion mechanisms, and electrochemical applications. *Adv Mater* 2019;31(11):1804903.

62. Dang S, Zhu Q-L, Xu Q. Nanomaterials derived from metal-organic frameworks. *Nat Rev Mater* 2017;3(1):17075.

63. Li Y, Xu Y, Yang W, Shen W, Xue H, Pang H. MOF-derived metal oxide composites for advanced electrochemical energy storage. *Small* 2018;14(25):1704435.

64. Xia W, Mahmood A, Zou R, Xu Q. Metal-organic frameworks and their derived nanostructures for electrochemical energy storage and conversion. *Energy Environ Sci* 2015;8(7):1837–66.

65. Wang L, Feng X, Ren L, Piao Q, Zhong J, Wang Y, et al. Flexible solid-state supercapacitor based on a metal-organic framework interwoven by electrochemically-deposited PANI. *J Am Chem Soc* 2015;137(15):4920–3.

66. Zhang Z, Chen Y, Xu X, Zhang J, Xiang G, He W, et al. Well-defined metal-organic framework hollow nanocages. *Angew Chem Int Ed* 2014;53(2):429–33.

67. Liu D, Wan J, Pang G, Tang Z. Hollow metal-organic framework micro/nanostructures and their derivatives: Emerging multifunctional materials. *Adv Mater* 2019;31(38):1803291.

68. Zhao M, Lu Q, Ma Q, Zhang H. Two-dimensional metal-organic framework nanosheets. *Small Methods* 2017;1(1–2):1600030.

69. Kumar P, Kim K-H, Bansal V, Kumar P. Nanostructured materials: A progressive assessment and future direction for energy device applications. *Coord Chem Rev* 2017;353:113–41.

70. Guan H-Y, LeBlanc RJ, Xie S-Y, Yue Y. Recent progress in the syntheses of mesoporous metal-organic framework materials. *Coord Chem Rev* 2018;369:76–90.

71. Bhardwaj SK, Bhardwaj N, Kaur R, Mehta J, Sharma AL, Kim K-H, et al. An overview of different strategies to introduce conductivity in metal-organic frameworks and miscellaneous applications thereof. *J Mater Chem A* 2018;6(31):14992–5009.

72. Bhadra BN, Vinu A, Serre C, Jhung SH. MOF-derived carbonaceous materials enriched with nitrogen: Preparation and applications in adsorption and catalysis. *Mater Today* 2019;25:88–111.

73. Guan BY, Yu XY, Wu HB, Lou XW. Complex nanostructures from materials based on metal--organic frameworks for electrochemical energy storage and conversion. *Adv Mater* 2017;29(47):1703614.

74. Lux L, Williams K, Ma S. Heat-treatment of metal-organic frameworks for green energy applications. *CrystEngComm* 2015;17(1):10–22.
75. Li X, Zheng S, Jin L, Li Y, Geng P, Xue H, et al. Metal-organic framework-derived carbons for battery applications. *Adv Energy Mater* 2018:1800716.
76. Yap MH, Fow KL, Chen GZ. Synthesis and applications of MOF-derived porous nanostructures. *Green Energy Environ* 2017;2(3):218–45.
77. Yang W, Li X, Li Y, Zhu R, Pang H. Applications of metal-organic-framework-derived carbon materials. *Adv Mater* 2019;31(6):1804740.
78. Salunkhe RR, Tang J, Kamachi Y, Nakato T, Kim JH, Yamauchi Y. Asymmetric supercapacitors using 3D nanoporous carbon and cobalt oxide electrodes synthesized from a single metal-organic framework. *ACS Nano* 2015;9(6):6288–96.
79. Guan C, Zhao W, Hu Y, Lai Z, Li X, Sun S, et al. Cobalt oxide and N-doped carbon nanosheets derived from a single two-dimensional metal-organic framework precursor and their application in flexible asymmetric supercapacitors. *Nanoscale Horizons* 2017;2(2):99–105.
80. Qu C, Liang Z, Jiao Y, Zhao B, Zhu B, Dang D, et al. "One-for-all" strategy in fast energy storage: Production of pillared MOF nanorod-templated positive/negative electrodes for the application of high-performance hybrid supercapacitor. *Small* 2018;14(23):1800285.
81. Dai E, Xu J, Qiu J, Liu S, Chen P, Liu Y. Co@carbon and Co_3O_4@carbon nanocomposites derived from a single MOF for supercapacitors. *Sci Rep* 2017;7(1):12588.
82. Bendi R, Kumar V, Bhavanasi V, Parida K, Lee PS. Metal organic framework-derived metal phosphates as electrode materials for supercapacitors. *Adv Energy Mater* 2016;6(3):1501833.
83. Kang W, Zhang Y, Fan L, Zhang L, Dai F, Wang R, et al. Metal-organic framework derived porous hollow Co_3O_4/N−C polyhedron composite with excellent energy storage capability. *ACS Appl Mater Interfaces* 2017;9(12):10602–9.
84. Zhu Z, Wang Z, Yan Z, Zhou R, Wang Z, Chen C. Facile synthesis of MOF-derived porous spinel zinc manganese oxide/carbon nanorods hybrid materials for supercapacitor application. *Ceram Int* 2018;44(16):20163–9.
85. Kong L, Chen Q, Shen X, Xu Z, Xu C, Ji Z, et al. MOF derived nitrogen-doped carbon polyhedrons decorated on graphitic carbon nitride sheets with enhanced electrochemical capacitive energy storage performance. *Electrochim Acta* 2018;265:651–61.
86. Khan IA, Badshah A, Khan I, Zhao D, Nadeem MA. Soft-template carbonization approach of MOF-5 to mesoporous carbon nanospheres as excellent electrode materials for supercapacitor. *Microporous Mesoporous Mater* 2017;253:169–76.
87. Tang Z, Zhang G, Zhang H, Wang L, Shi H, Wei D, et al. MOF-derived N-doped carbon bubbles on carbon tube arrays for flexible high-rate supercapacitors. *Energy Storage Mater* 2018;10:75–84.
88. Carrasco JA, Romero J, Abellán G, Hernández-Saz J, Molina SI, Martí-Gastaldo C, et al. Small-pore driven high capacitance in a hierarchical carbon via carbonization of Ni-MOF-74 at low temperatures. *Chem Commun* 2016;52(58):9141–4.
89. Salunkhe RR, Kaneti YV, Kim J, Kim JH, Yamauchi Y. Nanoarchitectures for metal-organic framework-derived nanoporous carbons toward supercapacitor applications. *Acc Chem Res* 2016;49(12):2796–806.
90. Ramachandran R, Zhao C, Luo D, Wang K, Wang F. Morphology-dependent electrochemical properties of cobalt-based metal-organic frameworks for supercapacitor electrode materials. *Electrochim Acta* 2018;267:170–80.
91. Gadipelli S, Li Z, Lu Y, Li J, Guo J, Skipper NT, et al. Size-related electrochemical performance in active carbon nanostructures: A MOFs-derived carbons case study. *Adv Sci* 2019;6(20):1901517.

92. Lim GJH, Liu X, Guan C, Wang J. Co/Zn bimetallic oxides derived from metal-organic frameworks for high performance electrochemical energy storage. *Electrochim Acta* 2018;291:177–87.

93. Long JY, Yan ZS, Gong Y, Lin JH. MOF-derived Cl/O-doped C/CoO and C nanoparticles for high performance supercapacitor. *Appl Surf Sci* 2018;448:50–63.

94. Chen Y, Ni D, Yang X, Liu C, Yin J, Cai K. Microwave-assisted synthesis of honeycomblike hierarchical spherical Zn-doped Ni-MOF as a high-performance battery-type supercapacitor electrode material. *Electrochim Acta* 2018;278:114–23.

95. Yu H, Xia H, Zhang J, He J, Guo S, Xu Q. Fabrication of Fe-doped Co-MOF with mesoporous structure for the optimization of supercapacitor performances. *Chin Chem Lett* 2018;29(6):834–6.

96. Liu S, Deng T, Hu X, Shi X, Wang H, Qin T, et al. Increasing surface active Co^{2+} sites of MOF-derived Co_3O_4 for enhanced supercapacitive performance via $NaBH_4$ reduction. *Electrochim Acta* 2018;289:319–23.

97. Yao M, Zhao X, Jin L, Zhao F, Zhang J, Dong J, et al. High energy density asymmetric supercapacitors based on MOF-derived nanoporous carbon/manganese dioxide hybrids. *Chem Eng J* 2017;322:582–9.

98. Zhao K, Lyu K, Liu S, Gan Q, He Z, Zhou Z. Ordered porous Mn_3O_4@N-doped carbon/graphene hybrids derived from metal-organic frameworks for supercapacitor electrodes. *J Mater Sci* 2017;52(1):446–57.

99. Zhao K, Xu Z, He Z, Ye G, Gan Q, Zhou Z, et al. Vertically aligned MnO_2 nanosheets coupled with carbon nanosheets derived from Mn-MOF nanosheets for supercapacitor electrodes. *J Mater Sci* 2018;53(18):13111–25.

100. Liu S, Zhou J, Cai Z, Fang G, Cai Y, Pan A, et al. Nb_2O_5 quantum dots embedded in MOF derived nitrogen-doped porous carbon for advanced hybrid supercapacitor applications. *J Mater Chem A* 2016;4(45):17838–47.

101. Zhou Q, Wang J, Zheng R, Gong Y, Lin J. One-step mild synthesis of Mn-based spinel $Mn^{II}Cr^{III}_2O_4/Mn^{II}Mn^{III}_2O_4$/C and Co-based spinel $CoCr_2O_4$/C nanoparticles as battery-type electrodes for high-performance supercapacitor application. *Electrochim Acta* 2018;283:197–211.

102. Yang Q, Liu Y, Yan M, Lei Y, Shi W. MOF-derived hierarchical nanosheet arrays constructed by interconnected NiCo-alloy@NiCo-sulfide core-shell nanoparticles for high-performance asymmetric supercapacitors. *Chem Eng J* 2019;370:666–76.

103. Samuel E, Joshi B, Kim M-W, Kim Y-I, Swihart MT, Yoon SS. Hierarchical zeolitic imidazolate framework-derived manganese-doped zinc oxide decorated carbon nanofiber electrodes for high performance flexible supercapacitors. *Chem Eng J* 2019;371:657–65.

104. Xia W, Qu C, Liang Z, Zhao B, Dai S, Qiu B, et al. High-performance energy storage and conversion materials derived from a single metal-organic framework/graphene aerogel composite. *Nano Lett* 2017;17(5):2788–95.

105. Han B, Cheng G, Zhang E, Zhang L, Wang X. Three dimensional hierarchically porous ZIF-8 derived carbon/LDH core-shell composite for high performance supercapacitors. *Electrochim Acta* 2018;263:391–9.

106. Sui Y, Zhang D, Han Y, Sun Z, Qi J, Wei F, et al. Effects of carbonization temperature on nature of nanostructured electrode materials derived from Fe-MOF for supercapacitors. *Electron Mater Lett* 2018;14(5):548–55.

107. Zou K-Y, Liu Y-C, Jiang Y-F, Yu C-Y, Yue M-L, Li Z-X. Benzoate acid-dependent lattice dimension of Co-MOFs and MOF-Derived CoS_2@CNTs with tunable pore diameters for supercapacitors. *Inorg Chem* 2017;56(11):6184–96.

108. Kim J, Young C, Lee J, Heo Y-U, Park M-S, Hossain MSA, et al. Nanoarchitecture of MOF-derived nanoporous functional composites for hybrid supercapacitors. *J Mater Chem A* 2017;5(29):15065–72.

109. Yu Z, Zhang X, Wei L, Guo X. MOF-derived porous hollow α-Fe$_2$O$_3$ microboxes modified by silver nanoclusters for enhanced pseudocapacitive storage. *Appl Surf Sci* 2019;463:616–25.
110. Wang Y, Huang J, Xiao Y, Peng Z, Yuan K, Tan L, et al. Hierarchical nickel cobalt sulfide nanosheet on MOF-derived carbon nanowall arrays with remarkable supercapacitive performance. *Carbon* 2019;147:146–53.
111. Li H, Lang J, Lei S, Chen J, Wang K, Liu L, et al. A High-performance sodium-ion hybrid capacitor constructed by metal-organic framework-derived anode and cathode materials. *Adv Funct Mater* 2018;28(30):1800757.
112. Meng X, Wan C, Jiang X, Ju X. Rodlike CeO$_2$/carbon nanocomposite derived from metal–organic frameworks for enhanced supercapacitor applications. *J Mater Sci* 2018;53(19):13966–75.
113. Chen S, Cai D, Yang X, Chen Q, Zhan H, Qu B, et al. Metal-organic frameworks derived nanocomposites of mixed-valent mnox nanoparticles in-situ grown on ultra-thin carbon sheets for high-performance supercapacitors and lithium-ion batteries. *Electrochim Acta* 2017;256:63–72.
114. Islam DA, Chakraborty A, Roy A, Das S, Acharya H. Fabrication of graphene-oxide (GO)-supported sheet-like CuO nanostructures derived from a metal-organic-framework template for high-performance hybrid supercapacitors. *ChemistrySelect* 2018;3(42):11816–23.
115. Borhani S, Moradi M, Kiani MA, Hajati S, Toth J. Co$_x$Zn$_{1-x}$ ZIF-derived binary Co$_3$O$_4$/ZnO wrapped by 3D reduced graphene oxide for asymmetric supercapacitor: Comparison of pure and heat-treated bimetallic MOF. *Ceram Int* 2017;43(16):14413–25.
116. Jayakumar A, Antony RP, Wang R, Lee J-M. MOF-derived hollow cage Ni$_x$Co$_{3-x}$O$_4$ and their synergy with graphene for outstanding supercapacitors. *Small* 2017;13(11):1603102.
117. He L, Liu J, Yang L, Song Y, Wang M, Peng D, et al. Copper metal-organic framework-derived CuO$_x$-coated three-dimensional reduced graphene oxide and polyaniline composite: Excellent candidate free-standing electrodes for high-performance supercapacitors. *Electrochim Acta* 2018;275:133–44.
118. Xu X, Shi W, Li P, Ye S, Ye C, Ye H, et al. Facile fabrication of three-dimensional graphene and metal-organic framework composites and their derivatives for flexible all-solid-state supercapacitors. *Chem Mater* 2017;29(14):6058–65.
119. Li X, Zhi L. Graphene hybridization for energy storage applications. *Chem Soc Rev* 2018;47(9):3189–216.
120. Chen K, Shi L, Zhang Y, Liu Z. Scalable chemical-vapour-deposition growth of three-dimensional graphene materials towards energy-related applications. *Chem Soc Rev* 2018;47(9):3018–36.
121. Wang P, Zhou H, Meng C, Wang Z, Akhtar K, Yuan A. Cyanometallic framework-derived hierarchical Co$_3$O$_4$-NiO/graphene foam as high-performance binder-free electrodes for supercapacitors. *Chem Eng J* 2019;369:57–63.
122. Tong M, Liu S, Zhang X, Wu T, Zhang H, Wang G, et al. Two-dimensional CoNi nanoparticles@S,N-doped carbon composites derived from S,N-containing Co/Ni MOFs for high performance supercapacitors. *J Mater Chem A* 2017;5(20):9873–81.
123. Wang Y, Chen B, Zhang Y, Fu L, Zhu Y, Zhang L, et al. ZIF-8@MWCNT-derived carbon composite as electrode of high performance for supercapacitor. *Electrochim Acta* 2016;213:260–9.
124. Zhang Q, Hu Z, Yang Y, Zhang Z, Wang X, Yang X, et al. Metal organic frameworks-derived porous carbons/ruthenium oxide composite and its application in supercapacitor. *J Alloys Compd* 2018;735:1673–81.
125. Zhang H, Wu L. Na+ intercalated manganese dioxide/MOF-derived nanoporous carbon hybrid electrodes for supercapacitors with high rate performance and cyclic stability. *J Electrochem Soc* 2018;165(11):A2815–A23.

126. Zhang Y, Lin B, Wang J, Tian J, Sun Y, Zhang X, et al. All-solid-state asymmetric supercapacitors based on ZnO quantum dots/carbon/CNT and porous N-doped carbon/CNT electrodes derived from a single ZIF-8/CNT template. *J Mater Chem A* 2016;4(26):10282–93.

127. Zeng W, Wang L, Shi H, Zhang G, Zhang K, Zhang H, et al. Metal-organic-framework-derived ZnO@C@NiCo$_2$O$_4$ core-shell structures as an advanced electrode for high-performance supercapacitors. *J Mater Chem A* 2016;4(21):8233–41.

128. Farisabadi A, Moradi M, Hajati S, Kiani MA, Espinos JP. Controlled thermolysis of MIL-101(Fe, Cr) for synthesis of Fe$_x$O$_y$/porous carbon as negative electrode and Cr$_2$O$_3$/porous carbon as positive electrode of supercapacitor. *Appl Surf Sci* 2019;469:192–203.

129. Haldorai Y, Choe SR, Huh YS, Han Y-K. Metal-organic framework derived nanoporous carbon/Co$_3$O$_4$ composite electrode as a sensing platform for the determination of glucose and high-performance supercapacitor. *Carbon* 2018;127:366–73.

130. Wang S, Wang T, Shi Y, Liu G, Li J. Mesoporous Co$_3$O$_4$@carbon composites derived from microporous cobalt-based porous coordination polymers for enhanced electrochemical properties in supercapacitors. *RSC Adv* 2016;6(22):18465–70.

131. Wu S-R, Liu J-B, Wang H, Yan H. NiO@ graphite carbon nanocomposites derived from Ni-MOFs as supercapacitor electrodes. *Ionics* 2019;25(1):1–8.

132. Yang Y, Yang F, Hu H, Lee S, Wang Y, Zhao H, et al. Dilute NiO/carbon nanofiber composites derived from metal-organic framework fibers as electrode materials for supercapacitors. *Chem Eng J* 2017;307:583–92.

133. Wang L, Jiao Y, Yao S, Li P, Wang R, Chen G. MOF-derived NiO/Ni architecture encapsulated into N-doped carbon nanotubes for advanced asymmetric supercapacitors. *Inorg Chem Front* 2019;6(6):1553–60.

134. Wang K, Wang H, Bi R, Chu Y, Wang Z, Wu H, et al. Controllable synthesis and electrochemical capacitor performance of MOF-derived MnO$_x$/N-doped carbon/MnO$_2$ composites. *Inorg Chem Front* 2019;6(10):2873–84.

135. Wang L, Tang P, Liu J, Geng A, Song C, Zhong Q, et al. Multifunctional ZnO-porous carbon composites derived from MOF-74(Zn) with ultrafast pollutant adsorption capacity and supercapacitance properties. *J Colloid Interface Sci* 2019;554:260–8.

136. Meng XH, Wan CB, Yu SY, Jiang XP, Ju X. Nickel/porous carbon composite derived from bimetallic MOFs for electrical double-layer supercapacitor application. *Int J Electrochem Sci* 2018;13(8):8179–88.

137. Kumar M, Kim MS, Jeong DI, Humayoun UB, Yoon DH. A core-shell assembly of hierarchical porous Ni@C nanospheres synthesized from metal-organic framework for electrochemical energy application. *Phys Status Solidi A* 2019;216(9):1800921.

138. Wang X, Yue L, Ai J, Shi Z, Lei X, Sun T, et al. A new Co-ZIF derived nanoporous cobalt-rich carbons with high-potential-window as high-performance electrodes for supercapacitors. *Int J Hydrogen Energy* 2019;44(16):8392–402.

139. Yang J, Zeng C, Wei F, Jiang J, Chen K, Lu S. Cobalt-carbon derived from zeolitic imidazolate framework on Ni foam as high-performance supercapacitor electrode material. *Mater Des* 2015;83:552–6.

140. Qiu J, Dai E, Xu J, Liu S, Liu Y. Functionalized MOFs-controlled formation of novel Ni-Co nanoheterostructure@carbon hybrid as the electrodes for supercapacitor. *Mater Lett* 2018;216:207–11.

141. Jayakumar A, Antony RP, Zhao J, Lee J-M. MOF-derived nickel and cobalt metal nanoparticles in a N-doped coral shaped carbon matrix of coconut leaf sheath origin for high performance supercapacitors and OER catalysis. *Electrochim Acta* 2018;265:336–47.

142. Yu F, Xiong X, Zhou L-Y, Li J-L, Liang J-Y, Hu S-Q, et al. Hierarchical nickel/phosphorus/nitrogen/carbon composites templated by one metal-organic framework as highly efficient supercapacitor electrode materials. *J Mater Chem A* 2019;7(6):2875–83.

143. Wang YC, Li WB, Zhao L, Xu BQ. MOF-derived binary mixed metal/metal oxide @carbon nanoporous materials and their novel supercapacitive performances. *Phys Chem Chem Phys* 2016;18(27):17941–8.

144. Díaz-Duran AK, Montiel G, Viva FA, Roncaroli F. Co,N-doped mesoporous carbons cobalt derived from coordination polymer as supercapacitors. *Electrochim Acta* 2019;299:987–98.

145. Lee G, Seo YD, Jang J. ZnO quantum dot-decorated carbon nanofibers derived from electrospun ZIF-8/PVA nanofibers for high-performance energy storage electrodes. *Chem Commun* 2017;53(83):11441–4.

146. Wang L, Yu J, Dong X, Li X, Xie Y, Chen S, et al. Three-dimensional macroporous carbon/Fe_3O_4-doped porous carbon nanorods for high-performance supercapacitor. *ACS Sustainable Chem Eng* 2016;4(3):1531–7.

147. Zhou Z, Zhang Q, Sun J, He B, Guo J, Li Q, et al. Metal-organic framework derived spindle-like carbon incorporated α-Fe_2O_3 grown on carbon nanotube fiber as anodes for high-performance wearable asymmetric supercapacitors. *ACS Nano* 2018;12(9):9333–41.

148. Dou Y, Zhou J, Yang F, Zhao M-J, Nie Z, Li J-R. Hierarchically structured layered-double-hydroxide@zeolitic-imidazolate-framework derivatives for high-performance electrochemical energy storage. *J Mater Chem A* 2016;4(32):12526–34.

149. Yi M, Zhang C, Cao C, Xu C, Sa B, Cai D, et al. MOF-derived hybrid hollow submicrospheres of nitrogen-doped carbon-encapsulated bimetallic Ni–Co–S nanoparticles for supercapacitors and lithium ion batteries. *Inorg Chem* 2019;58(6):3916–24.

150. Wu R, Wang DP, Kumar V, Zhou K, Law AWK, Lee PS, et al. MOFs-derived copper sulfides embedded within porous carbon octahedra for electrochemical capacitor applications. *Chem Commun* 2015;51(15):3109–12.

151. Li L, Liu Y, Han Y, Qi X, Li X, Fan H, et al. Metal-organic framework-derived carbon coated copper sulfide nanocomposites as a battery-type electrode for electrochemical capacitors. *Mater Lett* 2019;236:131–4.

152. Li Z-X, Yang B-L, Jiang Y-F, Yu C-Y, Zhang L. Metal-directed assembly of five 4-connected MOFs: One-pot syntheses of MOF-derived M_xS_y@C composites for photocatalytic degradation and supercapacitors. *Cryst Growth Des* 2018;18(2):979–92.

153. Qu C, Zhang L, Meng W, Liang Z, Zhu B, Dang D, et al. MOF-derived α-NiS nanorods on graphene as an electrode for high-energy-density supercapacitors. *J Mater Chem A* 2018;6(9):4003–12.

154. Tian D, Chen S, Zhu W, Wang C, Lu X. Metal–organic framework derived hierarchical Ni/Ni_3S_2 decorated carbon nanofibers for high-performance supercapacitors. *Mater Chem Front* 2019;3(8):1653–60.

155. Zhang S, Li D, Chen S, Yang X, Zhao X, Zhao Q, et al. Highly stable supercapacitors with MOF-derived Co_9S_8/carbon electrodes for high rate electrochemical energy storage. *J Mater Chem A* 2017;5(24):12453–61.

156. Cao F, Zhao M, Yu Y, Chen B, Huang Y, Yang J, et al. Synthesis of two-dimensional $CoS_{1.097}$/nitrogen-doped carbon nanocomposites using metal–organic framework nanosheets as precursors for supercapacitor application. *J Am Chem Soc* 2016;138(22):6924–7.

157. Wang Y, Chen B, Chang Z, Wang X, Wang F, Zhang L, et al. Enhancing performance of sandwich-like cobalt sulfide and carbon for quasi-solid-state hybrid electrochemical capacitors. *J Mater Chem A* 2017;5(19):8981–8.

158. Wang P, Li C, Wang W, Wang J, Zhu Y, Wu Y. Hollow Co_9S_8 from metal organic framework supported on rGO as electrode material for highly stable supercapacitors. *Chin Chem Lett* 2018;29(4):612–5.

159. Niu H, Zhang Y, Liu Y, Luo B, Xin N, Shi W. MOFs-derived Co_9S_8-embedded graphene/hollow carbon spheres film with macroporous frameworks for hybrid supercapacitors with superior volumetric energy density. *J Mater Chem A* 2019;7(14): 8503–9.

160. Sun S, Luo J, Qian Y, Jin Y, Liu Y, Qiu Y, et al. Metal-organic framework derived honeycomb Co_9S_8@C composites for high-performance supercapacitors. *Adv Energy Mater* 2018;8(25):1801080.

161. Wang Z, Zhu Z, Zhang Q, Zhai M, Gao J, Chen C, et al. Fabrication of N-doped carbon coated spinel copper cobalt sulfide hollow spheres to realize the improvement of electrochemical performance for supercapacitors. *Ceram Int* 2019;45(17, Part A):21286–92.

162. Liu H, Guo H, Yue L, Wu N, Li Q, Yao W, et al. Metal-organic frameworks-derived NiS_2/CoS_2/N-doped carbon composites as electrode materials for asymmetric supercapacitor. *ChemElectroChem* 2019;6(14):3764–73.

163. Chen C, Wu M-K, Tao K, Zhou J-J, Li Y-L, Han X, et al. Formation of bimetallic metal-organic framework nanosheets and their derived porous nickel–cobalt sulfides for supercapacitors. *Dalton Trans* 2018;47(16):5639–45.

164. Qiu J, Bai Z, Liu S, Liu Y. Formation of nickel–cobalt sulphide@graphene composites with enhanced electrochemical capacitive properties. *RSC Adv* 2019;9(12):6946–55.

165. Tang Q, Ma L, Yan F, Gan M, Li X, Cao F, et al. Designed cross-linking nanoporous $Zn_{0.76}Co_{0.24}S$ @C-ZIF-$Zn_{0.76}Co_{0.24}S$ core-shell nanosheet arrays on nickel foam for battery-type electrodes with high performance electrochemical energy storage. *Synth Met* 2019;250:136–45.

166. Liu Y-P, Qi X-H, Li L, Zhang S-H, Bi T. MOF-derived PPy/carbon-coated copper sulfide ceramic nanocomposite as high-performance electrode for supercapacitor. *Ceram Int* 2019;45(14):17216–23.

167. Wang Y, Du Q, Zhao H, Hou S, Shen Y, Li H, et al. Metal-organic framework derived leaf-like CoSNC nanocomposites for supercapacitor electrodes. *Nanoscale* 2018;10(37):17958–64.

168. Pham DT, Baboo JP, Song J, Kim S, Jo J, Mathew V, et al. Facile synthesis of pyrite (FeS_2/C) nanoparticles as an electrode material for non-aqueous hybrid electrochemical capacitors. *Nanoscale* 2018;10(13):5938–49.

169. Liu S, Tong M, Liu G, Zhang X, Wang Z, Wang G, et al. S,N-Containing Co-MOF derived Co_9S_8@S,N-doped carbon materials as efficient oxygen electrocatalysts and supercapacitor electrode materials. *Inorg Chem Front* 2017;4(3):491–8.

170. Yuan C, Wu HB, Xie Y, Lou XWD. Mixed transition-metal oxides: Design, synthesis, and energy-related applications. *Angew Chem Int Ed* 2014;53(6):1488–504.

171. Yu L, Wu HB, Lou XWD. Self-templated formation of hollow structures for electrochemical energy applications. *Acc Chem Res* 2017;50(2):293–301.

172. Yu L, Hu H, Wu HB, Lou XW. Complex hollow nanostructures: Synthesis and energy-related applications. *Adv Mater* 2017;29(15):1604563.

173. Yu X-Y, Yu L, Lou XW. Metal sulfide hollow nanostructures for electrochemical energy storage. *Adv Energy Mater* 2016;6(3):1501333.

174. Zhang X, Luo J, Tang P, Ye X, Peng X, Tang H, et al. A universal strategy for metal oxide anchored and binder-free carbon matrix electrode: A supercapacitor case with superior rate performance and high mass loading. *Nano Energy* 2017;31:311–21.

175. Liao J, Ni W, Wang C, Ma J. Layer-structured niobium oxides and their analogues for advanced hybrid capacitors. *Chem Eng J* 2020. doi: 10.1016/j.cej.2019.123489

176. Guan Q, Cheng J, Wang B, Ni W, Gu G, Li X, et al. Needle-like Co_3O_4 anchored on the graphene with enhanced electrochemical performance for aqueous supercapacitors. *ACS Appl Mater Interfaces* 2014;6(10):7626–32.

177. Zhang C, Xiao J, Lv X, Qian L, Yuan S, Wang S, et al. Hierarchically porous Co_3O_4/C nanowire arrays derived from a metal–organic framework for high performance supercapacitors and the oxygen evolution reaction. *J Mater Chem A* 2016;4(42):16516–23.

178. Deng X, Li J, Zhu S, He F, He C, Liu E, et al. Metal-organic frameworks-derived honeycomb-like Co_3O_4/three-dimensional graphene networks/Ni foam hybrid as a binder-free electrode for supercapacitors. *J Alloys Compd* 2017;693:16–24.

179. Devi B, Venkateswarulu M, Kushwaha HS, Halder A, Koner RR. A Polycarboxyl-decorated Fe^{III}-based xerogel-derived multifunctional composite (Fe_3O_4/Fe/C) as an efficient electrode material towards oxygen reduction reaction and supercapacitor application. *Chem Eur J* 2018;24(25):6586–94.

180. Mahmood A, Zou R, Wang Q, Xia W, Tabassum H, Qiu B, et al. Nanostructured electrode materials derived from metal–organic framework xerogels for high-energy-density asymmetric supercapacitor. *ACS Appl Mater Interfaces* 2016;8(3):2148–57.

181. Zhang Y, Chen H, Guan C, Wu Y, Yang C, Shen Z, et al. Energy-saving synthesis of MOF-derived hierarchical and hollow $Co(VO_3)_2$-$Co(OH)_2$ composite leaf arrays for supercapacitor electrode materials. *ACS Appl Mater Interfaces* 2018;10(22):18440–4.

182. Guan C, Liu X, Ren W, Li X, Cheng C, Wang J. Rational design of metal-organic framework derived hollow $NiCo_2O_4$ arrays for flexible supercapacitor and electrocatalysis. *Adv Energy Mater* 2017;7(12):1602391.

183. Li X, Wu H, Elshahawy AM, Wang L, Pennycook SJ, Guan C, et al. Cactus-like NiCoP/NiCo-OH 3D architecture with tunable composition for high-performance electrochemical capacitors. *Adv Funct Mater* 2018;28(20):1800036.

184. Yang Q, Liu Y, Xiao L, Yan M, Bai H, Zhu F, et al. Self-templated transformation of MOFs into layered double hydroxide nanoarrays with selectively formed Co_9S_8 for high-performance asymmetric supercapacitors. *Chem Eng J* 2018;354:716–26.

185. Zhao W, Zheng Y, Cui L, Jia D, Wei D, Zheng R, et al. MOF derived Ni-Co-S nanosheets on electrochemically activated carbon cloth via an etching/ion exchange method for wearable hybrid supercapacitors. *Chem Eng J* 2019;371:461–9.

186. Zhao Y, Dong H, He X, Yu J, Chen R, Liu Q, et al. Carbon cloth modified with metal-organic framework derived $CC@CoMoO_4$-$Co(OH)_2$ nanosheets array as a flexible energy-storage material. *ChemElectroChem* 2019;6(13):3355–66.

187. Dai S, Yuan Y, Yu J, Tang J, Zhou J, Tang W. Metal-organic framework-templated synthesis of sulfur-doped core–sheath nanoarrays and nanoporous carbon for flexible all-solid-state asymmetric supercapacitors. *Nanoscale* 2018;10(33):15454–61.

188. Wang Z, Cheng J, Guan Q, Huang H, Li Y, Zhou J, et al. All-in-one fiber for stretchable fiber-shaped tandem supercapacitors. *Nano Energy* 2018;45:210–9.

189. Yang D, Ni W, Cheng J, Wang Z, Li C, Zhang Y, et al. Omnidirectional porous fiber scrolls of polyaniline nanopillars array-N-doped carbon nanofibers for fiber-shaped supercapacitors. *Mater Today Energy* 2017;5:196–204.

190. Li X, Li X, Cheng J, Yuan D, Ni W, Guan Q, et al. Fiber-shaped solid-state supercapacitors based on molybdenum disulfide nanosheets for a self-powered photodetecting system. *Nano Energy* 2016;21:228–37.

191. Zhang Q, Zhou Z, Pan Z, Sun J, He B, Li Q, et al. All-metal-organic framework-derived battery materials on carbon nanotube fibers for wearable energy-storage device. *Adv Sci* 2018;5(12):1801462.

192. Zhao J, Li H, Li C, Zhang Q, Sun J, Wang X, et al. MOF for template-directed growth of well-oriented nanowire hybrid arrays on carbon nanotube fibers for wearable electronics integrated with triboelectric nanogenerators. *Nano Energy* 2018;45:420–31.

193. Jiao Y, Qu C, Zhao B, Liang Z, Chang H, Kumar S, et al. High-performance electrodes for a hybrid supercapacitor derived from a metal-organic framework/graphene composite. *ACS Appl Energy Mater* 2019;2(7):5029–38.

194. Bai X, Liu Q, Lu Z, Liu J, Chen R, Li R, et al. Rational design of sandwiched Ni-Co layered double hydroxides hollow nanocages/graphene derived from metal-organic framework for sustainable energy storage. *ACS Sustainable Chem Eng* 2017;5(11):9923–34.
195. Tabassum H, Mahmood A, Wang Q, Xia W, Liang Z, Qiu B, et al. Hierarchical cobalt hydroxide and B/N Co-doped graphene nanohybrids derived from metal-organic frameworks for high energy density asymmetric supercapacitors. *Sci Rep* 2017;7:43084.
196. Niu H, Zhang Y, Liu Y, Xin N, Shi W. NiCo-layered double-hydroxide and carbon nanosheets microarray derived from MOFs for high performance hybrid supercapacitors. *J Colloid Interface Sci* 2019;539:545–52.
197. Jin H, Yuan D, Zhu S, Zhu X, Zhu J. Ni–Co layered double hydroxide on carbon nanorods and graphene nanoribbons derived from MOFs for supercapacitors. *Dalton Trans* 2018;47(26):8706–15.
198. Xu X, Shi W, Liu W, Ye S, Yin R, Zhang L, et al. Preparation of two-dimensional assembled Ni–Mn–C ternary composites for high-performance all-solid-state flexible supercapacitors. *J Mater Chem A* 2018;6(47):24086–91.
199. Cheng C-F, Li X, Liu K, Zou F, Tung W-Y, Huang Y-F, et al. A high-performance lithium-ion capacitor with carbonized $NiCo_2O_4$ anode and vertically aligned carbon nanoflakes cathode. *Energy Storage Mater* 2019;22:265–74.
200. Shen L, Lv H, Chen S, Kopold P, van Aken PA, Wu X, et al. Peapod-like Li_3VO_4/n-doped carbon nanowires with pseudocapacitive properties as advanced materials for high-energy lithium-ion capacitors. *Adv Mater* 2017;29(27):1700142.
201. Wang R, Wang S, Jin D, Zhang Y, Cai Y, Ma J, et al. Engineering layer structure of MoS_2-graphene composites with robust and fast lithium storage for high-performance Li-ion capacitors. *Energy Storage Mater* 2017;9:195–205.

5 Photovoltaic Performance of Titanium Oxide/Metal-Organic Framework Nanocomposite

Phuti S. Ramaripa, Kerileng M. Molapo,
Thabang R. Somo, Malesela D. Teffu, Mpitloane J.
Hato, Manoko S. Maubane-Nkadimeng,
Katlego Makgopa, Emmanuel I. Iwouha, and
Kwena D. Modibane

CONTENTS

5.1   INTRODUCTION

Energy is of a great global concern with increasing population and industrialization resulting in an increase in demand [1–5]. The increase in energy demand accelerates the production of fossil fuels that are very harmful to the environment. Renewable energy sources are candidates to meet the increase in energy demand and are anticipated to play an important role in taking the world to an environmentally friendly, reliable, and sustainable energy system [2]. Photovoltaic (PV) technologies are known to convert sunlight radiation into electrical energy [1,5]. They are a type of energy which is renewable, clean, and sustainable. In addition, they were introduced in a quest to limit or eradicate the combustion of fossil fuels [1]. Currently, PVs are used in small-scale applications and show great potential to be used at a larger scale provided the power efficiency is above 10% [6]. The most commercialized PV solar cells are silicon-based due to their high efficiency of up to 25% [7]. Even though they show high efficiency, some disadvantages such as high-purity silicon single crystal manufacture uses expensive processes and restricts their widespread use [7]. Gong *et al.* [8] have also indicated that high production and environmental cost restricts the terrestrial PV market. Silicon-based PVs have also been shown to be expensive relative to other technologies and are generally used notwithstanding their changing cost factor and poor PV efficiency [7,8]. These disadvantages paved the way for intensive research in the quest to find an alternative in PV renewable energy production leading to the development of dye-sensitized solar cells (DSSC) [1]. The most attractive characteristic of DSSCs is its cost-effectiveness due to inexpensive materials and simple fabrication process [5]. DSSCs are good candidates to replace conventional silicon-based solar cells due to their good performance in diverse light conditions as well as being lightweight, flexible, and having low toxicity [1,5]. Daeneke *et al.* [9,10] have also indicated that DSSCs are potential PV devices, because they are easy to manufacture, and they have low-cost materials as well as mechanical flexibility. Moreover, it was shown that DSSCs are mainly composed of five components [1,5]. The photoanode is normally made of transparent conductive oxides where a semiconductor material may be coated to form film [1]. Currently, titanium dioxide (TiO_2) materials are used as photoanode materials due to their wide availability, low cost, and non-toxicity [11]. The TiO_2 materials have ability to adsorb dye sensitizer for harvesting of light and generation of photo-excited electrons [12]. In addition, DSSCs have an electrolyte containing a redox mediator, usually triiodide/iodine in an organic solvent, to transport electrons between photoanode and counter electrodes [5,12]. As a result, DSSCs based on TiO_2 semiconductors were extensively studied due to their advantage of a large surface area, and high chemical and optical stability [11,12]. However, TiO_2 has the larger band gap of 3.2 eV and low power conversion efficiency of 3.08% [13] that disadvantages it in harvesting more energy. It was seen that the n-type anatase TiO_2 may be doped with metal-organic frameworks (MOFs) to improve the harvesting of sunlight and reduces the band gap owing to their high surface area [14–18]. MOFs are known to be porous polymers which are made from a combination of metal cores and organic ligands through covalent bonds [2,3,16,17]. Therefore, this chapter reports on the recent position of improving the PV applications of TiO_2 with the addition of MOFs. This is accomplished by considering the fundamental mechanisms of photovoltaic cells (PVCs). Critical factors which are used to evaluate the PV properties of semiconductor materials are discussed. Furthermore, the background, structural properties, and synthetic methods of TiO_2 and MOFs are presented. To sum up, the challenges and approaches to address drawbacks of TiO_2/MOF composite as photoanode material in dye sensitized solar cells are outlined.

5.2 PHOTOVOLTAIC CELLS

5.2.1 BACKGROUND

The PV effect is a chemical process through which solar cells convert sunlight radiation into electrical energy [19,20]. It can occur in different kinds of solar cells (Figure 5.1a) such as nanocrystal (perovskite), organic (polymer), DSSCs, and concentrated solar cells [20–22]. In addition, Figure 5.1b presents a conversion efficiency as a function of years for perovskite, and dye-sensitized solar cells [21]. The figure shows that the efficiency progress of perovskite solar cells and other third-generation solar cells with an enhanced efficiency over the years [21,23,24].

5.2.2 KEY PARAMETERS TO EVALUATE PHOTOVOLTAIC CELLS

When a solar cell panel interacts with light, the photocurrent can be measured. There are many factors that affect the value of the photocurrent such as the quality of the device, incident wavelength, and the surface area of the device being illuminated. Therefore, different parameters are employed to determine the energy efficiency of the PVCs such as the maximum power point (Pmax), the energy conversion efficiency (η), and the fill factor (FF) [22].

The Upper Limit of Open Circuit Voltage

Open circuit voltage (Voc) represents the maximum voltage at zero current. It was seen that the Voc value rises logarithmically with increased sunlight radiation [1,5,22]. In addition, the band gap of the semiconductor material limits the increase in the Voc value and the maximum can only be achieved when the band gap is in the range 1.0–1.6 eV [25].

Short Circuit Current (Isc)

The short circuit current gives the maximum current density that can be obtained when the two leads are connected to each other. When analyzing a current-potential (I-V) curve, the Isc value is obtained at zero voltage as a result of generation and collection of carriers [22,25].

Fill Factor (FF)

The fill factor (FF) is defined as the ratio of the curve under the maximum power point of the cell (*Pmax* = Imax (maximum current) × Vmax (maximum potential))

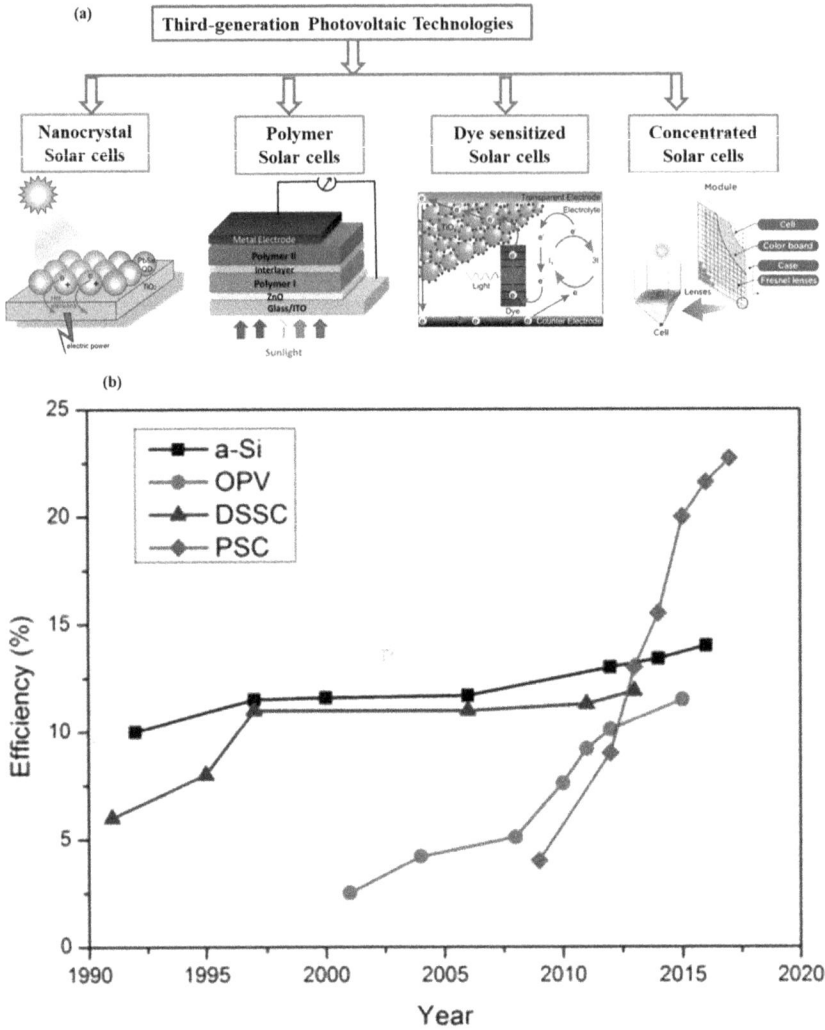

FIGURE 5.1 (a) Types of photovoltaic (PV) technologies and (b) the power conversion efficiency of different solar cells (a-Si: amorphous Si single layer, OPV: organic photovoltaic solar cells, DSSC: dye sensitized solar cells, and PSC: perovskite solar cells) over the years [21].

to the area associated with open and closed circuits (P = Isc × Voc) [22]. The FF is referred to as a quantifying unit for the squareness of the I-V curve and calculated using Equation 5.1 [22]. This is also influenced by the materials morphology and stability of the materials [1]. The theoretical limit of a FF is between 0.25 and 1 [1,5].

$$FF = \frac{(Pmax)(Vmax)}{(Isc)(Voc)} \tag{5.1}$$

Power Conversion Efficiency (η)

Power conversion efficiency (Equation 5.2) can be defined as the ratio between the maximum power output (Pmax) and the power that is from the incident light (Pin) [22]. As stated, the power conversion efficiency is greatly influenced by the Voc and Isc [5]. The power conversion efficiency is calculated as follows:

$$\eta = \frac{Pmax}{Pin} = \frac{(Isc)(Voc)(FF)}{Pin} \tag{5.2}$$

where Pin is the solar power input into the solar cell, and Voc is the voltage across the open circuit.

5.2.3 Perovskite Solar Cells

It was documented that perovskite solar cells (PSCs) are made of materials that are described by the formula ABX_3, where A represents organic material in a large quantity, B is a smaller amount of inorganic material (cations of different sizes), and X is an anion material, e.g., halogens [26]. In addition, it was shown that the PSCs exist in two forms as organic-inorganic and quantum dot sensitized PSCs [27]. The organic-inorganic cells have been used as a light harvesting sensitizer [26,27], whereas quantum dot sensitizer ones have been used as an electron transfer which is known as the third generation of solar cells due to their distinguished optical and electronic properties [27]. The advantage of quantum dot sensitizer solar cells (QDSSCs) is they have the probability of light harvesting materials as well as the low manufacturing cost [27,28]. The principles of QDSSCs is similar to that of DSSCs but differs with the size, which is larger than dye molecules [28]. Furthermore, lead-based material is usually known as organometallic hybrid PSCs [29]. These kinds of solar cells have brought much attention to the researchers due to their outstanding structural properties of lead-based materials [29,30]. The wide tunable band gap, high absorptivity coefficients, carrier mobility, and bipolar carrier transport make these solar cells good [30]. The lead-based materials are highly versatile materials and show a high efficiency in solar cells, however, they are toxic and have a long-term stability problem [31,32]. To solve the problem of toxicity, other technology such as DSSCs that are not harmful and qualify as clean energy should be employed. For example, Liu et al. [33], fabricated the PSCs by adding $CsPbI_3$ QDs as interface materials, as shown in Figure 5.2. They have observed the enhancement of the charge-transfer efficiency by $CsPbI_3$ QDs at the interface of the perovskite/hole-transporting materials (HTMs) layer. Their results predicted the energy level structure formed by interfacial material of the CsPbI3 QDs improved the hole extraction and that led to increases in FF, Isc, and PCE values. In addition, Yuan et al. [34] reported perovskite-based quantum dot (QD) solar cells with introduction of $CsPbI_3$ and a series of dopant-free polymeric HTMs, as shown in Figure 5.3. Their device possessed a significant power conversion efficiency of ~13% with low energy loss of 0.45 eV.

Moreover, the tunable band gap of the semiconductor materials improves the adsorption coefficient in QDSSCs that may lead to increases in the short circuit

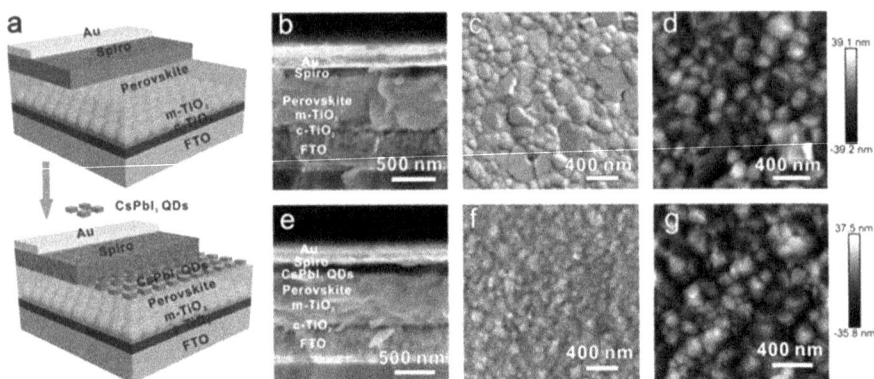

FIGURE 5.2 (a) Synthetic route of CsPbI$_3$ QD interface, scanning electron microgram (SEM) cross-section of (b) non-deposited perovskite solar cells and of (e) CsPbI$_3$ QD-deposited perovskite solar cells, upper-view SEM image of (c) non-deposited perovskite layer and of (f) CsPbI$_3$ QD-deposited layer, as well as AFM images of (d) non-deposited perovskite layer and of (g) CsPbI$_3$ QD-deposited layer [33].

current of the PVCs [35,36]. It was reported that QDSSCs fabrication uses inorganic materials containing chalcogenide such as CdS, CdSe, CdTe, PbS, and PbSe [36]. However, those materials are hazardous to health due to their high toxicity [37].

5.2.4 Organic Solar Cells

The fundamental properties of organic (as well as organometallic) semiconductors differ from those of their inorganic counterparts, hence, organic photovoltaic cells (OPCs) operate in a different manner to inorganic solar cells. For example, Figure 5.4 shows the chemical structures of typical organic solar cell donor and acceptor materials with their corresponding energy levels [38]. The OPC device needs a bulk heterojunction (BHJ) that can be prepared from a mixture of an electron-donor (p-type semiconductor) and electron-acceptor (n-type semiconductor) (Figure 5.4). The p-type semiconductor material is made of π-conjugated polymer and a small molecule semiconductor. The n-type semiconductor material is generally a fullerene derivative [38]. In Figure 5.4, P3HT (or PCBM), that is a solubility-enhanced fullerene, efficiently provides a BHJ [39]. It was shown in the figure that the MEH-CN-PPV provides a high open-circuit voltage with PCEs of 2.0% [38]. Moreover, it was reported that C70 derivative materials normally give higher power conversion efficiency as compared to C60 ones, due to high light absorption [40].

5.2.5 Dye-Sensitized Solar Cells

The sensitization and photoexcitation of light absorbing material (e.g., dye) is the key element for the operation of DSSCs. Vogel and Berlin were the first scientists to discover the dye-sensitization technique in 1873 [41]. Although, it took a century

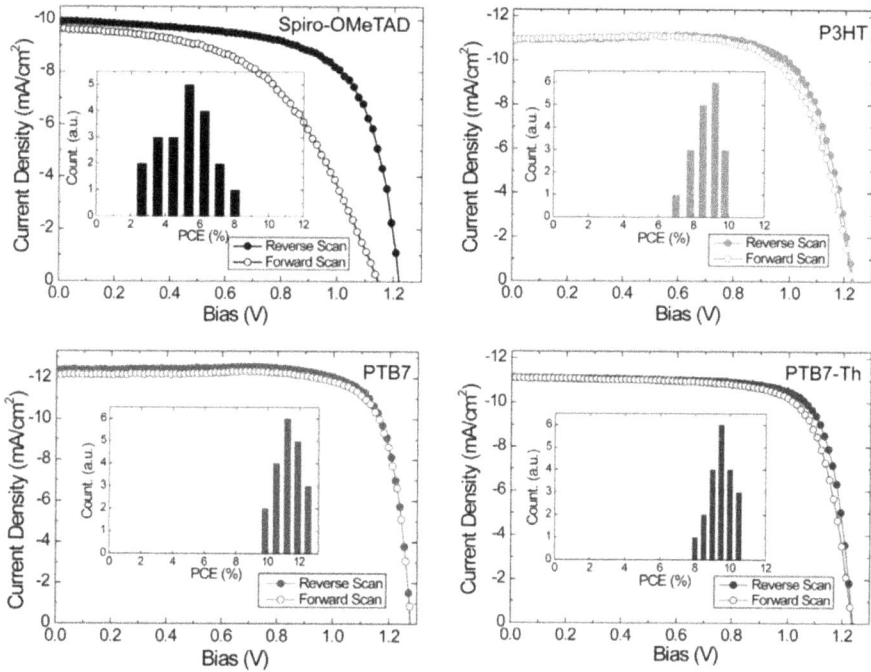

FIGURE 5.3 J-V curves from reverse and forward scan of optimized CsPbI₃ perovskite QD solar cells adopting 2,2,7,7-tetrakis-(N,N-di-p-methoxyphenylamine)-9,9-bifluorene (Spiro-OMeTAD); poly-3-hexylthiophene (P3HT); poly[4,8-bis[(2-ethylhexyl)oxy]benzo[1,2-b: 4,5-b′]dithiophene-2,6-diyl-alt-3-fluoro-2-[(2-ethylhexyl)carbonyl]thieno[3,4-b]thiophe ne-4,6-diyl] (PTB7); and poly[4,8-bis(5-(2-ethylhexyl)thiophen-2-yl)benzo[1,2-b; 4.5b′]dith iophene-2,6-diyl-alt-3-fluoro-2-[(2-ethyl-hexyl)-carbonyl]-thieno[3,4-b]thiophene-4,6-diyl] (PTB7-Th) as the hole-transporting material (HTM) [34].

to understand the DSSC mechanism which was reported in photo-electrochemical studies of dye-sensitized single-crystal electrodes in the 1970s [42]. Figure 5.5a shows the basic elements of DSSC, the components are conducting substrate, semiconductor active layer, dye sensitizer, electrolyte, and catalyst [43]. In this figure, the dye sensitizer molecule absorbs energy from light radiation and gets excited, passing electrons to the conduction band of semiconductor material. This passes the electron to the external circuit through conducting substrate to generate excited electrons from highest occupied molecular orbital (HOMO) to lowest unoccupied molecular orbital (LUMO) of the semiconductor [43]. Figure 5.5(b) shows the energy levels and the working principle of DSSC. It was seen that the dye's band energy level has to be higher than that of the semiconductor's conduction band [41–43]. On the other hand, electrolyte made of I⁻/I³⁻ receives an electron from the external circuit with the help of a photocatalyst and donates its electron to the excited dye [43]. The electrolyte plays important role as a bridge between photoanode and counter electrode where the migration of electrons takes place

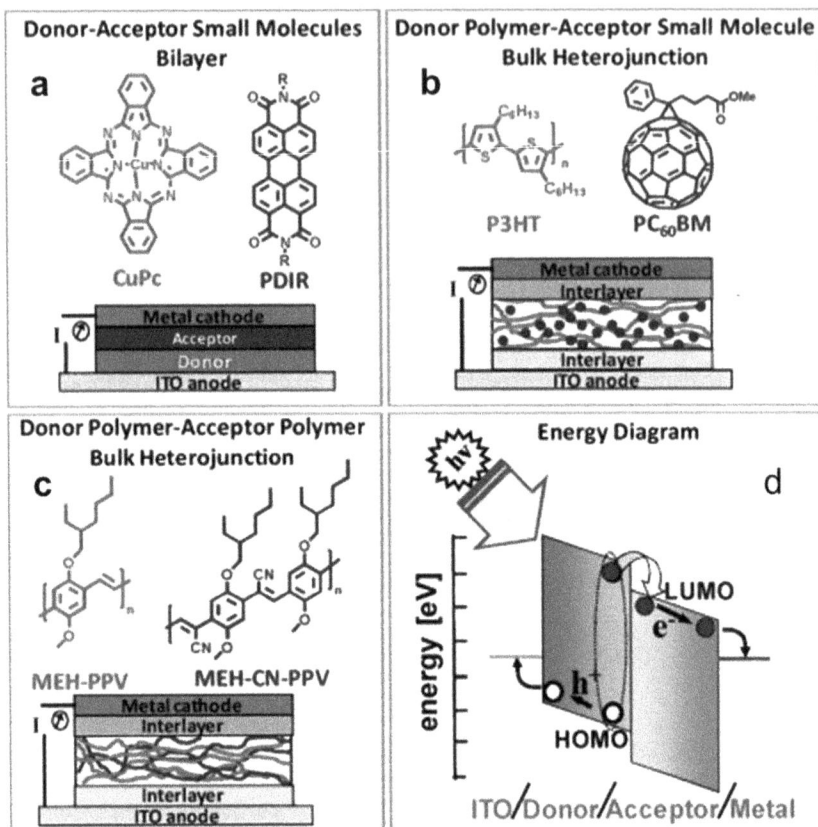

FIGURE 5.4 Structure and descriptive photoactive materials of (a) donor-acceptor small molecule single heterojunction, (b) donor polymer-acceptor small molecule bulk heterojunction, and (c) donor polymer-acceptor polymer bulk heterojunction photovoltaic cells. (d) Energy levels of the materials where light absorption/excitation dissociation/charge collection takes place [38].

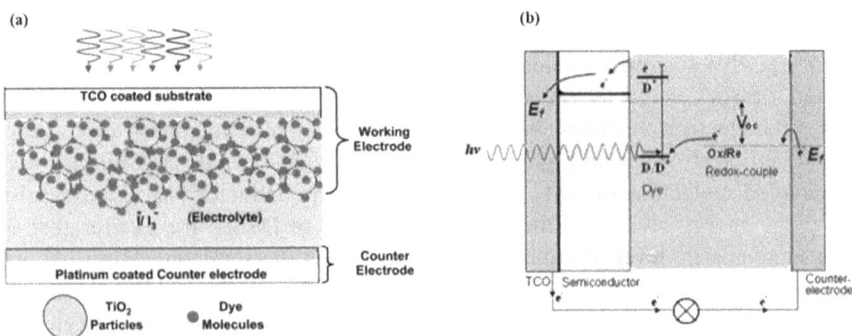

FIGURE 5.5 (a) Structure and (b) principle of a dye sensitized solar cell [43].

from the anode to cathode [42,43]. The photoanode electrode in DSSC consists of a semiconductor film of TiO_2, sensitized with a monolayer of dye (normally Ru-complex) [43]. Exhibiting some TiO_2-based materials is broadly used as photoanode materials in DSSCs due to their unique properties such as cost-efficiency, ecofriendliness, larger surface area for the flow of electrons, and a wide band gap of 3.2 eV [44].

5.3 TITANIUM DIOXIDE

5.3.1 BACKGROUND

Titanium dioxide (TiO_2) is a recognized and well investigated material owing to its remarkable properties such as high strength, toxicity-tolerance, biocompatibility, photosensitivity, and electrical conductivity [45]. TiO_2 exists primarily in three distinct crystallographic forms which are rutile, anatase, and brookite, as depicted in Figure 5.6 [46]. Rutile (Figure 5.6a) has proven to be the form of TiO_2 with the highest strength and was discovered by Wermer [47] (and the name rutile is derived from Latin meaning deep red color), and transmits light of the UV-visible spectroscopy. Anatase (Figure 5.6b) was earlier called octahedrite and named by Hauy in 1801 [47]. The word anatase comes from the Greek, anatasis, meaning extension [48]. Figure 5.6c) shows the brookite structure of TiO_2, and its name is in honor of the English mineralogist H.J. Brook in 1825 [49]. It can be seen that the crystal structure of brookite is dark brown to greenish black [49]. Crystal forms include the typical tubular to platy crystal with psuedohexagonal structure [45–49]. After heating, the anatase and brookite forms of TiO_2 are converted to a thermal stable rutile form [50]. Table 5.1 presents some of the physical and structural characteristics of the anatase and rutile forms of TiO_2 [51].

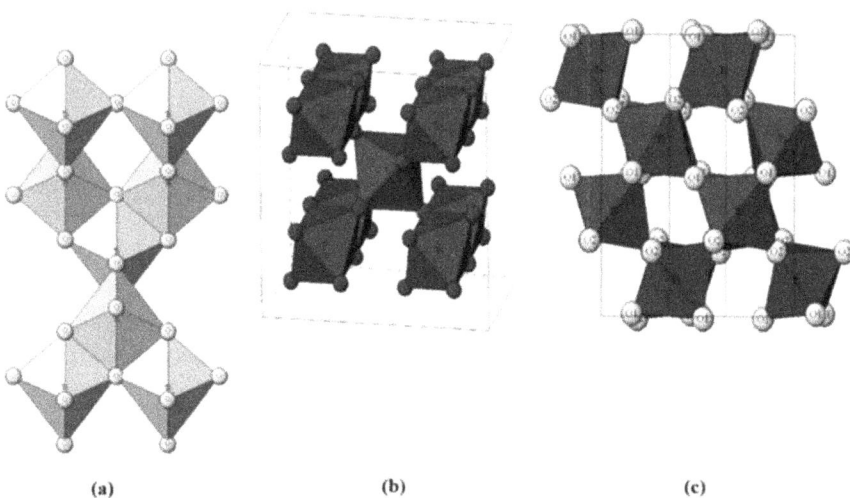

FIGURE 5.6 Crystalline structures of (a) anatase, (b) rutile, and (c) brookite forms of TiO_2 [49].

TABLE 5.1

Physical and Structural Characteristics of Anatase and Rutile Phase of TiO$_2$

Property	Rutile	Anatase
Molecular weight (g/mol)	79.88	79.88
Crystal structure	Tetragonal	Tetragonal
Density (g/cm^3)	4.13	3.79
Light absorption (nm)	$\lambda \leq 415$ nm	$\lambda \leq 385$ nm
Specific gravity	4	3.9
Mohr's hardness	6.5–7	5.5
Refractive index	2.75	2.55
Dielectric constant	114	31
Melting point (°C)	1825	1825
Boiling point (°C)	2500–3000	2500–3000
Lattice constants (Å)	a = 4.5936	a = 3.784
	c = 2.9587	c = 9.515
Ti—O bond length (Å)	1.949 (4)	1.937 (4)
	1.980 (2)	1.965 (2)

5.3.2 ELECTRONIC ABSORPTION OF TITANIUM DIOXIDE

Titanium dioxide is known to be a semiconductor; therefore, a suitable source of light can stimulate the transfer of electrons from the valence band to the conduction band [45]. The band gap energy is the energy required to encourage the photogeneration of electron-hole pairs [52]. The solar energy can be used as the primary source of radiation to activate the TiO$_2$ catalyst when the band gap energy is small enough [53]. On the other hand, ultra-violet (UV) radiation may be used in a high band gap. It was reported that band gap energy of TiO$_2$ is between 3 and 3.2 eV [52,53], hence, light radiation (wavelength <400 nm), known as ultraviolet A (UVA), will be required for its activation. Therefore, using TiO$_2$ as photocatalysts under sunlight radiation leads into low performance because natural light comprises 4–5% of UVA radiation [54]. As a result, the TiO$_2$ band gap must be reduced by using a suitable dopant in order to benefit from the remaining spectrum of sunlight radiation, the visible region in solar cells [53,54]. For example, the electronic absorption properties of the pure TiO$_2$ and Fe-doped TiO$_2$ materials and the optical energy band gap of these materials are presented in Figure 5.7 [55]. As expected, it was seen that the Fe-doped TiO$_2$ possessed an increased absorption with a red shift with respect to pure TiO$_2$. Additionally, the energy band gap (Eg) of the TiO$_2$ and Fe-doped TiO$_2$ was measured using the Kubelka-Munk equation [55], as shown in the inset of Figure 5.7. The Eg value of Fe-doped TiO$_2$ was close to 3.15 eV, which was smaller than 3.26 eV of pure TiO$_2$. These properties of red shift and smaller band gap of the Fe-doped TiO$_2$ material show that the absorption of light with a wider wavelength range ($\leq$394 nm) was observed using Fe-doped TiO$_2$ sample while pure TiO$_2$ absorbed at $\leq$380 nm [55].

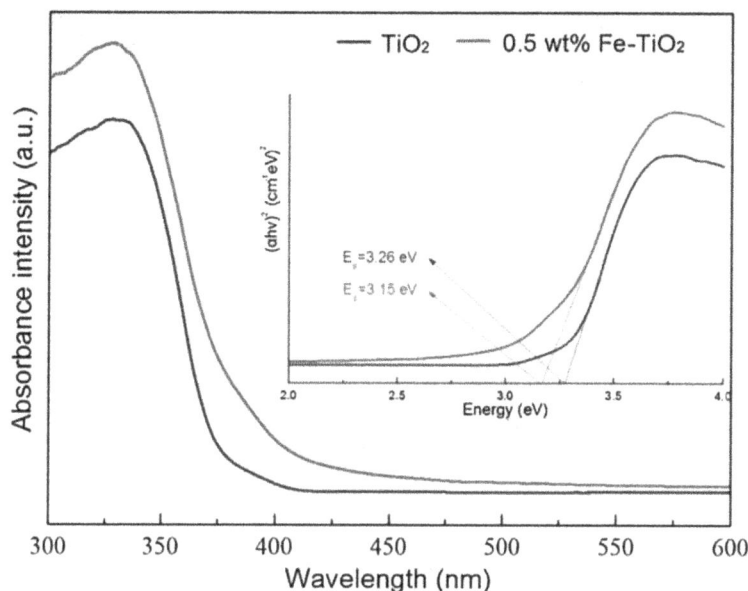

FIGURE 5.7 UV-vis spectra of TiO_2 and its doped derivative of 5% Fe-TiO_2. Inset: Kubełka-Munk Plot [55].

In addition, Momeni et al. [3] reduced the band gap of anatase TiO_2 by introducing Fe metal as an interesting dopant because its electronic configuration and ion radius is close to Ti^{4+}. They showed that Fe-doped TiO_2 decreased the band gap from 3.2 to 2.85 eV.

5.3.3 SYNTHESIS OF TITANIUM DIOXIDE

Most approaches for the synthesis of different phases of TiO_2 have been investigated [45–56]. It was that the gas phase (physical) and solution (chemical) synthetic methods are the common ones in making TiO_2 nanostructures [56]. Physical route yields enormous amounts of material; however, their resolution is much finer from micrometers upwards [56]. Physical vapor deposition (PVD) procedure is one of the physical methods in making TiO_2. It was reported in preparation of TiO_2 nanowires, a layer of TiO_2 nanopowders deposited on the substrate followed by growth of TiO_2 into nanowires [56]. For example, Figure 5.8 presents a model SEM image of nanowires of TiO_2 prepared via PVD method by Wu et al. [57]. They observed that TiO_2 nanowires with high density were uniformly grown over the entire substrate. Moreover, the image in the inset (Figure 5.8b) in Figure 5.8a indicates that the nanowires with diameters of 60–100 nm and lengths of 1–2 μm.

Chemical vapor deposit (CVD) is also as a physical method, which is the promising processes of synthesizing rutile TiO_2. Due to its high-quality film with controllability of micro and orientation at high deposit rates [56]. In addition, during the CVD

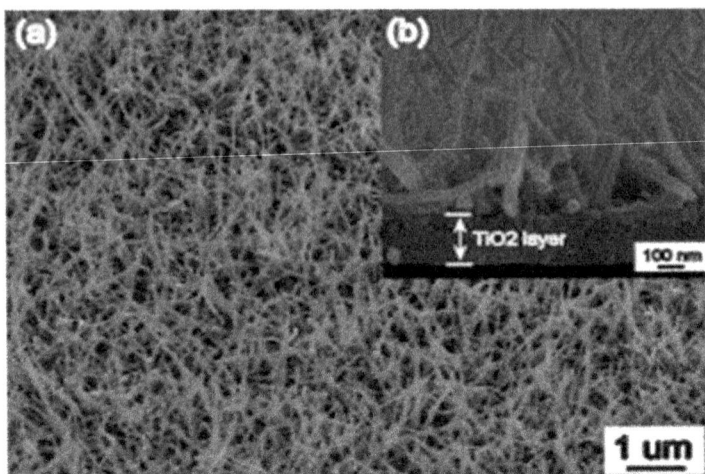

FIGURE 5.8 Morphological image of the TiO_2 nanowire arrays prepared by the PVD method observed by SEM technique [57].

process, the morphological orientation of the film is affected by various parameters such as laser power (P_l) and total pressure (P_{tot}) in the CVD chamber for the film with random orientation [56].

They discovered that there are two reaction mechanisms for titanium (iv) isopropoxide that occur in the reaction. The reactions are presented in Equation 5.3.

Pyrolysis

$$Ti(OC_3H_7)_4 \rightarrow TiO_2 + 4C_3H_6 + 2H_2O \tag{5.3}$$

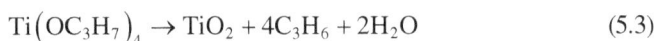

In this reaction, gases exposed to temperatures where the pyrolysis reaction can take place using liquid precursor and titanium tetraisopropoxide (TTIP) in a mixed gas (consisting of helium/oxygen atmosphere) leads to preparation of bushy crystalline of TiO_2 [56].

The chemical methods are capable of coating complex shapes, governing on stoichiometry, and synthesis of composite materials [56]. They can approach the atomic layer limit through the preparation of nanostructure materials (in <100 nm or <10 nm scale) [56]. Chemical methods are easy to prepare, cost effective, and environmentally friendly [51]. These methods include hydrothermal, solvothermal, and sol-gel methods [56]. Hydrothermal synthesis of TiO_2 is generally conducted in an autoclave of steel pressure vessels and sometimes in the presence of Teflon liners. The synthesis is done under the control temperature and pressure in an aqueous solution. The solvothermal method is similar to the hydrothermal one, however, the former uses non-aqueous solvent. In the solvothermal methods, greater temperatures than those used in hydrothermal technique are utilized. They control morphology, size distribution, and shape of TiO_2 nanoparticles a lot better than hydrothermal method [58].

Ramakrishnan et al. [58] synthesized TiO_2 nanoparticles using the modified solvo-thermal method for the photodegradation and PV applications (Figure 5.9). They showed that the synthesized TiO_2 nanostructures were used for photocatalytic and PV applications with photodegradation efficiency of $\leq 97\%$ for organic dyes, respectively overall power conversion efficiency of 5.92% in DSSC (Figure 5.9).

In sol-gel synthesizing methods, there are many sol-gel methods such as non-alk-oxide and alkoxide routes. These sol-gel methods have advantages over other fabrication techniques for high purity, homogeneity, felicity, and flexibility [59]. In addition, these routes are the methods that are simple, they are easy to introduce dopants in a larger concentration, have stoichiometric control, control over the composition, ability to coat on larger materials and complexes [56,59]. The synthesis involves the use of titanium tetra isopropoxide solution as a precursor in acidic catalyze hydrolysis (solvents such as hydrochloric, sulfuric, and glacial acetic acids) and in the presence of distilled water or ethanol in order to speed up the hydrolysis processes, as shown in Equation 5.4 and followed by condensation polymerization reactions to give gels as shown in Equation 5.5 [56].

Hydrolysis:

$$Ti(OR)_4 + 2H_2O \rightarrow Ti(OH)_4 + 4ROH \qquad (5.4)$$

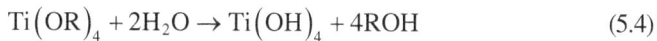

Condensation:

$$Ti(OH)_4 \rightarrow TiO_2 + 2H_2O \qquad (5.5)$$

It was seen that the sol particle size depends on the composition of solution, temperature and pH, and Ti-O-Ti chain development is preferred with a small amount of

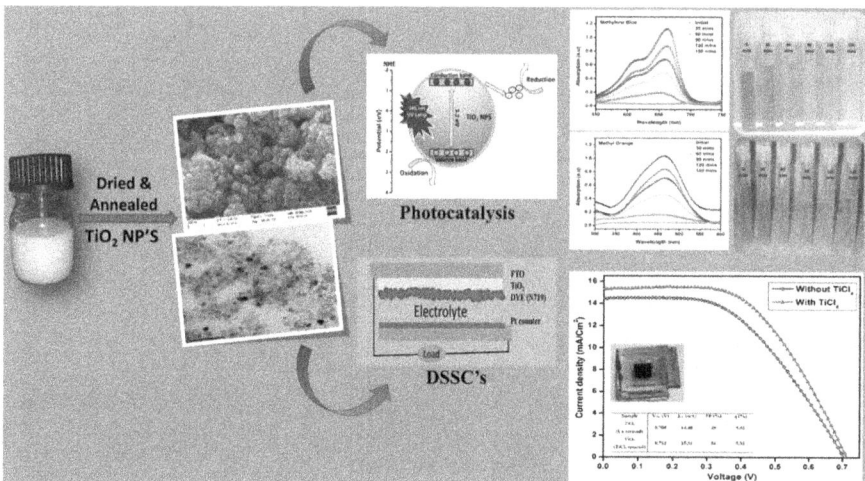

FIGURE 5.9 Synthesis and applications of TiO_2 using solvothermal method [58].

water, additional titanium alkoxide in the reaction combination, and stumpy hydrolysis rates [56].

5.3.4 TITANIUM DIOXIDE COMPOSITES

Doping TiO_2 using metals, non-metals, and metal complexes is another way of improving the performances on PV solar cells. Higashimoto et al. [20] investigated the doping of nitrogen molecules on TiO_2 (for the PV application). They discovered that the formation of vacancies is exhibited to the n-type [60]. Duarte et al. [61] also investigated the characterization of nitrogen-doped TiO_2 thin film for the PV application (Figure 5.10). They observed that the cells attained with conventional DSSCs exhibit shorter short-circuit photocurrent as compared to cells attained through doping of TiO_2 with nitrogen.

Furthermore, Wang et al. [62] reported on the influence of nitrogen doping on device preparation for TiO_2-based solid state DSSCs. They showed that the PCE % of the material doped with 0.1 wt.% nitrogen content was 4.1% (Table 5.2). On the other hand, carbon black dopants were used on TiO_2 to its abundance, low cost, and high catalytic activities with chemical stability against iodine redox couple [63]. Lim et al. [64] reported on the study of counter electrode material made of TiO_2/carbon black composition for DSSCs. They presented that the surface morphologies of the synthesized materials suggested more active sites and high porosity of 75 nm sizes. Their material with low carbon black content performed better for PV activity with the PCE of 7.4% (Table 5.2). In the case of metal and silicon dopants, the integration of transition metals and SiO_2 into TiO_2 contributes to the formation of new energy levels on the brink of the conduction band (CB) as a result of electrons in their d-orbitals [65,66]. The transition metals such as Mg, Nb, Y, Al, and Zr as dopants onto TiO_2 material have been used in solar cells [67]. The Mg-doping onto TiO_2 in DSSCs showed an improvement in V_{OC} value due to a higher CB energy and reduced recombination [68]. Nb-doping of TiO_2 solar cells led to an increment of electron

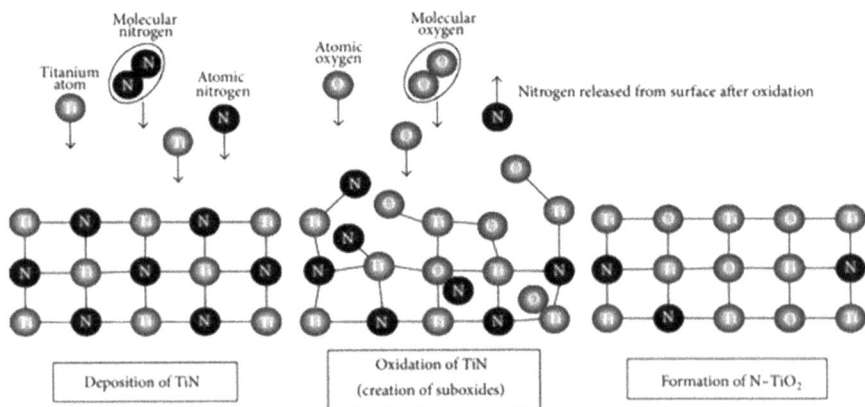

FIGURE 5.10   Tetrahedral structure of TiO_2 doped with nitrogen [61].

TABLE 5.2
Photovoltaic Parameters of TiO_2 Composites in DSSCs

Dopant	Isc (mA/cm²)	Voc (V)	FF (%)	η(%)	Ref.
N	8.31	0.79	62.0	4.1	[62]
C	15.5	0.71	67.4	7.4	[64]
Zr	19.35	1.02	76	15.1	[70]
Nb	17.7	0.70	63.0	7.8	[69]
Ga	13.4	0.76	79	8.1	[71]
Y	15.9	0.74	77	9.0	[71]
Bi	7.71	0.59	46.0	2.11	[73]
Ta	8.17	0.79	46.3	2.99	[74]
SiO_2	0.55	0.60	54	0.18	[66]
MgO	11.7	0.72	53.5	4.5	[75]
Nb_2O_5	9.32	0.71	68	4.5	[76]
$SrTiO_3$	10.2	0.71	58.4	4.39	[77]
$CaCO_3$	21.9	0.67	66.1	9.68	[78]
$Mg(OH)_2$	6.61	0.76	74.0	3.7	[79]
Graphite	12.6	0.82	54	5.76	[80]
Graphene oxide (GO)	13.7	0.69	69	6.49	[81]
Reduced GO	18.4	0.68	61	7.68	[82]
MWCNT	15.3	0.73	71	7.9	[83]

injection and transport, resulting in a higher Isc value with an overall 7.8% energy-conversion efficiency on 5.0 mol% Nb-dopedTiO_2 [69].

A high PCE of 15.1% with a good fill factor of 76% was achieved at 2 mol% Zr doping onto TiO_2 as compared to 12.3% of the pure TiO_2 [70]. The Mott-Schottky band gap showed that the elevated Voc value was owing to the upshifting of flat band potential of TiO_2 from −4.15 eV to −4.02 eV after doping [70]. The dopants of gallium doping and yttrium in the TiO_2 material showed a noticeable increment of the power conversion efficiency 8.1% for Ga doping and 9.0% for Y doping with simultaneous increases in other PV parameters (Table 5.2) [71]. Ko et al. [72] reported that Al-dopant has the ability to decrease the number of oxygen vacancies, and recombination as well as improving conductivity. They showed that introducing Al doping to TiO_2 DSSCs resulted in an overall increase in Isc values and also an enhancement of the long-term stability. An'Amt et al. [73] investigated the PV behavior of DSSCs based on Bi-TiO_2 materials. They obtained the V_{oc} values of 0.59 V with the Isc value of 7.71 mA/cm², and the efficiency of 2.11% as an increase of 77% with respect to pure TiO_2 materials. Under optimized conditions, 0.1 wt.% Ta doped TiO_2 nanotubes in DSSCs possessed a power efficiency of 2.99 as increased of 25% compared to non-doped TiO_2 nanotubes [74]. In addition, MgO coated on TiO_2 surface material showed increase in the power conversion efficiency of 4.5% (Table 5.2) [75]. This indicates a 45% increment as compared to the uncoated TiO_2 electrode due to improved specific surface area [75]. The Nb_2O_5 coated between FTO and TiO_2 nano-crystalline film as a potential blocking layer showed improvement of Voc that led to a better conversion efficiency of dye-sensitized TiO_2 solar cells [76]. The increment of PV parameters indicated that the Nb_2O_5 layer was a good blocking layer at FTO and TiO_2 interface to suppress the recombination processes [76]. The photovoltage current response of $SrTiO_3$-doped TiO_2 possessed an increase with overall conversion efficiency of the solar cell of 15% [77]. Lee et al. [78] reported that nanoporous $CaCO_3$-doped TiO_2 with a higher specific surface area led to the larger amount of dye adsorption that increased the PV parameter of the DSSCs (Table 5.2). Yum et al. [79] investigated the nature of the improvement of PV parameters on $Mg(OH)_2$-coated TiO_2 DSSCs, whereby little influence of $Mg(OH)_2$ coating on the diffusion coefficient was observed in relationship with the rise in Voc value to 0.76 V, and efficiency of 3.7%. Moreover, Hu et al. [80] presented the *current–potential (I-V)* curves of graphite-based TiO_2 DSSCs. They observed that the increment of Voc $_c$ value from 0.821 to 0.835 V with the addition of graphite from 0.00 to 0.02 wt% in the composite. The Isc of 12.59 mA/cm^2 was obtained only 0.01 wt% graphite content with power efficiency of 5.76 (Table 5.2) [80]. The effect of graphene oxide was also investigated on TiO_2, the photoelectric conversion efficiency of the DSSC based on graphene/TiO_2 composite film increased from 5.52 to 6.49%, with an increase of η by 17.6% [81]. This showed that graphene reduces the charge recombination rate and electrolyte-electrode interfacial resistance as well as improvement of the transport of electrons that led to rise in the power efficiency [81].

Cheng et al. [82] reported the preparation of anatase TiO_2@reduced graphene oxide hybrids for high performance of DSSC applications. They observed that the PCE became lower at the low reduced graphene content and showed a high PCE of 7.68% for the hybrid 1.6 wt.% reduced graphene content. Dembele et al. [83] investigated the double layer multiwall carbon nanotubes (MWCNTs)-TiO_2 composite for DSSCs. They observed a high PCE of 8.1% at 0.010 wt.% MWCTs concentration. An increase in concentration from 0.010 to 0.250 wt.%, resulted in decrease in PCE to 1.1% [83]. Moreover, it was reported that the conversion efficiency of MWCNT/TiO_2 DSSCs with conventional Pt counter electrode was increased from 6.51 to 7.00% as compared to pure TiO_2 [84]. On the other hand, MWCNT/TiO_2 mesoporous based DSSCs with FeS_2 counter electrode possessed a high solar conversion efficiency of 7.27% as a result of higher surface area and the good catalytic activity of FeS_2 as compared to the conventional Pt counter electrode [84]. As stated previously, a TiO_2-based photoanode has the disadvantages of weak interconnectivity and poor light-scattering capillarity [85]. This may overcome incorporation of metal organic frameworks (MOFs) into TiO_2 photoanode in order to improve structural properties [86].

5.4   METAL-ORGANIC FRAMEWORKS

5.4.1   BACKGROUND

MOFs are a class of inorganic-organic materials and made from metal cation and organic linkers [87–95]. As shown in Figure 5.11, the metal cores serve as connecting

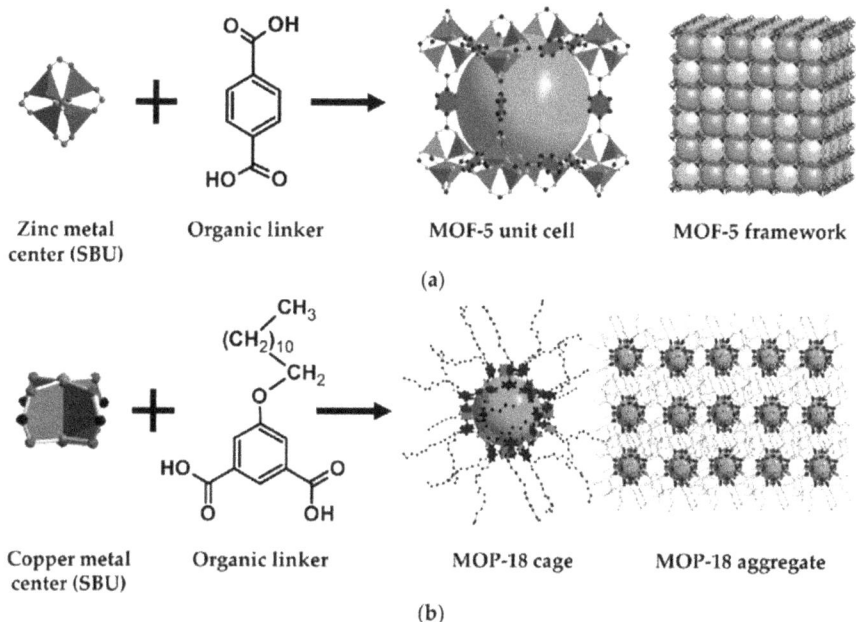

Zinc metal center (SBU) Organic linker MOF-5 unit cell MOF-5 framework

(a)

Copper metal center (SBU) Organic linker MOP-18 cage MOP-18 aggregate

(b)

FIGURE 5.11 (a) Preparation of metal-organic framework from a zinc metal salt and organic linker of terephthalic acid; (b) preparation of metal organic framework from a copper dimer and alkylated isophthalic acid [96].

points and the organic ligands function as bridging molecules [96]. These two constituents are connected by covalent bonds in the formation of the MOF structure [91–93]. The MOFs were first discovered through the study of zeolites [96]. The name metal-organic framework was given by Yaghi during synthesis of copper-4,4'-bipyridal compound that unveiled lengthy metal-organic interactions [96,97]. Basically, MOFs fit in to the family of coordination polymers. They are polymers of two- or three-dimensional crystallized networks and possess highly porous properties with respect to other coordination polymers [92,96]. It is well known that MOFs are porous solids and the precise positions of all atoms in the MOF structure can be recognized and associated with the dignified properties [92–96]. In porous solids, permanent pores are created, and they saturate the structure of the material. The pores are large enough to require guest molecules to diffuse into the structure or to be removed from the channels by heating or vacuum [98].

In addition, the MOF polymers are extremely tunable by altering the metal cores or organic linkers to attain an appropriate MOF material with the allowed structures and functionalities for definite applications [92]. It was seen that the porosity of the reported MOFs varies between 20 and 95% [98]. Figure 5.12 shows some MOF materials with different pores. Their porosity associated to high surface areas, permitting improved adsorption capacity of guest molecules. As a result, MOFs have attracted significant attention owing to structural properties [92, 98–100].

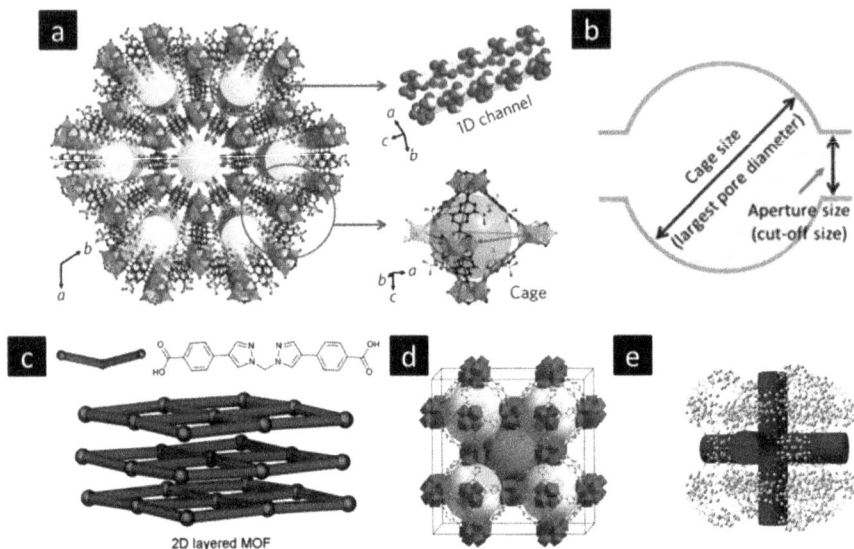

FIGURE 5.12 MOFs with different pore geometries: (a) BUT-8(M) contains the 0D cavity (light blue) and paralleled 1D channel (light green). (b) A schematic illustration for the 0D cavity showing the cavity diameter (i.e., cage size) and opening aperture size (i.e., window size). (c) A schematic diagram for 2D layered MOF showing interlayer spaces. (d) Two types of cavity (red and yellow) contained in a DUT-67 unit cell. (e) Intersecting channels (blue) along <100> in a DUT-67 unit cell [100].

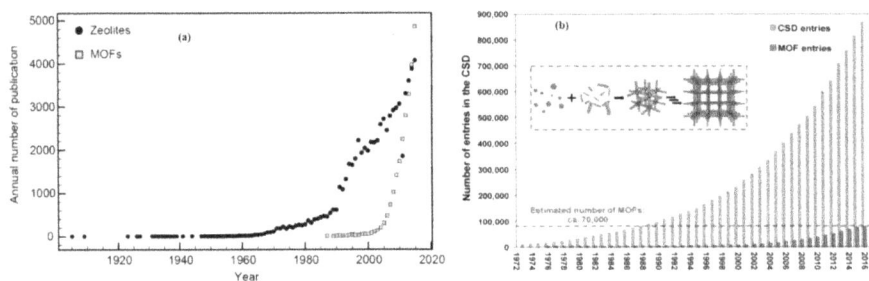

FIGURE 5.13 (a) Graph of number of MOFs (as compared to zeolites) published per year [101]; and (b) the reported MOF crystal structures in the CSD since 1970 [102].

At beginning of the 21st century, there is exponential growth in interest in the synthesis of MOFs and their publications are shown by the Figure 5.13a as compared with zeolitic materials [101]. These large numbers of reports, about 40,000 papers for MOFs and their rates of increase rationalize the significance of the overview based on bibliometrics in this report with respect to each aspect. Moreover, Figure 5.13b shows that over 69,000 MOFs have been reported in the Cambridge Structural Database (CSD) from 1970 to 2016 [102]. This intensive research interest concentrates on MOF design and applications such as hydrogen, gas sensing, and storage [87–95]. In the meantime, the pore dimension of MOFs can be bigger than inorganic nanoporous materials in general and approaches 100 Å [103] Amongst the synthesized MOFs, MOF-5, ZIF-8, HKUST-1, UiO-66, and NU-1000 have been widely known and extensively investigated due to their low cost and ease of preparation [104].

5.4.2 PROPERTIES OF MOFs

The unsaturated metal sites in MOF materials have remarkable influence on their structural properties. The metal elements as the center connector are customarily chosen from copper, zinc, manganese, chromium, cobalt, etc. [100]. The metal ions of MOFs act as Lewis acids and activate the attached organic substrate for successive organic transformation [105]. The partial positive charges of the metal cores in MOF structures have been shown to possess the ability to increase adsorption properties [101,102,105]. The organic linkers (Figures 5.14 and 5.15) used in the formation of MOFs are usually multidentate organic molecules with O- and N-donor atoms. It is unusual for linkers in coordination polymers not to contain aromatic rings [106]. It was seen that the aromatic rings based organic linkers in MOFs help to maintain the structural integrity of the complex and direct the geometry of the frameworks [106,107]. The surface area and pore volume of MOFs can be controlled by modifying organic ligands which act as spacers and create an open porous structure [107]. A recipe of inorganic and organic structures can lead to materials with unique properties [106]. Alternatively, the choice of the primary building units makes it possible to vary some parameters, including the pore size (to increase pore diameter to 98 Å), density (to decrease to 0.126 g cm^{-3}), as well as the specific surface area (up to $1000 \pm 10,000$ m^{2}g^{-1}), which opens up new ways to produce materials with tailored physical and chemical properties [100–106]. Based on porous structure, larger specific surface area, stability, and ability to be functionalized, MOFs are currently good candidates for practical applications such as gas separation, catalysis, and ion exchange reactions [87–95].

Moreover, secondary building units (SBUs) play a significant part as they dictate the final topology of MOFs. The geometry and chemical characteristics of the SBUs and organic ligands lead to the prediction of the design and preparation of MOFs [108]. Ren et al. [109] reported that under some conditions, multidentate linkers could aggregate and lock metal ions at certain positions, forming SBUs. The SBUs will be subsequently merged by rigid organic linkers to yield MOFs that exhibit high structural stability [108,109]. Zhao et al. [110] and Huang et al. [111] reported a comparative study of MOF nanosheets based on the design and theoretical calculations. Huang et al. [111] attained an MOF of high stability under the thermodynamic and kinetic conditions (Figure 5.16a). Figure 5.16b presents the electrostatic surface potential (ESP) diagrams calculated using density functional theory (DFT), the activities of the three-coordinated Cu (0.056 au) with more σ-holes were higher than those of pristine four-coordinated Cu (0.040 au) [111]. Furthermore, the frontier molecular orbitals (MOs) in Figure 5.16c show that the 3-coordinated Cu sites have the smallest band gap of 1.38 eV [111]. On the other hand, the 4-, 2-, and 1-coordinated Cu sites have band gaps of 1.43, 4.51, and 3.02 eV respectively [111].

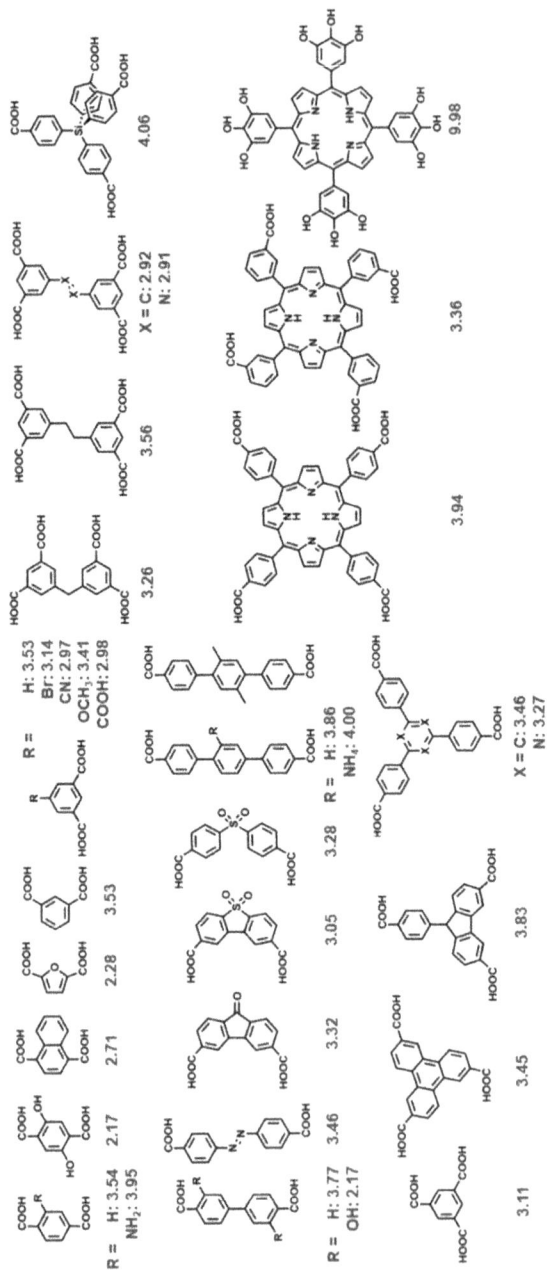

FIGURE 5.14   Some organic linkers used in the preparation of MOF with O donor ligands and their pKa values [106].

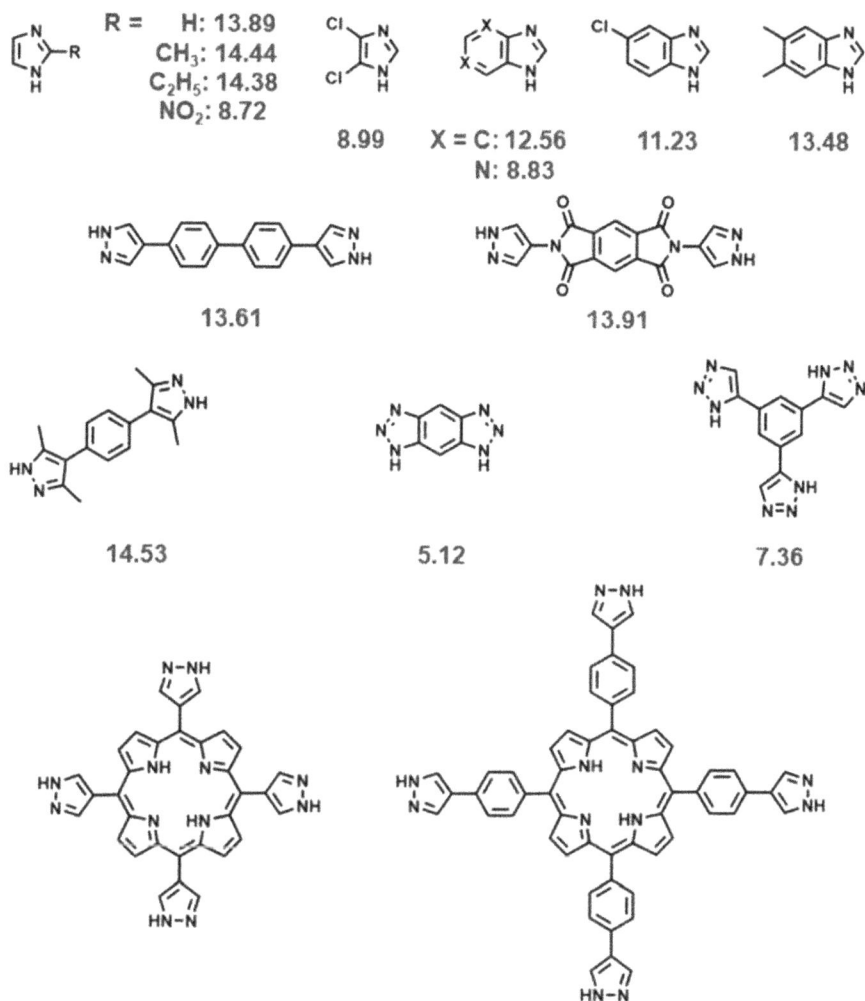

FIGURE 5.15 Some organic linkers used in the preparation of MOF with N donor ligands and their pKa values [106].

5.4.3 Synthesis of MOFs

Figure 5.17a illustrates a summary of several synthetic methods for MOF preparations. The main aspects of MOF preparation are to institute preparation conditions that can produce defined inorganic building blocks without decomposition of the organic linker [112]. In this chapter, the focus is on the selected preparation of the MOF structures illustrated in Figure 5.17b [113].

MOFs have been prepared effectively through the solvothermal method [114]. During the heating process, coordination of the primary carboxylate linkers and metal ions can result in many metal-oxygen-carbon clusters, known as SBUs, as

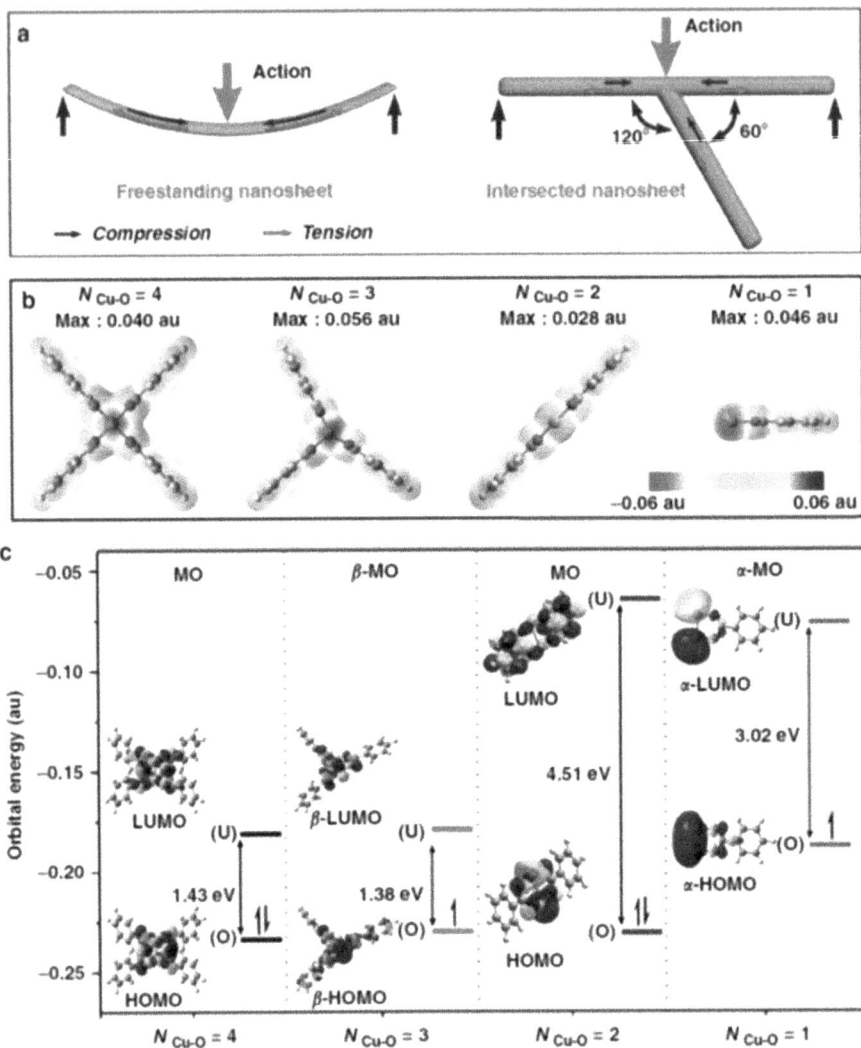

FIGURE 5.16 (a) Comparison between displacements for MOF nanosheets and intersected nanosheets when force is applied to them; (b) ESP maps of paddlewheel Cu_2 clusters in MOF structures where NCu-O stands for the paddlewheel Cu_2 clusters with different coordination numbers (CNs) of Cu-O (4, 3, 2, 1); and (c) the theoretical calculated frontier orbital energy levels and molecular orbital (MO) diagrams in MOF structure [111].

discussed above, and settle into final orderly MOF network [108]. The imperative parameters in the synthesis of MOF are concentration, temperature, metal salt, organic linker, and suitable solvents [114]. The metal cores are coordinated covalently through organic linkers to form a so-called paddle wheel unit in a three-dimensional porous cubic network [106]. The solvothermal method (a method of

Solvothermal
Energy: Thermal Energy
Time: 48-96 hours
Temperature: 353-453 K

Electrochemical
Energy: Electrical Energy
Time: 10-30 mins
Temperature: 273-303 K

Microwave
Energy: Microwave Radiation
Time: 4 mins - 4 hours
Temperature: 303-373 K

Metal Salt
+
Ligand
+
Solvent

Mechanochemical
Energy: Mechanical Energy
Time: 30 mins - 2 hours
Temperature: 298 K

Sonochemical
Energy: Ultrasonic Radiation
Time: 30-180 mins
Temperature: 273-313 K

Slow Evaporation
Energy: No External Energy
Time: 7 days - 7 months
Temperature: 298 K

(a)

(b)

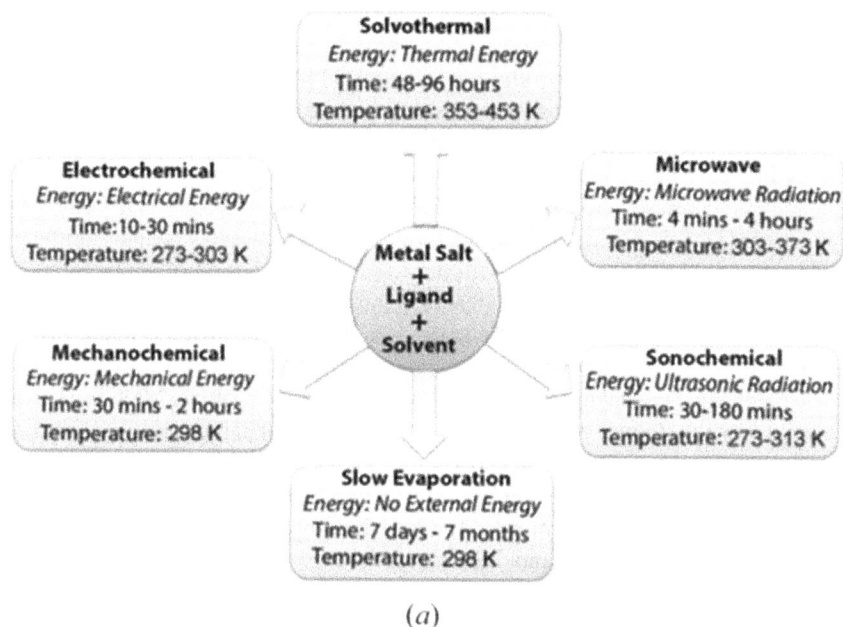

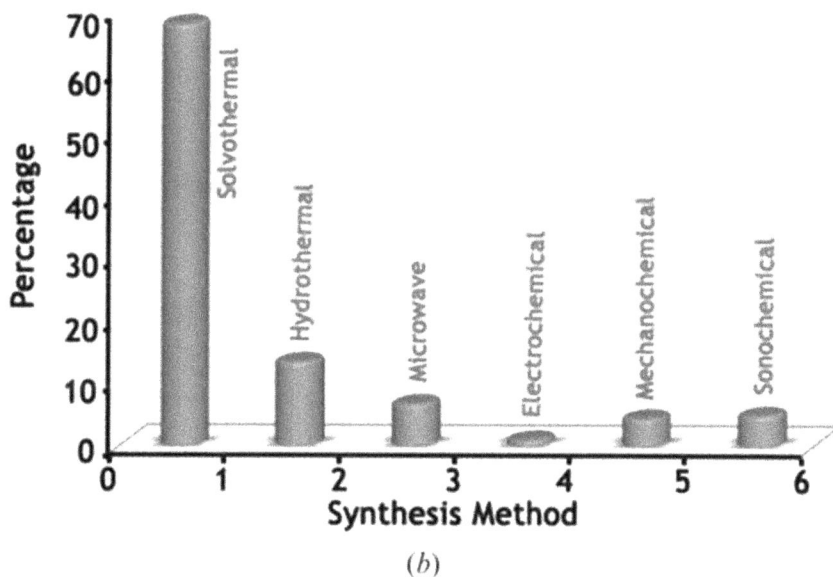

FIGURE 5.17   (a) Preparation conditions for synthesis of MOF and (b) percentage of MOFs synthesized using the various preparation methods [113], reproduced with permission of the International Union of Crystallography (http://journals.iucr.org/).

producing chemical compounds using a stainless steel autoclave) is presented in Figure 5.18a, it allows for precise control over the shape and size of the material to be synthesized, unlike the hydrothermal method [114].

The microwave- method has demonstrated to be an appealing route for quick preparation of nanoporous materials under hydrothermal conditions [114,115]. Besides fast crystallization and high efficiency, potential advantages of this technique include phase selectivity, narrow particle size distribution, and facile morphology control [114]. In the microwave method, the process involves applying heat to solution mixture of the substrates with microwaves for a period of about an hour to produce nanosized crystals (Figure 5.18b). This method produces excellence of the crystals which are normally the same as those by the regular solvothermal processes, but the synthesis is much quicker [115].

The mechanochemical method is a method which used to prepare MOF in a solvent-free environment (Figure 5.18c). Mechanical breakage of intramolecular bonds followed by chemical transformation takes place in a mechanochemical synthesis [116].

Quantitative yields of small MOF particles can be obtained in a short time, normally in the range of 10–60 min [114]. Occasionally, metal oxides are selected as starting materials in the mechanochemical method over metal salts due to the former

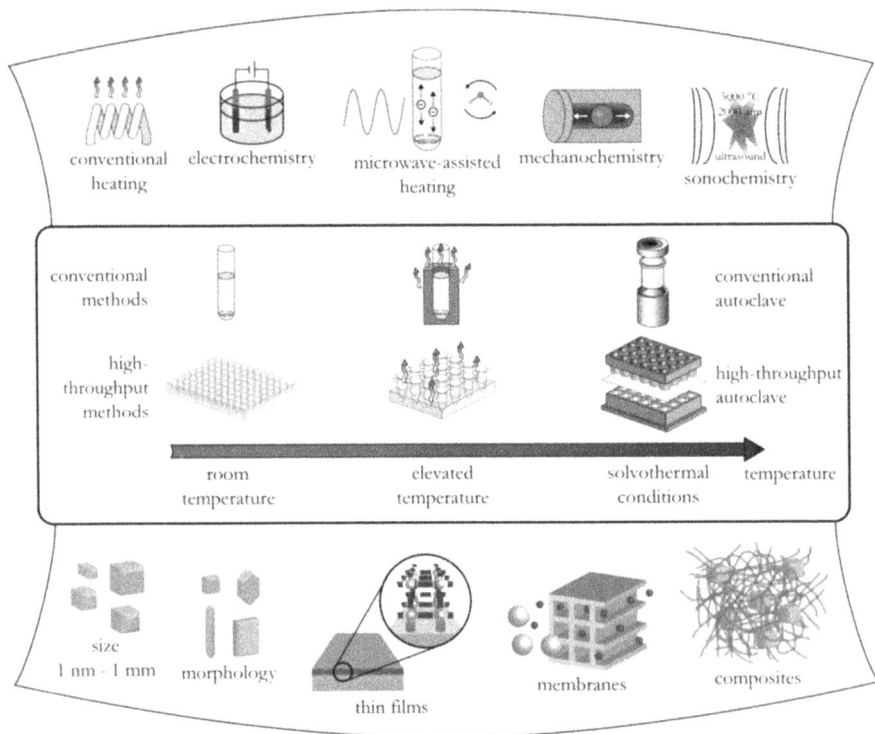

FIGURE 5.18 Various synthetic routes of MOF structures, operational temperatures related to each method and their final reaction products [114].

producing H_2O as a byproduct [114,116]. Currently, mechanochemical synthesis has been efficiently used for the rapid preparation of MOFs using liquid-assisted grinding (LAG), where a small amount of solvent is added into a solid reaction mixture [114,117]. This liquid could also work as a structure-directing agent, and ion-and-liquid-assisted grinding (ILAG) was used to be highly efficient for the selective construction of pillared-layered MOFs [118]. However, the mechanochemical preparation is limited to specific MOF types only and large amount of product is difficult to obtain [114]. The electrochemical preparation of MOFs (Figure 5.18d) uses metal ions continuously arranged as the anodes in an electrochemical cell with the organic linker dissolved in methanol as solvent and a metal cathode. In case of HKUST-1, trimesic acid dissolved in alcohol and copper cathode were used in electrochemical synthesis for period of 150 min at a voltage of 12–19 V and a current of 1.3 A, to attain a greenish blue precipitate [114]. After activation by heat, a dark blue-colored powder of octahedral crystals structure in nature with 0.5 to 5 μm size) possessing surface area of 1820 m^2/g was obtained [119]. Currently, HKUST-1, ZIF-8, Al-MIL-100, Al-MIL-53, and Al-MIL-53-NH_2 were reported to be prepared through anodic dissolution in an electrochemical cell [114]. The experimental parameters such as solvent, electrolyte, voltage-current density, and temperature on the preparation of MOF are vital in obtaining a desired structure [120].

5.4.4 TITANIUM OXIDE/METAL ORGANIC FRAMEWORK COMPOSITE FOR DSSC

It was seen that MOFs are promising crystalline materials and received attention in energy storage and conversion applications, because of their high surface area, variable pore sizes, and low density [13]. Currently, the concepts of MOF-based solar cells are emerging and some reports on applications of MOFs as candidate active materials in PVs can also be found [13,121–126]. However, a fully devised solar cell based on MOFs as active material for harvesting solar energy has been rarely investigated [123]. The performance of these cells based on commercial TiO_2/MOF and functionalized TiO_2/Al-MOF (BDC = organic linker) with 1,4-dimethoxybenzene (DMB) is summarized in Table 5.3. The device constructed with DMB@TiO_2/Al-MOF of lowest thickness, had an Isc value of 0.36 higher than TiO_2/Al-MOF cells prepared without DMB [126]. The feasibility of the Ru-MOF film as a sensitizer in a solar cell was investigated by Lee et al. [123] and it was demonstrated that the cell performance of Isc = 2.56 mA cm^{-2}, Voc = 0.63 V and efficiency = 1.22%. The thin layer of Cu-based MOFs (copper(II) benzene-1,3,5-tricarboxylate, HKUST-1) was fabricated using a layer-by-layer technique in TiO_2-based solar cells and it demonstrated the excellent performance with Voc = 0.49 V, Isc = 1.25 mA cm^{-2}, and power conversion efficiency = 0.26% [125]. Furthermore, Lee et al. [125] made a comparative study with the device without HKUST-1 sensitization under similar conditions, the device showed only 0.37 mA cm^{-2} of Isc and 0.06% of conversion efficiency. This was supported by EIS analysis, example in Figure 5.19a–d showing the Nyquist plots of the above cells under open circuit condition. The EIS revealed that iodine doping of MOF is very essential to reduce the charge-transfer resistance of the TiO_2/MOF/electrolyte interface, hence higher power efficiency was obtained [125].

TABLE 5.3

Photovoltaic Parameters of TiO$_2$/MOFComposites in DSSCs

Photoanode Material	Dye	Electrolyte	Voc (V)	Isc (mAcm^{-2})	η (%)	Ref.
TiO$_2$/Al-MOF	-		0.27	0.023	-	[126]
TiO$_2$/Al-MOF-DMB	-		0.36	0.036	-	[126]
TiO$_2$/Ru-MOF	-	I3/I_ redox	0.63	2.56	1.22	[123]
TiO$_2$/HKUST-1	-	I3/I_ redox	0.49	1.25	0.26	[125]
TiO$_2$/ZIF-8	N719	1DMPImI/LiI/ I2 /AN/TBP	0.75	10.28	5.34	[121]
TiO$_2$/ZIF-8	MK2	[Co(bpy)3](TFSI)3/LiTFSI/tert-butylpyridine (tBP) /AN	0.90	14.39	9.42	[122]
TiO$_2$/Zn-MOF	N719	PEGOPEG/NaI/I$_2$/EMITFMSI /AN	0.68	6.22	2.34	[13]
TiO$_2$/Ni-MOF	N719	PEGOPEG/NaI/I$_2$/EMITFMSI /AN	0.62	27.32	8.84	[127]
TiO$_2$/MIL-125	N219	PEGOPEG/NaI/I$_2$/EMITFMSI /AN	0.85	10.9	6.4	[128]

N719=D-tetrabutylammoniumcis-bis(isothiocyanato)bis(2,2'-bipyridyl-4,4'-dicarboxylato)ruthenium(II); MK2= 2-Cyano-3-{5'''-(9-ethyl-9H-carbazol-3-yl)-3',3'',3''',4-tetra-n-hexyl-[2,2':5',2'':5'',2''']-quater thiophen-5-yl] acrylic acid; AN=acetonitrile; TBP=4-tertbutylpyridine; DMPImI=1,2-dimethyl-3-n-propylimidazolium iodide; PEGOPEG=poly(ethylene oxide)–poly(ethylene glycol); and EMITFMSI= 1-ethyl-3-methylimidazolium bis(trifluoromethylsulphonyl)imide

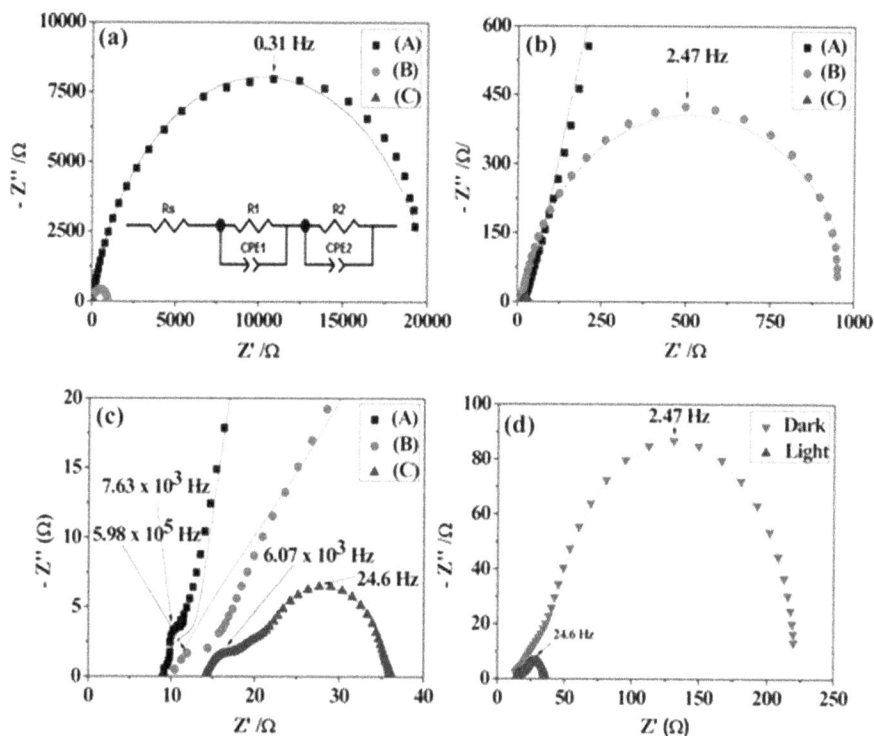

FIGURE 5.19 (a) Equivalent circuit model and the impedance spectra of the cell at open-circuit condition under illumination with simulated AM 1.5 solar radiations: (A) FTO/TiO$_2$/Cu-MOF undoped, (B) FTO/TiO$_2$, and (C) FTO/TiO$_2$/Cu-MOF iodine doped. (b) and (c) are zoomed impedance spectra of (a) at a higher frequency region. (d) Impedance spectra of the cell consisting of FTO/TiO$_2$/Cu-MOF iodine-doped photoanode under dark and illumination conditions. Squares, circles, and triangles are the experimental data, whereas the solid lines are the fitted curves [125].

On the other hand, Li et al. [121] reported fabrication of zeolite imidazolate framework (ZIF-8), a MOF reticulated using a nitrogen-containing ligand, 2-methylimidazole, and zinc metal, on top of the TiO$_2$ layer for DSSCs, for the first time. They controlled the growth time of the MOF in order to vary the thickness of the MOF layer which was demonstrated by the linear relationship of Voc and the thickness of the ZIF-8 layer on top of TiO$_2$. The I-V curves under dark conditions showed a higher voltage requirement for the motion of electrons for ZIF-8-coated TiO$_2$ electrodes [121]. This was supported by the charge transfer resistance determined using electrochemical impedance spectroscopy (EIS) possessing a higher resistance for TiO$_2$/ZIF-8 [121]. In addition, there was a decrease in both the Isc and η was achieved as the thickness of ZIF-8 layer increased even though the Voc was enhanced [121]. Gu et al. [122] prepared a thin blocking layer of a MOF ZIF-8 coated on the surface of mesoporous TiO$_2$ films for DSSCs using MK2 dye and tris(2,20-bipyridine) cobalt(II)/(III) redox in acetonitrile as an electrolyte. They observed that DSSCs with TiO_2/ZIF-8/ dye possessed increment of Voc value by 3% and Isc value by 21% with 9.42% power conversion efficiency (Table 3). Alwin et al. [13] reported for PV performance of TiO_2 aerogel-Zn-MOF $(Zn(1-L_{Cl})(Cl)/TiO_2)$ nanocomposite. The TiO_2 aerogel-Zn-MOF nanocomposite was used as photoanode in DSSCs and an overall power conversion efficiency 2.34% along with a short-circuit current density 6.22 mA cm^{-2} was achieved using N719 dye [13]. Ramasubbu et al. [127] fabricated DSSC based on three-dimensional TiO_2-Ni-MOF composite aerogels which was prepared through the sol-gel method. They observed an enhancement of PV performance of the TiO_2-Ni-MOF cell with respect to pure TiO_2-based QSDSCs using N719 dye (Figure 5.20). The maximum photo-conversion efficiency of the fabricated photoanode of 8.846% was obtained for 0.5% composite aerogel, which is ~30% higher than that of pure aerogel-based ones (6.805%) (Table 5.3) [127]. The increases in power efficiency enhancement were due to increased photocurrent density, reduced charge-transfer resistance, and suppressed electron recombination was confirmed by photocurrent

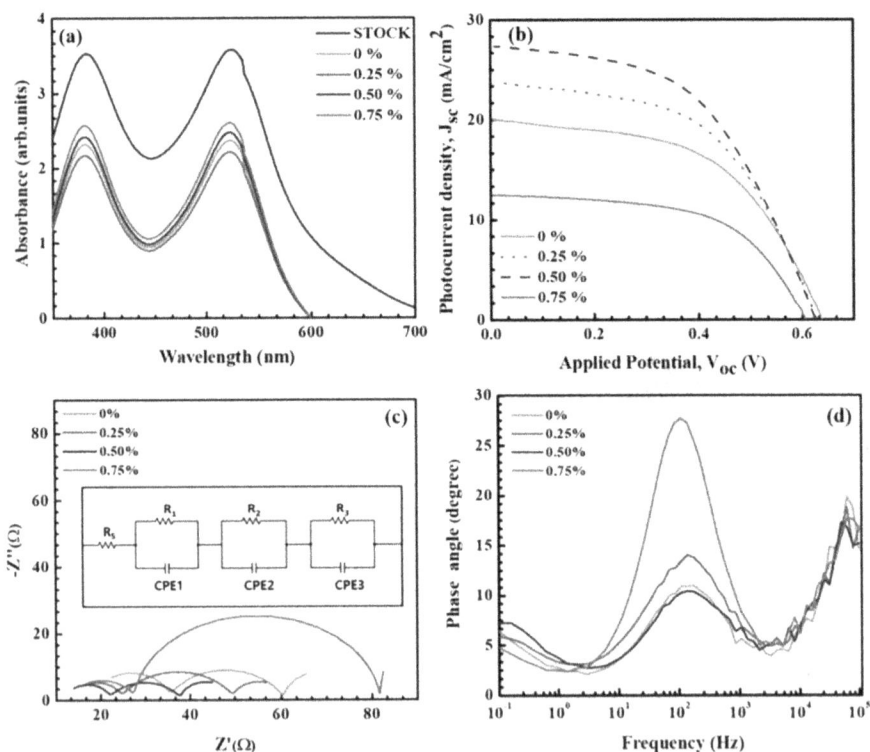

FIGURE 5.20 (a) Comparison between dye adsorption exhibited by pure TiO_2 aerogel, aerogel-coated photoanodes, and the blank. (b) Photocurrent density applied voltage curves (J-V), c) Nyquist plot of all the prepared QSDSCs, and (d) bode plot of all the prepared QSDSCs [127].

density-applied voltage curves and electrochemical impedance measurements as seen in Figure 5.20a–d. Furthermore, Vinogradov et al. [128] applied a single-step hydrothermal preparation for the first time to produce a MIL-125@TiO$_2$ composite for DSSC application. They have seen that the analysis of the photopolarization measurements for determination of optimal MOF concentration, the greatest photocurrent response of 39 µA cm^{-2} observed in 3% MOF-TiO$_2$ [128]. The MIL-125@TiO$_2$-based device showed high stability over time with a power efficiency of about 6.4% [128].

5.5   CONCLUSION

In conclusion, PVCs have demonstrated they are an alternative source of energy and used in many fields, such as mobile commerce, building integrated photovoltaics (BIPVs), and vehicles. It was seen that development of PVCs has reached efficiencies of over 14%, however, recombination losses, poor stability, and leakage of the electrolyte (in liquid-state devices) have hindered the enhancement of the device performance and commercialization. Recently, much hard work has been made to reduce the recombination losses within the PVC devices. Particularly, DSSCs based on TiO$_2$ have been shown to possess higher power conversion efficiency due to high surface area and also the ability of dye molecules to transport electrons quickly. However, TiO$_2$ materials have a large band gap value of 3.2 eV which is higher to absorb radiation at the visible region for practical usage which utilizes sunlight. In terms of low cost and light weight, organics, inorganics, and hybrid materials have brighter prospects than the semiconductors. In TiO$_2$ hybrid materials, PVCs have been prepared with the advantages of organics or inorganics selectively. However, it is not easy to increase the energy conversion efficiency. One potential solution is to use three-dimensional nanostructures, such as nanotubes, nanowires, composite films with metal, or MOFs. As of today, MOFs in PVCs still yield the higher energy conversion efficiency by supplying larger surface areas for the sensitizer adsorption. However, the electron transport and flexibility are sufficiently high in three-dimensional nanostructures. To improve the energy conversion efficiency of three-dimensional nanostructures based solar cells and take both advantages of organic- or inorganic-dye and MOFs on TiO$_2$, the light-harvesting ability must be reinforced. One way of achieving better light-harvesting ability is to look at other molecules such as phthlocyanine and polyaniline that can be adsorbed on the in the cavities of three-dimensional MOF structures and TiO$_2$ which is another currently active field of study.

ACKNOWLEDGMENT

MJH and KDM would like to thank the financial support from the National Research Foundation (NRF) under the Thuthuka programme (UID Nos. 117727 and 118113), Sasol Foundation for purchasing both STA and UV-vis instruments and University of Limpopo (Research Development Grants R202 and R232), South Africa.

REFERENCES

1. Sengupta D, Das P, Mondal B, Mukherjee K. Effects of doping, morphology, and film-thickness of photo-anode materials for dye sensitized solar cell application–A review. *Renew Sust Energy Rev* 2016;60:356–76.
2. Ramohlola KE, Hato MJ, Monama GR, Makhado E, Iwuoha EI, Modibane KD. State-of-the-Art advances and perspectives for electrocatalysis. In: Inamuddin, Boddula R, Asiri A (eds) *Methods for Electrocatalysis*. Springer: Cham, Switzerland. 2020. pp. 311–52.
3. Momeni MM. Dye-sensitized solar cell and photocatalytic performance of nano-composite photocatalyst prepared by electrochemical anodization. *B Mater Sci* 2016;39:1389–95.
4. Sánchez AS, Torres EA, Kalid RD. Renewable energy generation for the rural electrification of isolated communities in the Amazon Region. *Renew Sust Energ Rev* 2015;49:278–90.
5. Ye M, Wen X, Wang M, Iocozzia J, Zhang N, Lin C, et al. Recent advances in dye-sensitized solar cells: From photoanodes, sensitizers and electrolytes to counter electrodes. *Mater Today* 2015;18:155–62.
6. Ito S, Murakami TN, Comte P, Liska P, Grätzel C, Nazeeruddin MK, Grätzel M. Fabrication of thin film dye sensitized solar cells with solar to electric power conversion efficiency over 10%. *Thin Solid Films* 2008;516:4613–9.
7. Battaglia C, Cuevas A, De Wolf S. High-efficiency crystalline silicon solar cells: Status and perspectives. *Energy Environ Sci* 2016;9:1552–76.
8. Gong J, Darling SB, You F. Perovskite photovoltaics: Life-cycle assessment of energy and environmental impacts. *Energy Environ Sci* 2015;8:1953–68.
9. Daeneke T, Kwon TH, Holmes AB, Duffy NW, Bach U, Spiccia L. High-efficiency dye-sensitized solar cells with ferrocene-based electrolytes. *Nat Chem* 2011;3:211–5.
10. Daeneke T, Yu Z, Lee GP, Fu D, Duffy NW, Makuta S, et al. Dominating energy losses in NiO p-type dye-sensitized solar cells. *Adv Energy Mater* 2015;5:1401387.
11. Sun Z, Liao T, Sheng L, Kou L, Kim JH, Dou SX. Deliberate design of TiO_2 nanostructures towards superior photovoltaic cells. *Chem Eur J* 2016;22:11357–64.
12. Wang ZS, Kawauchi H, Kashima T, Arakawa H. Significant influence of TiO2 photo-electrode morphology on the energy conversion efficiency of N719 dye-sensitized solar cell. *Coord Chem Rev* 2004;248:1381–9.
13. Alwin S, Ramasubbu V, Shajan XS. TiO_2 aerogel-metal-organic framework nanocomposite: A new class of photoanode material for dye-sensitized solar cell applications. *B Mater Sci* 2018;41:27.
14. Chung HY, Lin CH, Prabu S, Wang HW. Perovskite solar cells using TiO_2 layers coated with metal-organic framework material ZIF-8. *J Chin Chem Soc* 2018;65:1476–81.
15. Chang TH, Kung CW, Chen HW, Huang TY, Kao SY, Lu HC, et al. Planar heterojunction perovskite solar cells incorporating metal-organic framework nanocrystals. *Adv Mater* 2015;27:7229–35.
16. Shen D, Pang A, Li Y, Dou J, Wei M. Metal-organic frameworks at interfaces of hybrid perovskite solar cells for enhanced photovoltaic properties. *Chem Comm* 2018;54:1253–6.
17. Li M, Wang J, Jiang A, Xia D, Du X, Dong Y, et al. Metal organic framework doped Spiro-OMeTAD with increased conductivity for improving perovskite solar cell performance. *Sol Energy* 2019;188:380–5.
18. Jin Z, Yan J, Huang X, Xu W, Yang S, Zhu D, et al. Solution-processed transparent coordination polymer electrode for photovoltaic solar cells. *Nano Energy* 2017;40:376–81.

19. Shastry TA, Balla I, Bergeron H, Amsterdam SH, Marks TJ, Hersam MC. Mutual photoluminescence quenching and photovoltaic effect in large-area single-layer MoS2–polymer heterojunctions. *ACS Nano* 2016;10:10573–9.

20. Bagher AM, Vahid MM, Mohsen M. Types of solar cells and application. *Am J Opt Photon* 2015;3:94.

21. Letcher TM, Fthenakis VM, editors. *A Comprehensive Guide to Solar Energy Systems: With Special Focus on Photovoltaic Systems.* Academic Press: Cambridge, MA. 2018.

22. Mohamed A, Selim Y. Factors affect dye sensitized solar cells performance. *Renew Sust Energy Rev* 2017;3:83–6.

23. Assadi MK, Bakhoda S, Saidur R, Hanaei H. Recent progress in perovskite solar cells. *Renew Sust Energy Rev* 2018;81:2812–22.

24. Sharma S, Jain KK, Sharma A. Solar cells: In research and applications—A review. *Mater Sci Appl* 2015;6:1145.

25. Nayak PK, Mahesh S, Snaith HJ, Cahen D. Photovoltaic solar cell technologies: Analysing the state of the art. *Nat Rev Mater* 2019;4:269.

26. Park NG. Perovskite solar cells: An emerging photovoltaic technology. *Mater Today* 2015;18:65–72.

27. Zhou S, Tang R, Yin L. Slow-photon-effect-induced photoelectrical-conversion efficiency enhancement for carbon-quantum-dot-sensitized inorganic CsPbBr3 inverse opal perovskite solar cells. Adv Mater. 2017;29:1703682.

28. Albero J, Clifford JN, Palomares E. Quantum dot based molecular solar cells. Coord Chem Rev 2014;263:53–64.

29. Kang R, Kim JE, Yeo JS, Lee S, Jeon YJ, Kim DY. Optimized organometal halide perovskite planar hybrid solar cells via control of solvent evaporation rate. *J Phys Chem C* 2014;118:26513–20.

30. Lin Q, Armin A, Nagiri RC, Burn PL, Meredith P. Electro-optics of perovskite solar cells. *Nat Photon* 2015;9:106.

31. Qiu L, Ono LK, Qi Y. Advances and challenges to the commercialization of organic–inorganic halide perovskite solar cell technology. *Mat Today Energy* 2018;7:169–89.

32. Shi Z, Jayatissa AH. Perovskites-based solar cells: A review of recent progress, materials, and processing methods. *Materials* 2018;11:729.

33. Liu C, Hu M, Zhou X, Wu J, Zhang L, Kong W, et al. Efficiency and stability enhancement of perovskite solar cells by introducing CsPbI$_3$ quantum dots as an interface engineering layer. *NPG Asia Mater* 2018;10:552–61.

34. Yuan J, Ling X, Yang D, Li F, Zhou S, Shi J, et al. Band-aligned polymeric hole transport materials for extremely low energy loss α-CsPbI3 perovskite nanocrystal solar cells. *Joule* 2018;2:2450–63.

35. Omrani MK, Minbashi M, Memarian N, Kim DH. Improve the performance of CZTSSe solar cells by applying a SnS BSF layer. *Solid State Electron* 2018;141:50–7.

36. Raj CJ, Karthick SN, Park S, Hemalatha KV, Kim SK, Prabakar K, et al. Improved photovoltaic performance of CdSe/CdS/PbS quantum dot sensitized ZnO nanorod array solar cell. *J Power Sources* 2014;248:439–46.

37. Huy BT, Kim Phuong NT, Nguyen TT, Lee YI. Photoluminescence spectroscopy of Cd-based quantum dots for optosensing biochemical molecules. *Appl Spectrosc Rev* 2018;53:313–32.

38. Facchetti A. Polymer donor–polymer acceptor (all-polymer) solar cells. *Mater Today* 2013;16:123–32.

39. Jung B, Kim K, Eom Y, Kim W. High-pressure solvent vapor annealing with a benign solvent to rapidly enhance the performance of organic photovoltaics. *ACS Appl Mater Interfaces* 2015;7:13342–9.

40. Wang X, Perzon E, Oswald F, Langa F, Admassie S, Andersson MR, et al. Enhanced photocurrent spectral response in low-bandgap polyfluorene and C70-derivative-based solar cells. *Adv Funct Mater* 2005;15:1665–70.

41. Grätzel M. Perspectives for dye-sensitized nanocrystalline solar cells. *Prog Photovolt* 2000;8:171–85.

42. Beranek R. (Photo) electrochemical methods for the determination of the band edge positions of TiO_2-based nanomaterials. *Adv Phys Chem* 2011;2011.

43. Weerasinghe HC, Huang F, Cheng YB. Fabrication of flexible dye sensitized solar cells on plastic substrates. *Nano Energy* 2013;2:174–89.

44. Semalti P, Sharma SN. Dye sensitized solar cells (DSSCs) electrolytes and natural photo-sensitizers: A review. *J Nanosci Nanotechnol* 2020;20:3647–58.

45. Selloni A. Titania and its outstanding properties: Insights from first principles calculations. In: Andreoni W, Yip S (eds) *Handbook of Materials Modeling: Applications: Current and Emerging Materials*. 2020, pp. 29–51.

46. Alemany LJ, Bañares MA, Pardo E, Martın-Jiménez F, Blasco JM. Morphological and structural characterization of a titanium dioxide system. *Mater Charact* 2000;44:271–5.

47. Khataee AR, Aleboyeh H, Aleboyeh A. Crystallite phase-controlled preparation, characterisation and photocatalytic properties of titanium dioxide nanoparticles. *J Exp Nanosci* 2009;4:121–37.

48. Carp O, Huisman CL, Reller A. Photoinduced reactivity of titanium dioxide. *Prog Solid State Ch* 2004;32:33–177.

49. Pelaez M, Nolan NT, Pillai SC, Seery MK, Falaras P, Kontos AG, et al. A review on the visible light active titanium dioxide photocatalysts for environmental applications. *Appl Catal B* 2012;125:331–49.

50. Zhang H, Banfield JF. Understanding polymorphic phase transformation behavior during growth of nanocrystalline aggregates: Insights from TiO2. *J Phys Chem B* 2000;104:3481–7.

51. Prieto-Mahaney OO, Murakami N, Abe R, Ohtani B. Correlation between photocatalytic activities and structural and physical properties of titanium (IV) oxide powders. *Chem Lett* 2009;38:238–9.

52. Kim CH, Kim BH, Yang KS. TiO_2 nanoparticles loaded on graphene/carbon composite nanofibers by electrospinning for increased photocatalysis. *Carbon* 2012;50:2472–81.

53. Sakthivel S, Neppolian B, Shankar MV, Arabindoo B, Palanichamy M, Murugesan V. Solar photocatalytic degradation of azo dye: Comparison of photocatalytic efficiency of ZnO and TiO_2. *Sol Energy Mat Sol C* 2003;77:65–82.

54. Malato S, Blanco J, Cáceres J, Fernández-Alba AR, Agüera A, Rodrıguez A. Photocatalytic treatment of water-soluble pesticides by photo-Fenton and TiO_2 using solar energy. *Catal Today* 2002;76:209–20.

55. Xu C, Zhang Y, Chen J, Lin J, Zhang X, Wang Z, et al. Enhanced mechanism of the photo-thermochemical cycle based on effective Fe-doping TiO_2 films and DFT calculations. *Appl Catal B* 2017;204:324–34.

56. Ali I, Suhail M, Alothman ZA, Alwarthan A. Recent advances in syntheses, properties, and applications of TiO_2 nanostructures. *RSC Adv* 2018;8:30125–47.

57. Wu JM, Shih HC, Wu WT. Electron field emission from single crystalline TiO_2 nanowires prepared by thermal evaporation. *Chem Phys Lett* 2005;413:490–4.

58. Ramakrishnan VM, Natarajan M, Santhanam A, Asokan V, Velauthapillai D. Size controlled synthesis of TiO_2 nanoparticles by modified solvothermal method towards effective photo catalytic and photovoltaic applications. *Mater Res Bull* 2018;97:351–60.

59. Karaagac H, Yengel E, Islam MS. Physical properties and heterojunction device demonstration of aluminum-doped ZnO thin films synthesized at room ambient via sol-gel method. *J Alloys Compd* 2012;521:155–62.

60. Higashimoto S, Azuma M. Photo-induced charging effect and electron transfer to the redox species on nitrogen-doped TiO_2 under visible light irradiation. *Appl Catal B* 2009;89:557–62.

61. Duarte DA, Massi M, da Silva Sobrinho AS. Development of dye-sensitized solar cells with sputtered N-doped thin films: From modeling the growth mechanism of the films to fabrication of the solar cells. *Int J Photoenergy* 2014;2014:1–13.

62. Wang J, Tapio K, Habert A, Sorgues S, Colbeau-Justin C, Ratier B, et al. Influence of nitrogen doping on device operation for TiO_2-based solid-state dye-sensitized solar cells: Photo-physics from materials to devices. *Nanomaterials* 2016;6:35.

63. Ramasamy E, Lee WJ, Lee DY, Song JS. Nanocarbon counterelectrode for dye sensitized solar cells. *Appl Phys Lett* 2007;90:173103.

64. Lim J, Ryu SY, Kim J, Jun Y. A study of TiO_2/carbon black composition as counter electrode materials for dye-sensitized solar cells. *Nanoscale Res Lett* 2013;8:1–5.

65. Umebayashi T, Yamaki T, Itoh H, Asai K. Analysis of electronic structures of 3d transition metal-doped TiO_2 based on band calculations. *J Phys Chem Solids* 2002;63:1909–20.

66. Nguyen TV, Lee HC, Khan MA, Yang OB. Electrodeposition of TiO_2/SiO_2 nanocomposite for dye-sensitized solar cell. *Sol Energy* 2007;81:529–34.

67. Lee S, Noh JH, Han HS, Yim DK, Kim DH, Lee JK, et al. Nb-doped TiO_2: A new compact layer material for TiO_2 dye-sensitized solar cells. *J Phys Chem C* 2009;113:6878–82.

68. Zhang C, Chen S, Mo LE, Huang Y, Tian H, Hu L, Huo Z, Dai S, Kong F, Pan X. Charge recombination and band-edge shift in the dye-sensitized Mg^{2+}-doped TiO_2 solar cells. *J Phys Chem C* 2011;115:16418–24.

69. Lü X, Mou X, Wu J, Zhang D, Zhang L, Huang F, Xu F, Huang S. Improved-performance dye-sensitized solar cells using Nb-doped TiO_2 electrodes: Efficient electron injection and transfer. *Adv Funct Mater* 2010;20:509–15.

70. Bu T, Wen M, Zou H, Wu J, Zhou P, Li W, et al. Humidity controlled sol-gel Zr/TiO_2 with optimized band alignment for efficient planar perovskite solar cells. *Sol Energy.* 2016;139:290–6.

71. Chandiran AK, Sauvage F, Etgar L, Graetzel M. Ga^{3+} and Y^{3+} cationic substitution in mesoporous TiO_2 photoanodes for photovoltaic applications. *J Phys Chem C* 2011;115:9232–40.

72. Ko KH, Lee YC, Jung YJ. Enhanced efficiency of dye-sensitized TiO_2 solar cells (DSSC) by doping of metal ions. *J Colloid Interface Sci* 2005;283:482–7.

73. An'Amt MN, Radiman S, Huang NM, Yarmo MA, Ariyanto NP, Lim HN, et al. Sol–gel hydrothermal synthesis of bismuth–TiO_2 nanocubes for dye-sensitized solar cell. *Ceram Int* 2010;36:2215–20.

74. Lee K, Schmuki P. Ta doping for an enhanced efficiency of TiO_2 nanotube-based dye-sensitized solar cells. *Electrochem Commun* 2012;25:11–4.

75. Jung HS, Lee JK, Nastasi M, Lee SW, Kim JY, Park JS, et al. Preparation of nano-porous MgO-coated TiO_2 nanoparticles and their application to the electrode of dye-sensitized solar cells. *Langmuir* 2005;21:10332–5.

76. Xia J, Masaki N, Jiang K, Yanagida S. Sputtered Nb_2O_5 as a novel blocking layer at conducting glass/TiO_2 interfaces in dye-sensitized ionic liquid solar cells. *J Phys Chem C* 2007;111:8092–7.

77. Diamant Y, Chen SG, Melamed O, Zaban A. Core–shell nanoporous electrode for dye sensitized solar cells: The effect of the $SrTiO_3$ shell on the electronic properties of the TiO_2 core. *J Phys Chem B* 2003;107:1977–81.

78. Lee S, Kim JY, Youn SH, Park M, Hong KS, Jung HS, et al. Preparation of a nano-porous $CaCO_3$-coated TiO_2 electrode and its application to a dye-sensitized solar cell. *Langmuir* 2007;23:11907–10.

79. Yum JH, Nakade S, Kim DY, Yanagida S. Improved performance in dye-sensitized solar cells employing TiO$_2$ photoelectrodes coated with metal hydroxides. *J Phys Chem B* 2006;110:3215–9.

80. Hu X, Huang K, Fang D, Liu S. Enhanced performances of dye-sensitized solar cells based on graphite–TiO$_2$ composites. *Mat Sci Eng B Adv* 2011;176:431–5.

81. Zhang H, Wang W, Liu H, Wang R, Chen Y, Wang Z. Effects of TiO$_2$ film thickness on photovoltaic properties of dye-sensitized solar cell and its enhanced performance by graphene combination. *Mater Res Bull* 2014;49:126–31.

82. Cheng G, Akhtar MS, Yang OB, Stadler FJ. Novel preparation of anatase TiO2@ reduced graphene oxide hybrids for high-performance dye-sensitized solar cells. *ACS Appl Mater Interfaces* 2013;5:6635–42.

83. Dembele KT, Selopal GS, Soldano C, Nechache R, Rimada JC, Concina I, et al. Hybrid carbon nanotubes-TiO$_2$ photoanodes for high efficiency dye-sensitized solar cells. *J Phys Chem C* 2013;117:14510–7.

84. Kilic B, Turkdogan S, Astam A, Ozer OC, Asgin M, Cebeci H, et al. Preparation of carbon nanotube/TiO$_2$ mesoporous hybrid photoanode with iron pyrite (FeS$_2$) thin films counter electrodes for dye-sensitized solar cell. Sci Rep 2016;6:27052.

85. Xu F, Zhang X, Wu Y, Wu D, Gao Z, Jiang K. Facile synthesis of TiO$_2$ hierarchical microspheres assembled by ultrathin nanosheets for dye-sensitized solar cells. *J Alloys Compd* 2013;574:227–32.

86. Dou J, Li Y, Xie F, Ding X, Wei M. Metal-organic framework derived hierarchical porous anatase TiO2 as a photoanode for dye-sensitized solar cell. *Cryst Growth Des* 2016;16:121–5.

87. Ramohlola KE, Monana GR, Hato MJ, Modibane KD, Molapo KM, et al. Polyaniline-metal organic framework nanocomposite as an efficient electrocatalyst for hydrogen evolution reaction. *Compos B Eng* 2018;137:129–39.

88. Monama GR, Mdluli SB, Mashao G, Makhafola MD, Ramohlola KE, Molapo KM, et al. Palladium deposition on copper (II) phthalocyanine/metal organic framework composite and electrocatalytic activity of the modified electrode towards the hydrogen evolution reaction. *Renew Energy* 2018;119:62–72.

89. Monama GR, Modibane KD, Ramohlola KE, Molapo KM, Hato MJ, Makhafola MD, et al. Copper (II) phthalocyanine/metal organic framework electrocatalyst for hydrogen evolution reaction application. *Int J Hydrogen Energy* 2019;44:18891–902.

90. Makhafola MD, Ramohlola KE, Maponya TC, Somo TR, Iwuoha EI, Makgopa K, et al. Electrocatalytic activity of graphene oxide/metal organic framework hybrid composite on hydrogen evolution reaction properties. *Int J Electrochem Sci* 2020, 15:4884–99.

91. Monama GR, Hato MJ, Ramohlola KE, Maponya TC, Mdluli SB, Molapo KM, Modibane KD, et al. Hierarchiral 4-tetranitro copper (II) phthalocyanine based metal organic framework hybrid composite with improved electrocatalytic efficiency towards hydrogen evolution reaction. *Results Phys* 2019;15:102564.

92. Chughtai AH, Ahmad N, Younus HA, Laypkov A, Verpoort F. Metal-organic frameworks: Versatile heterogeneous catalysts for efficient catalytic organic transformations. *Chem Soc Rev* 2015;44(19):6804–49.

93. Ramohlola KE, Masikini M, Mdluli SB, Monama GR, Hato MJ, Molapo KM, et al. Electrocatalytic hydrogen production properties of polyaniline doped with metal-organic frameworks. In *Carbon-related Materials in Recognition of Nobel Lectures by Prof. Akira Suzuki in ICCE*. Springer: Cham, Switzerland. 2017. pp. 373–89.

94. Mashao G, Ramohlola KE, Mdluli SB, Monama GR, Hato MJ, Makgopa K, Molapo KM, Ramoroka ME, Iwuoha EI, Modibane KD. Zinc-based zeolitic benzimidazolate framework/polyaniline nanocomposite for electrochemical sensing of hydrogen gas. *Mater Chem Phys* 2019;230:287–98.

95. Mashao G, Modibane KD, Mdluli SB, Iwuoha EI, Hato MJ, Makgopa K, et al. Polyaniline-cobalt benzimidazolate zeolitic metal-organic framework composite material for electrochemical hydrogen gas sensing. *Electrocatalysis* 2019;10:406–19.

96. Perez EV, Karunaweera C, Musselman IH, Balkus KJ, Ferraris JP. Origins and evolution of inorganic-based and MOF-based mixed-matrix membranes for gas separations. *Processes* 2016;4:32.

97. Eddaoudi M, Li H, Yaghi OM. Highly porous and stable metal-organic frameworks: Structure design and sorption properties. *J Am Chem Soc* 2000;122:1391–7.

98. Kawano M, Fujita M. Direct observation of crystalline-state guest exchange in coordination networks. *Coord Chem Rev* 2007;251:2592–605.

99. Yuan S, Zou L, Qin JS, Li J, Huang L, Feng L, Wang X, et al. Construction of hierarchically porous metal–organic frameworks through linker labilization. *Nat Commun* 2017;8:15356.

100. [100] Wang T. Host-*Guest Systems and Their Derivatives Based on Metal-Organic Frameworks* (Doctoral dissertation, University of Cambridge). 2019.

101. Ogawa T, Iyoki K, Fukushima T, Kajikawa Y. Landscape of research areas for zeolites and metal-organic frameworks using computational classification based on citation networks. *Materials* 2017;10:1428.

102. Moghadam PZ, Li A, Wiggin SB, Tao A, Maloney AG, Wood PA, et al. Development of a Cambridge Structural Database subset: A collection of metal-organic frameworks for past, present, and future. Chem Mater 2017;29:2618–25.

103. Hong DY, Hwang YK, Serre C, Ferey G, Chang JS. Porous chromium terephthalate MIL-101 with coordinatively unsaturated sites: Surface functionalization, encapsulation, sorption and catalysis. *Adv Funct Mater* 2009;19:1537–52.

104. Wu H, Yildirim T, Zhou W. Exceptional mechanical stability of highly porous zirconium metal-organic framework UiO-66 and its important implications. *J Phys Chem Lett* 2013;4:925–30.

105. Jiang HL, Makal TA, Zhou HC. Interpenetration control in metal-organic frameworks for functional applications. *Coord Chem Rev* 2013;257:2232–49.

106. Ding M, Cai X, Jiang HL. Improving MOF stability: Approaches and applications. *Chem Sci* 2019;10:10209–30.

107. Lu W, Wei Z, Gu ZY, Liu TF, Park J, Park J, et al. Tuning the structure and function of metal-organic frameworks via linker design. *Chem Soc Rev* 2014;43:5561–93.

108. Kitagawa S. Metal–organic frameworks (MOFs). *Chem Soc Rev* 2014;43:5415–8.

109. Ren Y, Chia GH, Gao Z. Metal-organic frameworks in fuel cell technologies. *Nano Today* 2013;8:577–97.

110. Zhao M, Huang Y, Peng Y, Huang Z, Ma Q, Zhang H. Two-dimensional metal-organic framework nanosheets: Synthesis and applications. *Chem Soc Rev* 2018;47:6267–95.

111. Huang C, Dong J, Sun W, Xue Z, Ma J, Zheng L, et al. Coordination mode engineering in stacked-nanosheet metal–organic frameworks to enhance catalytic reactivity and structural robustness. *Nat Commun* 2019;10:2779.

112. Ansari A, Siddiqui VU, Khan I, Akram MK, Ahmad W, Siddiqi AK, et al. Metal-Organic-Frameworks (MOFs) for industrial wastewater treatment. In: Khan A, Abu-Zied BM, Hussein MA, Asiri AM, Azam (eds) *Metal-Organic Framework Composites*: *Volume I*. Materials Research Foundations 2019;53:1–28.

113. Dey C, Kundu T, Biswal BP, Mallick A, Banerjee R. Crystalline metal-organic frameworks (MOFs): Synthesis, structure and function. *Acta Cryst B* 2014;70:3–10.

114. Stock N, Biswas S. Synthesis of metal-organic frameworks (MOFs): Routes to various MOF topologies, morphologies, and composites. *Chem Rev* 2012;112:933–69.

115. Kim DS, Chang JS, Hwang JS, Park SE, Kim JM. Synthesis of zeolite beta in fluoride media under microwave irradiation. *Micropor Mesopor Mat* 2004;68:77–82.

116. Lee YR, Kim J, Ahn WS. Synthesis of metal-organic frameworks: A mini review. *Korean J Chem Eng* 2013;30:1667–80.
117. Friščić T, Fábián L. Mechanochemical conversion of a metal oxide into coordination polymers and porous frameworks using liquid-assisted grinding (LAG). Cryst Eng Comm 2009;11:743–5.
118. Friščić T, Reid DG, Halasz I, Stein RS, Dinnebier RE, Duer MJ. Ion-and liquid-assisted grinding: Improved mechanochemical synthesis of metal–organic frameworks reveals salt inclusion and anion templating. *Angew Chem Int Ed* 2010;49:712–5.
119. Mueller U, Schubert M, Teich F, Puetter H, Schierle-Arndt K, Pastre J. Metal-organic frameworks—prospective industrial applications. *J Mater Chem* 2006;16 :626–36.
120. Al-Kutubi H, Gascon J, Sudhölter EJ, Rassaei L. Electrosynthesis of metal-organic frameworks: Challenges and opportunities. *Chem Electro Chem* 2015;2:462–74.
121. Li Y, Pang A, Wang C, Wei M. Metal–organic frameworks: Promising materials for improving the open circuit voltage of dye-sensitized solar cells. *J Mater Chem* 2011;21:17259–64.
122. Gu A, Xiang W, Wang T, Gu S, Zhao X. Enhance photovoltaic performance of tris (2, 2'-bipyridine) cobalt (II)/(III) based dye-sensitized solar cells via modifying TiO$_2$ surface with metal-organic frameworks. *Sol Energy* 2017;147:126–32.
123. Lee DY, Kim EK, Shin CY, Shinde DV, Lee W, Shrestha NK, et al. Layer-by-layer deposition and photovoltaic property of Ru-based metal–organic frameworks. *RSC Adv* 2014;4:12037–42.
124. Feldblyum JI, Keenan EA, Matzger AJ, Maldonado S. Photoresponse characteristics of archetypal metal–organic frameworks. *J Phys Chem C* 2012;116:3112–21.
125. Lee DY, Shinde DV, Yoon SJ, Cho KN, Lee W, Shrestha NK, et al. Cu-based metal-organic frameworks for photovoltaic application. *J Phys Chem C* 2014;118:16328–34.
126. Lopez HA, Dhakshinamoorthy A, Ferrer B, Atienzar P, Alvaro M, Garcia H. Photochemical response of commercial MOFs: Al$_2$(BDC)$_3$ and its use as active material in photovoltaic devices. *J Phys Chem C* 2011;115:22200–6.
127. Ramasubbu V, Kumar PR, Mothi EM, Karuppasamy K, Kim HS, Maiyalagan T, et al. Highly interconnected porous TiO$_2$-Ni-MOF composite aerogel photoanodes for high power conversion efficiency in quasi-solid dye-sensitized solar cells. *Appl Surf Sci* 2019;496:143646.
128. Vinogradov AV, Zaake-Hertling H, Hey-Hawkins E, Agafonov AV, Seisenbaeva GA, Kessler VG, Vinogradov VV. The first depleted heterojunction TiO$_2$-MOF-based solar cell. *Chem Commun* 2014;50:10210–3.

6 Bio-Based Magnetic Metal-Organic Framework Nanocomposites

Manickam Ramesh and
Mayakrishnan Muthukrishnan

CONTENTS

6.1 INTRODUCTION

Metal-organic frameworks (MOFs), also known as porous polymers (PCPs), have appeared in the last two decades as a new category of porous and crystalline material. MOFs are a group of open crystalline framework compounds which consist of inorganic subunits like clusters of metal ions linking chemically to organic molecules like phosphonates, azolates, carboxylates, etc. They may be organic-inorganic porous materials called linker molecules with regular array of positively charged metal ions that outline nodes and attach the arms of linkers to form a recurring cage-like structure. The properties of MOFs vary widely with respect to the independent properties of inorganic subunits and the linked organic molecules called organic linkers. In general, all MOFs have common properties of open crystalline frameworks. MOFs can be developed as highly diversified 1D, 2D, and 3D structures and can provide a variety of MOF architecture by chemical design. They are porous in nature with a wide range of pore sizes. Thus, by suitable modification of the functional nodes or molecules and organic linkers, several properties like electronic, electrical, magnetic, optical properties, etc., can be introduced to facilitate their use in a wide variety of applications. The advantages of MOFs are that, i) they provide scope for endless combinations and in recent research advancements, development of more than 80,000 MOFs are reported [1], ii) better control of primary building blocks allows researchers to develop MOFs which add on different properties with complex architectures, and iii) these complex architectures with different physical and chemical properties provide scope in numerous application fields like catalysis, photoluminescence, energy storage, photo-magnetism, storage of gases, thin film membranes, and in various biomedical applications like drug delivery, etc. (Figure 6.1).

MOFs are widely used in gas purification, where due to their porous nature they allow the desired gas to pass through. They are also used for gas separation whereby the varying pore sizes of the MOF membrane allow mass transport of molecules to pass through. Thus, MOFs as a membrane material use continuous membrane-based separations, membrane reactors, etc. MOFs have high adsorption enthalpy properties which are used for storing molecules like methane, oxygen, and carbon dioxide.

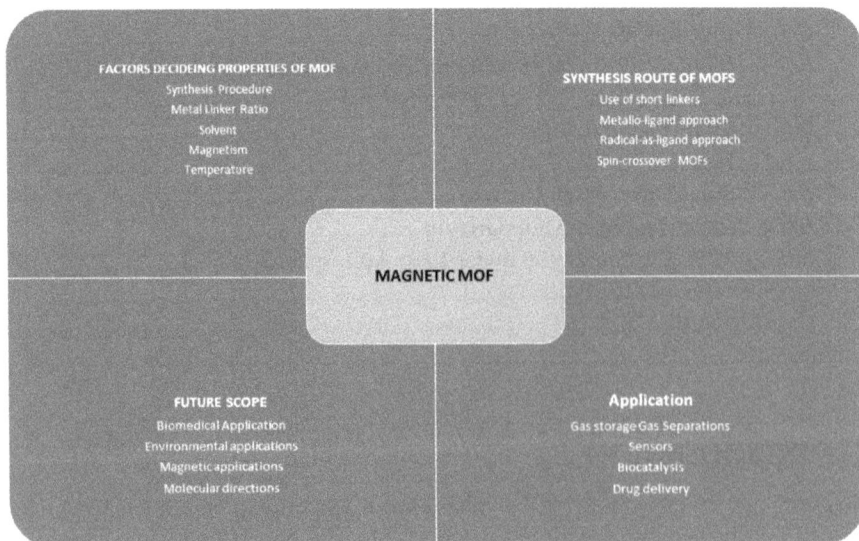

FIGURE 6.1 Fundamentals of magnetic MOFs.

Owing to their size and shape, they are also widely used as a catalyst. MOFs are also susceptible to cracks and fractures and are extremely sensitive to moisture. Thus, the challenge before designing any MOFs is to make them low cost, crack free, and have continuous MOF membranes. Modern studies in chemistry have combined MOFs with biomacromolecules to synthesize new biocomposites where MOFs will act as porous matrices to encapsulate enzymes, nucleotides, and many complex structures like viruses, bacteria, etc. Such an approach of preparation of bio-MOFs is called as biomimetic mineralization where it mimics the natural biomineralization process.

MOFs, owing to their excellent catalytic efficiency, better thermal stability, easy accessibility to dynamic sites, and effective enzyme-loading capacity are used as porous support in the immobilization of enzymes. But the challenges in developing bioenzyme-MOF composites like handling, separation, and low dispersion of enzyme-MOF composites can be overcome by adding magnetic properties which provide a larger surface area, easy loading, and speedy collection [2]. Since most of the MOF materials in general are insulators, introduction of addition properties like electrical conductivity, optical, and magnetic are a relatively unexplored area and profound research interests in the exploitation of MOFs have been seen recently [3]. The primary building units used in the formation of 3D MOFs are metal ion connectors that comprise Cr_3^+, Fe_3^+, Co_2^+, and Zn_2^+ [4]. Also, alkali metal ions and rare earth metal ions (Cs, K, Ba, or Sr_2^+) [5] are also used as connectors. With respect to developing porous magnets, transition metals like V, Cr, Mn, Fe, Co, Ni, and Cu, and lanthanides (La, Pr, Ce, Sm, Tb, Eu) are used [6]. In most of the synthetic routes to prepare MOFs, metal oxides, nitrate, sulphate, acetate, and chloride are used as a precursor. These metal ion connectors are capable of functional coordination bonds like amine, nitrile, sulfonate, phosphate, carboxylate, etc. On the other hand, MOF topology with intrinsic geometric properties is determined by secondary building units where organic linkers instead of metal ions, are connected through metal oxygen carbon clusters. For example, when metal salts are heated with carboxylic acid, they form intrinsic clusters units of MOF form different geometric properties as shown in Figure 6.2 and Figure 6.3.

MOFs based on the structural composition are classified into i) rigid frameworks where the resulting frameworks possess a robust MOF porous structure. They retain their stability during adsorption and desorption of guest molecules, ii) dynamic frameworks which are flexible in nature while adding or removing the guest molecules without affecting their porosity. Some of the flexible MOFs like MIL-5, MIL-8, and SNU-M10 exhibit a breathing effect during adsorption and desorption of guest molecules that affect the pore volume capacity of the MOFs [8–11], and iii) surface functionalized frameworks where the surface functionality of certain functional MOF groups like arylaminealkylamine, hydroxyl where modified for specific adsorption, or desorption [12–14] (Figure 6.4).

6.2 MOLECULAR DESIGN OF MAGNETIC MOFs

Magnetic MOFs are distinguished based on individual molecule and extended influence within the framework as i) MOFs exhibiting cooperative spin crossover and ii)

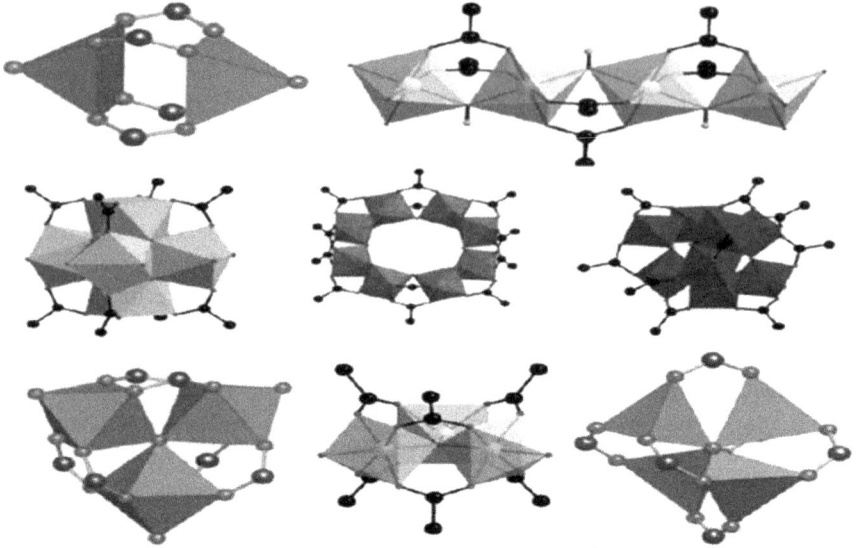

FIGURE 6.2 Secondary building units of MOFs [7].

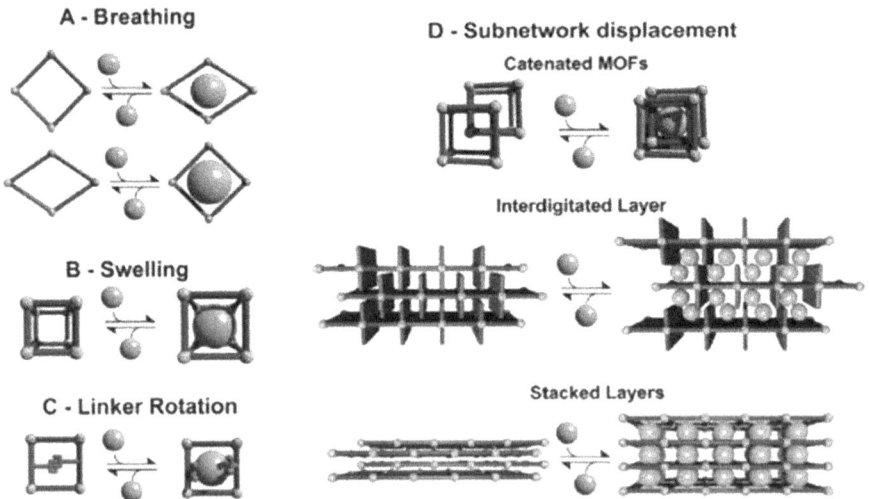

FIGURE 6.3 Some schematic representations of flexible framework of MOFs [7].

single or individual molecule magnetism. In the field of molecular design of magnetic MOFs, coordination polymers are used to develop cooperative magnetic properties. The challenge is to develop organic molecular magnetic MOF materials that exhibit effective magnetism above room temperature [16] which would have better structural rigidity and versatile electronic properties to accommodate two or more functional properties. Also, such materials should have sufficient catalytic efficiency

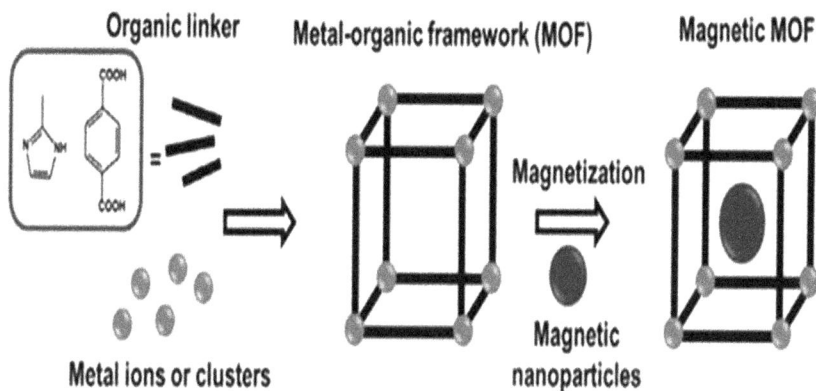

FIGURE 6.4 Schematic representation of magnetic MOFs [15].

to be chemically versatile. In conclusion, a biomagnetic MOF should have the primary property of magnetism concomitant with the secondary properties like electrical conductivity, thermal conductivity, bioluminescence, better structural stability, porosity, etc.

Depending on the type of application, biomagnetic MOFs can be designed to have i) two independent network materials where multiple properties like magnetism, conductivity, or luminescence are integrated into a crystal [17,18] or ii) one network material where two different properties are coupled to form a stimuli-responsive coordination material. The best example of the above is the spin crossover complexes (SCO) like the light-induced excited spin state trapping (LIESTT) effect, piezo-magnetism, etc. [19,20]. Thus, the effect of SCO in MOF magnetism is coupled with luminescence, pressure, and temperature which responds to external stimuli to tune their spin state. Therefore, magnetic coordination polymers can be used to develop multifunctional materials by increasing the number of pores wherein magnetism can be tuned for different applications. With an increase in the number of pores, additional molecules can be accommodated to have guest framework intermolecular interactions that respond to the magnetic stimuli. Also, the presence of magnetic centers at the pores or nodes will aid in designing organized MOF structures with better spacing.

6.3   MOFs BASED ON MAGNETIC FRAMEWORKS

As discussed earlier, properties like magnetism and porosity coexist to allow additional guest molecules to act as external chemical stimuli that respond to magnetism [21]. Based on the source of the magnetic property, MOFs are classified to four different types such as i) magnetic MOFs, where ligands are used for magnetism, ii) spin-crossover (SCO) MOFs where all nodes in the MOF clusters are used for this magnetic phenomenon, iii) MOFs with magnetic relaxation, where a single molecule in a cluster group will be in possession of magnetic behavior, and iv) MOFs with a magneto-caloric effect, where the nodes or clusters will have an isotropic spin ground state.

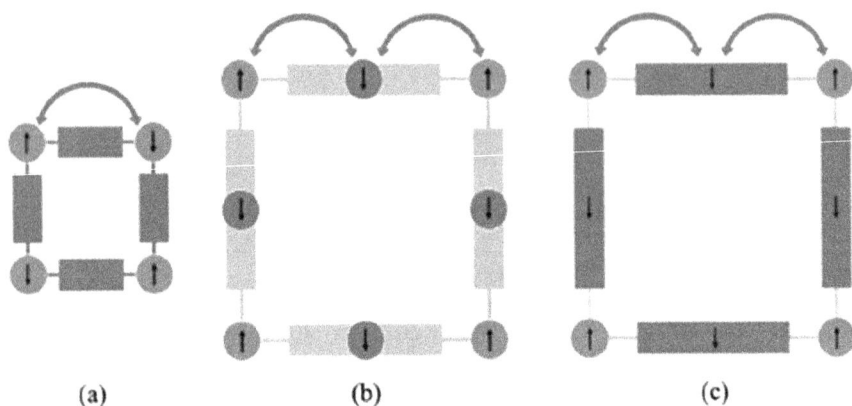

FIGURE 6.5 Schematic representation of various synthesis routes of preparation of magnetic MOFs, a) short linker usage, b) metallo-ligand, and c) use of radicals as ligands [22].

The challenging aspect of designing magnetic MOFs is to understand the fundamental idea that magnetism and porosity are unfavorable to one another. It is a known fact that any interactions through magnetic exchange between the metal centers require short distances, whereas for better porosity long linkers are mostly preferred [22,23]. These long linkers affect the magnetism at the nodes or at magnetic centers above absolute zero temperature. To overcome the above difficulties, several synthetic routes like short linker usage, the metallo-ligand approach and use of radicals as ligands as shown in the Figure 6.5 are suggested to design porous magnetic materials without affecting their individual nature.

6.3.1 USE OF SHORT LINKERS

In order to have better magnetic coupling between the metal centers, generally short linkers are preferred, but they affect the porosity of the MOF. The solution for the above problem is to have denser short linkers, and long-range magnetic order can emerge with chemisorptions of gases. For example, MOFs of Cu coordination polymers can switch from antiferromagnetism to ferromagnetism upon chemical gas absorption of gaseous HCl molecules [24]. The advantage of this method is that the gaseous molecules are integrated directly into the framework rather than through the pores without affecting the porosity and stability of the structures. Ammonium MOFs [(CH$_3$)$_2$NH$_2$] [M(HCOO)$_3$] and [NH$_4$][M(HCOO)$_3$] (M = Mn, Fe, Co, and Ni) are developed based on short linkers in higher dimensional systems where magnetic and electrical properties coexist [17,18]. Also, these MOFs exhibit phase transitions from paramagnetic to ferromagnetic properties at elevated temperature between 160 and 254 K [25,26] (Figure 6.6).

A 3D magnetic coordination polymer of cyanide forms the smallest bridging linker to exhibit magnetic properties which is why its family of bimetallic frameworks is called Prussian blue analogues (PBAs). PBAs are photo-magnetic materials

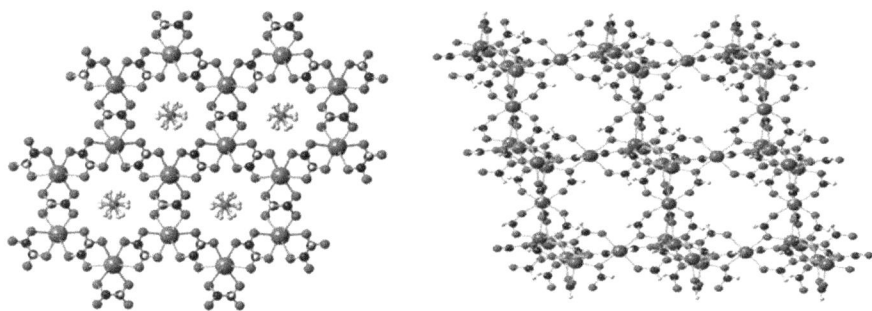

FIGURE 6.6 MOFs with short linkers, (a) crystal structure of $[NH_4][M(HCOO)_3$[27] and (b) crystal structure of $M_3(HCOO)_6$ [22].

with the basic chemical formula $[M_{12x}Co_{1+x}[Fe(CN)_6]\bullet zH_2O]$ where Fe, Co, and CN form the base materials, x and z represent variables and M represents alkali. They are widely used in energy storage devices because of their unique properties of frameworks that facilitate a high rate of conduction [27] (Figure 6.7).

Furthermore, $Co_3[Co(CN)_5]_2$ is the first family of magnetic coordination polymers to have a better combination of high range magnetic ordering and porosity. Another approach is that in place of ammonium MOFs, bulky amines are used to improve porosity. For example, amines act as a structuring agent in $M_3(HCOO)_6$ (M = Mn, Fe, Co, Ni) which forms an independent family of 3D porous magnets [28–30] where Fe and Mn form ferromagnets with critical temperatures between 8–16 K, Co forms an antiferromagnet below 1.8 K and Ni forms an extensive array of ferromagnetic order of 2.7 K. Similarly, azolates form another family of coordination polymers that is widely used in developing magnetic MOFs owing to its strong coordination bonds

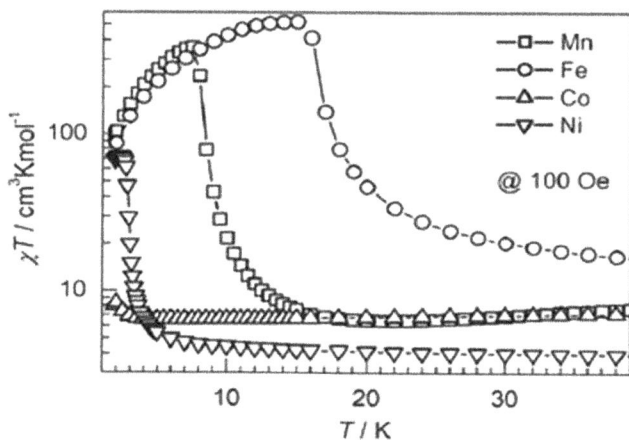

FIGURE 6.7 Temperature dependence susceptibility of MOFM$_3$(HCOO)$_6$ (M = Mn, Fe, Co, Ni) [22].

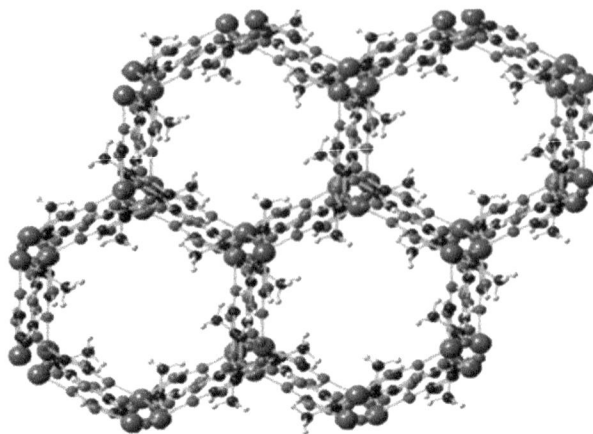

FIGURE 6.8   Pentacoordinated cocenters [22].

between metal ions [31]. Functional imidazolates like Co(II)-Imidazolate-4-amide-5-imidate based MOFs with pentacoordinated cocenters with imidazolate-amide-imidate as linkers are also used for formation of magnetic MOFs. The Cu-based MOF of formula [Cu(F-pymo)2(H_2O)1.25]n (F-pymo = 5-fluoropyrimidin-2-olate) forms a set of MOFs to respond to magnetism under a conducive environment of gases [32]. This gas absorption capacity has increased its potential in storing more volume in gas tankers. These MOFs release H_2O gases resulting in empty channels where different gases like CO_2 can be integrated (Figure 6.8).

6.3.2   METALLO-LIGAND APPROACH

The metallo-ligand approach is an alternative approach in synthesizing MOFs. Metallo-ligands are molecular complexes containing two or more electron pair donors, i.e., Lewis base sites which aid in coordination between other meta [33,34]. In general, a metallo-ligand comprises two coordination sites, primary, and secondary functional groups. The primary functional groups comprise carboxylates and pyridine derivatives along with a secondary functional group in forming active MOFs. The secondary functional group reacts with active metal centers to produce sites for guest molecule interactions. The metallo-ligand approach of oxamato-based dinuclear Cu_{II} metallacyclic complexes which exhibit weak ferromagnetic coupling between Cu_{II} ions. These complexes can also coordinate with Mn_{II} ions and yield 3D MOFs by carbonyl oxygen atoms [35] (Figure 6.9).

6.3.3   RADICAL-AS-LIGAND APPROACH

The radical-as-ligand approach refers to the addition of spin centers into organic linkers. The stronger coupling between the metal centers and radical ligand spin can be achieved by connecting paramagnetic metal centers by paramagnetic radical bridging

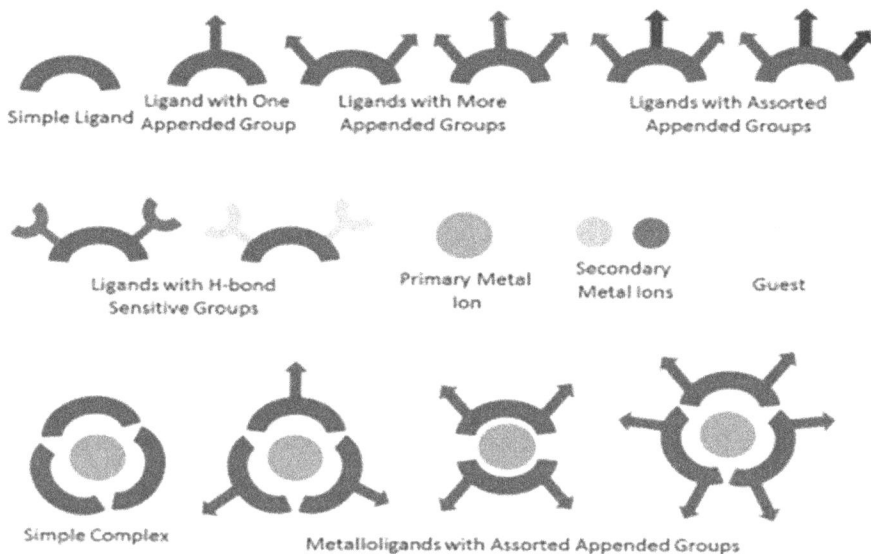

FIGURE 6.9 Metallo-ligand structure with primary and secondary functional groups [33].

ligands [36]. For example, 2,5-dichloro-3,6- dihydroxy-1,4-benzoquinone (Cl_2dhbq) is a radical form of chloranilic acid and is bridged by metallic center Fe III to form a 2D honeycomb-like crystalline solid $(Me_2NH_2)_2$- [$Fe_2(Cl_2$dhbq)3]_$2H_2O$_6DMF where, Me_2NH_2 forms counter cations [37]. In this case, it is revealed on a magnetic susceptibility test that magnetization occurs below Tc = 100 K and on activation of metal centers, the temperature is reduced to 30 K. Thus, the radical ligand approach is favorable for synthesis of magnetic MOFs for coupling between metal nodes, and radical ligands are widely spaced without affecting their stability. Also, the radical ligand approach is suitable for developing magnetic MOFs having high Tc.

6.4 SPIN-CROSSOVER MOFs

Spin crossover is one of the interesting phenomena found due to the movement of transition metal ions in magnetic MOFs with variable pore sizes. It is also called spin transition or spin equilibrium behavior which results in response to external stimuli by switching electronic properties of transition metal ions' atomic orbitals from a high spin (HS) state to low spin (LS) state [38]. The HS or LS electron configuration is determined by the magnitude of the ligand field splitting in association with the pairing strength of the configuration complexes whereas a HS state occurs when the complex's pairing energy is greater than the ligand splitting and is a favorable process for spin crossover [39]. Some of the external stimuli are magnetic field, light, temperature, pressure, soft X-ray, guest molecule inclusion, chemical environments' electric field, etc., which result in light-induced excited spin-state trapping (LIESST), ligand-driven light-induced spin change (LD-LISC), charge transfer-induced spin transition (CTIST), soft X-ray-induced excited spin state trapping (SOXIESST), etc. Spin crossover eliminates the necessity of designing MOFs with the required interactions through interfaces between adjacent magnetic centers. Instead, the necessary ligand field can be achieved using appropriate first-row transition metals that respond to the required transition from HS to LS. The advantages are i) the length of the ligand field is not constrained and ii) no metal center connectivity. The SCO phenomenon associated with inorganic chemical materials provides wider applications in the research areas of memory devices, displays, sensors, and artificial muscle actuators. Similarly, multifunctional SCO material like cobalt (II) terpyridin-4'-yl nitroxide complex is one of the MOFs with successful magnetic exchange coupling interactions [40]. A dithio oxalato-bridged iron mixed-valence complex also reported for its affirmative Fe_{II} and Fe_{III} sites for electron ions phase transfer [41]. For smart materials, octahedral Fe(II) systems are widely investigated. Fe(2-pytrz)2{Pt(CN)4}]_3H$_2$O is proven to exhibit HS spin behavior at high temperature.

6.5 MAGNETIC MOFs IN BIOMEDICINE

The human body comprises numerous cells and constant movement of ions within and outside cell membranes and results in electrical activity which forms the basis of biomagnetic fields that can be measured by external instruments placed outside our body. Biomagnetism involves the study and manipulation of these biomagnetic fields. Generally, prior to usage of magnetic MOFs, magnetic materials are widely used in the medical field in the areas like cell separation and magnetic resonance imaging (MRI). Some of the prominent magnetic biomaterials are magnetic alloys of Fe, Co, Ni, Nd-Fe-B, etc. The choice of the materials is limited by biocompatibility issues and toxicity limits. However, biocompatibility of these materials can be enhanced by encapsulation of these magnetic alloys as magnetic metal cores and form shells using biocompatible polymers. These core-shell combinations which possess coercivity, high magnetization, and better susceptibility are suitable for developing magnetic MOFs which are place in the medical field owing to their molecular storage capabilities, separation, delivery, and enzymatic biocatalysis. Thus, by manipulation of the above properties, biomagnetism can be achieved by altering the composition of the core-shell combination, temperature, crystal structure of MOFs, pressure, and by varying the size of the microporous material. Recent researchers have tuned MOF pore environments for the adsorption of biomolecules into a porous framework of MOF crystals and modified it as a protein-enabled MOF surface. It also paves way for effective drug delivery systems by host guest encapsulation and other interactions with effective responsiveness.

Magnetic MOFs have proven potential applications as molecular magnets, magnetic molecular sensors, and in the molecular biomedicine field. The coating of the magnetic materials by biocompatible polymers, gold, activated carbon, or silica reduces the aggravated effects of the core materials and prevents it from direct exposure to the body. In the same way polymer-coated MOFs will have colloidal stability, targeting ability, augmented biocompatibility, better encapsulation efficiency, longer circulation, and leakage prevention. Heparin, poly(N-vinylpyrrolidone) (PVP), chitosan, and poly(sodium 4-styrenesulfonate)(PSS) are some of the polymer-coating derivatives for their excellent biocompatibility and aqueous solubility. Polyvinyl alcohols, phospholipids, dextran, polyethylene glycol (PEG) are few of the derivatives used for coating magnetic materials [42–44].

Generally, prior to usage of magnetic MOFs, magnetic materials are widely used in the medical field in areas like cell separation and magnetic resonance imaging. MOFs have found their place in the medical field owing to their molecular storage, separation, delivery, and enzymatic biocatalysis. Recent researchers have tuned MOF pore environment for adsorption of biomolecules into the porous framework of MOF crystals and modified as a protein-enabled MOF surface. It also paves way for effective drug delivery systems by host guest encapsulation and other interactions with effective responsiveness. Magnetic MOFs have proven potential applications as molecular magnets, magnetic molecular sensors, and in the molecular biomedicine field. Elsaidi et al. [45] synthesize functionalized magnetic MOFs with Fe_3O_4 as the core which resembles a microsphere and MIL-101-SO_3 as the shell. Poly (sodium 4-styrenesulfonate) (PSS) is added as a linker to promote MOF shell growth over Fe_3O_4 microsphere cores. Also, PSS aids in absorbing metal ions but also helps in balancing the charges of Na^+ and H^+ cations by providing sulfonic acid (SO_3 functional groups) and keeps the material stable in the aqueous solution. Initially the diameter of the Fe_3O_4 core is 300 to 500 nm [46]. With the addition of PSS as linker, the diameter of the MOF shell is enlarged to 800–900 nm as it grows around the surface of the Fe_3O_4 without affecting the spherical shape. The schematic graphical representation of the polymer@MOF of Fe_3O_4 is shown in Figure 6.10a and the resultant scanning electron microscope (SEM) images are shown in Figure 6.10b. These MOFs are used in the extraction of rare earth ions in the aqueous solution owing to their function of ease of separation.

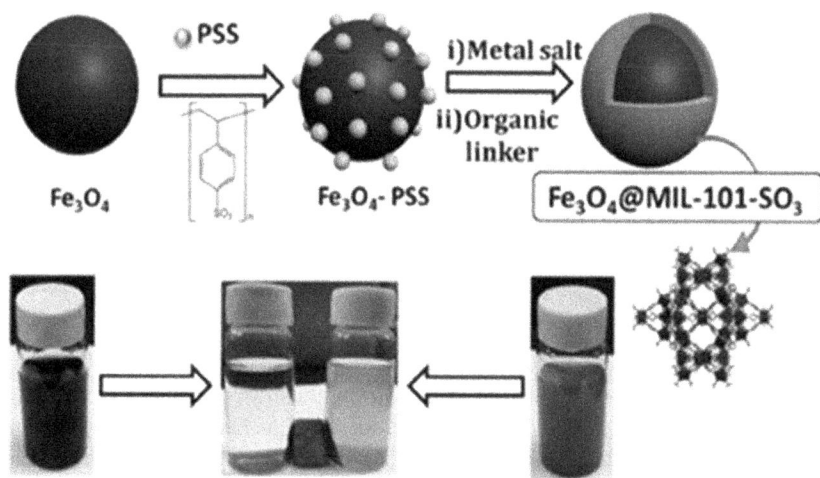

FIGURE 6.10A   Graphical Illustration of the Fe_3O_4@MIL-101-SO3 core formation [45].

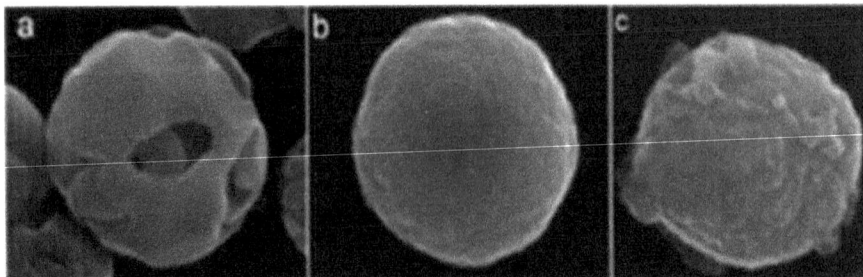

FIGURE 6.10B   SEM images of Fe_3O_4, Fe_3O_4-PSS, and Fe_3O_4@MIL-101-SO_3 [45].

Hidalgo et al. [47] have developed a biocompatible oral carrier, polymer MOF, chitosan-coated mesoporous MIL-100 (Fe) nanoparticles using 1,3,5-benzene tricarboxylic acid (BTC) as linker. Bellido et al. [46] engineered mesoporous Fe MoF nanoparticles using BTC as linkers for stealth drug nanocarriers. Polyacrylic acid bridged gadolinium MOF-gold nanoparticle composites using 1,4-benzene dicarboxylic acid (BDC) as linkers is analyzed by Tian et al. [48] as contrast agents in magnetic resonance biomedical imaging. UiO-66-L1 nanoparticles loaded with dichloroacetate using BDC as linkers is developed by Lazaro et al. [49] for anticancer cytotoxicity and immunity systems. For effective drug delivery into the blood stream, gene delivery, and hyperthermia, the drug-coated magnetic particles will be very small in size, called, in general, magnetic nanoparticles (MNPs) and are generally in powder form. For example, in targeted drug delivery to the site of disease, drug-coated magnetic MOFs are introduced into the blood vessel. Then, by giving an appropriate external magnetic field, particles are drawn towards the disease site and are retained (Figure 6.11).

Many researchers have focused on drug delivery through the intravenous route that would inflict more pain and needless complications. Hidalgo et al. [47] proposed an oral delivery drug system based on biocompatible mesoporous iron (III) trimethylsulfonate NPs that are coated with bioadhesive polysachharide chitosan. These NPs are biocompatible that do not pose issues during adsorption or during drug release. Ke et al. [51] designed magnetic core shell NPs of Fe_3O_4@MIL-100(Fe) loaded along with the anti-inflammatory drug Ibuprofen (IBU) and it took seven hours to completely release IBU in the aqueous buffer solution.

6.5.1   MAGNETIC HYPERTHERMIA

Magnetic hyperthermia is a concept of combining the effect of magnetic energy and thermal energy which paves way for the hyperthermia therapy wherein particular tissue with carcinogenic cells can be targeted and killed by exposing the tissue to high temperature using electromagnetic energy. Among various MNPs, iron oxide

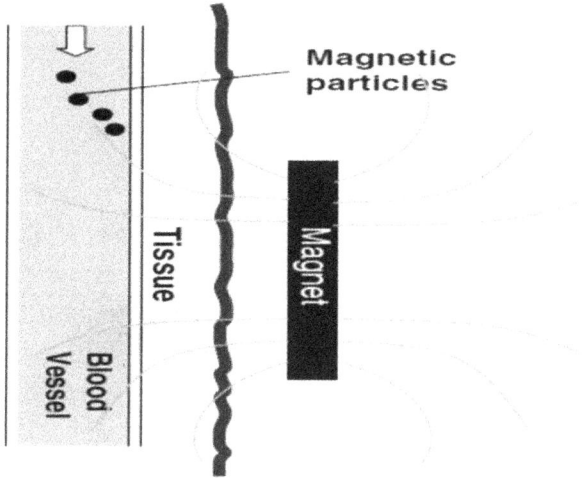

FIGURE 6.11 Motion of biomagnetic particles by external magnetic effect [50].

nanoparticles (IONPs) are widely used MNPs owing to their tunable magnetic properties. IONPs are the preferred material for hyperthermia due to their unique properties of converting electromagnetic energy to heat energy in targeting the cancer cells. The development of heating mechanism in magnetic hyperthermia is achieved through i) eddy current heating by alternating pulsing magnetic field, ii) frictional heating by interface between IONPs and the surrounding medium, and iii) hysteric losses of IONPs [52] (Figures 6.12 and 6.13).

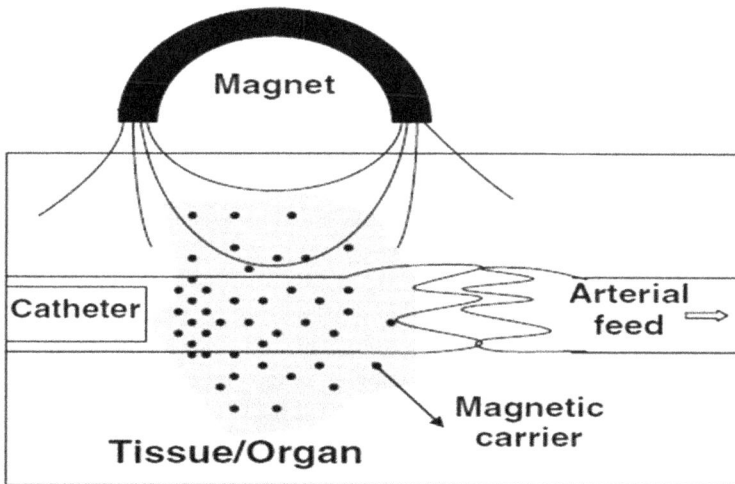

FIGURE 6.12 Targeting drugs using magnetic hyperthermia [52].

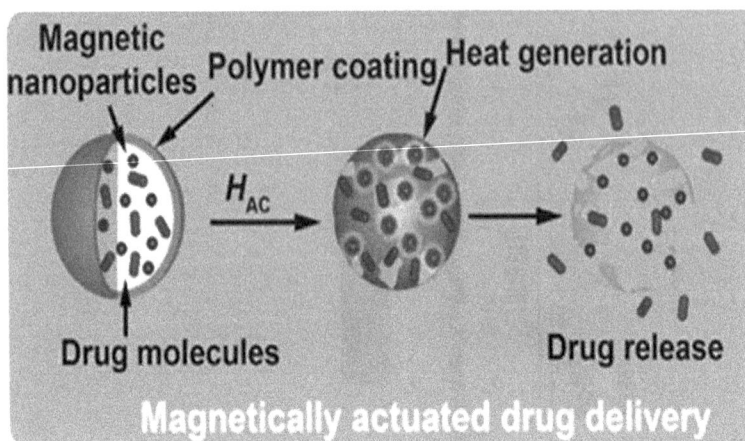

FIGURE 6.13   Magnetic actuated drug delivery [53].

6.5.2   BIOCATALYSIS

It is an important process that uses natural resources like enzymes to speed up the rate of the biological process by lowering the activation energy. Enzymes are non-living things fundamentally made of protein and the study of these macromolecules is important in the food, beverage, and pharmaceutical industries. Protein engineering of enzymes involves designing enzymes to react with the cell membranes or to settle on solids for better interaction. However, the issues faced are that they react selectively under mild conditions, are unpredictable in reactions with organic solvents, have low stability, and are difficult to recover as they get lost due to leakage from microcapsules during the interaction process. Also, the handling and separation of enzyme-MOF composites is challenging owing to their low density and high dispersion [54].

In biocatalysis, immobilization techniques or encapsulation of enzymes are widely demonstrated to overcome these limitations. Enzyme immobilization is the detention of enzymes to a phase that is different from that of substrates and products. Many researchers have identified a diverse range of materials to augment stable enzyme activity for longer periods. However, in biotechnology, the drive for identifying innovative methods for synthesizing hybrid enzymes with structured shells, better porous support, easy accessibility to active sites, potential recyclability, and elevated enzyme loading and controllable compositions for wider areas are currently ongoing. Here, magnetic MOFs are identified as potential candidates for enzyme immobilization with their distinguished characteristics such as manipulative composition, large surface area, uncomplicated loading, and swift collection improve the activity of enzymes and stabilizes enzymes during reactions [55].

The enzyme immobilization onto magnetic-MOFs can be strategies split into four parts: (i) physical adsorption onto magnetic MOFs, (ii) covalent/co-ordination bonding with magnetic MOFs, (iii) metal-ion affinity co-ordination bonding, and (iv) de novo encapsulation method [56, 57]. Adsorption is one of the widely used immobilization techniques which is based on van der Waals forces where there is physical interaction between the solid carrier and the required enzyme form weak bonds. Here the solid carriers are MOFs on which enzymes interact for a fixed period of time under favorable conditions. It is a low-cost process and it does not disturb the active sites during binding and maintains stability and favors reusability. On the other hand, the weak bonds between the carrier and enzyme can be altered by the pH in the aqueous medium, temperature, and variation in the strength of ions may affect the adsorption ability and lead to desorption.

In covalent bonding, enzymes through side chain functional groups support the solid carriers. The chain functional groups are carboxyl groups and amino groups (lysine, arginine, aspartic acid, and histidine) [58,59]. The inherent activities of the enzymes are independent of these functional groups as they are not involved in catalytic activity [60,61].The enzyme linkage improves enzyme activity and provides structural and thermo stability and it aids in increasing enzyme half-life [62,63]. NPs like magnetic cellulose, commercially called ECR8285 and MANAE agarose, are used for lipase immobilization for their improved performance in enzyme stability [64,65]. MOF materials like ZIF-8, Td-BDC, MIL-88, and HKUST-1 have ultrahigh porosity, variable functions, and higher surface properties have found their place in enzyme immobilization [66,67] (Figure 6.14).

6.5.3 Methods of Enzyme Immobilization of MOFs

Enzyme immobilization on MOF methods follows a combination of adsorption and covalent bonding methods. Generally, the methods of immobilization follow these following steps: i) MOF surface immobilization, ii) enzyme diffusion into MOFs, and iii) in situ encapsulation [69]. In the MOF surface immobilization method, a

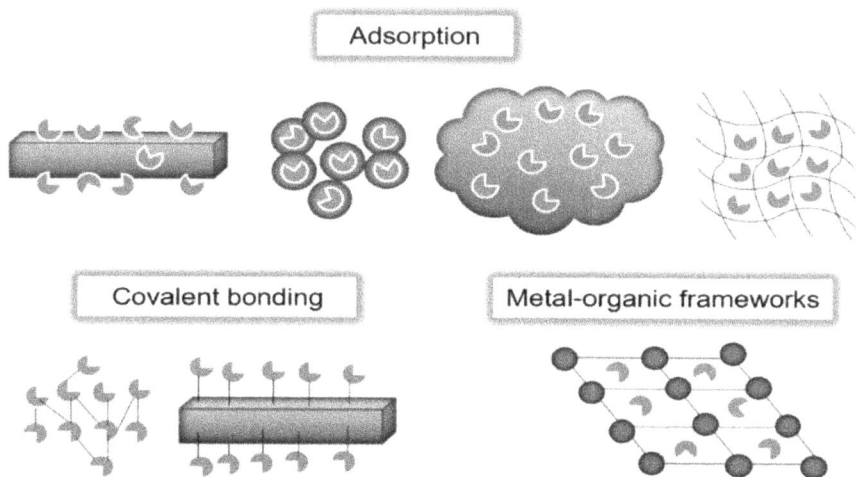

FIGURE 6.14   Graphical representation various immobilization techniques [68].

combination of enzyme-SNF-IF-8) and PGA forms a hybrid material with tunable uniform pores with thermal and water stability [70]. In the second method, the diffusion rate of enzyme activity is improved by diffusing microperoxidase enzymes into mesoporous Tb-TATB MOF [71]. Similarly, catalytic efficiency is improved by immobilizing OPAA with Nu-1003 MOF [72]. In the third method of in situ encapsulation, simultaneous synthesis of MOF or NPs is made possible from an enzyme or from any organic ligands [73]. Investigations on applied MgAl-LDH in monometallics like ZIF-8, ZIF-67, and Cu-BTC and in bimetallics like CoZn-ZIF MOFs revealed that both MgAl-LDH/ZIF-8 and MgAl-LDH/Cu have responded with better activity and recyclability [74]. The post-synthesis and in situ approaches with enzyme immobilization for β-glucosidase enzyme in two different MOF components at normal temperature and in solvent are investigated and the results are compared [75].

6.6 MAGNETIC SOLID-PHASE EXTRACTION USING MOFs

Sorbent is a substance that has the ability to collect or filter the materials through adsorption. Thus, a sorbent material can be used to absorb or adsorb solids, liquids, and even gases through attraction. Magnetic solid-phase extraction (MSPE) is a useful technique to absorb the dispersed sorbents and extract them in the large sample volume using magnetic retrieval methods. MSPE is one of the effective techniques used in sample preparation to extract solid sorbents because of its simple procedure, effective pore size variability, enabling reusability of the sorbents, etc. [76,77]. Compared with conventional solid phase extraction methods MSPE overcomes the associated problems due to non-spherical and uneven pore size which result in high back pressures, sorbent packing, or packed bed clogging [78,79].

Generally, sorbent materials used for MSPE are basically comprised of MOFs of magnetite (Fe_3O_4) NPs that are coated with silica, organic or inorganic polymers, or carbon nanostructures. As there are no commercially available magnetic MOFs, conventional MOFs are magnetized using several methods like i) direct MOF magnetization is achieved by direct mixing of the MOF and the magnetic NPs in the sample solution under electrostatic interactions through the sonication method, ii) in situ growth of magnetic MOFs, iii) single-step MOF coating, and iv) layer-by-layer MOF growth [80] (Figure 6.15).

6.6.1 MOF MAGNETIZATION

Direct MOF magnetization is achieved by direct mixing of the MOF and the magnetic NPs in the sample solution under electrostatic interactions through the sonication method which enables the magnetic NPs to attach to the MOFs. These magnetized MOFs can be retrieved by a magnetic retrieval process [81] (Figure 6.16).

6.6.2 IN SITU GROWTH OF MAGNETIC NPs

In situ growth of magnetic NPs involves dispersion of MOFs in a mixture containing the reagents for the synthesis of the magnetic MOFs. Thus, magnetic NPs (Fe_3O_4) grow in the presence of MOF crystals and can be separated by the centrifugal process [82].

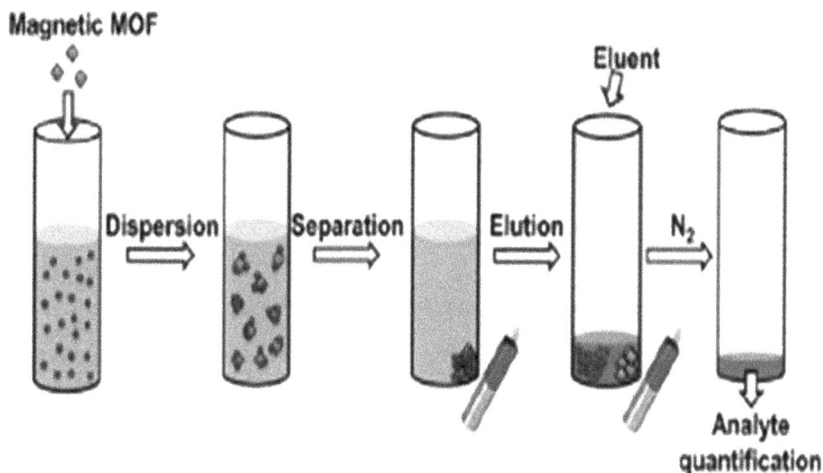

FIGURE 6.15 Schematic representation of MSPE using MOFs [15].

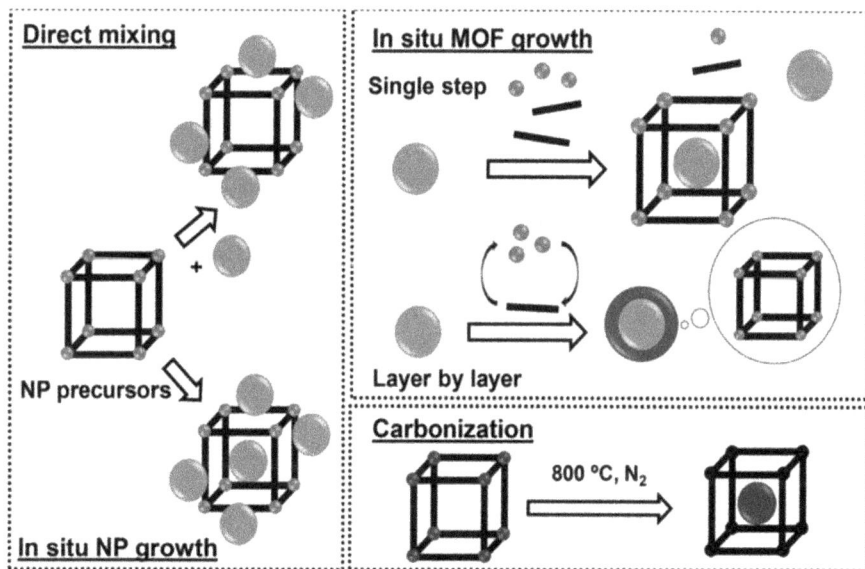

FIGURE 6.16 Strategies for the magnetization of MOFs for MSPE [15].

6.6.3 SINGLE-STEP MOF COATING

Single-step MOF coating involves discrete thin coating of magnetic NPs encapsulated in large MOF crystals. For example, ZIF-8 MOF crystals are polymer-coated magnetic NPs encapsulated in a time-dependent manner [83]. Just by controlling the rate of encapsulation of magnetic NPs in MOFs, spatial resolution of NPs and addition of different NPs or multiple applications can be generated.

6.6.4 Layer-by-Layer MOF Growth

This method is a time-consuming and highly labor-intensive method of generating thin films by depositing alternating layers of opposite charged magnetic NPs that are washed with pure solvent and any non-retained materials will be removed [84]. This cycle repeats with every layer using organic linkers for additional increases in thickness, but now attempts have been made to be automated using flow-based techniques [85].

6.6.5 MOF Carbonization under Inert Atmosphere

In this method, the precursor MOFs that contain metal ions are carbonized at high temperature at inert atmosphere to form magnetic NPs. The source of carbonization is an organic linker that encapsulates NPs by forming a porous carbon structure [86]. For example, an organic linker ZIF-67 which has a composition of Co(II)/2-methylimidazole, Co(II) aggregates to form Co NPs and 2-methylimidazole is encapsulated in N-doped porous carbon..

6.7 APPLICATIONS OF MSPE

MIL-101(Fe) MOF [87] is used for extraction of pesticides from samples of urine and human hair. MIL-101(Fe) MOF structure comprises encapsulated magnetic NPs functionalized in ethylenediamine is also used for extracting some metallic species like Cd(II), Pb(II), Zn(II), Hg(II), and Cr(III) from agricultural samples [88]. Pyrocatechol-embedded magnetic NPs are used for identifying flight/fight hormones like dopamine, epinephrine, and norepinephrine from human urine and serum samples [89,90]. These hormones are released from the adrenal medulla and nervous system when the body is under extreme stress conditions. ZIF-8 MOF using different kinds of magnetic NP sorbents for the extraction of fungicides [91] or phthalate esters [92] prior to chromatographic separation. Magnetic ZIF-8@Fe_3O_4 sorbent material has been successful in the determination of arsenic in water and milk samples and more than 90% of MSPE samples are recovered.

6.8 CONCLUSION

In conclusion, this chapter provided an overview of the bio-based magnetic MOF nanocomposites including coating, cross-linking MOF to polymer, and metal polymer ligands, with excellent physicochemical properties to their biomedical applicability. However, due to their excellent water stability, most of the bio-based MOFs currently used are Fe-based MOFs and Zr-based MOFs, and the use of other bio-based MOFs is rarely reported. Though bio-based magnetic MOFs have demonstrated excellent performance efficiency in enhancing the biocompatibility of MOFs with colloidal stability, these composites still have limitations. The biggest problem at the moment is that the incorporation of nanoparticles with MOF composites. The encapsulation of MOFs in polymeric composites, have been addressed. In addition to significantly high preconcentration capabilities, some of the advantages including environmentally friendly equipment, low detection limits, usability, and speed of analysis are also discussed.

REFERENCES

1. Moghadam PZ, Li A, Wiggin SB, Tao A, Maloney AGP, Wood PA, et al. Development of a Cambridge structural database subset: A collection of metal-organic frameworks for past, present, and future. *Chem Mater* 2017;29:2618–25.
2. Nadar SS, Rathod VK. Magnetic-metal organic framework (magnetic-MOF): A novel platform for enzyme immobilization and nanozyme applications. *Int J Biol Macromol* 2018;120:2293–302.
3. Park SS, Hontz ER, Sun L, Hendon CH, Walsh A, VoorhisTV, et al. Cation-dependent intrinsic electrical conductivity in isostructural tetrathiafulvalene-based microporous metal-organic frameworks. *J Am Chem Soc* 2015;137:1774–7.
4. Wu H, Zhou W, Yildirim T. High-capacity methane storage in metal-organic frameworksM_2(dhtp): The important role of open metal sites. *J Am Chem Soc* 2009;131:4995–5000.
5. Zacher D, Shekhah O, Woll W, Fischer RA. Thin films of metal-organic frameworks. *Chem Soc Rev* 2009;38:1418–29.
6. Kurmoo M. Magnetic metal-organic frameworks. *Chem Soc Rev* 2009;38:1353–79.
7. Soni S, Bajpai PK, Arora C. A review on metal-organic framework: Synthesis, properties and application. *Charact Appl Nanomater* 2019;2:doi:10.24294/can.v2i2.551
8. Sabouni R. *Carbon dioxide adsorption by metal organic frameworks* (Synthesis, testing and modeling) Ph.D. Thesis, University of Western Ontario, 2013:28–9.
9. Hamon L, Llewellyn PL, Devic T, Ghoufi A, Clet G, Guillerm V, et al. Co-adsorption and separation of CO_2-CH_4 mixtures in the highly flexible MIL-53(Cr)MOF. *J Am Chem Soc* 2009;131:17490–9.
10. Llewellyn P, BourrellyS, Serre C, Filinchuk Y, Férey G. How hydration drastically improves adsorption selectivity for CO_2 over CH_4 in the flexible chromium terephthalate MIL-53. *Angew Chem Int Ed* 2006;45:7751–4.
11. Serre C, Mellot C, Surblé S, Audebrand N, Filinchuk Y, Férey G. Role of solvent-host interactions that lead to very large swelling of hybrid frameworks. *Science* 2007;315:1828–31.
12. Millward A, Yaghi OM. Metal-organic frameworks with exceptionally high capacity for storage of carbon dioxide at room temperature. *J Am Chem Soc* 2005;127:17998–9.
13. Demessence A, D'Alessandro D, Foo M, Long J. Strong CO_2 binding in a water stable, triazolate-bridged metal-organic framework functionalized with ethylenediamine. *J Am Chem Soc* 2009;131:8784–6.
14. Serre C, Bourrelly S, Vimont A, Ramsahye NA, Maurin G, Llewellyn P, et al. An explanation for the very large breathing effect of a metal-organic framework during CO_2 adsorption. *Adv Mater* 2007;19:2246–51.
15. Maya F, Cabello CP, Frizzarin RM, Estela JM, Palomino GT, CerdAV. Magnetic solid-phase extraction using metal-organic frameworks (MOFs) and their derived carbons. *Trend Analyt Chem* 2017;90:142–52.
16. Manriquez JM, Yee GT, McLean RS, Epstein AJ, Miller JS. A room temperature molecular/organic-based magnet. *Science* 1991;252:1415–7.
17. Clemente-Leon M, Coronado E, Marti-Gastaldo C, RomeroFM. Multifunctionality in hybrid magnetic materials based on bimetallic oxalate complexes. *Chem Soc Rev* 2011;40:473–97.

18. Coronado E, Galan-Mascaros JR, Gomez-GarciaCJ, Laukhin V. Molecule-based magnets formed by bimetallic three-dimensional oxalate networks and chiral tris(bipyridyl) complex cations. The series $[Z^{II}(bpy)_3][ClO_4][M^{II}\ Cr^{III}(ox)_3]$ (Z^{II} = Ru, Fe, Co, and Ni; M^{II} = Mn, Fe, Co, Ni, Cu, and Zn; ox = oxalate dianion). *Nature* 2000;408:447–9.

19. Coronado E, Day P. Magnetic molecular conductors. *Chem Rev* 2004;104:5419–48.

20. Guetlich P, Gaspar AB, Beilstein YG. Spin state switching in iron coordination compounds. *J Org Chem* 2013;9:342–91.

21. Nuida T, Matsuda T, Tokoro H, Sakurai S, Hashimoto K, Ohkoshi S. Nonlinear magnetooptical effects caused by piezoelectric ferromagnetism in $F\bar{4}3m$-type prussian blue analogues. *J Am Chem Soc* 2005;127:11604–5.

22. Espallargas GM, Coronado E. Magnetic functionalities in MOFs: From the framework to the pore. *Chem Soc Rev* 2018;47:533–57.

23. Coronado E, Espallargas GM. Dynamic magnetic MOFs. *Chem Soc Rev* 2013;42:1525–39.

24. Coronado E, Marques MG, Espallargas GM, Brammer L. Tuning the magneto-structural properties of non-porous coordination polymers by HCl chemisorptions. *Nat Commun* 2012;3:828.

25. Jain P, Ramachandran V, Clark RJ, Zhou HD, Toby BH, Dalal NS, et al. Ultiferroicbehavior associated with an order-disorder hydrogen bonding transition in metal-organic frameworks (MOFs) with theperovskite ABX3 architecture. *J Am Chem Soc* 2009;131:13625–7.

26. Xu GC, Zhang W, Ma XM, Chen YH, Zhang L, Cai HL, et al. Coexistence of magnetic and electric orderings in the metal–formate frameworks of $[NH_4][M(HCOO)_3]$. *J Am Chem Soc* 2011;133:14948–51.

27. Wang Z, Zhang Y, Liu T, Kurmoo M, Gao S. $[Fe_3(HCOO)_6]$: A permanent porous diamond framework displaying H_2/N_2 adsorption, guest inclusion, and guest-dependent magnetism. *Adv Funct Mater* 2007;17:1523–36.

28. Wang Z, Hu K, Gao S, Kobayashi H. Formate-based magnetic metal-organic frameworks templated by protonated amines. *Adv Mater* 2010;22:1526–33.

29. Wang ZM, Zhang B, Fujiwara H, Kobayashi H, Kurmoo M. $Mn_3(HCOO)_6$: A 3D porous magnet of diamond framework with nodes of Mn-centered $MnMn_4$ tetrahedron and guest-modulated ordering temperature. *Chem Commun* 2004;4:416–7.

30. Zhang B, Wang ZM, Kurmoo M, Gao S, Inoue K, Kobayashi H. Guest-induced chirality in the ferromagnetic nanoporous diamond framework $Mn_3(HCOO)_6$. *Adv Funct Mater* 2007;17:577–84.

31. Zhang JP, Zhang YB, Lin JB, Chen XM. Metal azolate frameworks: From crystal engineering to functional materials. *Chem Rev* 2012;112:1001–33.

32. Mondal SS, Bhunia A, Demeshko S, KellingA,Schilde U, Janiak C, et al. Synthesis of a Co(ii)-imidazolate framework from an anionic linker precursor: Gas-sorption and magnetic properties. *Cryst Eng Comm* 2014;16:39–42.

33. Kumar G, Gupta R. Molecularly designed architectures—the metallo-ligand way. *J Chem Soc Rev* 2013;42:9403–53.

34. Evans JD, Sumby CJ, Doonan C. Post-synthetic metalation of metal-organic frameworks. *J Chem Soc Rev* 2014;43:5933–51.

35. Kleij AW. New templating strategies with salen scaffolds (Salen=N,N'-Bis(salicylidene) ethylenediamine Dianion). *Chem Eur J* 2008;14:10520–9.

36. Demarteau J, Debuigne A, Detrembleur C. Organocobalt complexes as sources of carbon-centered radicals for organic and polymer chemistries. *Chem Rev* 2019;119(12):6906–55.

37. Jeon IR, Negru B, Van Duyne RP, Harris TD. A 2D semiquinone radical-containing microporous magnet with solvent-induced switching from T_c = 26 to 80 K. *J Am Chem Soc* 2015;137:15699–702.

38. Bousseksou A, Molnar G, Salmon L, Nicolazzi W. Molecular spin crossover phenomenon: Recent achievements and prospects. *Chem Soc Rev* 2011;40:3313–35.
39. Cotton FA, Wilkinson G, Gaus PL. *Basic Inorganic Chemistry* (3rd ed.). Wiley. 1995.
40. Ondo A, Ishida T. Cobalt(II) Terpyridin-4′-yl Nitroxidecomplex as an exchange-coupled spin-crossover material. *Crystals* 2018;8(4):155.
41. Enomoto M, Ida H, Okazawa A, Kojima N. Effect of transition metal substitution on the charge-transfer phase transition and ferromagnetism of dithiooxalato-bridged hetero metal complexes (n-C$_3$H$_7$)$_4$N[Fe^{II}$_{1-x}$Mn^{II}$_x$Fe^{III}(dto)$_3$]. *Crystals* 2018;8(12):446.
42. Berry CC, Curtis ASG. Functionalisation of magnetic nanoparticles for applications in biomedicine. *J Phys D: Appl Phys* 2003;36:R198–R206.
43. BahadurD, Giri J. Biomaterials and magnetism. *Sadhana* 2003;28:639–56.
44. Shinkai M. Functional magnetic particles for medical application. *J Biosci Bioeng* 2002;94:606–13.
45. Elsaidi SK, Sinnwell MA, Banerjee D, Devaraj A, Kukkadapu RK, Droubay TC, et al. Reduced magnetism in core–shell magnetite@MOF composites. *Reduced Nano Lett* 2017;17:6968–73.
46. Bellido E, Hidalgo T, Lozano MV, Guillevic, M, Vazquez RS, Ortega MJS, et al. Heparin-engineered mesoporous from iron metal organic framework nanoparticles: Toward stealth drug nanocarriers. *Adv Healthcare Mater* 2015;4:1246–57.
47. Hidalgo T, Gimenez-Marques M, Bellido E, Avila J, Asensio MC, Salles F, et al. Chitosan-coated mesoporous MIL-100(Fe) nanoparticles as improved bio-compatible oral nanocarriers. *Sci Rep* 2017;7:43099.
48. Tian C, Zhu L, Lin F, Boyes SG. Poly(acrylic acid) bridged gadolinium metal-organic framework-gold nanoparticle composites as contrast agents for computed tomography and magnetic resonance bimodal imaging. *ACS Appl Mater Interf* 2015;7:17765–75.
49. Abanades Lazaro I, Haddad S, Rodrigo-Muñoz JM, Orellana-Tavra C, del Pozo V, Fairen-Jimenez D, et al. Mechanistic investigation into the selective anticancer cytotoxicity and immune system response of surface-functionalized, dichloroacetate-loaded UiO-66 nanoparticles. *ACS Appl Mater Interf* 2018;10:5255–68.
50. Pankhurst QA, Connolly J, Jones SK, Dobson J. Applications of magnetic nanoparticles in biomedicine. *J Phys D Appl Phys* 2003;36:R167–R168.
51. Ke X, Song X, Qin N, Cai Y, Kei F. Rational synthesis of magnetic Fe$_3$O$_4$@MOF nanoparticles for sustained drug delivery. *J Porous Mat* 2019;26(3):813–8.
52. Tartaj P, Morales MP, Verdaguer SV, Carreno TG, Serna CJ. The preparation of magnetic nanoparticles for applications in biomedicine. *J Phys D: Appl Phys* 2003;36: R182–97.
53. Abenojara EC, Wickramasinghe S, Concepcion JB, Samia ACS. Structural effects on the magnetic hyperthermia properties of iron oxide nanoparticles. *Prog Nat Sci: Mater Int* 2016;26:440–8.
54. Giustini AJ, Petryk AA, Casssim SM, Tate JA, Baker I, Hoopes PJ. Magnetic nanoparticle hyperthermia in cancer treatment. *Nano Life* 2010;1:17–32.
55. Hanefeld U, Gardossi L, Magner E. Understanding enzyme immobilisation. *Chem Soc Rev* 2009;38:453–68.
56. Lerin LA, Loss RA, Remonatto D, Zenevicz MC, Balen M, NettoVO, et al. A review on lipase-catalyzed reactions in ultrasound-assisted systems. *Bioprocess Biosyst Eng* 2014;37:2381–94.
57. Mohamad NR, Marzuki NHC, Buang NA, Huyop F, Wahab RA. An overview of technologies for immobilization of enzymes and surface analysis techniques for immobilized enzymes. *Biotechnol Biotechnol Equip* 2015;29:205–20.
58. Sirisha VL, Jain A, Jain A. An overview of technologies for immobilization of enzymes and surface analysis techniques for immobilized enzymes. *Adv Food Nutr Res* 2016;79:179–211.

59. Morawietz T, Singraber A, Dellago C, Behler. How van der Waals interactions determine the unique properties of water. *J Proc Natl* Acad *Sci* 2016;113:8368–73.
60. Liu CH, Li XQ, Jiang XP, Zhuang MY, Zhang JX, Bao CH, et al. Preparation of functionalized graphene oxide nanocomposites for covalent immobilization of NADH oxidase. *Nanosci Nanotechnol Lett* 2016;8:164–7.
61. Jiang XP, Lu TT, Liu CH, Ling XM, Zhuang MY, Zhang JX, et al. Immobilization of dehydrogenase onto epoxy-functionalized nanoparticles for synthesis of (*R*)-mandelic acid. *Int J Boil Macromol* 2016;88:9–17.
62. Mateo C, Abian O, Fernandez-Lafuente R, Guisan M. Increase in conformational stability of enzymes immobilized on epoxy-activated supports by favoring additional multipoint covalent attachment. *J Enzym Microb Technol* 2000;26:509–15.
63. Guzik U, Hupert-Kocurek K, Wojcieszynska. Immobilization as a strategy for improving enzyme properties-application to oxidoreductases. *Molecules* 2014;19:8995–9018.
64. Ispas C, Sokolov I, Andreescu S. Enzyme-functionalized mesoporous silica for bioanalytical applications. *Anal Bioanal Chem* 2009;393:543–54.
65. Feng D, Liu TF, Su J, Bosch M, Wei Z, Wan W, et al. Stable metal-organic frameworks containing single-molecule traps for enzyme encapsulation. *Nat Commun* 2015;6:5979.
66. Yi S, Dai F, Zhao C, Si Y. A reverse micelle strategy for fabricating magnetic lipase-immobilized nanoparticles with robust enzymatic activity. *Sci Rep* 2017;7:9806.
67. Cao SL, Huang YM, Li XH, Xu P, Wu H, Li N, et al. Preparation and characterization of immobilized lipase from pseudomonas cepacia onto magnetic cellulose nanocrystals. *Sci Rep* 2016;6:20420.
68. Zhao Z, Zhou MC, Liu RL. Recent developments in carriers and non-aqueous solvents for enzyme immobilization. *Catalysts* 2019;9:647.
69. Wang C, Liu D, Lin W. Metal-organic frameworks as a tunable platform for designing functional molecular materials. *J Am Chem Soc* 2013;135:13222–34.
70. Xiong F, Hu K, Yu H, Zhou L, Song L, Zhang Y, et al. Functional iron oxide nanoparticles modified with PLA-PEG-DG as tumor-targeted MRI contrast agent. *Pharm Res* 2017;34:1683–92.
71. Li P, Modica JA, Howarth AJ, Vargas E, Moghadam PZ, Snurr RQ, et al. Toward design rules for enzyme immobilization in hierarchical mesoporous metal-organic frameworks. *Chem* 2016;1:154–69.
72. Lawson S, Ali AH, Rezaei ARF. MOF immobilization on the surface of polymer-cordierite composite monoliths through *in-situ* crystal growth. *Sep Purif Technol* 2017;183(7):173–80.
73. Du Y, Gao J, Liu H, Zhou L, Ma L, He Y, et al. Enzyme@silicananoflower@metal-organic framework hybrids: A novel type of integrated nanobiocatalysts with improved stability. *Nano Res* 2018;11:4380–9.
74. Banerjee R. *Functional Supramolecular Materials: From Surfaces to MOFs*. Royal Society of Chemistry: London, UK, 2017.
75. Li P, Moon S-Y, Guelta MA, Lin L, Gomez-Gualdron DA, Snurr RQ, et al. Nanosizing a metal-organic framework enzyme carrier for accelerating nerve agent hydrolysis. *ACS Nano* 2016;10:9174–82.
76. Zeng HC, Li P. Immobilization of metal-organic framework nanocrystals for advanced design of supported nanocatalysts. *ACS Appl Mater Interf* 2016;8:29551–64.
77. Sanchez-Sanchez M, Getachew N, Díaz K, Diaz-Garcia M, Chebude Y, Díaz I. Synthesis of metal-organic frameworks in water at room temperature: Salts as linker sources. *Green Chem* 2015;17:1500–9.

78. Lopez de Alda MJ, Barcelo D. Determination of steroid sex hormones and related synthetic compounds considered as endocrine disrupters in water by fully automated online solid-phase extraction-liquid chromatography-diode array detection. *J Chromatogr A* 2001;911:203–10.

79. Gago-Ferrero P, Mastroianni N, Díaz-Cruz MS, Barcelo D. Fully automated determination of nine ultraviolet filters and transformation products in natural waters and wastewaters by on-line solid phase extraction-liquid chromatography-tandem mass spectrometry. *J Chromatogr A* 2013;1294:106–16.

80. Aghaei E, Alorro R, Encila AN, Yoo K. Magnetic adsorbents for the recovery of precious metals from leach solutions and wastewater. *Metals* 2017;7:529. doi:10.3390/met7120529

81. Safarıkova M, Safarık I. Magnetic solid-phase extraction. *J Magn Mater* 1999;194:108–12.

82. Lu G, Li S, Guo Z, Farha OK, Hauser BG, Qi X, et al. Imparting functionality to a metal-organic framework material by controlled nanoparticle encapsulation. *Nat Chem* 2012;4:310–6.

83. Huo SH, Yan XP. Facile magnetization of metal-organic framework MIL-101 for magnetic solid-phase extraction of polycyclic aromatic hydrocarbons in environmental water samples. *Analyst* 2012;137:3445–51.

84. Saikia M, Bhuyan D, SaikiaL. Facile synthesis of Fe_3O_4 nanoparticles on metalorganic framework MIL-101(Cr): Characterization and catalytic activity. *New J Chem* 2015;39:64–7.

85. Sorribas S, Zornoza B, Tellez C, Coronas J. Ordered mesoporous silica–(ZIF-8) coreshell spheres. *Chem Commun* 2012;48:9388–90.

86. Torad NL, Hu M, Ishihara S, Sukegawa H, Belik AA, Imura M, et al. Direct synthesis of MOF-derived nanoporous carbon with magnetic co nanoparticles toward efficient water treatment. *Small* 2014;10:2096–107.

87. Zhang S, Jiao Z, Yao W. A simple solvothermal process for fabrication of a metalorganic framework with an iron oxide enclosure for the determination of organophosphorus pesticides in biological samples. *J Chromatogr A* 2014;1371:74–81.

88. Babazadeh M, Khanmiri RH, Abolhasani J, Kalhor EG, Hassanpour A. Speciation analysis of inorganic arsenic in food and water samples by electrothermal atomic absorption spectrometry after magnetic solid phase extraction by a novel MOF-199/modified magnetite nanoparticle composite. *RSC Adv* 2015;5:19884–92.

89. Khezeli T, Daneshfar A. Dispersive micro-solid-phase extraction of dopamine, epinephrine and norepinephrine from biological samples based on green deep eutectic solvents and Fe_3O_4@MIL-100 (Fe) core–shell nanoparticles grafted with pyrocatechol. *RSC Adv* 2015;5:65264–73.

90. Du F, Qin Q, Deng J, Ruan G, Yang X, Li L, et al. Magnetic metal-organic framework MIL-100(Fe) microspheres for the magnetic solid-phase extraction of trace polycyclic aromatic hydrocarbons from water samples. *J Sep Sci* 2016;39:2356–64.

91. Su H, Lin Y, Wang Z, Wong YLE, Chen X., Chan TWD. Magnetic metal-organic framework–titanium dioxide nanocomposite as adsorbent in the magnetic solid-phase extraction of fungicides from environmental water samples. *J Chromatogr A* 2016;1466:21–8.

92. Liu X, Sun Z, Chen G, Zhang W, Cai Y, Kong R, Wang X, et al. Patterns of use of angiotensin-converting enzyme inhibitors/angiotensin receptor blockers among patients with acute myocardial infarction in China from 2001 to 2011: China PEACE-Retrospective AMI Study. *J Chromatogr A* 2015;1409:46–52.

7 Synthesis of Metal-Organic Framework Hybrid Composites Based on Graphene Oxide and Carbon Nanotubes

Nasani Rajendar, Bhagwati Sharma,
Tridib K. Sarma, and Anish Khan

CONTENTS

7.1 INTRODUCTION

Metal-organic frameworks (MOFs) are a class of versatile hybrid organic-inorganic porous materials and have attracted immense research interest in the last 20 years [1–4]. Owing to their exceptional properties such as large surface area, tunable structure, and high porosity, they are considered superior to other porous materials such as zeolites and silica [5,6]. MOFs have shown enormous potential in a myriad of applications such as gas separation and storage, sensing, catalysis, biomedicine, energy and environment, supercapacitors, and batteries [7–12]. Despite their considerable advancement in several applications, there are certain issues that need to be addressed to utilize MOFs to their true potential [12,13]. Firstly, MOFs in general are electrically non-conductive and unstable against high temperature and moisture, as well as strong electron beams [13]. Upon exposure to moisture, the surface area of MOFs decreases drastically [12]. Thus, the poor stability of the MOFs restricts their use in large-scale industrial applications, which require harsh conditions. Secondly, the MOF crystals are typically obtained in the form of powder which makes their subsequent processability a tedious job [13]. Furthermore, the narrow micropores of MOFs restrict the rapid diffusion of gas inside the pores [12]. Therefore, it is of utmost importance to address these issues to utilize the full potential of MOFs.

To overcome the above-mentioned challenges associated with MOFs, researchers developed three general strategies [13]: (i) *pre synthetic* approach, which can involve doping metal ions or functionalization of the ligands, or selection of specific building blocks, (ii) *post synthetic* approach which might involve exchanging ligands and/ or metal ions or grafting active groups after synthesis of MOFs, and (iii) *composite* formation, which involves hybridization of the MOFs with some other material to develop a new hybrid functional material with superior properties. This latter strategy has recently attracted a lot of attention compared to the other two strategies discussed.

MOF composites comprise the benefits of both the MOFs as well as the other functional material [12–33]. Moreover, fabrication of MOF composites is relatively easy. The composites can integrate the advantages of both the components and at the same time alleviate the disadvantages associated with each of the single components [13]. Thus, MOF composite possess superior properties due to the synergistic effects of the individual components. Composites of MOFs with various other functional materials have been developed. MOF-metal nanoparticles/quantum dot composites [14–16], MOF-silica composites [17–19], MOF-polymer composites [20–22], MOF-polyoxometalates [23,24] and MOF-carbon composites [13,25–32] are widely studied. Among all these, MOF-carbon composites (composites of MOFs with carbon-based materials, such as carbon dots, carbon nanotubes, graphene, and graphene oxide) are special as MOFs are regarded as kind of state-of-the-art materials and carbon-based materials are known to be somewhat classical [13]. These carbon-based materials are all unique in the sense that they possess several interesting properties such as high mechanical as well as elastic strength, excellent thermal and chemical stability, unique electronic and photophysical properties, low weight, biocompatibility as well as low cost. Such unique properties of the carbon-based materials have resulted in their use in sensing and catalysis, as well as energy and environmental applications [25–32]. Therefore, fabricating MOF-carbon composites will not only alleviate the disadvantages of MOFs but will also endow them with several novel functionalities such as improved thermal conductivity, stability, etc. [13]. In this chapter, however, we will restrict ourselves to only MOF-graphene oxide (MOF-GO) and MOF-carbon nanotube (MOF-CNT) composites. CNTs are highly conductive 1D materials and composite formation of MOFs with CNTs can assist in the transportation of electrons from the outer circuit to the inner surface of the MOFs. GO, on the other hand, is a 2D material and it can function as a support to anchor and decorate the MOF nanocrystals on its surface enabling them to be employed in a variety of other applications. Both these carbon-based materials can suppress MOF aggregation and can control the properties, structure and morphology of MOFs.

7.2 SYNTHESIS OF MOF-CARBON COMPOSITE

Composites are materials which consist of two or more components which work synergistically [13]. They are hierarchical architectures which are endowed with new and enhanced properties without losing the intrinsic nature of individual components. In this chapter we discuss most generalized synthetic procedures used for MOF-carbon material composite fabrication which are suitable for both MOF-GO and MOF-CNT systems. Several studies have been performed on these composites and it has been found that these hybrids could be generated either by growing MOF crystals on carbon scaffolds, direct mixing of MOFs with carbon materials, or incorporating MOFs in the carbon matrices during synthesis [13,26]. For convenience, the synthesis methods have been divided into two major classes, namely (i) in situ fabrication, which involves one-pot synthesis as well as stepwise synthesis and (ii) ex situ fabrication, which involves hybridization after individual synthesis. A third method which involves rarely used procedures has also been included under the title *miscellaneous approaches*.

7.2.1 IN SITU SYNTHETIC APPROACH

It is one of the most commonly employed methods to fabricate MOF-carbon composite and is pretty straightforward and easy to handle. It basically involves three steps. (i) The first step involves dispersing the carbon material into the solution containing MOF precursor components (metal ions and ligands). (ii) The second step involves proper mixing of all the precursors using techniques such as stirring or ultra-sonication followed by allowing them to react under conditions similar to those for pristine MOF synthesis. (iii) The third and final step involves separation of the formed composite by filtration or centrifugation followed by washing and drying. The functional groups and defects on the carbon material allow close contact between the composite components, which enhances the synergistic effect between the two, together with an increase in the porosity of the component. The in situ synthesis can be further divided into one-pot synthesis and stepwise synthetic approach.

7.2.1.1 One-Pot Synthesis

The one-pot synthesis method involves subjecting the reactants to successive chemical reactions, all in a single reaction vessel. Using this method of synthesis, tedious and time-consuming purification and separation steps of intermediates can be avoided. This method is suitable for the synthesis of almost all types of MOF-carbon composites. It involves the addition of carbon-based material to the solution containing MOF precursors which are then subjected to various reaction methods such as ambient room temperature agitation, conventional heating, solvothermal or hydrothermal reactions, ultra-sonication, and/or microwave irradiation, etc. (Scheme 1).

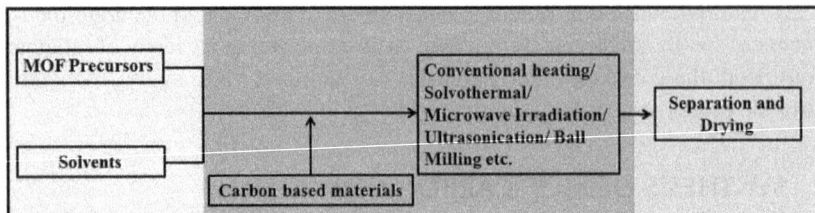

SCHEME 1   Schematic representation of one-pot synthesis of MOF-carbon nanocomposite. Reproduced with permission from ref. 29. Copyright 2015 Elsevier.

7.2.1.1.1   In Situ Synthesis in Ambient Conditions

This method is a low energy-consuming in situ method for fabrication of MOF-carbon nanomaterial composite [34–38]. The synthesis is generally performed at room temperatures, without the addition of any catalyst. Guo et al. [34] successfully fabricated ZIF-CNT composite for the electrocatalytic reduction of CO_2 to CO using this approach. The synthetic methodology involved the dropwise addition of 80 mL of a methanolic solution of $Zn(NO_3)_2$ (5.65 mmol of $Zn(NO_3)_2.6H_2O$) to another methanolic solution (80 mL) containing 3.7 g of 2-methylimidazole and 0.1 g of MWCNTs under vigorous stirring. The stirring was continued for 24 h at room temperature. The composite product was separated using centrifugation, followed by washing with ethanol. The authors then loaded furfuryl alcohol (FA) into the pores of ZIF-MWCNT composites by stirring the ZIF-MWCNTs in FA. The FA-loaded composite was then subjected to pyrolysis at 950°C, which resulted in the formation of a brown powder. This pyrolyzed composite ZIF-CNT-FA-p was used as an electrocatalyst for the reduction of CO_2 to CO (Figure 7.1a). It was observed that using ZIF-CNT-FA-p as catalyst, about 100% faradaic efficiency could be achieved, due to an enhanced electron transport on the MWCNT network as compared to the catalyst without CNTs. Using a similar approach Abdi et al. [35] synthesized ZIF-8@GO and ZIF-8@CNT hybrid nanocomposites at ambient conditions for removal of dyes. Xiong et al. [36] also utilized a similar ambient condition approach for fabricating salicylaldehyde@ZIF-8-GO hybrid composite for improving the corrosion protection properties of PVP coating. In their synthesis, in addition to the precursors for ZIF-8 and GO dispersion, salicyladehyde was also added to the reaction mixture (Figure 7.1b).

7.2.1.1.2   Conventional Heating

This method is one of the less energy-consuming approaches where reaction components are heated up or refluxed until it produces desired product. Anastasiou et al. [39] produced a MOF-GO composite by means of in situ ZIF-8 growth on GO layers. During fabrication, GO dispersed aqueous mixture was added to a clear solution holding ZIF-8 reactants (namely $Zn(NO_3)_2.6H_2O$ and 2-methyl-imidazole) and the resultant mixture was stirred and sonicated until it produced a well dispersed reaction system. Then the reaction mixture was allowed to reflux at 140°C for 24 h. Once the reaction was stopped and cooled down to room temperature, the product was

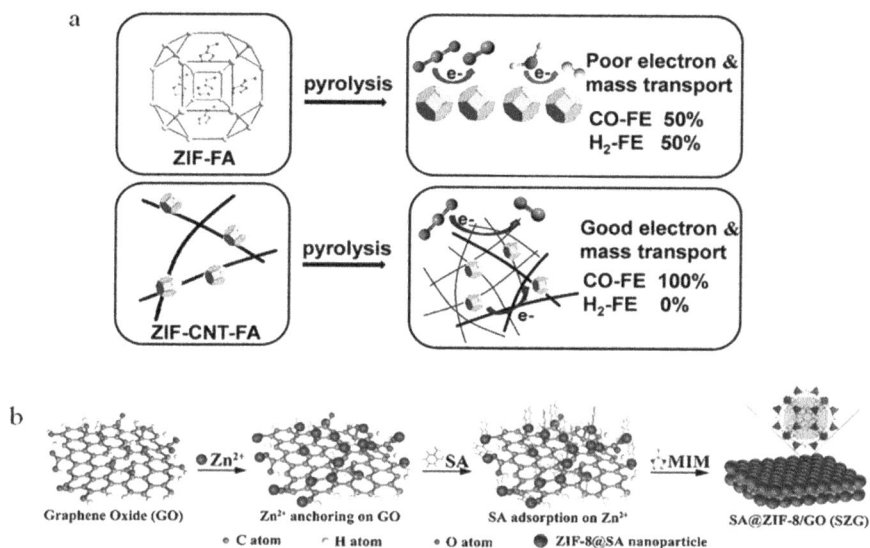

FIGURE 7.1 (a) Schematic representation showing that MWCNT support enhances the interparticle conductivity and mass transport of pyrolyzed ZIFs for reduction of CO_2. Reproduced with permission from ref. 34. Copyright 2017 Royal Society of Chemistry. (b) Schematic representation for the formation of salicylaldehyde@ZIF-8-GO composite. Reproduced with permission from ref. 36. Copyright 2019 Elsevier.

filtered off and washed successively with $CHCl_3$, DMF followed by DCM. The product was then dried overnight under vacuum at 116°C. In later studies the ZIF-8-GO composite was used as filler in polysulfone (PSF) matrix to fabricate PSF+ZIF-8-GO mixed matrix membranes for selective gas separation.

Using this methodology Pokhrel et al. [40] synthesized composites of UiO-66 and HKUST-1 with GO for adsorption of acid gases at ambient temperature. For the synthesis of UiO-66-GO composite, two solutions in DMF were prepared. The first solution was prepared by dispersing $ZrCl_4$ and terephthalic acid in 40 mL DMF under stirring and sonication, while the second solution was a dispersion of dry GO in DMF, prepared under the same stirring and sonication conditions. Both these solutions were then mixed under sonication to yield a well-dispersed mixture, which was heated at 120°C for 24 h. The as-obtained product was washed with DMF and methanol and dried at 105°C. The formation of HKUST-1-GO composite was accomplished by mixing 15 mL of DMF solution containing 0.6 g $Cu(NO_3)_2$ and 0.3 g trimesic acid and an aqueous solution containing 0.5 g of GO under stirring and sonication. The resultant mixture was then stirred at 85°C for 12 h, resulting in HKUST-1-GO composite. The obtained composite was washed several times with DCM. The FTIR spectrum of UiO-66-GO composite showed peaks corresponding to the oxygen functionalities present in GO, in addition to the characteristic peaks of UiO-66, confirming the formation of its composite with GO. On the other hand, in case of HKUST-1-GO composite, the peaks due to oxygen functionalities in GO could not be observed, which was ascribed to the high dispersion of GO flakes. The formation of composite was also ascertained using SEM. Although, the UiO-66-GO composite did not show GO layers, due to the relatively lower loading of GO (10%), HKUST-1-GO composite showed a uniform distribution of the particles of HKUST-1 over the GO flakes, indicating homogeneous growth of MOF over GO. The PXRD pattern confirmed that the crystals retained their identity in the composite (Figure 7.2). Several other groups have utilized this strategy to fabricate MOF-GO or MOF-CNT hybrid composites for various applications [31,41–43].

7.2.1.1.3   Hydrothermal (or Solvothermal) Assisted In Situ Synthesis

It is the most commonly used method for the in situ fabrication of MOF-carbon composites and involves mixing of the carbon material and MOF precursor, followed by subjecting the reaction mixture to high temperatures, where reactions are carried out at self-generated higher pressures. Several groups have successfully synthesized MOF-GO or MOF-CNT composites using this approach [44–51].

Using a hydrothermal approach, Pang and coworkers [44] produced $[Zn_2(ATA)_3(ATA)_{2/2}]_n$ in two polymorphic forms [MOF(ATA-a) and MOF(ATA-b)] by reaction between $Zn(NO_3)_2.6H_2O$ (for [MOF (ATA-a)]) or $ZnClO_4.6H_2O$ (for [MOF (ATA-b)] and 5-amino-1-H-tetrazole (HATA) at 160°C for 48 h. They found that [MOF(ATA-a)] was unstable and readily converted to stable form [MOF(ATA-b)] upon heating at 120°C or standing at room temperature for three months. To improve the stability of the MOF, the authors prepared a hybrid composite of [MOF(ATA-a)] and GO by an in situ approach. The formation of the MOF-GO composite was same as that of pure MOF, except that GO powder was also added to the aqueous solution of $ZnNO_3.6H_2O$ and HATA to yield black massive crystals of [GO⊂MOF(ATA-a)]. The resulting crystals were filtered, washed, and dried and the physical analysis indicated that it had superior chemical and thermal stability along with improved BET surface area compared to pure MOF. Mechanistic studies indicated that the stability of the metastable polymorph [MOF(ATA-a)] might be a result of the extensive coordination bonds between GO and the MOF. It was also interesting to note that the GO stabilized metastable [MOF(ATA-a)] exhibited a higher onset decomposition temperature (377.4°C) which makes it an excellent candidate for its usage as heat-resistant explosive.

Recently, Wang et al. [45] synthesized flower string-like NiCo MOF-MWCNT composite for application as a high-performance supercapacitor electrode material. The composite was synthesized in four steps (Figure 7.3). In the first step, the MWCNTs were functionalized by refluxing the CNTs in a mixture of H_2SO_4 and HNO_3 for 5 h at 110°C, followed by drying overnight. In the next step, the carboxylated MWCNTs were dispersed in 5 mL DMF by ultrasonication to obtain a homogeneous solution. Then $Ni(NO_3)_2.6H_2O$ and $Co(NO_3)_2.6H_2O$ were added to the above CNT dispersion. In the third step, a solution of 4,4'-biphenydicarboxylic acid (BPDC) in DMF-ethanol mixture was prepared by dissolving 0.125 mmol of the ligand in 17 mL of the mixture solvent. In the final step, both the MWCNT-metal mixture and ligand solution were slowly mixed under agitation and the final mixture was transferred to a stainless steel autoclave and heated at 170°C for 12 h. The MOF-MWCNT

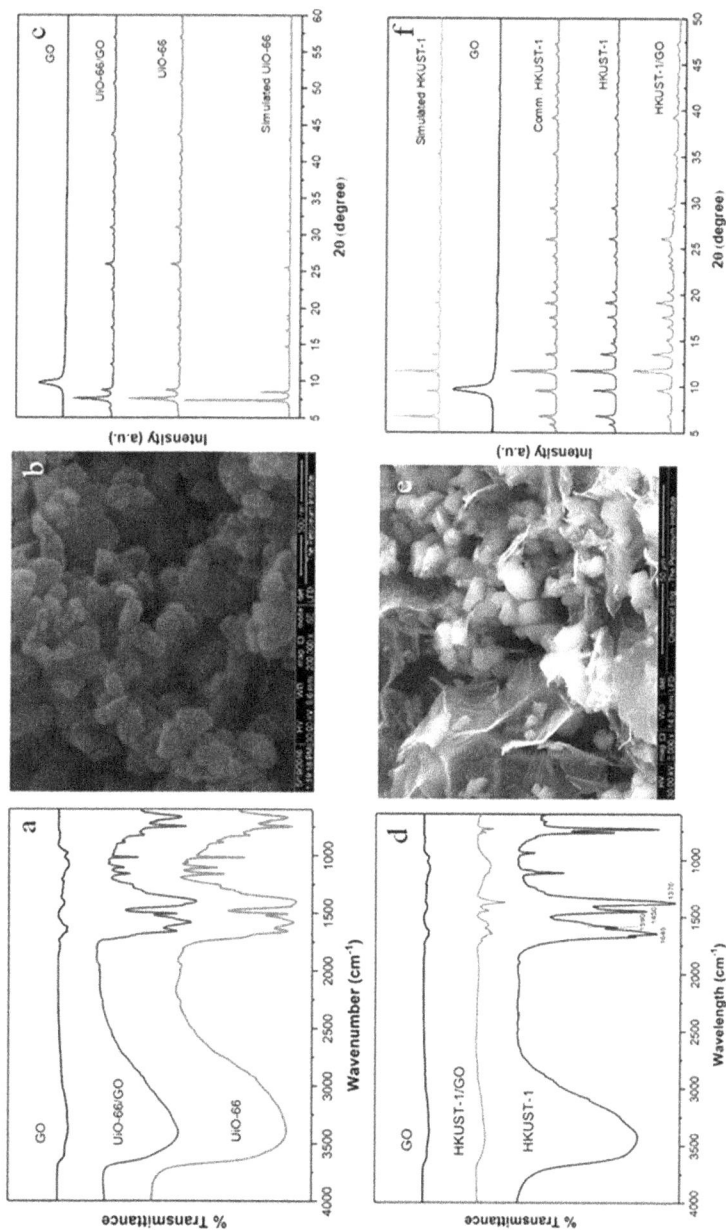

FIGURE 7.2 (a) FTIR spectrum of GO, UiO-66 and UiO-66/GO composite, (b) SEM image of UiO-66-GO composite, (c) PXRD pattern of GO, UiO-66, and UiO-66-GO composite compared with simulated UiO-66, (d) FTIR spectrum of GO, HKUST-1 and HKUST-1/GO composite, (e) SEM image of HKUST-1-GO composite and (f) PXRD pattern cf GO, HKUST-1 and HKUST-1-GO composite compared with simulated HKUST-1 and commercially available HKUST-1. Reproduced with permission from ref. 40. Copyright 2018 Elsevier.

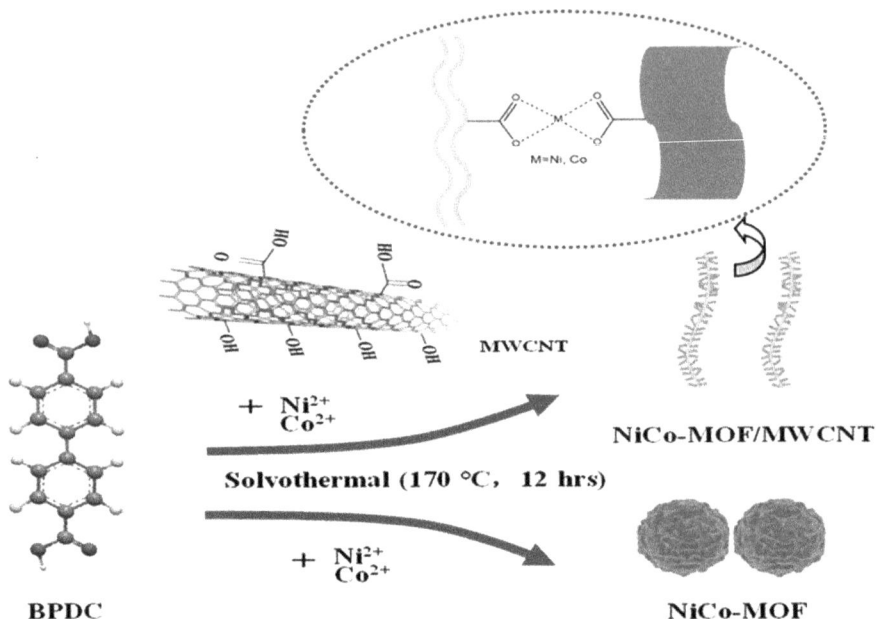

FIGURE 7.3 Schematic representation for the formation of NiCo-MOF and NiCo-MOF-MWCNT composite. Reproduced with permission from ref. 45. Copyright 2019 Elsevier.

composite was obtained as a black powder upon cooling, which was washed with DMF and ethanol and finally dried overnight at 60°C under vacuum. Microscopic studies revealed that the NiCo-MOF grew mostly around the MWCNT to give flower-like strings, whereby the MWCNT were wrapped with some NiCo-MOF nanosheets.

7.2.1.1.4   Ultrasonication

Ultrasonication has been a long-used method for the synthesis of various nanomaterials. It is a simple, environmentally friendly, and bulk synthesis method. Recently, this technique has also been utilized in preparing composites of MOFs with carbon nanomaterials [52–56] due to its simple operational procedure. Tanhahei et al. [52] used ultrasonication for the facile synthesis of two-fold interpenetrated Zn based MOF-GO composite (TMU-16-GO$_n$) (n = the percentage of GO) at ambient temperature and pressure. In a typical procedure, Zn(OAc)$_2$·2H$_2$O was added to GO suspension in DMF which was well dispersed by mild ultra-sonication, and the sonication was continued for 15 more minutes to make sure that complete interaction of Zn^{+2} ions with carboxylic acid functionalities over the GO surface took place. It was followed by addition of solution of H$_2$BDC (1,4-benzenedicarboxylic acid) and 4-bpdh (2,5-bis(4-pyridyl)-3,4-diaza-2,4-hexadiene) in DMF and the resultant reaction components were kept under sonication for an additional hour. The solid product obtained was centrifuged out (at 6000 rpm) and washed with DMF followed by drying in a vacuum oven and was characterized by common analytical techniques like FTIR, SEM, PXRD, Raman spectroscopy, etc. In a similar manner, Zhao [55] and Safarifard [53,54] have also synthesized various MOF-carbon composite materials. Ansari et al. [56] synthesized a MOF-CNT composite by simple blending of the MOF IITI-1 and CNTs under sonication for 1 h and utilized the composite for application as supercapacitor. They first synthesized the MOF (IITI-1) by using an organic linker H_2L as the ligand and $Cu(NO_3)_2$ as the metal salt in a mixed solvent system consisting of DMF and H_2O. The composite of IITI-1 with CNTs was prepared by dispersing 1:1 ratio (wt%) of the MOF and CNTs in ethanol, followed by ultrasonication for 60 minutes. The black product was dried on a rotary evaporator and named IITI-1-CNT. The formation of the composite was ascertained from electron microscopic studies, which showed uniform wrapping of CNTs by MOF.

7.2.1.1.5 Microwave-Assisted

Microwave (MW)-assisted MOF synthesis is well reported in literature because of its added advantage in terms of lowering reaction time and low energy consumption along with production of MOFs with uniform crystal size and high reproducibility. As far as the synthesis of MOF-carbon material using MW is concerned [57–61], it is believed that MW irradiation produces super-hot spots (SHS) on the surface of the carbon material and these SHS enhance the formation of small nanocrystals of MOF along with a boosting of the interaction between the MOF and carbon materials [60].

Han et al. [58] reported a simplified and efficient synthetic method for graphene oxide-MIL-101(Fe) composite fabrication. GO-MIL-101 (Fe) composite was fabricated by adopting a microwave-assisted in situ growth approach. In a typical procedure, GO and $FeCl_3.6H_2O$ solution in DMF were dispersed. Another solution containing NH_2-H_2BDC and PVP mixture was also prepared in DMF. Both these solutions were mixed in the next step and transferred to a MW vessel and subjected to MW irradiation at 160°C for 10 min. The final product was separated by centrifugation followed by washing with DMF and ethanol and was vacuum-dried overnight. The SEM images confirmed the good distribution of MIL-101 MOF on both sides of the GO nanoplatelets.

Zhao et al. reported, for the first time, in 2018 the fabrication of a GO-MOF composite using a unique method of microwave irradiation-assisted ball milling, in which a ball milling instrument was directly kept in a microwave oven [59]. This combined technique has advantages in the sense that microwave heating helped in enhancing the rate of the chemical reaction and ball milling helped in refining the particle size by suppressing grain growth. Varying amounts of GO were placed in a tetrafluoroethylene milling pot along with nickel acetate, trimesic acid, and stainless steel balls with deionized water. Both the systems (microwave and ball mill) were switched simultaneously and the speed of the ball mill was kept at 200 rpm. After 30 minutes of reaction, light blue colored powder was formed, which was filtered and kept for drying.

7.2.1.2 Stepwise Synthesis

Although an in situ synthetic approach offers several advantages such as short preparation processes and easy separation and purification of intermediates, it is associated with several disadvantages such as the lack of control over ratios of components in products and inability to obtain defect-free materials, as well as difficulty in modifications on prepared composites [13]. In order to circumvent these drawbacks, stepwise synthesis methods such as the seeded growth method, transfer from metal oxides/carbon-based materials, and layer-by-layer assembly have been developed.

7.2.1.2.1 Seeded Growth Method

The seeded growth method, also known as the secondary growth method, allows for a more effective control of orientation and is one of the most efficient procedures for fabrication of crack- and defect-free membranes for purification, adsorption, and separation applications. This method involves the deposition of the preformed MOF crystals on the supports as seeds, which greatly assists in further crystallization of the MOFs [62–67].

Using the seeded growth method, Wang et al. [62] synthesized a 100 nm thick defect-free ZIF-8-GO membrane using ZIF-8-GO hybrid nanosheets as seeds. The 2D ZIF-8/-GO seeds were prepared by adding GO solution (methanol:water mixture 4:1) to a mixture of methylimidazole and $Zn(NO_3)_2$ in methanol followed by stirring at different times. The seeds were collected by centrifugation and washing. Porous anodic aluminum oxide disks were used as supports for the formation of ZIF-8-GO membrane. The seeds were spin coated onto the support for 30 s at 1000 rpm to obtain a thin layer of the ZIF-8-GO seed on the support. The layer was then dried for 2 h at 50°C, followed by spin coating of another layer, which was then finally dried overnight. Next, the coated support was mounted on a home-made setup, where the metal ion (Zn^{2+}) and ligand (methylimidazole) were separated by the support, with the seeded layer facing towards methanolic Zn^{2+} solution. The experimental setup was kept at room temperature for reactants to diffuse into confined spaces of the seeded layers and grown into well-dispersed ZIF-8 crystals in membranes. Finally, membranes were taken out and rinsed with fresh methanol followed by drying at 50°C overnight. (Figure 7.4).

Dumèe et al. employed the seeded growth method for the fabrication of a dense ZIF-8 membrane on porous CNT bucky paper [63]. The CNT bucky paper substrate was fabricated by vacuum filtration of the suspension of CNTs onto a poly(ethersulfone) membrane till it was completely dry. The as-formed CNT bucky paper was placed into a methanolic solution containing the precursors for ZIF-8 formation at room temperature to anchor the nanocrystal seeds. The modified bucky paper was then transferred to a 200 mL Teflon autoclave and solvothermally heated at 100°C for 24 hours, where the ZIF-8 seeds grew homogeneously within the pores of the bucky paper and packed themselves into a continuous, dense ZIF-8 network. Using a similar strategy, Zhang and coworkers [67] fabricated a continuous and low-defect ZIF-8 membrane on macroporous CNTs. First, composites of graphite and ZIF-8 were dip coated and rubbed on the surfaces of the CNTs, producing a seed layer. Graphite was used in the seed layer to anchor and stabilize the ZIF-8 seeds on the CNT support, as well as to fill up the pores and polish the surface of the CNTs. The ZIF-8 membrane was then produced using a simple solvothermal approach, in which the precursors of ZIF-8 in methanol were mixed and transferred to a stainless steel autoclave. The substrate with the seed layer was then placed

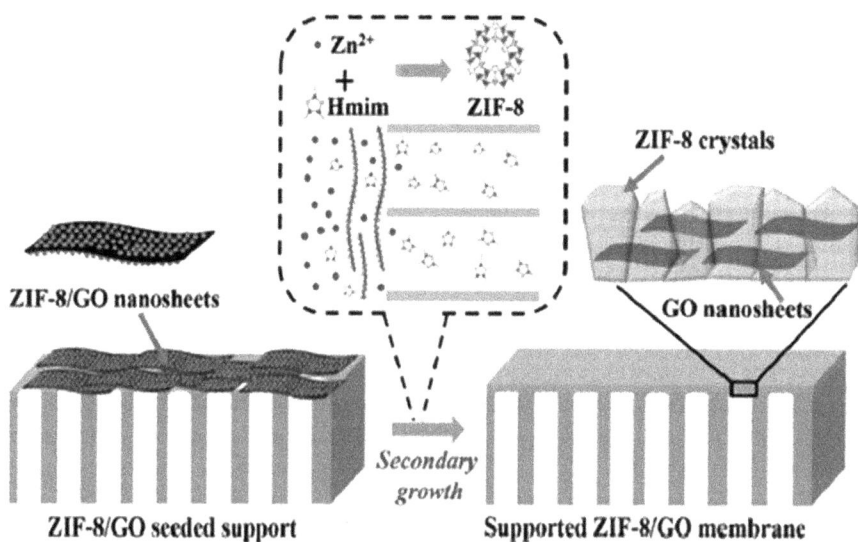

FIGURE 7.4 Schematic representation for the formation of ZIF-8-GO membrane using the seed growth method. Reproduced with permission from ref. 62. Copyright 2016 John Wiley and Sons.

vertically in the reaction vessel, which was then subjected to solvothermal heating for 4 h at 120°C, after which the autoclave was allowed to cool naturally to room temperature.

7.2.1.2.2 Transfer from Metal Oxides/Carbon-Based Materials

The synthesis of MOF-carbon composites can also be accomplished using metal oxides/carbon materials, whereby the metal oxide can be converted into MOF. Dai and coworkers [68] fabricated a ZIF-8/MWCNTs from CNT-attached ZnO nanoparticle-assisted growth of ZIF-8 crystal over CNTs. The ZnO nanoparticles were converted to ZIF-8 upon treatment with 2-methylimidazole. ZIF-8 crystals nucleated in situ around the MWCNTs, wherever the ZnO nanoparticles were located. The process could be divided into two steps. In the first step, MWCNTs were treated with ZnO nanoparticles in methanol to form a ZnO-CNT composite, whereas the second step involved the addition of 2-methylimidazole to the composite, followed by room temperature stirring for 48 h. The ZIF-8-CNT was obtained in the form of a grey powder, which was washed and finally dried. Using a similar process, Bechelany and coworkers [69] fabricated ZIF-8/PAN nanofibers and MIL-53-NH_2-PAN nanofiber mats.

7.2.1.2.3 Layer-by-Layer Assembly

Layer-by-layer assembly is one of the most systematic and straightforward methods for the fabrication of MOF-carbon composites and provides excellent control over structure penetration, composition, and layer thickness of MOFs [70–73]. Yu et al.

reported a facile and easy-to-handle procedure to produce an ordered MOF-GO composite with layered architecture [70]. HKUST-1- GO, ZIF-67-GO, and ZIF-8-GO were fabricated by systematic treatment of metal hydroxide nanowires. Copper hydroxide nanowires (Cu-NWs), cobalt carbonate hydroxide nanowires (Co-NWs) and zinc nitrate hydroxide nanoflakes (Zn-NFs) were used as the metal precursors for the formation of HKUST-1, ZIF-67, and ZIF-8, respectively. All the metal sources (nanowires and nanoflakes) were then mixed with GO by ultrasonic treatment. The nanomaterials not only acted as metal source for MOF formation on treatment with ligands, but also functioned as excellent spacers which keep GO layers separated and allow the MOF to grow freely in confined spaces between GO layers. The metal source-GO hybrid layer-by-layer film was prepared and exposed to the linker solution which makes space for the linker (1,3,5-benzenetricarboxylic acid and 2-methylimidazole) to interact with metal precursors. It is assumed that the metal precursors form plenty of interconnected open channels which can allow the ligand solution to flow through. Once the solution enters the spaces the ligands with dissociable protons can easily break the bond in metal hydroxides and react with free metal centers. It is expected that the process continues till the whole metal is released and no metal-hydroxide is left. The growth of MOF material spreads into all open spaces results in MOF-GO composite with layer-by-layer architecture (Figure 7.5).

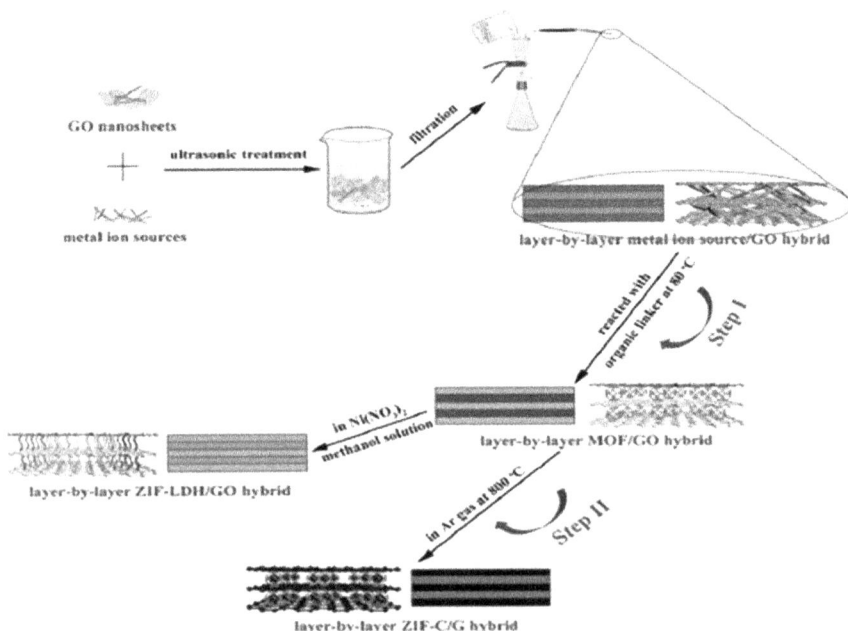

FIGURE 7.5   Schematic representation showing the synthesis of ZIF-GO hybrid films using layer by layer assembly and their supercapacitor electrode material. Reproduced with permission from ref. 70. Copyright 2017Royal Society of Chemistry.

7.2.2 Ex Situ Synthesis Approach

The excessive interference of carbon material during MOF synthesis due to the improper quantity of carbon-based materials in some cases of in situ synthesis can result in low performance or may interfere with the structural features of the MOF [13]. Therefore, under such circumstances, ex situ approaches, which involve the integration of preformed MOFs with the carbon material becomes important for fabrication of ideal composites. The generally used ex situ approaches involve direct mixing, self-assembly approach, or other mechanical methods.

7.2.2.1 Direct Mixing

The direct mixing method generally involves the simple mixing of the MOFs with carbon materials to yield the composite material [74–76]. Bao et al. used the direct mixing approach to generate a MOF@rGO/sulfur multi composite for application in lithium sulfur batteries [74]. The synthesis involved the fine dispersion of GO in water under sonication, which yielded a homogeneous GO suspension. The pre-formed MOF, MIL-101(Cr), was then dispersed into the GO dispersion in a mass ratio of 1:1, followed by room temperature stirring, resulting in the formation of MIL-101 (Cr)-GO composite. The reduction of GO was then performed using hydrazine hydrate at 95°C for 12 h, which yielded a black-colored MIL-101 (Cr)-rGO composite. Finally, sulfur was loaded onto the composite by liquid phase infiltration yielding MIL-101 (Cr)@rGO/S multi composite.

Ben-Mansour et al. [75] reported a solid-state direct mixing method to fabricate MOF-CNT composites in which MWCNTs in different ratios were incorporated into Mg-MOF-74 and MIL-100(Fe) materials to yield MWCNT/Mg-MOF-74 and MWCNT/MIL-100(Fe). In the first step MOFs were prepared solvothermally from their respective reaction precursors (2,5-dihydroxyterephthalic acid and 1.4 g $Mg(NO_3)_2 \cdot 6H_2O$ for Mg-MOF-74 and for MIL-100(Fe) $Fe(NO_3)_3 \cdot 9H_2O$ and BTC). After completion of the reactions, the MOFs were purified by multi-step washing. The MOF-CNT composite was synthesized by direct physical mixing of MWCNTs in powder form with solid powder of the MOFs at different weight percentage of MWCNTs.

7.2.2.2 Self-Assembly Method

It is a special type of solvent-assisted direct mixing method in which pre-synthesized MOF is directly mixed with carbon material in solution/dispersed form. Secondary forces such as electrostatic interactions, π-π stacking, hydrogen bonding, van der Waals forces etc., play key roles in holding the composite components together to yield MOF-C hybrid composites [77–82]. Following this route Cheng et al. [77] synthesized GO/Co-MOF composite via an electrostatic self-assembly. The synthesis was achieved by dropwise addition of a freshly exfoliated GO suspension in DMF to freshly synthesized Co-MOF dispersed in DMF under stirring. The stirring process was continued for an additional hour after complete addition of the GO suspension. The resulting uniform mixture was filtered through organic membrane filter and solid layer left on the membrane was peeled off and left for vacuum drying at 60°C for 24 h. The zeta potential measurement showed that the surface of GO was electronegative (measured zeta potential −52.0 mV), which was attributed to the presence of plenty of residual oxygen functionalities. On the other hand, Co-MOF was positive in nature (zeta potential of +59.1 mV). When these oppositely charged nanomaterials come together and interact, they form a composite due to electrostatic self-assembly. Using a similar electrostatic self-assembly approach, Shen et al. [78] fabricated a NH_2-mediated UiO-66-rGO composite, where UiO-66 (NH_2) was positively charged, while GO was negatively charged. Lin and coworkers [79] also employed the self-assembly approach for the fabrication of ZIF-67-GO composites, which were later used as precursors for the fabrication of a new Co-reduced graphene composite, which was magnetic in nature. The ZIF-67-GO composite was carbonized under nitrogen at 600°C for 6 h to yield the magnetic Co-graphene composite (Figure 7.6).

Li et al. [80] recently investigated the interactions of GO with a family of Zr-MOFs at different pH. Depending upon the pH, the MOFs possessed a net positive or negative charge, while the GO was negatively charged. When MOFs were positively charged, they electrostatically interacted with the negatively charged GO to yield MOF-GO composites. On the other hand, when the MOF was negatively charged, a composite formation was not expected due to the repulsion between both the materials of the same charge. Nevertheless, adsorption of GO on the MOFs was observed, which the authors attributed to other non-covalent interactions such as π-π stacking and acid-base interaction, which can overcome the effect of electrostatic repulsion.

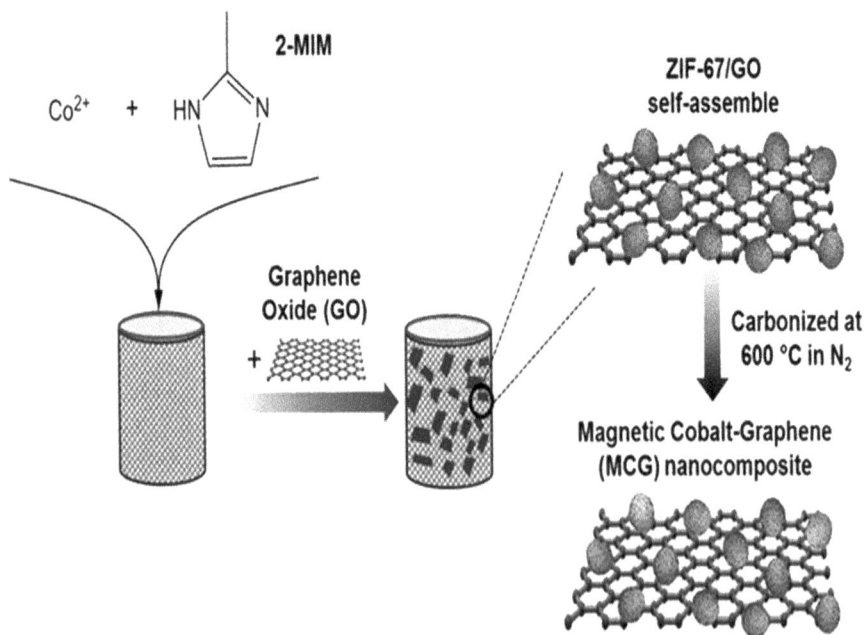

FIGURE 7.6 Schematic representation for the formation of magnetic cobalt-graphene nanocomposite using self-assembly approach. Reproduced with permission from ref. 79. Copyright 2015 Royal Society of Chemistry.

7.2.3 Miscellaneous Approach

In addition to the in situ and ex situ synthetic approaches, a few methods which are less common but effective for synthesis of composites are also used.

7.2.3.1 Pickering Emulsion-Induced Growth

Pickering emulsion is a kind of system where solid particles occupy the interphase of a two phase and stabilizes it. If solid particles are added to the emulsifying mixture like water-oil emulsion, the solid particles accumulate at the water-oil interphase and stabilize the emulsion by decreasing the energy of the system. In recent years MOFs and GO have also been found as emerging emulsifiers owing to their amphiphilic nature [83,84] (MOF: hydrophilic metal/metal cluster and hydrophobic aromatic ligands, GO: hydrophobic carbon ring system and hydrophilic oxygen surface functional groups). Because of these unique characteristics they are suitable candidates for stabilizing Pickering emulsions. Using the Pickering emulsion method, Bian et al. [85] synthesized $Cu_3(BTC)_2$-GO composite. In their studies, trimesic acid in octanol formed the oil phase, while the $Cu_3(BTC)_2$ and GO in water formed the aqueous phase. Both the solutions were mixed and emulsified using a B25 emulsifier at a speed of 10,000 rpm for 300 s followed by heating at 60°C for 1 h, which resulted in the formation of the MOF-GO composite at the interface, which was separated and vacuum dried at 60°C for 24 h. CO_2 uptake studies were performed and the composite formed by the Pickering emulsion method showed better activity than the one obtained by simple mixing the MOF and GO. (Figure 7.7).

Zhang et al. [86] fabricated Zr-BDC-NO_2-GO composite using a similar Pickering emulsion method using a two-step process. In the first step the emulsion was prepared by addition of cyclohexane to the slurry of ultrasonically dispersed Zr-BDC-NO_2 and GO in water. Emulsification was performed and controlled by ultrasonication to result in the formation of well-dispersed water/cyclohexane/ Zr-BDC-NO_2-GO Pickering emulsion where water/cyclohexane boundaries were well protected MOF-GO composite components. This procedure was followed by second step in which the emulsion was solidified by liquid nitrogen and the solvent systems were removed by freeze drying which yielded solid material of Zr-BDC-NO_2-GO composite.

7.2.3.2 In Situ Polymerization Method

This method involves the formation of a covalent metal-polymeric framework over the carbon materials by in situ polymerization of metal complex monomers to yield covalent-metal polymeric framework-CNT composites. Du and coworkers developed a multilayered covalent cobalt framework wrapped MWCNT composite ((CoP)n-MWCNTs) by in situ polymerization of cobalt porphyrin monomer (CoP) in the presence of MWCNTs [87]. The covalent metal-polymer-CNT composite was synthesized by sonication and room temperature stirring of a mixture of CoP and MWCNTS in NMP in the presence of tetramethylethylenediamine (TMEDA) and CuCl as catalysts and bubbling of O_2 for additional 48 h. In a similar manner, Yang et al. produced a CNTs coated by cationic Zn-porphyrin polymer-CNT composite using the in situ polymerization method [88].

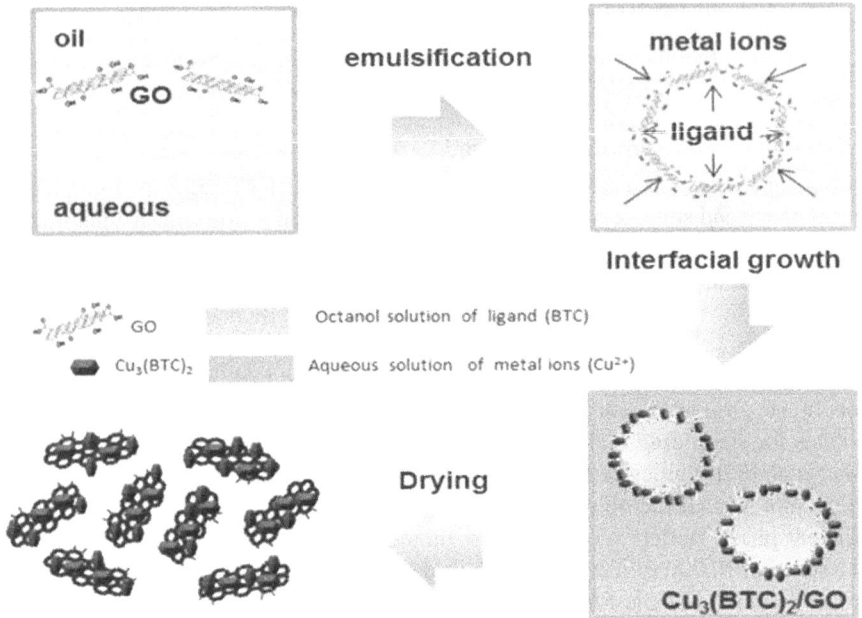

FIGURE 7.7 Schematic representation for the formation of Cu3(BTC)2-GO nanocomposite using Pickering emulsion induced growth. Reproduced with permission from ref. 85. Copyright 2015 American Chemical Society.

7.3 CONCLUSION

The development of hybrid MOF-carbon composites has improved the stabilities of the MOFs and has expanded the range of their applications. The fabrication of MOF-carbon composites is a step towards exploiting the full potential of MOFs. Combining MOFs with carbon materials brings about several newer functionalities such as enhanced stabilities and structural changes, which are crucial in order to realize the applications of MOFs for practical purposes. The formation of MOF-carbon composites such as MOF-GO and MOF-CNTs has improved several properties as compared to bare MOFs. Indeed, the composites have shown much better mechanical and chemical stability and conductivities, as well as adsorption properties compared to bare MOFs.

Various methods of synthesis of MOF-carbon composites are important as the properties of the composites are highly dependent upon the methods of synthesis. Different strategies involved for the synthesis of MOF-GO and MOF-CNT nanocomposites have been discussed in detail in the chapter. While in situ fabrication methods are relatively simple and easy to handle and involve the addition of carbon materials to the MOF precursors followed by synthesis of MOFs, ex situ synthesis methods integrate the preformed MOF with GO or CNTs mainly via the self-assembly approach, involving various non-covalent interactions. Although these synthesis methods have proved to be highly successful, the specific function-led synthesis of these composites is still in its infancy. More scalable and facile synthetic strategies are highly desired in order to take complete advantage of the MOF-carbon composite materials.

ABBREVIATIONS

MOF	Metal-organic framework
GO	Graphene oxide
CNT	Carbon nanotube
MWCNT	Multi-wall carbon nanotubes
FA	Furfuryl alcohol
HATA	5-amino-1-*H*-tetrazole
BPDC	4,4′-biphenyldicarboxylic acid
H$_2$BDC	1,4-benzenedicarboxylic acid
4-bdph	(2,5-bis(4-pyridyl)-3,4-diaza-2,4-hexadiene)
MW	Microwave
rGO	Reduced graphene oxide
BTC	Trimesic acid
PVP	Polyvinyl pyrollidone
DMF	Dimethylformamide
DCM	Dichloromethane
PXRD	Powder X-ray diffraction
FTIR	Fourier Transform Infrared Spectroscopy
SEM	Scanning Electron Microscopy

REFERENCES

1. H.-C. Zhou, S. Kitagawa. Metal–organic frameworks (MOFs). *Chem. Soc. Rev.* 43 (2014) 5415–5418.
2. S. Kitagawa, R. Kitaura, S. -I. Noro. Functional porous coordination polymers. *Angew. Chem. Int. Ed.* 43 (2004) 2334–2375.
3. A. Y. Robin, K. M. Fromm. Coordination polymer networks with O- and N-donors: What they are, why and how they are made. *Coord. Chem. Rev.* 250 (2006) 2127–2157.
4. S. Yuan, L. Feng, K. Wang, J. Pang, M. Bosch, C. Lollar, Y. Sun, J. Qin, X. Yang, P. Zhang, Q. Wang, L. Zou, Y. Zhang, L. Zhang, Y. Fang, J. Li, H. -C. Zhou. Stable metal-organic frameworks: Design, synthesis, and applications. *Adv. Mater.* 30 (2018) 1704303.
5. I. T. Hillman, A. Laybourn, C. Dodds, S. W. Kingman. Realising the environmental benefits of metal-organic frameworks: Recent advances in microwave synthesis. *J. Mater. Chem. A* 6 (2018) 11564–11581.
6. H. -C. Zhou, J. R. Long, O. M. Yaghi. Introduction to metal–organic frameworks. *Chem. Rev.* 112 (2012) 673–674.
7. L. E. Kreno, K. Leong, O. K. Farha, M. Allendorf, R. P. Van Duyne, J. T. Hupp. Metal-organic framework materials as chemical sensors. *Chem. Rev.* 112 (2012) 1105–1125.
8. K. Sumida, D. L. Rogow, J. A. Mason, T. M. McDonald, E. D. Bloch, Z. R. Herm, T.-H. Bae, J. R. Long. Carbon dioxide capture in metal-organic frameworks. *Chem. Rev.* 112 (2012) 724–781.

9. P. Horcajada, R. Gref, T. Baati, P. K. Allan, G. Maurin, P. Couvreur, G. Fèrey, R. E. Morris, C. Serre. Metal–organic frameworks in biomedicine. *Chem. Rev.* 112, (2012) 1232–1268.

10. T. Zhang, W. Lin. Metal–organic frameworks for artificial photosynthesis and photocatalysis. *Chem. Soc. Rev.* 43 (2014) 5982–5993.

11. L. Yang, Y. Han, X. Feng, J. Zhou, P. Qi, B. Wang. Metal-organic frameworks for energy storage: Batteries and supercapacitors. *Coord. Chem. Rev.* 307 (2016) 361–381.

12. Y. Xue, S. Zheng, H. Xue, H. Pang. Metal–organic framework composites and their electrochemical applications. *J. Mater. Chem. A* 7 (2019) 7301–7327.

13. X. -W. Liu, T. -J. Sun, J. -L. Hu, S. -D. Wang. Composites of metal-organic frameworks and carbon-based materials: Preparations, functionalities, and applications. *J. Mater. Chem. A* 4 (2016) 3584–3616.

14. Q. Yang, Q. Xu, H. -L. Jiang. Metal–organic frameworks meet metal nanoparticles: Synergistic effect for enhanced catalysis. *Chem. Soc. Rev.* 46 (2017) 4774–4808.

15. H. R. Moon, D.W. Lim, M. P. Suh. Fabrication of metal nanoparticles in metal-organic frameworks. *Chem. Soc. Rev.* 42 (2013) 1807–1824.

16. J. A. –Sigalat, D. Bradshaw. Synthesis and applications of metal-organic framework-quantum dot (QD@MOF) composites. *Coord. Chem. Rev.* 307 (2016) 267–291.

17. J. Gorka, P. F. Fulvio, S. Pikus, M. Jaroniec. Mesoporous metal-organic framework-boehmite and silica composites. *Chem. Commun.* 46 (2010) 6798–6800.

18. S. Sorribas, B. Zornoza, C. Tellez and J. Coronas. Ordered mesoporous silica-(ZIF-8) core-shell spheres. *Chem. Commun.* 48 (2012) 9388–9390.

19. Z. Karimi, A. Morsali. Modulated formation of metal-organic frameworks by oriented growth over mesoporous silica. *J. Mater. Chem. A* 1 (2013) 3047–3054.

20. Z. Zhang, H. T. H. Nguyen, S. A. Miller, S. M. Cohen, polyMOFs: A class of interconvertible polymer-metal-organic-framework hybrid materials. Angew. Chem. Int. Ed. 54 (2015), 6152–6157.

21. J. Huo, M. Marcello, A. Garai, D. Bradshaw. MOF-polymer composite microcapsules derived from Pickering emulsions. *Adv. Mater.* 25 (2013) 2717–2722.

22. A. J. Brown, J. R. Johnson, M. E. Lydon, W. J. Koros, C. W. Jones, S. Nair. Continuous polycrystalline zeolitic imidazolate framework-90 membranes on polymeric hollow fibers. *Angew. Chem. Int. Ed.* 51 (2012) 10615–10618.

23. J. Juan-Alcaniz, J. Gascon, F. Kapteijn. Metal–organic frameworks as scaffolds for the encapsulation of active species: State of the art and future perspectives. *J. Mater. Chem.* 22 (2012) 10102–10118.

24. L. E. Lange, S. K. Obendorf. Functionalization of cotton fiber by partial etherification and self-assembly of polyoxometalate encapsulated in $Cu_3(BTC)_2$ metal-organic framework. *ACS Appl. Mater. Interfaces* 7 (2015) 3974–3980.

25. L. Zhu, L. Meng, J. Shi, J. Li, X. Zhang, M. Feng. Metal-organic frameworks/carbon-based materials for environmental remediation: A state-of-the-art mini-review. *J. Environ. Manag.* 232 (2019) 964–997.

26. M. Muschi, C. Serre. Progress and challenges of graphene oxide/metal-organic composites. *Coord. Chem. Rev.* 387 (2019) 262–272.

27. Y. Zheng, S. Zheng, H. Xue, H. Pang. Metal-organic frameworks/graphene-based materials: Preparations and applications. Adv. Funct. Mater. 28 (2018) 1804950.

28. S. Sundriyal, H. Kaur, S. K. Bhardwaj, S. Mishra, K. -H. Kim, A. Deep. Metal-organic frameworks and their composites as efficient electrodes for supercapacitor applications. *Coord. Chem. Rev.* 369 (2018) 15–38.

29. C. Petit, T. J. Bandosz. Engineering the surface of a new class of adsorbents: Metal–organic framework/graphite oxide composites. *J. Colloid. Interface Sci.* 447 (2015) 139–151.

30. C. Petit, T. J. Bandosz. Exploring the coordination chemistry of MOF-graphite oxide composites and their applications as adsorbents. *Dalton Trans.* 41 (2012) 4027–4035.

31. C. Petit, T. J. Bandosz. MOF-graphite oxide composites: Combining the uniqueness of graphene layers and metal–organic frameworks. *Adv. Mater.* 21 (2009) 4753–4757.
32. L. Xu, G. Fang, J. Liu, M. Pan, R. Wang, S. Wang. One-pot synthesis of nanoscale carbon dots-embedded metal–organic frameworks at room temperature for enhanced chemical sensing. *J. Mater. Chem. A* 4 (2016) 15880–15887.
33. S. Li, F. Huo. Metal–organic framework composites: From fundamentals to applications. *Nanoscale* 7 (2015) 7482–7501.
34. Y. Guo, H. Yang, X. Zhou, K. Liu, C. Zhang, Z. Zhou, C. Wang, W. Lin. Electrocatalytic reduction of CO_2 to CO with 100% faradaic efficiency by using pyrolyzed zeolitic imidazolate frameworks supported on carbon nanotube networks. *J. Mater. Chem. A* 5 (2017) 24867–24873.
35. J. Abdi, M. Vossoughi, N. M. Mahmoodi, I. Alemzadeh. Synthesis of metal-organic framework hybrid nanocomposites based on GO and CNT with high adsorption capacity for dye removal. *Chem. Eng. J.* 326 (2017) 1145–1158.
36. L. Xiong, J. Liu, M. Yu, S. Li. Improving the corrosion protection properties of PVB coating by using salicylaldehyde@ZIF-8/graphene oxide two-dimensional nanocomposites. *Corros. Sci.* 146 (2019) 70–79.
37. J. Li, Z. Wua, Q. Duana, X. Li, X. Tan, A. Alsaedi, T. Hayat, C. Chen. Mutual effects behind the simultaneous U(VI) and humic acid adsorption by hierarchical MWCNT/ZIF-8 composites. *J. Mol. Liq.* 288 (2019) 110971.
38. X. Chen, D. Liu, G. Cao, Y. Tang, C. Wu. In situ synthesis of a sandwich-like graphene@zif-67 heterostructure for highly sensitive nonenzymatic glucose sensing in human serums. *ACS Appl. Mater. Interfaces* 11 (2019) 9374–9384.
39. S. Anastasiou, N. Bhoria, J. Pokhrel, K. S. K. Reddy, C. Srinivasakannan, K. Wang, G. N. Karanikolos. Metal-organic framework/graphene oxide composite fillers in mixed matrix membranes for CO_2 separation. *Mater. Chem. Phys.* 212 (2018) 513–522.
40. J. Pokhrel, N. Bhoria, C. Wua, K. S. K. Reddy, H. Margetis, S. Anastasiou, G. George, V. Mittal, G. Romanos, D. Karonis, G. N. Karanikolos. Cu– and Zr-based metal organic frameworks and their composites with graphene oxide for capture of acid gases at ambient temperature. *J. Solid State Chem.* 266, (2018) 233–243.
41. S. Zhang, Q. Yang, Z. Li, W. Wang, X. Zang, C. Wang, Z. Wang. Solid phase microextraction of phthalic acid esters from vegetable oils using iron (III)-based metal-organic framework/graphene oxide coating. *Food Chem.* 263 (2018) 258–264.
42. N. Wei, X. Zheng, H. Ou, P. Yu, Q. Lia, S. Fengab. Fabrication of an amine-modified ZIF-8@GO membrane for high-efficiency adsorption of copper ions. *New J. Chem.* 43 (2019) 5603–5610.
43. P. K. Samantaray, G. Madras, S Bose. Water remediation aided by a graphene-oxide-anchored metal organic framework through pore- and charge-based sieving of ions. *ACS Sustain. Chem. Eng.* 7 (2019) 1580–1590.
44. H. Su, Y. Du, J. Zhang, P. Peng, S. Li, P. Chen, M. Gozin, S. Pang. Stabilizing metastable polymorphs of metal-organic frameworks via encapsulation of graphene oxide and mechanistic studies. *ACS Appl. Mater. Interfaces* 10 (2018) 32828–32837.
45. X. Wang, N. Yang, Q. Li, F. He, Y. Yang, B. Wu, J. Chu, A. Zhou, S. Xiong. Solvothermal synthesis of flower-string-like NiCo-MOF/MWCNT composites as a high-performance supercapacitor electrode material. *J. Solid State Chem.* 277 (2019) 575–586.
46. M. Oveisi, M. A. Asli, N. M. Mahmoodi. Carbon nanotube-based metal-organic framework nanocomposites: Synthesis and their photocatalytic activity for decolorization of colored wastewater. *Inorganica Chim. Acta* 487 (2019) 169–176.
47. E. Akbarzadeh, H. Z. Soheili, M. R. Gholami. Novel Cu_2O/Cu-MOF/rGO is reported as highly efficient catalyst for reduction of 4-nitrophenol. *Mater. Chem. Phys.* 237 (2019) 121846.

48. J. Meng, Q. Chen, J. Lu, H. Liu, ZScheme photocatalytic CO_2 reduction on a hetero-structure of oxygen-defective Zno/reduced graphene oxide/UiO-66-NH_2 under visible light. *ACS Appl. Mater. Interfaces* 11 (2019) 550–562.

49. X. -Y. Hou, X. -L. Yan, X. Wang, X. Li, Y. Jiang, M. Hu, Q. -G. Zhai. Excellent super-capacitor performance of robust nickel–organic framework materials achieved by tun-able porosity, inner-cluster redox, and in situ fabrication with graphene oxide. *Cryst. Growth Des.* 18 (2018) 6035–6045.

50. S. Luo, J. Wang. MOF/graphene oxide composite as an efficient adsorbent for the removal of organic dyes from aqueous solution. *Environ. Sci. Pollut. Res.* 25 (2018) 5521–5528.

51. C. Petit, T. J. Bandosz. Synthesis, characterization, and ammonia adsorption proper-ties of mesoporous metal-organic framework (MIL(Fe))-graphite oxide composites: Exploring the limits of materials fabrication. *Adv. Funct. Mater.* 21 (2011) 2108–2117.

52. M. Tanhaei, A. R. Mahjoub, V. Safarifard. Energy-efficient sonochemical approach for the preparation of nanohybrid composites from graphene oxide and metal-organic framework. *Inorg. Chem. Commun.* 102 (2019) 185–191.

53. M. Tanhaei, A. R. Mahjoub, V. Safarifard. Ultrasonic-assisted synthesis and charac-terization of nanocomposites from azine-decorated metal-organic framework and gra-phene oxide layers. *Mater. Lett.* 227 (2018) 318–321.

54. M. Tanhaei, A. R. Mahjoub, V. Safarifard. Sonochemical synthesis of amide-function-alized metal-organic framework/graphene oxide nanocomposite for the adsorption of methylene blue from aqueous solution. *Ultrasonics—Sonochem.* 41 (2018) 189–195.

55. S. Zhao, D. Chen, F. Wei, N. Chen, Z. Liang, Y. Luo. Removal of Congo red dye from aqueous solution with nickel-based metal-organic framework/graphene oxide com-posites prepared by ultrasonic wave-assisted ball milling. *Ultrasonics - Sonochem.* 39 (2017) 845–852.

56. S. N. Ansari, M. Saraf, A. Gupta, S. M. Mobin. Functionalized Cu-MOF@CNT hybrid: Synthesis, crystal structure and applicability in supercapacitors. *Chem. Asian. J.* 14 (2019) 3566–3571.

57. A. Dastbaz, J. Karimi-Sabet, M. A. Moosavian. Intensification of hydrogen adsorption by novel Cu-BDC@rGO composite material synthesized in a microwave-assisted cir-cular micro-channel. *Chem. Eng. Process Process Intensification* 135 (2019) 245–257.

58. B. Han, E. Zhang, G. Cheng. Facile preparation of graphene oxide-MIL-101(Fe) com-posite for the efficient capture of uranium. *Appl. Sci.* 8 (2018) 2270.

59. S. Zhao, D. Chen, F. Wei, N. Chen, Z. Liang, Y. Luo. Synthesis of graphene oxide/metal-organic frameworks hybrid materials for enhanced removal of Methylene blue in acidic and alkaline solutions. *J. Chem. Technol. Biotechnol.* 93 (2018) 698–709.

60. X. Wang, X. Zhao, D. Zhang, G. Li, H. Li. Microwave irradiation induced UiO-66-NH_2 anchored on graphene with high activity for photocatalytic reduction of CO_2. *Appl. Catal. B: Environ.* 228 (2018) 47–53.

61. X. Li, Z. Le, X. Chen, Z. Li, W. Wang, X. Liu, A. Wu, P. Xu, D. Zhang. Graphene oxide enhanced amine-functionalized titanium metal organic framework for visible-light-driven photocatalytic oxidation of gaseous pollutants. *Appl. Catal. B: Environ.* 236 (2018) 501–508.

62. Y. Hu, J. Wei, Y. Liang, H. Zhang, X. Zhang, W. Shen, H. Wang. Zeolitic imidazo-late framework/graphene oxide hybrid nanosheets as seeds for the growth of ultrathin molecular sieving membranes. *Angew. Chem. Int. Ed.* 55 (2016) 2048–2052.

63. L. Dumèe, L. He, M. Hill, B. Zhu, M. Duke, J. Schütz, F. She, H. Wang, S. Gray, P. Hodgson, L. Kong. Seeded growth of ZIF-8 on the surface of carbon nanotubes towards self-supporting gas separation membranes. *J. Mater. Chem. A* 1 (2013) 9208–9214.

64. A. Gholidoust, J. W. Maina, A. Merenda, J. A. Schütz, L. Kong, Z. Hashisho, L. F. Dumée. CO₂ sponge from plasma enhanced seeded growth of metal organic frameworks across carbon nanotube bucky-papers. *Sep. Purif. Technol.* 209 (2019) 571–579.

65. Y. Hu, Y. Wu, C. Devendran, J. Wei, Y. Liang, M. Matsukata, W. Shen, A. Neild, H. Huang, H. Wang. Preparation of nanoporous graphene oxide by nanocrystal-masked etching: Toward a nacre mimetic metal-organic framework molecular sieving membrane. *J. Mater. Chem. A* 5 (2017) 16255–16262.

66. E. Shamsaei, X. Lin, L. Wan, Y. Tong, H. Wang. A one-dimensional material as a nano-scaffold and a pseudo-seed for facilitated growth of ultrathin, mechanically reinforced molecular sieving membranes. *Chem. Commun.* 52 (2016) 13764–13767.

67. L. Kong, X. Zhang, H. Liu, T. wang, J. Qiu. Preparation of ZIF-8 membranes supported on macroporous carbon tubes via a dipcoating-rubbing method. *J. Phys. Chem. Solids* 77 (2015) 23–29.

68. Y. Yue, B. Guo, Z-A. Qiao, P. F. Fulvio, J. Chen, A. J. Binder, C. Tian, S. Dai. Multi-wall carbon nanotube@zeolite imidazolate framework composite from a nanoscale zinc oxide precursor. *Micropor. Mesopor. Mat.* 198 (2014) 139–143.

69. M. Bechelany, M. Drobek, C. Vallicari, A. Abou Chaaya, A. Julbe, P. Miele. Highly crystalline MOF-based materials grown on electrospun nanofibers. *Nanoscale* 7 (2015) 5794–5802.

70. D. Yu, L. Ge, X. Wei, B. Wu, J. Ran, H. Wang. T. Xu. A general route to the synthesis of layer-by-layer structured metal organic framework/graphene oxide hybrid films for high-performance supercapacitor electrodes. J. Mater. Chem. A 5 (2017) 16865–16872.

71. L. Wu, Z. Lu, J. Ye. Enzyme-free glucose sensor based on layer-by-layer electrodeposition of multilayer films of multi-walled carbon nanotubes and Cu-based metal framework modified glassy carbon electrode. *Biosens. Bioelectron.* 135 (2019) 45–49.

72. Y. Huang, Y. Xiao, H. Huang, Z. Liu, D. Liu, Q. Yang, C. Zhong. Ionic liquid functionalized multi-walled carbon nanotubes/zeolitic imidazolate framework hybrid membranes for efficient H₂/CO₂ separation. *Chem. Commun.* 51 (2015) 17281–17284.

73. A. Huang, Q. Liu, N. Wang, Y. Zhu, J. Caro. Bicontinuous zeolitic imidazolate framework ZIF-8@GO membrane with enhanced hydrogen selectivity. *J. Am. Chem. Soc.* 136 (2014) 14686–14689.

74. W. Bao, Z. Zhang, Y. Qu, C. Zhou, X. Wang, J. Li. Confine sulfur in mesoporous metal-organic framework @ reduced graphene oxide for lithium sulfur battery. *J. Alloys Compd.* 582 (2014) 334–340.

75. R. BenMansour, N. A. A. Qasem, M. A. Habib. Adsorption characterization and CO₂ breakthrough of MWCNT/MgMOF74 and MWCNT/MIL100(Fe) composites. *Int. J. Energy Environ. Eng.* 9 (2018) 169–185.

76. X. Zhou, W. Huang, J. Miao, Q. Xia, Z. Zhang, H. Wang, Z. Li. Enhanced separation performance of a novel composite material GrO@MIL-101 for CO₂/CH₄ binary mixture. *Chem. Eng. J.* 266 (2014) 339–344.

77. J. Cheng, J. Liang, L. Dong, J. Chai, N. Zhao, S. Ullah, H. Wang, D. Zhang, S. Imtiaz, G. Shan, G. Zheng. Self-assembly of 2D-metal-organic framework/graphene oxide membranes as highly efficient adsorbents for the removal of Cs⁺ from aqueous solutions. *RSC Adv.* 8 (2018) 40813–40822.

78. L. Shen, L. Huang, S. Liang, R. Liang, N. Qin, L. Wu. Electrostatically derived self-assembly of NH₂-mediated zirconium MOFs with graphene for photocatalytic reduction of Cr(VI). *RSC Adv.* 4 (2014) 2546–2549.

79. K. -Y. A. Lin, F. -K. Hsu, W. -D. Lee. Magnetic cobalt-graphene nanocomposite derived from self-assembly of MOFs with graphene oxide as an activator for peroxymonosulfate. *J. Mater Chem. A* 3 (2015) 9480–9490.

80. J. Li, Q. Wu, X. Wang, Z. Chai, W. Shi, J. Hou, T. Hayat, A. Alsaedi, X. Wang. Heteroaggregation behavior of graphene oxide on Zr-based metal-organic frameworks in aqueous solutions: A combined experimental and theoretical study. *J. Mater. Chem. A* 5 (2017) 20398–20406.

81. J. Mao, M. Ge, J. Huang, Y. Lai, C. Lin, K. Zhang, K. Meng, Y. Tang. Constructing multifunctional MOF@rGO hydro-/aerogels by the self-assembly process for customized water remediation. *J. Mater. Chem. A* 5 (2017) 11873–11881.

82. F. Yang, M. Wu, Y. Wang, S. Ashtiani, H. Jiang. A GO-induced assembly strategy to repair MOF nanosheet-based membrane for efficient H_2/CO_2 separation. *ACS Appl. Mater. Interfaces* 11 (2019) 990–997.

83. S. Cui, M. Qin, A. Marandi, V. Steggles, S. Wang, X. Feng, F. Nouar, C. Serre. Metal-organic frameworks as advanced moisture sorbents for energy-efcient high temperature cooling. *Sci. Rep.* 8 (2018) 15284.

84. A. J. P. Neto, E. E. Fileti. Elucidating the amphiphilic character of graphene oxide. *Phys. Chem. Chem. Phys.* 20 (2018) 9507–9515.

85. Z. Bian, J. Xu, S. Zhang, X. Zhu, H. Liu, J. Hu. Interfacial growth of metal organic framework/graphite oxide composites through Pickering emulsion and their CO_2 capture performance in the presence of humidity. *Langmuir* 31 (2015) 7410–7417.

86. F. Zhang, L. Liu, X. Tan, X. Sang, J. Zhang, C. Liu, B. Zhang, B. Han, G. Yang. Pickering emulsions stabilized by a metal-organic framework (MOF) and graphene oxide (GO) for producing MOF/GO composites. *Soft Matter* 13 (2017) 7365–7370.

87. H. Jia, Z. Sun, D. Jiang, P. Du. Covalent cobalt porphyrin framework on multiwalled carbon nanotubes for efficient water oxidation at low overpotential. *Chem. Mater.* 27 (2015) 4586–4593.

88. S. Jayakumar, H Li, J. Chen, Q. Yang. Cationic Zn-porphyrin polymer coated onto CNTs as a cooperative catalyst for the synthesis of cyclic carbonates. *ACS Appl. Mater. Interfaces* 10 (2018) 2546–2555.

8 Application of Nanoscale Metal-Organic Frameworks for Phototherapy of Cancer

Bhagwati Sharma, Tridib K. Sarma, and Anish Khan

CONTENTS

8.1 INTRODUCTION

Cancer, being one of the deadliest diseases, has been a threat to humans for several decades and has attracted immense research interest [1–4]. The most widely used methods to treat cancer involve radiotherapy, surgery, and chemotherapy. Each of these methods, however, suffers from certain limitations [5,6]. For instance, radiotherapy is known to cause toxic side effects such as damage to the normal tissues near the radiotherapy site as well as systemic immune suppression. Surgical treatment, on the other hand, suffers from its inability to treat metastasis tumors and tumors growing in sensitive parts of the body. Further, in most of the cases, the tumor cells are not completely removed. Chemotherapy uses chemical drugs to control and kill the cancer cells. These drugs can effectively be circulated throughout the human body and are effective towards systemic treatment. However, they have poor targeting ability and hence cause serious side effects to normal tissues. Therefore, there has been an urge to develop alternative cancer treatment methods that are not only safe and cost-effective, but also have the potential to completely remove/kill the cancer cells without any side effects to other normal tissues. In this regard, cancer therapy by the use of light, commonly known as phototherapy has recently gained immense attention [2,3,7–10]. Phototherapy, which employs radiation energy, is a non-invasive clinical approach with marginal side effects, used for the treatment of tumor cells. It makes use of near infrared light of wavelength 650–1300 nm for excitation of the photosensitizers (PS) present in the tumor sites to generate chemically reactive species (photodynamic therapy) or sufficient heat in the body (photothermal therapy), capable of killing tumor cells. Photodynamic therapy (PDT) involves the administration of PS to the tumor cells, followed by illumination of the region using infrared light of suitable wavelength. This leads to the excitation of the PS, which then transfer their excess energy to molecular oxygen, generating highly cytotoxic reactive oxygen species (ROS) such as singlet oxygen ($^1O_2$), which consequently leads to cell death and tissue destruction [11,12]. In the case of photothermal therapy (PTT), PS are excited by the absorption of light, and the excess energy is released in the form of heat which elevates the temperature of the region above 40°C, leading to cell death [13–15].

Over the past two decades there has been a rapid development in the area of nanoscience and nanotechnology. The tunable size, good biocompatibility, and the ability to favorably deposit in the tumor via enhanced permeation and retention effect [16,17] has led to the use of nanoparticles for biomedical applications such as sensing, drug delivery, cancer imaging, and phototherapy, etc. [18–21]. Nanomaterials are broadly classified into three different types, viz., purely organic, purely inorganic, and hybrid nanomaterials [22]. While purely organic nanoparticles constitute micelles, dendrimers, liposomes, and polymeric hydrogel nanoparticles, purely inorganic nanoparticles constitute metal nanoparticles, metal oxide nanoparticles, quantum dots, etc. On the other hand, the hybrid nanomaterials constitute nanoscale metal-organic frameworks and coordination polymer particles.

Metal-organic frameworks (MOFs), a class of hybrid, porous material, can be built using diverse combinations of metal ions and organic linkers [23,24]. The properties of MOFs can easily be tuned as a result of the diversity of their constitutional unit. MOFs offer several advantages such as high porosity, tunable pore size, and good biocompatibility when compared to the traditional porous materials such as zeolites and carbon-based materials [25,26]. As a result of these advantages, MOFs have been employed in a variety of important applications such as sensing, gas separation and storage, energy and environment, and catalysis as well as biomedicine [27–30]. The last decade has witnessed the rise of miniaturized nanoscale MOFs (NMOFs) for applications in nanomedicine and cancer therapy [31]. Due to the regularity and tunability in structure, high porosity as well as biodegradability, NMOFs hold promise for application in phototherapy [22,32]. In the case of PDT, the tunability in the composition of the NMOFs permits the incorporation of several PS, leading to a high loading of the PS. On the other hand, the structural regularity prevents the PS from self-quenching, by keeping them isolated from one another. Similarly, the highly porous nature of the NMOFs helps in the movement of ROS to exert cytotoxic effects and the biodegradable nature of NMOFs decreases the concerns related to their long-term toxicity. Moreover, the nanometer dimension of the NMOFs enhances the solubility and leads to an increase in the cellular uptake of the PS [32,33]. In the case of PTT, the versatile tunability of the NMOFs allows the functionalization of both the metal-cluster secondary building units and bridging linkers such that the efficiency of photothermal conversion can be increased. This chapter deals with the use of various NMOFs for PDT and PTT of cancer as well as a combination of both and summarizes the recent advances in the particular field.

8.2   NMOFs FOR PHOTODYNAMIC THERAPY

The tunable, porous, and crystalline structure of MOFs makes them impressive nanocarriers for delivery of contrast agents and therapeutic cargos. The optimal size of the nanocarriers must be 10–200 nm such that rapid renal clearance can be avoided and an enhanced permeation and retention (EPR) effect can be observed [16,17]. Several factors such as concentration of the precursors, solvents, reaction temperatures, etc. play a key role in controlling the size and shape of the NMOFs. The porous and crystalline nature of NMOFs prevents self-quenching of the PS by keeping them isolated from one another and the biodegradable nature of the NMOFs ensures that long-term toxicity is alleviated [22]. The first report on the use of NMOFs for PDT was reported by the Lin group in 2014 [34] and since then several other papers have been published, mostly by the Lin group [34–40], indicating the importance of this class of nanomaterials for PDT. Basically, there are three types of PS that are employed in PDT by NMOFs, viz., porphyrin- and chlorin-based PS [34–47], phthalocyanine-based PS [48, 49] and BODIPY-based PS [50,51]. However, most of the reported methods of PDT employ porphyrins as PS. The reports on all these three types of PS are discussed in the subsequent pages.

8.2.1   PORPHYRIN-BASED NMOFs FOR PDT

Porphyrin derivatives are an effective photosensitizer capable of producing toxic ROS. However, their hydrophobic nature and reduced selectivity towards the malignant tissues prevents their delivery to the target tumor cells. The miniaturization of MOFs into functional nanomaterials has provided ample opportunities to mitigate the disadvantages associated with the porphyrin derivatives. By incorporating the porphyrin-based derivatives into the MOF structure, a high loading of the porphyrin can be achieved. The porphyrin-incorporated MOFs show good water dispersibility and resistance towards aggregation and self-quenching, often associated with porphyrin molecules.

8.2.1.1   Porphyrins as MOF Organic Linkers

Lin et al. first reported the use of a Hf-porphyrin NMOF (DBP-UiO) as PS for PDT of head and neck cancer by using Hf^{4+} as metal ion and 5,15-di(p-benzoato) porphyrin (H_2DBP) as organic linker [34]. The NMOF was synthesized employing the solvothermal approach, whereby $HfCl_4$ was treated with H_2DBP in DMF at 90°C for three days in an oven. Electron microscopic studies suggested the formation of DBP-UiO NMOFs with a plate-like morphology having a diameter of approximately 100 nm and thickness of 10 nm. The thickness of the nanoplates (10 nm) was suited for diffusion of $^1O_2$ from the interior of the NMOF to the cell cytoplasm to exert cytotoxic effects. A porphyrin loading of 77 wt% was observed in the NMOF and it exhibited at least twice the $^1O_2$ generation than H_2DBP alone as observed from the singlet oxygen sensor green (SOSG) assay. The DBP-UiO NMOF was highly stable in physiological conditions and prevented the self-quenching of PS and the porous structure of the NMOF ensured the proper transport of the generated $^1O_2$. The PDT efficiency of the DBP-UiO NMOF was investigated by incubating the NMOF with SQ20B head and neck cancer cells. They observed that upon incubation for 4 h, up to 30% of the NMOF was taken up by the tumor cells, indicating that a high PS concentration could be achieved in the tumor cells. The PDT efficacy was studied by illuminating the cancer cells incubated with H_2DBP alone and DBP-UiO NMOF at different concentrations under 640 nm light for 15 or 30 minutes. Excellent PDT efficacy was observed for NMOF-treated cancer cells, while only a moderate PDT efficacy was observed for the cells treated only with the ligand. In vivo studies were also performed using the NMOF using a mouse model. A significant PDT effect was observed after a single illumination with a 640 nm LED (100 mW/cm^2) for 30 minutes as indicated by the complete eradication in half of the mice with a single dose of DBP-UiO (3.5 mg/kg) (Figure 8.1).

The lowest energy absorption for DBP-UiO had a relatively small extinction coefficient of 2200 M^{-1}cm^{-1} at 634 nm, which is close to the high energy edge of the tissue penetrating window. Lin and coworkers reduced the porphyrin H_2DBP to chlorin H_2DBC using toluenesulfonhydrazide [35], which led to a red shift in the lowest energy absorption from 634 to 646 nm. The extinction coefficient for the chlorin system also increased 11-fold with a value of 2200 M^{-1}cm^{-1}. Using the earlier method, they synthesized Hf-DBC NMOF with similar plate-like morphology. The diameter of these nanoplates was in the range of 100–200 nm, while their thickness was in the range of 3.3–7.5 nm. The smaller thickness of DBC-Hf as compared to DBP-UiO (10 nm), suggested that DBC-Hf NMOFs could facilitate the ROS diffusion better than the DBP-Hf NMOFs during PDT. As expected, DBC-Hf was found to generate $^1O_2$ oxygen three times more than that of DBP-UiO NMOF. The PDT efficacy of DBC-Hf was studied by incubating CT26 and HT29 colon cancer mouse models with DBC-Hf followed by irradiation with a light (650 nm) at 0.1 W/cm^2 for four cycles, each of 15 minutes duration. The effect of the PDT treatment was analyzed by flow cytometry using an Alexa Fluor 488 Annexin V/dead cell apoptosis kit. While no apoptosis was observed for cells treated with DBC-Hf in the dark, significant apoptosis was observed for cells treated with Hf-DBC upon illumination with light, clearly indicating the PDT efficacy of the NMOFs. Using a ligand dose of 3.5 mg/kg, DBC-Hf could eradicate tumors in a single treatment in HT29 model, while two treatments were required for the eradication in CT26 model.

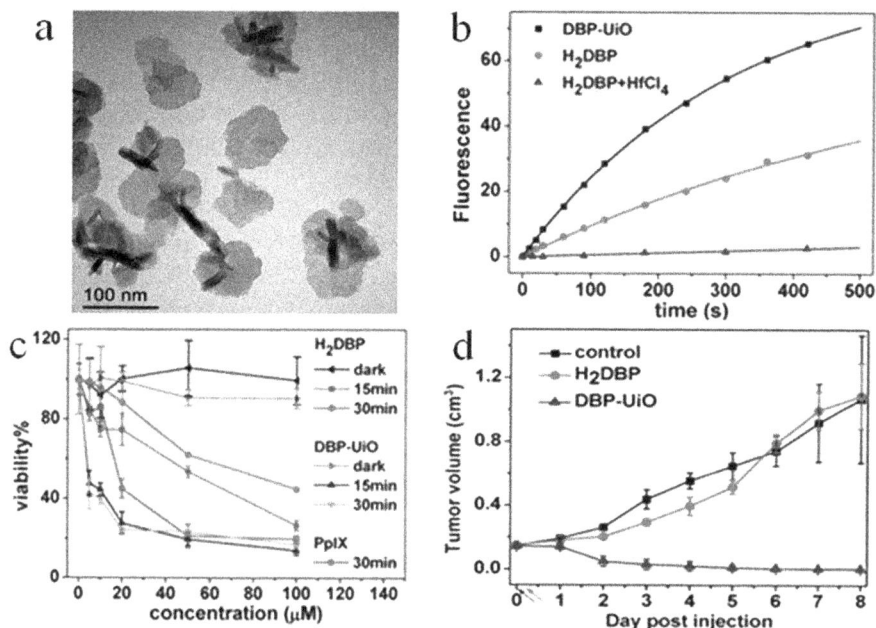

FIGURE 8.1 (a) TEM image of DBP-UIO NMOFs showing plate-like morphology, (b) SOSG assay showing the singlet oxygen generation efficiency of DBP-UiO, H$_2$BDP and H$_2$BDP+HfCl$_4$, (c) in vitro PDT cytotoxicity of H$_2$DBP, DBP–UiO, and PpIX at different PS concentrations and irradiation times, and (d) in vivo tumor growth inhibition curve of DBP-Hf and H$_2$BDP. Black and red arrows on X-axis refer to the time of injection and light irradiation respectively. Reproduced with permission from ref. 34. Copyright 2014 American Chemical Society.

In 2019 the same group also developed a Ti-based NMOF (Ti-TBP) by solvothermal reaction between TiCl$_4$.2THF and 5, 10, 15, 20-tetra(p-benzoato) porphyrin (H$_4$TBP) in DMF for seven days in the presence of acetic acid as modulator [38]. TEM images showed the formation of square-shaped nanoplates with an average size of 150 nm. The in vivo PDT efficacy of the system was studied on a colorectal adenocarcinoma model of CT26 tumor bearing BALB/c mice. Ti-TBP was intratumorally injected using a TBP dose of 0.2 μmole, followed by illumination with light of wavelength 650 nm (180 J/cm^2). The PDT treatment resulted in effective tumor regression of 98.4% in volume with a cure rate of 60%, in comparison to PBS in dark on day 20.

8.2.1.2 One Pot in Situ Synthesis Involving Porphyrin

In early 2018, Dong and coworkers reported the introduction of an iodine-attached Zn(II)-porphyrinic dicarboxylic building block (Zn-DTPP-I$_2$-2H) into UiO-66 NMOF [46]. The introduction of the Zn-DTPP-I$_2$-2H into the framework of UiO-66 did not alter its structural features but provided added advantages. Firstly, the iodine-attached porphyrin PS were dispersed well in the matrix of UiO-66, which avoided the aggregation induced quenching. Secondly, the introduction of heavy atom (iodine) helped promoting the intersystem crossing (ISC) process and enhanced the singlet oxygen generation capability and thirdly, due to the porous nature of the MOF, the generated singlet oxygen could easily be released. The NMOF was prepared by a solvothermal approach, wherein $ZrCl_4$, p-phthalic acid, Zn-DTPP-I_2-2H and acetic acid in DMF were heated for 24 h at 120°C. The NMOFs formed had spherical morphology with an average diameter of 110 nm. The PDT efficacy of the NMOFs was studied against HEPG2 cell line using standard MTT assay using different concentrations of NMOFs after illumination with a green LED of 540 nm wavelength and power density 20 mW/cm². The results showed that while 87% cell viability was observed in the absence of light, the illumination of the cells with NMOFs (40 μg/mL) led to a decline in cell viability to less than 25%, indicating the excellent PDT efficacy of the synthesized NMOFs (Figure 8.2).

8.2.2 PHTHALOCYANINE CONTAINING NMOFs FOR PDT

Phthalocyanine-based PS can be introduced as guest species into the matrix of the MOFs by using an in situ encapsulation approach. Using the coprecipitation method, Ma et al. successfully encapsulated hydrophobic Zinc(II) phthalocyanine (ZnPc) PS into zeolitic imidazolate framework (ZIF-8) [48]. The ZnPc molecules were encapsulated in the micropores of ZIF-8, which helped in maintaining the ZnPc in a monomeric state and prevented self-aggregation, thus enabling it to produce $^1O_2$ upon irradiation with a 650 nm light in an aqueous medium. The PDT efficacy of the ZnPc@ZIF-8 was studied in vitro against HEPG-2 cells. Nearly 80% of the cancer cells were destroyed using 50 μg/mL of the ZnPc@ZIF-8 NMOFs, upon irradiation

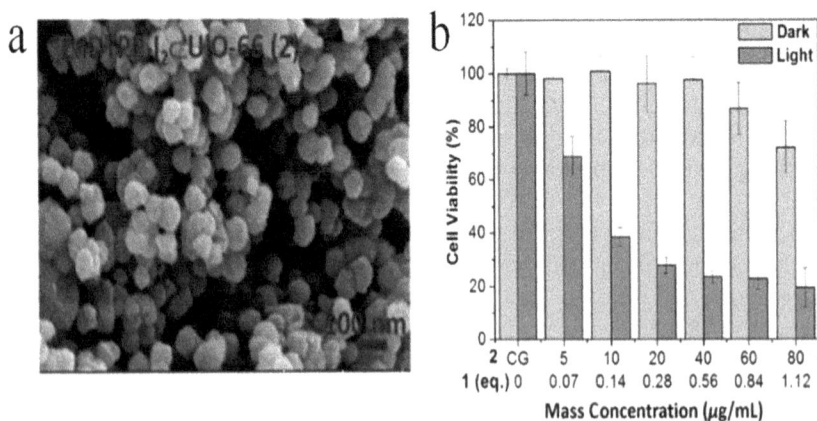

FIGURE 8.2 (a) SEM image of Zn-DTPP-I_2-2H-UiO-66 NMOFs showing particles with spherical morphology and (b) MTT assays for HepG2 cells incubated with Zn-DTPP-I_2-UiO-66 of different concentrations (blue) and without (red) upon irradiation with light (540 nm, 20 mW/cm²) for 10 min. Reproduced with permission from ref. 46. Copyright 2018 American Chemical Society.

with light for ten minutes, indicating excellent PDT efficiency of the NMOFs (Figure 8.3). Furthermore, the sensitivity of ZIF-8 towards acidic pH was also taken advantage of for the complete degradation of ZnPc@ZIF-8 after PDT, which could easily be monitored by the self-quenching of emission due to ZnPc. A similar study was performed by Song et al. where the ZnPc@ZIF-8 nanospheres functioned as pH responsive drug delivery systems for effective PDT [49].

8.2.3 BODIPY-CONTAINING NMOFs FOR PDT

In addition to porphyrins and phthalocyanines, diiodo-substituted BODIPYs (I_2-BDPs) can also function as effective PS due to their high extinction coefficient and marginal dark toxicities. Xie et al. used a solvent-assisted ligand exchange (SALE) method to incorporate carboxyl-modified I_2-BDP within UiO NMOFs to afford the light-active UiO-PDT [50]. UV-visible spectrum, EDS mapping, and NMR studies confirmed the introduction of I_2-BDP within UiO NMOFs by SALE. The powder XRD pattern of the UiO and UiO-PDT showed identical Bragg diffraction peaks, indicating retention of the structural features of the NMOFs after incorporation of I_2-BDP. Microscopic images of UiO NMOFs before and after incorporation of I_2-BDP were also similar and showed octahedral morphology with an average size of 70 nm. Confocal laser scanning microscopy (CLSM) studies indicated that both UiO-PDT NMOFs as well as I_2-BDP could pass across the cell membrane and accumulate in

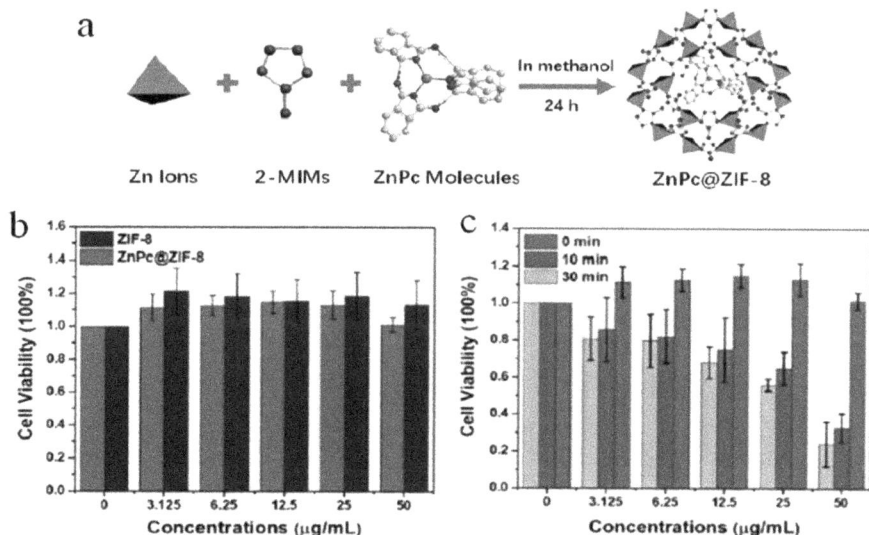

FIGURE 8.3 (a) Schematic illustration for the one-step synthesis of ZnPc@ZIF-8 NMOFs, (b) cell viability of HepG-2 cancer cells after incubation with ZIF-8 and ZnPc@ZIF-8 for 24 h in dark, and (c) photodynamic cytotoxicity of ZnPc@ZIF-8 under irradiation (650 nm) for 0 min, 10 min, and 30 min, respectively and further incubation for 24 h. Reproduced with permission from ref. 48. Copyright 2019 American Chemical Society.

the cytoplasm. Flow cytometry studies indicated that the cellular uptake of UiO-PDT was 1.54 times higher than the I_2-BDP after incubation for 2 h. The $^1O_2$ generation ability of both the UiO-PDT and free I_2-BDP was studied by using 1,3-diphenyl-isobenzofuran (DPBF) as a detector under light illumination. The results showed that the ability of I_2-BDP to generate $^1O_2$ was better than the UiO-PDT, due to the heterogeneous nature of UiO-PDT. In vitro PDT studies were performed for both UiO-PDT and I_2-BDP against B16F10, CT26, and C26 cell lines under light irradiation (power 80 mW/cm^2) for 10 min. Both UiO-PDT and free I_2-BDP exhibited good cytotoxicity against all the three cell lines and inhibited the cell growth by more than 80%. Although the PDT efficacy of UiO-PDT was not higher than the free I_2-BDP, this work paved a newer method for the introduction of PDT agents in the NMOF via SALE method (Figure 8.4).

Dong et al. also reported a new NMOF system based on ZIF-90 by using host-guest chemistry, where the guest, 2I-BODIPY-PhNO$_2$ (PS) was encapsulated into the host (ZIF-90) employing a one-pot in situ method to give photosensitive 2I-BODIPY-PhNO$_2$@ZIF-90 NMOFs [51]. The NMOFs showed a spherical morphology with diameters less than 80 nm and the PXRD pattern of the ZIF-90 before and after encapsulation of the PS were identical, indicating no changes in the crystal lattice due to encapsulation of the PS. The NMOFs showed low cytotoxicity and good

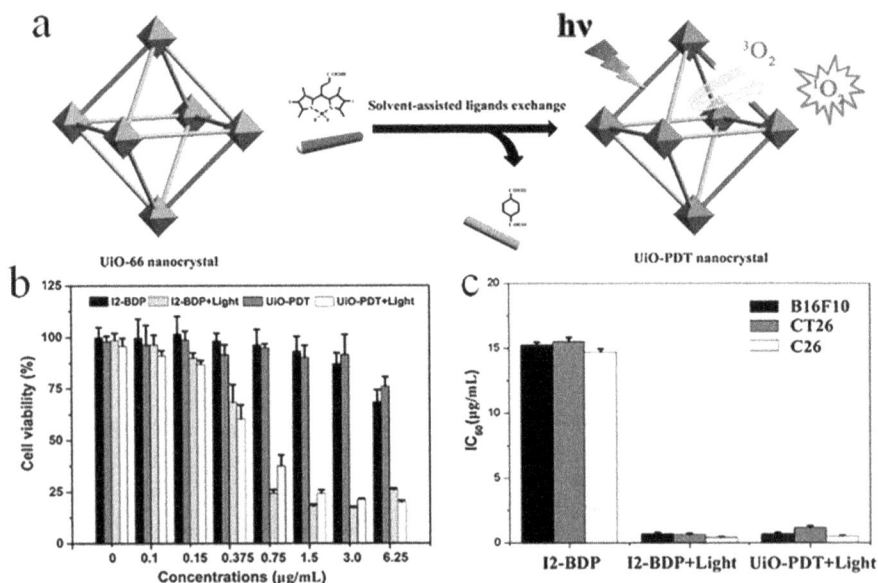

FIGURE 8.4 (a) Schematic illustration for the synthesis of UiO-PDT NMOFs and their use in PDT, (b) in vitro cytotoxicity of free I_2-BDP and UiO-PDT nanocrystal against B16F10 cells before and after irradiation by visible light at a power density of 80 mW cm^{-2} for 10 min, and (c) half maximal inhibitory concentration (IC50) of I_2-BDP and UiO-PDT against the three cells with and without light irradiation. Reproduced with permission from ref. 50. Copyright 2016 The Royal Society of Chemistry.

biocompatibility in the dark as evidenced by more than 80% cell viability upon incubation with HepG2 cells. In addition, the NMOFs exhibited a pH-driven selective uptake and release of cancer cells and mitochondria targeting ability as well as highly efficient $^1O_2$ generation ability. Therefore, it was reasonable to speculate that the BODIPY-containing ZIF-90 system would function as an efficient PDT agent. The PDT efficacy was studied by illumination of the NMOF incubated HepG2 cells. It was observed that less than 10% of the cancer cells were viable after illumination with 540 nm green LED light (20 mW/cm²) for ten minutes. Moreover, the NMOFs showed a distinct selectivity towards cancer cells over normal cells as more than 60% of the normal HL-7702 liver cells were viable under the same PDT conditions.

8.3   SURFACE MODIFICATION OF NMOFs

Singlet oxygen is highly reactive and has a half-life of 3.3 ± 0.5 µs and a diffusion coefficient, $D = 2 \times 10^{-5}$ cm²s⁻¹, making the effective range of $^1O_2$ very limited [52]. As a result of this, only those tumor cells which are immediately close to the $^1O_2$ generation area are affected. Therefore, the exact location of the PS in the NMOF system is crucial to achieve effective PDT and surface modification of the NMOFs becomes important.

Liu and coworkers synthesized Hf-TCPP NMOFs (TCPP = 5,10,15,20-tetrakis(4-carboxyphenyl)-porphyrin) employing a solvothermal approach, wherein Hf⁴⁺ ions were heated with TCPP in DMF at 80°C for over 24 h [53]. The purple-colored product, when observed under TEM showed the formation of spherical nanoparticles with size ranging from 80 to 150 nm. These particles were hydrophobic in nature and the average hydrodynamic diameter of the particles, as estimated using Dynamic Light Scattering (DLS), was 130 nm. To make the particles hydrophilic, surface modification of the NMOFs was performed by treating as synthesized NMOFs in chloroform with polyethylene glycol grafted poly(maleicanhydride-alt-1-octdecene) (PEG-C18PMH) in chloroform under stirring for 12 h. The Hf-TCPP-PEG obtained showed good water dispersibility and stability at physiological conditions. It showed good $^1O_2$ generation ability and enhanced PDT efficiency both in vitro and in vivo against 4T1 tumor cells upon exposure to 661 nm light. The in vivo studies were performed by intravenous injection of Hf-TCPP-PEG in healthy Balb/c mice (Hf = 12.5 mg/kg, TCPP = 24 mg/kg). Blood was extracted at different time intervals from the mice and the concentration of the NMOF was determined by the quantification of fluorescence due to TCPP. The NMOF had a blood circulation half-life of 3.27 h, which was attributed to the successful surface modification of the NMOFs. Due to the EPR effect, the NMOFs were gradually accumulated in the tumor tissues during circulation of blood. The PDT efficacy was studied by irradiating the tumor sites 20 h post intravenous injection of NMOFs with a 661 nm laser light (5 mW/cm²) for 60 minutes, which showed partial suppression in the tumor growth.

Chen et al. synthesized folic acid-nanoscale gadolinium porphyrin MOFs (FA-NPMOFs) which had a spherical morphology with sizes around 200 nm [54]. Folic acid was attached to the surface of the NMOFs by EDC:NHS coupling using the polyethyleneimine on the surface of the NMOFs. The NMOFs exhibited good dispersion and water solubility. The NMOFs showed bright red fluorescence and were good candidates for both in vitro and in vivo magnetic resonance imaging by virtue of their low biotoxicity as studied against HepG2 cells and embryonic and larval zebrafish. The FA-NPMOFs also demonstrated strong affinity towards folate receptor (FR)-expressing cells and could be delivered in a targeted fashion. The PDT efficacy of the FA-NPMOFs was studied in transgenic zebrafish with deoxycycline-induced hepatocellular carcinoma (HCC) cells.

Pang et al. successfully introduced TCPP into NU-1000 by a post synthetic ligand exchange [55]. NU-1000 was synthesized by the reaction between $ZrOCl_4$ and 1, 3, 6, 8-tetrakis(p-benzoic acid)pyrene (H4TBAPy). The ligand exchange could be performed by simple stirring of NU-1000 nanorods in DMF with TCPP in DMF at 40°C for 12 h. The ligand exchange was ascertained by observing a change in color from yellow to purple. While pure NU-1000 was incapable of $^1O_2$ generation, the ligand exchanged NMOFs (NT-1000 nanorods) showed effective $^1O_2$ generation upon illumination with a 650 nm light. In vitro cell toxicity was studied against HeLa cells and it was found that irradiation of the cells incubated with NMOFs (650 nm, 50 mW/cm^2) for five minutes led to the death of a significant number of cells, indicating the excellent PDT activity of NT-1000 nanorods.

8.4 TARGETED DELIVERY OF NMOFs FOR PDT

Since ROS cannot discriminate between normal and tumor cells, it is of utmost importance that the PS be selectively accumulated in the tumor cells only, so that normal cells are not killed. In 2016 Zhou and coworkers reported that modification of the surface of a Zr-based porphyrinic NMOF, PCN-224, (PCN = porous coordination network) with folic acid could lead to an increased cellular uptake of NMOFs, together with an increased PDT efficiency in vitro [41]. PCN-224 nanospheres were synthesized by solvothermal reaction between $ZrOCl_2$ and H_2TCPP in the presence of benzoic acid in DMF at 120°C for 24 h. The diameter of the nanospheres could be controlled from 24 to 232 nm by increasing the concentration of benzoic acid from 22 to 33 mg/mL. Cytotoxicity studies performed using HeLa cells using particles of sizes 30, 60, 90, 140, and 190 nm for 12 h, in the concentration range of 0.5–40 μM of TCPP indicated that 90 nm PCN-224 nanoparticles showed highest cellular uptake. In vitro PDT studies against HeLa cells were performed by irradiating the NMOFs (20 μM) using a 420 nm light. The best PDT efficacy was observed with the PCN-224 nanoparticles with a diameter of 90 nm, which showed 81% cytotoxicity, while the 190 nm PCN-224 nanoparticles showed the least PDT efficacy (49% cytotoxicity). The authors then modified the PCN-224 nanoparticles with a diameter of 90 nm with folic acid by coordination interaction between the –COO$^-$ group of folic acid and Zr_6SBUs. It was observed that the folic acid modified PCN-224 nanoparticles showed better cytotoxicity (90%) than the unmodified PCN-224 nanoparticles (81%) against FR-positive HeLa cells. The cytotoxicity studies against FR-negative HeLa cells did not show any appreciable increase for the folic acid-modified nanoparticles in comparison to the unmodified nanoparticles (Figure 8.5).

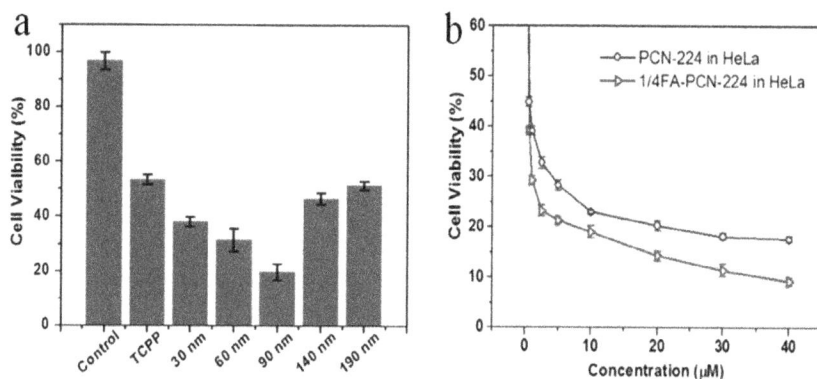

FIGURE 8.5 (a) Comparison of PDT efficacy between different sized PCN-224 nanoparticles and free TCPP molecules, and (b) comparison of in vitro PDT efficacy of pristine PCN-224 and 1/4FA-PCN-224 in HeLa cells. Reproduced with permission from ref. 41. Copyright 2016 American Chemical Society.

Lang et al. synthesized 50–70 nm PCN-222 nanoparticles with hexagonal morphology [42]. Upon photoexcitation, the PCN-222 nanoparticles induced apoptosis of cancer cells due to generation of ROS. Experiments showed that PCN-222 nanoparticles were localized within the endosomes and could not escape into cytoplasm. The results indicated that targeting the endosomes resulted in oxidative stress in the cancer cells and accelerated the apoptosis. Over the course of time, the endosomes gradually develop into lysosomes and the PCN-222 nanoparticles are degraded by the enzymes present in lysosome.

8.5 NMOFs FOR PHOTOTHERMAL THERAPY

Thermal therapy involves increasing the temperature of the whole body or a part of the body above the normal body temperature (37°C) and has emerged as an effective method for treatment of cancer due to its simple operative methods [14,56]. In the past few years, there have been substantial efforts for the development of simple and efficient techniques for the controlled and localized heating of the cells in the body and understanding the mechanisms of cell destruction using temperature [14,56,57]. The energy sources commonly employed for thermal therapy include radiofrequency pulse and microwave irradiation as well as ultrasound waves [14,56,57]. Recently, PTT has emerged as a fascinating method of thermal treatment, where near infrared light (NIR) is used to increase the temperature of the body. PTT is an oxygen-independent method of cancer treatment mediated by photothermal agents. In PTT, the PS are excited by NIR light and during relaxation to ground state, the PS release vibrational energy which generates heat [58,59]. The suddenly generated heat can regress tumors by increasing the temperature locally. Several nanoparticles such as gold nanospheres, nanorods, nanoshells, nanostars and nanocages, carbon nanotubes, graphene, iron oxide nanoparticles, etc. [60–67] have been shown to exhibit good photothermal activity. The high light absorptivity and efficient photothermal conversion make nanoparticle PTT agents selective [22], whereby the temperature increases in the tumor areas much more than normal tissues. The immense use of NMOFs for biomedical applications has led to increased research on the applications of NMOFs for PTT and PTT-based combination therapies.

Due to the inhomogeneous heat distribution in the cancer cells, PTT itself is generally incapable in the complete eradication of tumors. Therefore, there has been development of several new strategies (mainly three) to increase the PTT efficiency [22].

(i) Firstly, PTT agents which have superior light absorption in the NIR region, good blood circulation time, enhanced tumor uptake and high photothermal conversion are being developed.
(ii) Secondly, PTT is being combined with other modalities for enhanced tumor regression.
(iii) Thirdly, image-guided PTT with functional nanomaterials which function as theranostic agents can also increase the PTT efficacy.

All these three strategies can be employed in NMOFs for enhanced PTT. For instance, both the metal-cluster structural building units as well as bridging ligands can readily be functionalized to improve photothermal conversion efficiency [22]. Combination therapies can easily be employed using NMOFs as functional moieties can easily be incorporated into NMOFs by post synthetic modifications. Functional components can also be easily encapsulated into the cavities of NMOFs to further enhance the PTT efficacy as well as PTT-based combination therapies. The next few pages discuss the use of various NMOFs as PTT agents (NMOF enabled PTT), combination of NMOFs with other modalities (NMOF-combined PTT), and as hybrid nanomaterials for combined PTT and PDT (nMOFs for PTT + PDT).

8.5.1 NMOF-ENABLED PTT

Prussian blue nanoparticles (PB NPs) are a type of ancient dye used as a typical drug in clinic. PB is a MOF which lacks large pores but shows high photothermal efficiency and stability as well as good biocompatibility [68]. Yue and coworkers first employed the superior NIR photothermal conversion of PB NPs [69]. The synthesis was performed by simple mixing of aqueous solutions of $FeCl_3$ and $K_4[Fe(CN)_6]$ with citric acid acting as the stabilizer. The TEM images showed the formation of PB nanocubes with average diameter of 42 nm. The size of the nanocubes could be tuned from 10 to 50 nm by varying the concentration of the stabilizer. The PB nanocubes obtained were highly dispersible and stable in both water as well as biological environments like blood serum. The photothermal efficiency of the PB nanocubes was studied by exposing the aqueous dispersion of the nanocubes to a 808 nm laser with an output of 2 W. Using a concentration of 500 ppm and irradiation for 600 s, the temperature of the dispersion reached to more than 60°C, suggesting the excellent photothermal efficiency of the PB nanocubes. HeLa cells stained with fluorescent calcein AM were then incubated with PB nanocubes in a six-well plate and irradiated with 808 nm light for 4 h to study the PTT effect of the PB nanocubes. The HeLa cells were then observed under a fluorescence microscope which showed a dark region in the plate which was incubated with PB nanocubes and irradiated with a laser light of 808 nm. The results indicated that cells present in dark region were killed due to the PTT effect of PB nanocubes, while green fluorescence could still be observed from the cells outside, which indicated the strong PTT efficiency of the PB nanocubes (Figure 8.6).

Chen and coworkers used a simple coprecipitation method to synthesize a core-shell composite of Prussian blue analog, $Mn_3[Co(CN)_6]_2@SiO_2@Ag$ [70]. The hydrodynamic radius of the nanocomposite was found to be 125 nm. The loading of Ag nanoparticles on the composite was ascertained from UV-visible studies, which showed a band at 420 nm, characteristic of surface plasmon resonance band of Ag nanoparticles. The nanocomposite was found to possess T_1-T_2 dual-modal magnetic resonance imaging (MRI) capability. Due to the presence of the Prussian blue analog $Mn_3[Co(CN)_6]_2$, the nanocomposite exhibited brilliant two photon fluorescence (TPF) imaging upon excitation at 730 nm. Comparison between the TPF imaging intensities showed that upon loading the Ag nanoparticles on the $Mn_3[Co(CN)_6]_2$, the TPF imaging intensity increased 1.85 fold. They also utilized the nanocomposite as a drug carrier and loaded the anticancer drug Doxorubin (DOX) on the nanocomposite which showed a drug loading capacity of 600 mg/g. The drug could be released from the nanocomposite upon excitation with 808 nm light was 30% higher than that in the dark. The PTT efficiency was studied by exposing the aqueous solutions containing various concentrations of the nanocomposite to 808 nm laser light for ten minutes. It was observed that using concentrations of 0.05, 0.1, 0.25, and 0.5 mg/mL, the temperature of the solution could be increased to 43.2°C, while control experiments without the nanocomposite showed only an increase of 5°C, indicating excellent PTT efficacy of the nanocomposite. The drug-loaded nanocomposite was utilized for a combined chemotherapy-PTT application against lung cancer A549 cells, which showed excellent anticancer activity.

A similar strategy was employed by Zhang et al. who reported $PB@mSiO_2$-PEG nanocomposite for efficient loading of DOX [71]. The presence of $mSiO_2$ and PEG improved the photostability and biocompatibility of the nanocomposite and PB was used both for PTT and as photoacoustic agent. The DOX-loaded nanocomposite was a highly effective anticancer agent, as it combined chemotherapy and PTT.

In 2016 Chen et al. coated PB nanocubes with MIL-100 (Fe) shells to obtain core-shell like NMOFs and utilized it for combined MRI and PTT [72]. The hydrodynamic radius of the core-shell NMOFs was found to be 210 nm using (DLS). The hybrid NMOFs showed T_1-T_2 dual- modal MRI contrast and the presence of PB in the core and MIL-101 (Fe) in the shell ensured that they can also be utilized for fluorescence optical imaging. They also showed that the NMOFs could be utilized for efficient loading and release of anticancer drugs. Due to the presence of PB in the core, the NMOF showed strong NIR absorption, which could be utilized for PTT. In vivo and in vitro studies performed showed that the core-shell NMOFs loaded with an anticancer drug (artemisinin) possessed high antitumor efficiency due to a

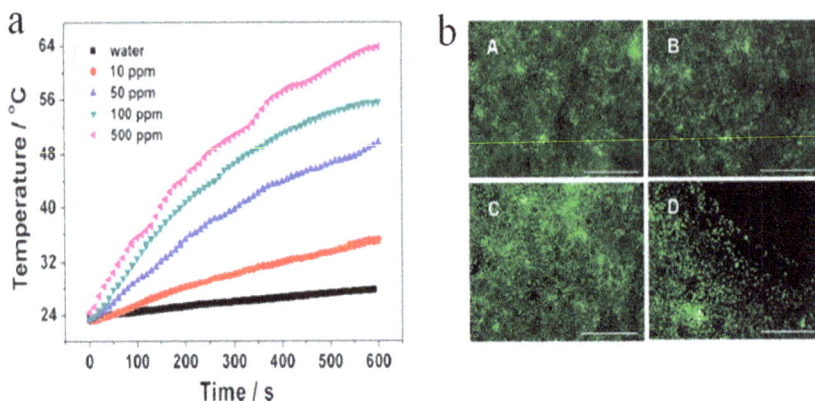

FIGURE 8.6   (a) Increase in temperatures of aqueous dispersions of PB NP with different concentrations under NIR laser irradiation (808 nm, 2 W) for 10 min. The temperature was measured every 10 s with a digital thermometer, and (b) fluorescence microscopy images of HeLa cells with different mode of treatments stained with calcein AM: (A) without PB NPs and without laser irradiation; (B) only with laser irradiation; (C) PB NPs (25 ppm) only; (D) with both PB NPs (25 ppm) and laser irradiation (scale bars: 500 mm). Reproduced with permission from ref. 69. Copyright 2012 The Royal Society of Chemistry.

combined effect of chemotherapy from the released drug and PTT due to NIR irradiation. The low toxicity of both the NMOF and artemisinin ensured that the system was not toxic to the major organs of the tested animal model.

Au nanostructures are biocompatible and can absorb the NIR radiation, as a result of which it can function as a CT contrast agent. However, due to the low photostability, it cannot be used for photoacoustic imaging. To overcome this problem, Dai and coworkers synthesized Au@PB core-shell nanoparticles for both photoacoustic imaging and PTT [73]. The synthesis involved slow mixing of the precursors required for the formation of PB nanoparticles to a preformed aqueous solution of Au nanostructures. The average diameter of the core-shell nanoparticles as observed under TEM was 17.8 ± 2.3 nm. The obtained core-shell nanoparticles were found to be an excellent photoabsorbing agent for both PTT and photoacoustic imaging because of the high stability and high molar extinction coefficient in the NIR region. The presence of Au core with a size of 9.1 ± 0.64 nm ensured that contrast enhancement was obtained for CT imaging. The PTT studies performed on nude mice showed that through a single treatment, 100 mm^3 sized tumors in the mice could completely be eradicated without any recurrence after laser irradiation of intravenously injected Au@PB NPs.

8.5.2   NMOF-COMBINED PTT

Excellent photothermal activity in NMOFs can be achieved by incorporating photothermal agents into the NMOFs. The pores present in the NMOFs can easily be used for the encapsulation of PTT agents. Furthermore, the PTT agents can also be hybridized with the NMOF to give a core-shell like structure, where the PTT agents can serve as either the core or shell of the hybrid composite [22].

In 2016 Tian et al. developed a one-pot synthesis of ZIF-8/graphene quantum dot composite for combined PTT-chemotherapy [74]. Due to the weak coordination interactions between the zinc ions in ZIF-8 and an anticancer drug (DOX), they were able to encapsulate DOX in the pores of ZIF-8 during its synthesis. The large number of oxygen-containing functional group such as hydroxyl, carboxyl, and epoxy groups in the graphene quantum dots (GQD) could form H-bonding with the imidazolate group in the ZIF-8 to give a ZIF-8/DOX/GQD nanocomposite having a spherical morphology with sizes between 50–100 nm. Using breast cancer $4T_1$ cells as model system, the combined chemo and PTT effect of the composite was found to exhibit a significant synergistic effect, resulting in higher cytotoxicity to cancer cells than chemotherapy or PTT alone (Figure 8.7).

In early 2017, Wang and coworkers developed a novel ZIF-8-based MOF nanocomposite which encapsulated Pd@Au nanoparticles and DOX and used it for pH and NIR-triggered PTT-chemotherapy [75]. Pd nanocubes were first synthesized by

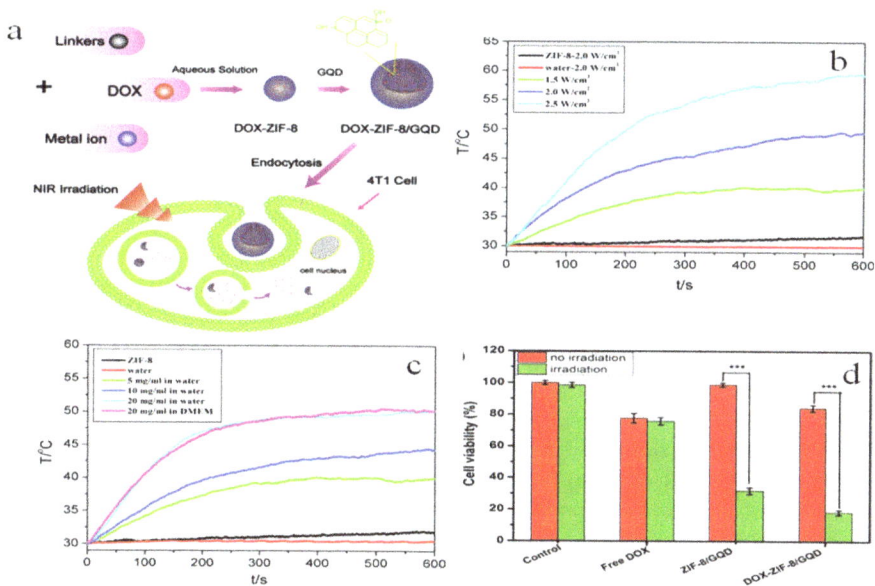

FIGURE 8.7 (a) Schematic representation for the synthesis of ZIF-8/GQD-DOX nanocomposite and synergistic delivery of DOX and PTT, (b) PTT efficacy of DOX-ZIF-8/GQD nanoparticles measured by noting increase in temperature using NIR irradiation ($\lambda = 808$ nm) with different laser intensities for 10 min; (c) PTT efficiency of DOX-ZIF-8/GQD nanoparticles in water and DMEM with different concentrations using 808 nm laser irradiation at 1.5 W/cm² (efficiency was measured with increase in temperature upon irradiation). (d) Cell viability of 4T1 cells after 8 h of incubation without and with free DOX solution, ZIF-8/GQD, and DOX-ZIF-8/GQD suspensions (DOX: 5 µg/mL, ZIF-8/ GQDs: 100 µg/mL) and without and with 3 min NIR irradiation. Reproduced with permission from ref. 74. Copyright 2017 American Chemical Society.

reduction of the palladium salt with ascorbic acid, which were then covered with nanosheets of gold to yield Pd@Au nanoparticles as photothermal agent. ZIF-8 was then used to encapsulate the Pd@Au nanoparticles and DOX to afford Pd@Au/DOX@ZIF-8 nanocomposite. The photothermal efficiency of the nanocomposite was studied by irradiating the solution containing the nanocomposite and studying the rise in temperature. Upon irradiation with a NIR light of 780 nm, it was observed that the nanocomposite not only converted the light energy into heat energy, but also assisted the release of DOX from the MOF making it a system for combined PTT-chemotherapy. In vivo cytotoxic studies indicated that upon incubating the cancer cells with the nanocomposite for 24 h, nearly 50% of cell death occurred due to intracellular acidic pH that led to the release of DOX from the composite. However, when these cells were irradiated with a NIR laser for ten minutes, almost all the cells died, clearly indicating the superior treatment effect of PTT-chemotherapy due to synergistic effect.

Chen et al. developed a facile method to fabricate polypyrolle@MOF core-shell nanocomposite for dual modal imaging and combined chemotherapy-PTT, where polypyrolle (PPy) served as the core, while MIL-100(Fe) was the shell [76]. PPy core was used as photothermal agent and an organic photoacoustic imaging agent for deep tissue imaging, while the MIL-100 shell was used for loading of DOX as well as for MRI T_2 contrast imaging. The PPy@MIL-100 core-shell nanocomposite enabled dual-modal imaging as well as light and pH mediated release of the anticancer drug, DOX and efficient PTT-chemotherapy treatment.

8.5.3   NMOFs for Combination of PDT and PTT

Combination of two therapies for cancer treatment is always expected to be better than monotherapy as the efficiency and potency of the therapy will increase upon combination with a different therapy due to the synergistic effects. NMOFs have been studied and employed as an efficient medium for combination of PDT and PTT [77–82]. Yang et al. synthesized a hybrid NMOF composite consisting of Fe_3O_4, ZIF-8, and Au_{25} nanoclusters by a green synthesis method [77]. The NMOFs were found to have an average size of 100 nm and it not only showed temperature elevation upon irradiation with NIR light (PTT) of wavelength 808 nm for five minutes, but could also be used for magnetic targeting and MRI due to the presence of Fe_3O_4 in the core. Additionally, the Au_{25} clusters (about 2.5 nm) present in the NMOF were capable of generating the highly reactive singlet oxygen ($^1O_2$) in the tissues upon irradiation with NIR light causing effective photodynamic effects. Furthermore, the Au_{25} clusters assisted the hyperthermia effect due to Fe_3O_4 in the core. The combined PDT-PTT efficacy of the nanocomposite was studied in vivo, and a higher anti-tumor efficiency was observed than individual monotherapies.

Yin and coworkers prepared a biocompatible porphyrin-based metal-organic framework (PMOF) nanocomposite for dual-modality imaging-guided PTT-PDT combination therapy [78]. The nanocomposite had a core-shell like morphology, where Fe_3O_4@C served as the core, while PMOF formed the shell. The Fe_3O_4@C@PMOF was synthesized by a hydrothermal procedure by heating $ZrCl_4$ and TCPP in the presence of Fe_3O_4@C, whereby in situ self-assembly of PMOF took place on the surface of Fe_3O_4@C giving a core-shell like morphology. The average diameter of the nanocomposite was found to be 95 nm. The Fe_3O_4@C core was utilized for both T2 weighted MRI as well as photothermal activity. The in vivo synergetic PDT and PTT therapy performed on MCF-7 tumor-bearing mice suggested that the PTT-PDT co-therapy group showed the highest therapeutic efficiency compared to that of the single PTT or PDT groups.

8.6   SUMMARY AND OUTLOOK

In summary, the potential and use of NMOFs for phototherapy of cancer using NIR light is discussed. Although NMOFs have been in use for applications in biomedical imaging and drug delivery for more than 12 years now, it has been nearly seven years that they have been in use for phototherapy of cancer. Since the first publication by Lin and coworkers in 2014, NMOFs have shown their capability as efficient PS for PDT. This chapter focused on the PS used for PDT, the main strategies and the advances in sensitizing PDT. NMOFs have shown several advantages for PDT compared to conventional PS: (i) they can show high loading of PS and can avoid self-quenching and aggregation, (ii) NMOFs can be used for the loading of variety of therapeutics for synergistic therapies, and (iii) they can easily be surface functionalized to endow them with biocompatibility. Nevertheless, the research on PDT using NMOFs is still in its infancy and requires extensive realistic in vivo studies to assess the clinical applications of NMOFs for PDT. Further, the synthesis of NMOF-based PDT systems is complicated, which makes their synthesis in bulk quantities quite difficult, and their accumulation in tumor tissues is still insignificant. Although targeted strategies have increased their accumulation in tumor tissues, further optimization of the accumulation of NMOFs in tumor sites is essential. All these issues need to be addressed for effective use of NMOFs in PDT in real life systems.

PTT, which leads to an increase in the temperature of the region irradiated by NIR light due to absorption of light by PS, has not been studied in depth using NMOFs alone. It has been explored in combination with other therapies such as magnetic resonance imaging, fluorescence, chemo or radiotherapy, and PDT. As multiple therapies are involved in studies involving PTT using NMOFs, it is difficult to understand the exact effect of PTT in cancer therapy. The light dose required for PTT is several times higher than the accepted clinical light dose, and although several therapeutic techniques have been combined with PTT, it is not clear how such systems can be used for clinical trials. This issue needs to be addressed before using NMOF-based PTT systems for clinical trials.

Lastly, the combination of PDT and PTT using NMOFs has been discussed. Due to their tunability in size and shape, the NMOFs can be combined for PDT and PTT to achieve superior anticancer efficacy. Although, the studies are still in their infancy, the results obtained so far indicate that NMOFs can prove to be a crucial factor in the phototherapy of cancer. It is expected that NMOFs will soon find application for phototherapy of cancer in clinics.

ABBREVIATIONS

PDT	Photodynamic therapy
PTT	Photothermal therapy
ROS	Reactive oxygen species
PS	Photosensitizer
MOF	Metal-organic framework
NMOF	Nanoscale metal-organic framework
EPR	Enhanced permeation and retention
DBP	5,15-di(p-benzoato) porphyrin
DBC	5,15-di(p-benzoato) chlorin
TBP	5,10,15,20-tetra(p-benzoato) porphyrin
ZnPc	Zn(II) phthalocyanine
ZIF	Zinc imidazolate framework
CLSM	Confocal laser scanning microscopy
DPBF	1,3-diphenylisobenzofuran
BDP	BODIPYs
SALE	Solvent assisted ligand exchange
TCPP	5,10,15,20-tetrakis (4-carboxyphenyl)-porphyrin
EDC	1-Ethyl-3-(3-dimethylaminopropyl)carbodiimide
NHS	N-hydroxysuccinimide
TBAPy	1,3,6,8-tetrakis(p-benzoic acid pyrene)
PCN	Porous coordination network
PB	Prussian blue
DOX	Doxorubicin
GQD	Graphene quantum dots
DMF	Dimethylformamide
XRD	X-ray diffraction

REFERENCES

1. M. Ferrari. Cancer nanotechnology: Opportunities and challenges. *Nat. Rev. Cancer* 5 (2005) 161–171.
2. S. S. Lucky, K. C. Soo, Y. Zhang. Nanoparticles in photodynamic therapy. *Chem. Rev.* 115 (2015) 1990–2042.
3. L. Cheng, C. Wang, L. Feng, K. Yang, Z. Liu. Functional nanomaterials for phototherapies of cancer. *Chem. Rev.* 114 (2014) 10869–10939.
4. C. B. Blackadar. Historical review of cancer. *World J. Clin. Oncol.* 7 (2016) 54–86.
5. D. Peer, J. M. Karp, S. Hong, O. C. Farokhzad, R. Margalit, R. Langer. Nanocarriers as an emerging platform for cancer therapy. *Nat Nanotechnol.* 2 (2007) 751–760.
6. J. Cao, Z. Chen, J. Chi, Y. Sun, Y. Sun. Recent progress in synergistic chemotherapy and phototherapy by targeted drug delivery systems for cancer treatment. *Artificial Cells Nanomed. Biotechnol.* 46 (2018) 817–830.
7. C. W. Ng, J. Li, K. Pu. Recent progresses in phototherapy-synergized cancer immunotherapy. *Adv. Funct. Mat.* 28 (2018) 1804688.
8. C. Liang, L. Xu, G. Song, Z. Liu. Emerging nanomedicine approaches fighting tumor metastasis: Animal models, metastasis-targeted drug delivery, phototherapy, and immunotherapy. *Chem. Soc. Rev.* 45 (2016) 6250–6269.

9. R. Bonnett. Photosensitizers of the porphyrin and phthalocyanine series for photodynamic therapy. *Chem. Soc. Rev.* 24 (1995) 19–33.

10. X. Li, S. Lee, J. Yoon. Supramolecular photosensitizers rejuvenate photodynamic therapy. *Chem. Soc. Rev.* 47 (2018) 1174–1188.

11. B. W. Henderson, T. J. Dougherty. How does photodynamic therapy work? *Photochem. Photobiol.* 55 (1992) 145–157.

12. W. M. Sharman, C. M. Allen, J. E. van Lier. Photodynamic therapeutics: Basic principles and clinical applications. *Drug Discovery Today* 4 (1999) 504–517.

13. M. R. K. Ali, Y. Wu, M. A. El-Sayed. Gold-nanoparticle-assisted plasmonic photothermal therapy advances toward clinical application. *J. Phys. Chem. C* 123 (2019) 15375–15393.

14. Y. Liu, P. Bhattarai, Z. Dai, X. Chen. Photothermal therapy and photoacoustic imaging via nanotheranostics in fighting cancer. *Chem. Soc. Rev.* 48 (2019) 2053–2108.

15. H. S. Jung, P. Verwilst, A. Sharma, J. Shin, J. S. Sessler, J. S. Kim. Organic molecule-based photothermal agents: An expanding photothermal therapy universe. *Chem. Soc. Rev.* 47 (2018) 2280–2297.

16. H. Maeda, J. Wu, T. Sawa, Y. Matsumura, K. Hori. Tumor vascular permeability and the EPR effect in macromolecular therapeutics: A review. *J. Control. Release* 65 (2000) 271–284.

17. A. Albanese, P. S. Tang, W. C. W. Chan. The effect of nanoparticle size, shape, and surface chemistry on biological systems. *Annu. Rev. Biomed. Eng.* 14 (2012) 1–16.

18. K. Saha, S. S. Agasti, C. Kim, X. Li, V. M. Rotello. Gold nanoparticles in chemical and biological sensing. *Chem. Rev.* 112 (2012) 2739–2779.

19. X. Chen, W. Zhang. Diamond nanostructures for drug delivery, bioimaging, and biosensing. *Chem. Soc. Rev.* 46 (2017) 734–760.

20. Y. Hu, S. Mignani, J. -P. Majoral, M. Shen, X. Shi. Construction of iron oxide nanoparticle-based hybrid platforms for tumor imaging and therapy. *Chem. Soc. Rev.* 47 (2018) 1874–1900.

21. E. -K. Lim, T. Kim, S. Paik, S. Haam, Y. -M. Huh, K. Lee. Nanomaterials for theranostics: Recent advances and future challenges. *Chem. Rev.* 115 (2015) 327–394.

22. G. Lan, K. Ni, W. Lin. Nanoscale metal-organic frameworks for phototherapy of cancer. *Coord. Chem. Rev.* 379 (2019) 65–81.

23. H.-C. Zhou, S. Kitagawa. Metal-organic frameworks (MOFs). *Chem. Soc. Rev.* 43 (2014) 5415–5418.

24. S. Yuan, L. Feng, K.Wang, J. Pang,M. Bosch, C. Lollar, Y. Sun, J. Qin, X. Yang, P. Zhang, Q. Wang, L. Zou, Y. Zhang, L. Zhang, Y. Fang, J. Li, H. -C. Zhou. Stable metal-organic frameworks: Design, synthesis, and applications. *Adv. Mater.* 30 (2018) 1704303

25. I. T. Hillman, A. Laybourn, C. Dodds, S. W. Kingman. Realising the environmental benefits of metal-organic frameworks: Recent advances in microwave synthesis. *J. Mater. Chem. A* 6 (2018) 11564–11581.

26. H. -C. Zhou, J. R. Long, O. M. Yaghi. Introduction to metal-organic frameworks. *Chem. Rev.* 112 (2012) 673–674.

27. L. E. Kreno, K. Leong, O. K. Farha, M. Allendorf, R. P. Van Duyne, J. T. Hupp. Metal–organic framework materials as chemical sensors. *Chem. Rev.* 112 (2012) 1105–1125.

28. K. Sumida, D. L. Rogow, J. A. Mason, T. M. McDonald, E. D. Bloch, Z. R. Herm, T.-H. Bae, J. R. Long. Carbon dioxide capture in metal-organic frameworks. *Chem. Rev.* 112 (2012) 724–781.

29. P. Horcajada, R. Gref, T. Baati, P. K. Allan, G. Maurin, P. Couvreur, G. Fèrey, R. E. Morris, C. Serre. Metal-organic frameworks in biomedicine. *Chem. Rev.* 112, (2012) 1232–1268.

30. T. Zhang, W. Lin. Metal–organic frameworks for artificial photosynthesis and photocatalysis. *Chem. Soc. Rev.* 43 (2014) 5982–5993.
31. L. He, Y. Liu, J. Lau, W. Fan, Q. Li, C. Zhang, P. Huang, X. Chen. Recent progress in nanoscale metal-organic frameworks for drug release and cancer therapy. *Nanomedicine* 14 (2019) 1343–1365.
32. C. He, D. Liu, W. Lin. Nanomedicine applications of hybrid nanomaterials built from metal-ligand coordination bonds: Nanoscale metal-organic frameworks and nanoscale coordination polymers. *Chem. Rev.* 115 (2015) 11079–11108.
33. Q. Guan, Y. -A. Li, W. -Y. Li, Y. -B. Dong. Photodynamic therapy based on nanoscale metal-organic frameworks: From material design to cancer nanotherapeutics. *Chem. Asian. J.* 13 (2018) 3122–3149.
34. K. Lu, C. He, W. Lin. Nanoscale Metal-organic framework for highly effective photodynamic therapy of resistant head and neck cancer. *J. Am. Chem. Soc.* 136 (2014) 16712–16715.
35. K. Lu, C. He, W. Lin. Chlorin-based nanoscale metal-organic framework for photodynamic therapy of colon cancers. *J. Am. Chem. Soc.* 137 (2015) 7600–7603.
36. K. Lu, C. He, N. Guo, C. Chan, K. Ni, R. R. Weichselbaum, W. Lin. Chlorin-based nanoscale metal-organic framework systemically rejects colorectal cancers via synergistic photodynamic therapy and checkpoint blockade immunotherapy. *J. Am. Chem. Soc.* 138 (2016) 12502–12510.
37. G. Lan, K. Ni, Z. Xu, S. S. Veroneau, Y. Song, W. Lin. Nanoscale metal-organic framework overcomes hypoxia for photodynamic therapy primed cancer immunotherapy. *J. Am. Chem. Soc.* 140 (2018) 5670–5673.
38. G. Lan, K. Ni, Z. Xu, S. S. Veroneau, X. Feng, G. T. Nash, T. Luo, Z. Xu, W. Lin. Titanium-based nanoscale metal-organic framework for type I photodynamic therapy. *J. Am. Chem. Soc.* 141 (2019) 4204–4208.
39. G. Lan, K. Ni, R. Xu, K. Lu, Z. Lin, C. Chan, W. Lin. Nanoscale metal-organic layers for deeply penetrating x-ray-induced photodynamic therapy. Angew. Chem. Int. Ed. 56 (2017) 12102–12106.
40. K. Ni, T. Aung, S. Li, N. Fatuzzo, X. Liang, W. Lin. Nanoscale metal-organic framework mediates radical therapy to enhance cancer immunotherapy. *Chem* 5 (2019) 1892–1913.
41. J. Park, Q. Jiang, D. Feng, L. Mao, H. -C. Zhou. Size-controlled synthesis of porphyrinic metal-organic framework and functionalization for targeted photodynamic therapy. *J. Am. Chem. Soc.* 138 (2016) 3518–3525.
42. D. Bůžek, J. Zelenka, P. Ulbrich, T. Ruml, I. Křížová, J. Lang, P. Kubát, J. Demel, K. Kirakci, K. Lang. Nanoscaled porphyrinic metal-organic frameworks: Photosensitizer delivery systems for photodynamic therapy. *J. Mater. Chem. B* 5 (2017) 1815–1821.
43. M. Liu, L. Wang, X. Zheng, S. Liu, Z. Xie. Hypoxia-triggered nanoscale metal-organic frameworks for enhanced anticancer activity. *ACS Appl. Mater. Interfaces* 10 (2018) 24638–24647.
44. J. Liu, Y. Yang, W. Zhu, X. Yi, Z. Dong, X. Xu, M. Chen, K.Yang, G. Lu, L. Jiang, Z. Liu. Nanoscale metal-organic frameworks for combined photodynamic & radiation therapy in cancer treatment. *Biomaterials* 97 (2016) 1–9.
45. L. Zhang, J. Lei, F. Ma, P. Ling, J. Liu, H. Ju. A porphyrin photosensitized metal-organic framework for cancer cell apoptosis and caspase responsive theranostics. Chem. Commun. 51 (2015) 10831–10834.
46. L. -L. Zhou, Q. Guan, Y. -A. Li, Y. Zhou, Y. -B. Xin, Y.-B. Dong. One-pot synthetic approach toward porphyrinatozinc and heavy-atom involved Zr-NMOF and its application in photodynamic therapy. *Inorg. Chem.* 57 (2018) 3169–3176.

47. X. Zhao, Z. Zhang, X. Cai, B. Ding, C. Sun, Z. Liu. C. Hu, S. Shao, M. Pang. Postsynthetic ligand exchange of metal–organic framework for photodynamic therapy. *ACS Appl. Mater. Interfaces* 11 (2019) 7884–7892.

48. D. Xu, Y. You, F. Zeng, Y. Wang, C. Liang, H. Feng, X. Ma. Disassembly of hydrophobic photosensitizer by biodegradable zeolitic imidazolate framework-8 for photodynamic cancer therapy. *ACS Appl. Mater. Interfaces* 10 (2018) 15517–15523.

49. M. -R. Song, D. -Y. Li, F. -Y. Nian, J. -P. Xue, J. -J. Chen. Zeolitic imidazolate metal organic framework-8 as an efficient pH-controlled delivery vehicle for zinc phthalocyanine in photodynamic therapy. *J. Mater. Sci.* 53 (2018) 2351–2361.

50. W. Wang, L. Wang, Z. Li, Z. Xie. BODIPY-containing nanoscale metal-organic frameworks for photodynamic therapy. *Chem. Commun.* 52 (2016) 5402–5405.

51. Q. Guan, L. -L. Zhou, Y. -A. Li, Y. –B. Dong. Diiodo-Bodipy-encapsulated nanoscale metal-organic framework for pH-driven selective and mitochondria targeted photodynamic therapy. *Inorg. Chem.* 57 (2018) 10137–10145.

52. E. Boix-Garriga, B. Rodriguez-Amigo, O. Planas, S. Nonell, Properties of singlet oxygen. In S. Nonell, C. Flors (Eds.), *Singlet Oxygen: Applications in Biosciences and Nanosciences* Vol. 1. Cambridge, UK: The Royal Society of Chemistry. 2016, pp. 23–46.

53. J. Liu, Y. Yang, W. Zhu, X. Yi, Z. Dong, X. Xu, M. Chen, K. Yang, G. Lu, L. Jiang, Z. Liu. Nanoscale metal-organic frameworks for combined photodynamic & radiation therapy in cancer treatment. *Biomaterials* 97 (2016) 1–9.

54. Y. Chen, W. Liu, Y. Shang, P. Cao, J. Cui, Z. Li, X. Yin, Y. Li. Folic acid-nanoscale gadolinium-porphyrin metal-organic frameworks: Fluorescence and magnetic resonance dual-modality imaging and photodynamic therapy in hepatocellular carcinoma. *Int. J. Nanomed.* 14 (2019) 57–74.

55. X. Zhao, Z. Zhang, X. Cai, B. Ding, C Sun, G. Liu, C. Hu, S. Shao, M. Pang. Postsynthetic ligand exchange of metal–organic framework for photodynamic therapy. *ACS Appl. Mater. Interfaces* 11 (2019) 7884–7892.

56. D. Jaque, L. Martínez Maestro, B. del Rosal, P. Haro-Gonzalez, A. Benayas, J. L. Plaza, E. Martín Rodríguez, J. García Solé. Nanoparticles for photothermal therapies. *Nanoscale* 6 (2014) 9494–9530.

57. A. C. V. Doughty, A. R. Hoover, E. Layton, C. K. Murray, E. W. Howard, W. R. Chen. Nanomaterial applications in photothermal therapy for cancer. *Materials* 12 (2019) 779.

58. J. Li, K. Pu. Development of organic semiconducting materials for deep-tissue optical imaging, phototherapy and photoactivation. *Chem. Soc. Rev.* 48 (2019) 38–71.

59. E. A. Hussein, M. M. Zagho, G. K. Nasrallah, A. A. Elzatahry. Recent advances in functional nanostructures as cancer photothermal therapy. *Int. J. Nanomed.* 13 (2018) 2897–2906.

60. M. Aioub, M. A. El-Sayed. A real-time surface enhanced raman spectroscopy study of plasmonic photothermal cell death using targeted gold nanoparticles. J. Am. Chem. Soc. 138 (2016) 1258–1264.

61. Z. Zhang, L. Wang, J. Wang, X. Jiang, X. Li, Z. Hu, Y. Ji, X. Wu, C. Chen. Mesoporous silica-coated gold nanorods as a light-mediated multifunctional theranostic platform for cancer treatment. *Adv. Mater.* 24 (2012) 1418–1423.

62. R. Bardhan, W. Chen, C. Perez-Torres, M. Bartels, R. M. Huschka, L. L. Zhao, E. Morosan, R. G. Pautler, A. Joshi, N. J. Halas. Nanoshells with targeted simultaneous enhancement of magnetic and optical imaging and photothermal therapeutic response. *Adv. Funct. Mater.* 19 (2009) 3901–3909.

63. S. Wang, P. Huang, L. Nie, R. Xing, D. Liu, Z. Wang, J. Lin, S. Chen, G. Niu, G. Lu. Single continuous wave laser induced photodynamic/plasmonic photothermal therapy using photosensitizer-functionalized gold nanostars. *Adv. Mater.* 25 (2013) 3055–3061.

64. Y. Xia, W. Li, C. M. Cobley, J. Chen, X. Xia, Q. Zhang, M. Yang, E. C. Cho, P. K. Brown. Gold nanocages: From synthesis to theranostic applications. *Acc. Chem. Res.* 44 (2011) 914–924.

65. J. Song, F. Wang, X. Yang, B. Ning, M. G. Harp, S. H. Culp, S. Hu, P. Huang, L. Nie, J. Chen. Gold nanoparticle coated carbon nanotube ring with enhanced raman scattering and photothermal conversion property for theranostic applications. *J. Am. Chem. Soc.* 138 (2016) 7005–7015.

66. Y.-W. Chen, Y.-L. Su, S.-H. Hu, S.-Y. Chen. Functionalized graphene nanocomposites for enhancing photothermal therapy in tumor treatment. *Adv. Drug Deliv. Rev.* 105 (2016) 190–204.

67. S. Shen, S. Wang, R. Zheng, X. Zhu, X. Jiang, D. Fu, W. Yang. Magnetic nanoparticle clusters for photothermal therapy with near-infrared irradiation. *Biomaterials* 39 (2015) 67–74.

68. M. B. Zakaria, T. Chikyow. Recent advances in Prussian blue and Prussian blue analogues: Synthesis and thermal treatments. *Coord. Chem. Rev.* 352 (2017) 328–345.

69. G. Fu, W. Liu, S. Feng, X. Yue. Prussian blue nanoparticles operate as a new generation of photothermal ablation agents for cancer therapy. *Chem. Commun.* 48 (2012) 11567–11569.

70. D. Wang, Z. Guo, J. Zhou, J. Chen, G. Zhao, R. Chen, M. He, Z. Liu, H. Wang, Q. Chen. Novel $Mn_3[Co(CN)_6]_2$@SiO_2@Ag core-shell nanocube: Enhanced two-photon fluorescence and magnetic resonance dual-modal imaging-guided photothermal and chemo-therapy. *Small* 11 (2015) 5956–5967.

71. Y. Y. Su, Z. Teng, H. Yao, S. J. Wang, Y. Tian, Y. L. Zhang, W. F. Liu, W. Tian, L. J. Zheng, N. Lu, Q. Q. Ni, X. D. Su, Y. X. Tang, J. Sun, Y. Liu, J. Wu, G. F. Yang, G. M. Lu, L. J. Zhang. A Multifunctional PB@$mSiO_2$–PEG/DOX nanoplatform for combined photothermal-chemotherapy of tumor. *ACS Appl. Mater. Interfaces* 8 (2016) 17038–17046.

72. D. Wang, J. Zhou, R. Chen, R. Shi, G. Zhao, G. Xia, R. Li, Z. Liu, J. Tian, H. Wang, Z. Guo, H. Wang, Q. Chen. Controllable synthesis of dual-MOFs nanostructures for pH-responsive artemisinin delivery, magnetic resonance and optical dual-model imaging-guided chemo/photothermal combinational cancer therapy. *Biomaterials* 100 (2016) 27–40.

73. L. Jing, X. Liang, Z. Deng, S. Feng, X. Li, M. Huang, C. Li, Z. Dai. Prussian blue coated gold nanoparticles for simultaneous photoacoustic/CT bimodal imaging and photothermal ablation of cancer. *Biomaterials* 35 (2014) 5814–5821.

74. Z. Tian, X. Yao, K. Ma, X. Niu, J. Grothe, Q. Xu, L. Liu, S. Kaskel, Y. Zhu. Metal-organic framework/graphene quantum dot nanoparticles used for synergistic chemo- and photothermal therapy. *ACS Omega* 2 (2017) 1249–1258.

75. X. Yang, L. Li, D. He, L. Hai, J. Tang, H. Li, X. He, K. Wang. A metal-organic framework-based nanocomposite with co-encapsulation of Pd@Au nanoparticles and doxorubicin for pH- and NIR-triggered synergistic chemo-photothermal treatment of cancer cells. *J. Mater. Chem. B* 5 (2017) 4648–4659.

76. X. Chen, M. Zhang, S. Li, L. Li, L. Zhang, T. Wang, M. Yu, Z. Mou, C. Wang. Facile synthesis of polypyrrole@metal-organic framework core-shell nanocomposites for dual-mode imaging and synergistic chemo-photothermal therapy of cancer cells. *J. Mater. Chem. B* 5 (2017) 1772–1778.

77. D. Yang, G. Yang, S. Gai, F. He, G. An, Y. Dai, R. Lv, P. Yang. Au25 cluster functionalized metal-organic nanostructures for magnetically targeted photodynamic/photothermal therapy triggered by single wavelength 808 nm near-infrared light. *Nanoscale* 7 (2015) 19568–19578.

78. H. Zhang, Y.-H. Li, Y. Chen, M.-M. Wang, X.-S. Wang, X.-B. Yin. Fluorescence and magnetic resonance dual-modality imaging-guided photothermal and photodynamic dual-therapy with magnetic porphyrin-metal organic framework nanocomposites. *Sci. Rep.* 7 (2017) 44153.

79. X. Cai, B. Liu, M. Pang, J. Lin. Interfacially synthesized Fe-soc-MOF nanoparticles combined with ICG for photothermal/photodynamic therapy. *Dalton Trans.* 47 (2018) 16329–16336.

80. Y. Wang, X. Pang, J. Wang, Y. Cheng, Y. Song, Q. Sun, Q. You, F. Tan, J. Li, N. Li. Magnetically targeted and near infrared fluorescence/magnetic resonance/photoacoustic imaging-guided combinational anti-tumor phototherapy based on polydopamine-capped magnetic Prussian blue nanoparticles. *J. Mater. Chem. B* 6 (2018) 2460–2473.

81. B. Zhou, B.-P. Jiang, W. Sun, F.-M. Wei, Y. He, H. Liang, X.-C. Shen. Water-dispersible Prussian blue hyaluronic acid nanocubes with near-infrared photoinduced singlet oxygen production and photothermal activities for cancer theranostics. *ACS Appl. Mater. Interfaces* 10 (2018) 18036–18049.

82. J.-Y. Zeng, M.-K. Zhang, M.-Y. Peng, D. Gong, X.-Z. Zhang. Porphyrinic metal-organic frameworks coated gold nanorods as a versatile nanoplatform for combined photodynamic/photothermal/chemotherapy of tumor. *Adv. Funct. Mater.* 28 (2018) 1705451.

9 Carbon Nanotube-Based Metal-Organic Framework Nanocomposites

Manickam Ramesh and Arivumani Ravanan

CONTENTS

9.1 INTRODUCTION

Metal-organic frameworks (MOFs) have developed as a comprehensive category of crystalline materials. Many analysts have given more attention to the favorable characteristics at greater level. MoFs comprise extraordinarily highly porous spacious capacity and tremendous surface areas. The desirable properties and special attributes of porous materials have led to rapid growth, especially in the nanoporous materials, to a broad range of applications as engineering ceramics in the last 25 years.

9.1.1 MOFs

When compared with other recently developed porous materials, MOFs are significant and different in nature as well as quality due to their attractive thermal stability and higher degree of porosity. As such, characteristics combined with a broad level and quality of variety of constituents subjected to structural aspects of organic and inorganic compounds, cause MOFs to be attractive for application in the environment, energy, chemical, and bio fields. Furthermore, MOFs are playing more and more crucial roles by applying to the preparation of membranes, biomedical images, thin films, and catalysis (Figure 9.1).

MOFs constitute a set of porous materials which are constructed through firm binding of metal ions and organic linkers. The strategic analysis and attentive selection of components provide the MOFs with superior chemical stability, tremendous porous volume, and a wide surface area. Analysis of synthesis, structures, and characteristics of different MOFs has indicated them as well-suited materials and they can also be directly used for storage purposes, sensing activities, and catalysis (Figure 9.2).

Put another way, for nanomaterials, MOFs are applied as support substrates. In order to produce as many different kinds of functional nanostructures, this task is to use MOFs as sacrificial templates and precursors. It has been considered that one of the efficient ways to produce porous carbons, metal, etc. is the one-step pyrolysis method.

9.1.2 CNTs

Iijima developed CNTs in 1991 [1]. The exceptional characteristics of CNTs draw the attention of the research world and keep the focus on the development and analysis of CNTs to employ them in all related applications. Because of the bad transformation of nano-graded inbuilt characteristics of CNTs into macro-sized dimensions such as powder, stacks, and film layers, it's not possible to use them in bulk shapes except in a few cases. Hence, in the majority of usages, CNTs have been applied in terms of blends, alloys, and hybridized components with others in the systematic proportions [2]. When combining with other materials, their fundamental properties provide the synergistic integration of adaptability, lighter weight, and adroit processing performance with extraordinary mechanical, thermal, and electrical behavior; also introducing their other additional electromagnetic and thermal features, thus broadening their sectors of application (Figure 9.3).

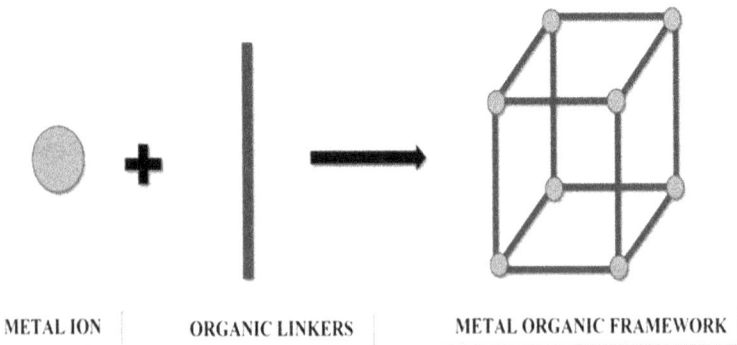

METAL ION   ORGANIC LINKERS   METAL ORGANIC FRAMEWORK

FIGURE 9.1 Basic MOF structure.

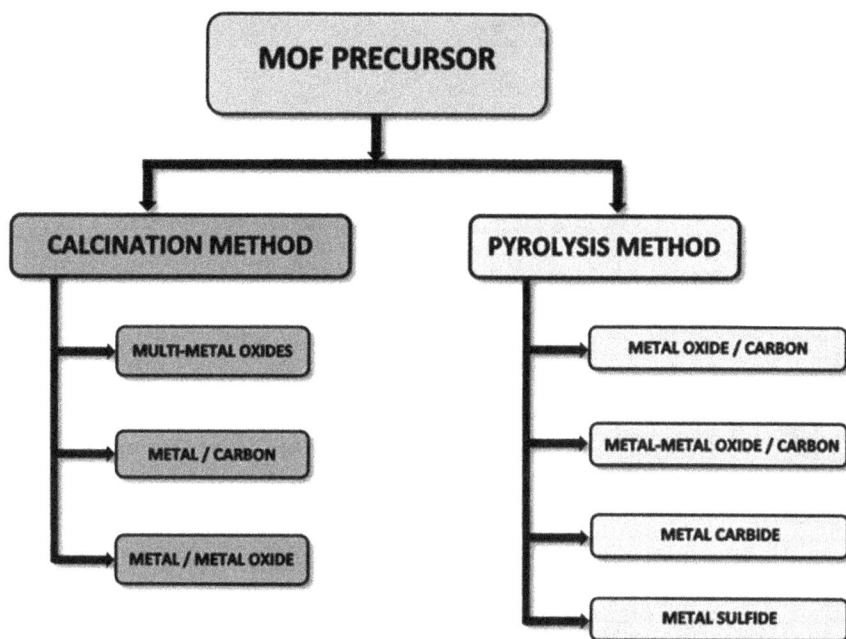

FIGURE 9.2 Various nanostructures developed from MOF precursor.

$1s^2$, $2s^2$, and $2p^2$ are the electron specifications of carbon which reveal that it possess two potent electrons in first "s" level and four valance electrons that are comparatively powerless bound electrons in next two orbits. The little variation in the energy of the last two orbits permits the carbon to exist in many hybridizing conditions. The adaptability of hybridization allows each atomic band of carbon to array itself in construction in which CNTs come under a one-dimensional range. The structure of CNTs is of hollow forms of graphene with tremendous electronic attributes and mechanical features. And there are two types CNTs, one graphene sheet is composed as CNTs, which are called single-walled CNTs and the coaxial arrangement of more nanotubes is called multi-walled CNTs [3,4].

9.1.3 SYNTHESIS

CNTs are one of the significant carbon-based materials to enhance the functioning process of MOFs owing to their distinctive electrical conductivity and greater thermal stabilizing attributes with larger surface areas. The CNTs could synthesize with MOFs to construct nanocomposites with custom-made electrical and chemical attributes [5], i.e., applying the CNTs as inorganic metal ions with organic ligands, provides the newly designed material with flexibility. It contributed to the tailor-made materials and the design has been achieved by using different construction blocks through functioning, surface area, and porosity. In order to employ energy and cleansing applications in the MOFs, porous carbon, metals, metal oxides, and

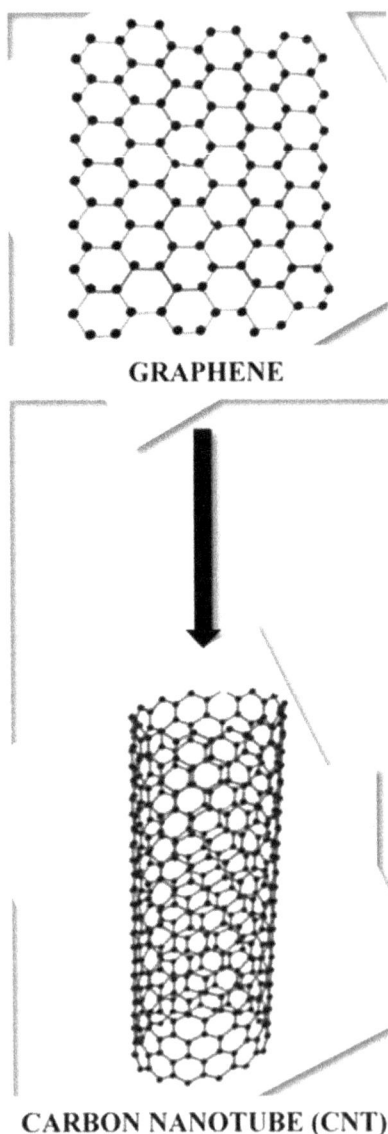

GRAPHENE

CARBON NANOTUBE (CNT)

FIGURE 9.3   CNT from graphene.

their hybrids are used as inorganic components. Porous materials are produced using various synthesis actions based on the required function and outcome. There are two major actions namely, hard templating and soft templating (Figure 9.4).

The quick growth in the field of synthesizing of nanomaterial leads to the feasibility of specific construction and morphology. Amongst such techniques, the hard template approach is highly beneficial in combining specific nanomaterials due to

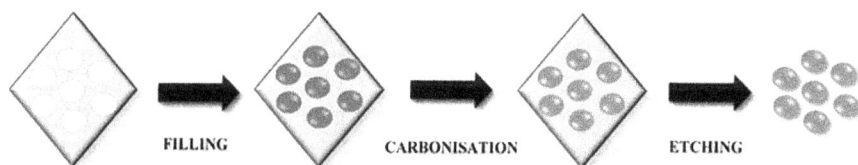

FIGURE 9.4   General process of hard template approach.

their attributes of tunable pore dimensions and ease of process. In hard templating, the set of cavities existing in the well-formed solid template are fulfilled by the framework precursors where at the end of the synthesizing process the template is eliminated. CNTs are very frequently applied as hard templates due to their strong ability to determine the pore sizes at nano-level. Also, it is highly suitable for mass production, hence it is employed in industrial applications. This approach has greater reproducibility and stability. Nevertheless, the detachment of template and product would harm the construction of nanotubes and hollow spheres. Also, semi-filled and partly filled components cause defects in the pores and discontinuity in construction. The delimited facilities of template materials controls and stops the comprehensive applications of the hard template approach [6] (Figure 9.5).

In the soft templating approach, a high degree of accurate micro-fine physical, and chemical actions occur mutually and reciprocally amongst the templates and the origins of the framework; it conducts the auto-assemblage process to initiate greater determination of the material characterization [7]. Static electricity, chemical bonding, and hydrogen bonding are known as intramolecular interaction forces. Generally, a soft template has a loose non-rigid structure. Because of the structure, it possesses either intermolecular force or intramolecular force. Better repeatability, ease of procedure, and no need to eliminate the silicon are the major merits of soft templates, hence they have been applied in a wide range of developments. If relating both of these approaches, soft templating is favorable for synthesizing the arranged and disarranged poriferous matrices [8].

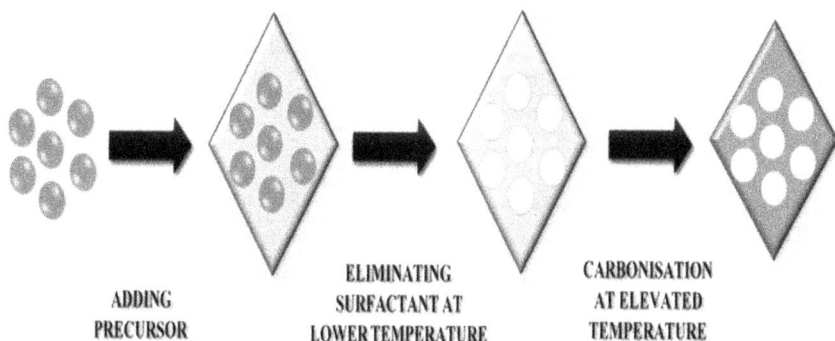

FIGURE 9.5   General process of soft template approach.

A hard template is produced earlier than the reactions, while on the contrary, a soft template is produced between the reactions. It indicates the simplicity in constructing and eliminating when compared to the hard template. So, no need for complicated tools and no disciplined production specifications and it has better control in response. The soft template approach is generally applicable for fabricating the different dimensions and distinct nanostructure. Nevertheless, the soft template approach has certain demerits. The stability of this approach depends upon particular synthetic systems, which is not advisable. It produces only an organized-mesoscopic structure when the template has powerful mutual action with a precursor. It leads to a mechanically strengthened independent structure which causes non-collapsible construction at the time of eliminating the template. In addition, there is no coordinated methodology of soft template approach for synthesizing nanomaterials [6].

Solvothermal is one of the techniques applied in order to produce the chemical compounds of porous substances. It's a very desirable technique because of its key features such as relative simplicity and scalability, which contribute this technique towards a greater volume of production. The solvothermal technique has been employed for the process of synthesizing s as well as MOFs, with a high success rate. Synthesizing the single crystal based on the solubility of minerals is known as the hydro-thermal method. This process takes place under heavy pressure and the solubility is tested in hot water. Producing the crystalline phases is one of the merits of this method whereas it's unstable at melting point. MOFs are performing as a precursor under the activity of producing a variety of nanostructures, including CNTs, from MOFs. Herein, metal components yield the intrinsic origin of metals in order to obtain the nanostructures. Either metal oxides, metal, or self-sacrificing templates are the derived nano-contents. In the same way, organic components are engaged as the origin of the carbon material to produce nanoporous carbon such as CNTs [9] (Figure 9.6).

Commonly, particular types of MOFs like Cu-BTC and MOF-5 would break down metaphorically while carbonization; at the same time, other MOFs' structure, for instance, ZIF-8 and ZIF-67, behave well in giving a template for directing the establishment of pores by permitting vaporization from the governable moisturized organics during pyrolysis, resulting in the spongy-pore arrangement. Very close dispersion of pore volumes, and a wider surface area are the major merits of such MOF-derived nanostructures. Moreover, a tremendous variety of morphology is an added benefit when compared to other methods of obtaining nanostructures; this technique renders more advantageous structures such as hollow-shaped cylinders, spheres, polyhedrons, and nanocages, etc. Ignorance of decomposition mechanisms, difficulties in controlling the dimensions of micropores, and demanding comparatively higher temperatures during calcination are the demerits during the process of MOF-derived nanostructures. Specifically, when a variety of precursors of MOF structures are subjected to various temperatures of calcination, which results in nanostructures with dissimilar topologies, crystalline phases, and porous properties. Ion exchange is one of the best techniques and can replace the thermal treatment in the fabrication of a variety of nanostructures out of MOF precursors. Despite the involvement of so many chemicals and observational procedures, ion exchange methods evidenced some remarkable characteristics [11–13].

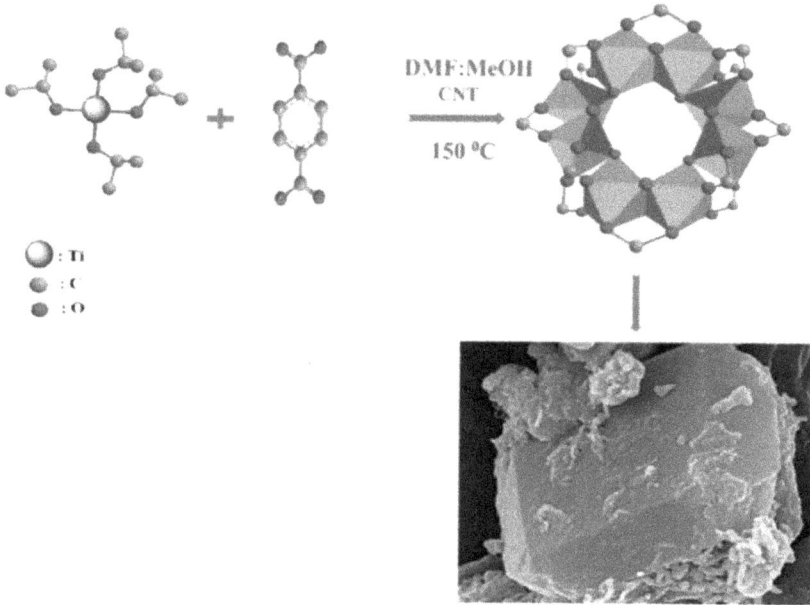

FIGURE 9.6   Example for synthesizing of a MOF and CNT as nanocomposite [10].

9.2   APPLICATIONS OF CNT-BASED MOF NANOCOMPOSITES

This chapter deals with the role of CNT-based MOF nanocomposites applied in various fields. Analysis of construction, preparation, and characteristics of different MOFs has indicated that MOFs are excellent materials for diverse applications, for instance, energy/gas storage, sensing, chemical/electrochemical, environmental, biological, and medical applications. Physical and chemical techniques are the two approaches involved in the fabrication of nanomaterials. Physical pulverization, coacervation, and spraying are the three methods mainly applied physical in approaches. Chemical vapor deposition, template, sol-gel, hydrothermal, and chemical precipitation are the chemical-approached methods in the fabrication of nanomaterials. In order to produce the different operational nanostructures, MOFs can be used directly and as substrates for nanostructured-materials and as sacrificial templates/precursors (Figure 9.7).

9.2.1   Chemical/Electrochemical Sensing Applications

MOFs are frequently synthesized to conduct capable components especially with CNTs. Because of the extradentary aspect ratio, better mechanical attributes, and extremely appreciable electrical behaviors, CNTs have been broadly employed as electrode candidates. Apart from that, they ensure an enormous specific surface area; approximately, in single walled CNTs it's greater than 600 m^2/g, and in multi-walled CNTs it's about more than 430 m^2/g [14]. Nanoparticles of composites and

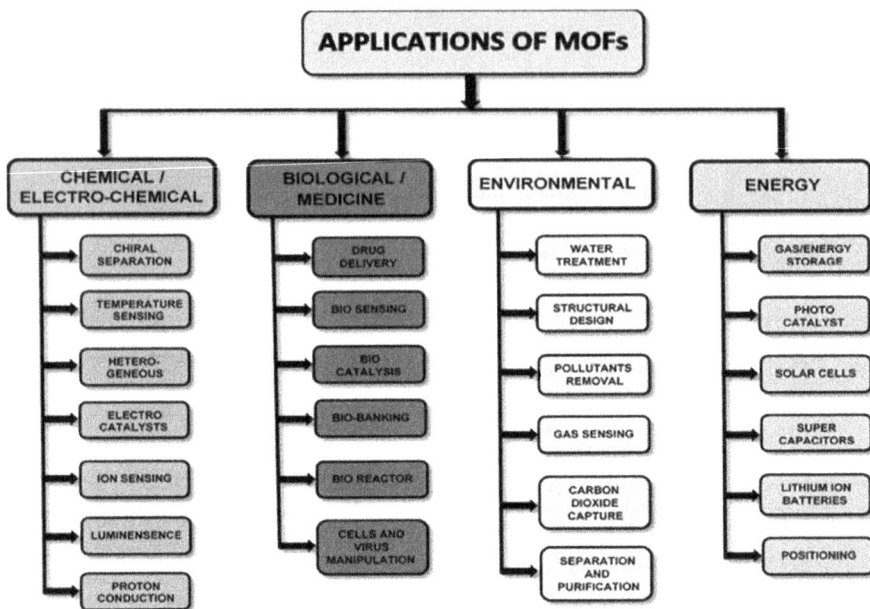

FIGURE 9.7   Overview of general applications of MOFs.

MOFs have attained a greater level in the field of electrochemical sensing due to the substantial development of nanoscience and nanotechnology, significantly, conductive nanoparticles such as CNTs and metal nan-particles. A nanocomposite of multi-walled CNTs and $Cu_3(BTC)_2$ MOFs was fabricated to change the electrode. It was done for determining the remnant stages of lead by gathering the lead over the faces of the electrode followed by quantitating the same in terms of concentration range [15].

In an electrochemical sensor system, in order to identify the high toxin pollutants of hydroquinone and catechol simultaneously, a nanocomposite of MOF-199/HKUST-1/single-walled CNTs was suggested. A HKUST-1 structure is also understood as $Cu_3(BTC)_2$, and hydroquinone and catechol are the abbreviated as HQ and CT respectively. The modifying process was accomplished by casting the single-walled CNTs on a plain glassy carbon electrode. And then the surfaces of the single-walled CNT-altered electrode were electrodeposited by Cu-MOF-199. The composite electrode showed the finest electrocatalytic activity. Also, because of the interactions of Cu-MOF-199 and single-walled CNTs, electrochemical signals for the oxidation of HQ and CT were increased. Furthermore, significant reproducibility and attractive anti-interference behavior was displayed by the altered electrode. The amperometry method was also applied on the altered electrode to analyze the feasible interference by a few ions. The strong, influential, interrupting compounds had no impact in determining the HQ and CT. Hence, this nanocomposite-altered electrode can be said to be as highly qualified as HQ and CT. This electrochemical sensor arrangement was tested with the sampling of spiked tap and river water to determine the recovery rates of HQ and CT; and received highly positive outcomes [16].

Synthesized nickel and MOFs were fixed firmly on carboxyl CNTs through the solvothermal technique. Nanoparticles of the synthesized nanocomposite were homogeneously distributed over the outer-face of the CNTs. It permitted the microcrystalline of the MOFs to be connected to the current collector through the CNT conductance. When comparing prepared (Ni-MOF/CNT) composites with bare Ni-MOF altered composite, the outcome hybridized nanostructure candidate registered more prominent sensitivity and required stability for the recognition of hydrogen peroxide because of the interactive integrated catalytic actions of the nanocomposite [17].

Aijaz and team registered an attractive nanocomposite with the support of ZIF-67 MOF, which was developed through the core-shell construction of Co@Co_3O_4 nanoelements firmly placed in the NCNT-engrafted carbon polyhedral. This novel formulated hybridized composite was combined through reductive carbonization in the inert condition, also accompanied by oxidation. Pt/C, IrO_2, and RuO_2 are all known as well accurate and effective Oxygen evolution reaction (OER) catalysts. When relating the prepared economy material with those traditional catalysts, it registered a favorable result on very lower reversible over voltage against a reversible hydrogen electrode. Investigations about the differences revealed the significances about the approaches of MOFs with other effective CNT combinations in the development of influenced constituents for electrocatalysis. The appreciable task is accompanied by the hollow nanostructure and specific characterized components comprised by the carbon shells of many layers for the core components of N-doped CNTs and cobalt nanosized interfacial layers [18] (Figure 9.8).

Wang et al. conducted a comparative analysis of pure MOFs, multi-walled CNTs/GCE, Ni-MOF/GCE, and Ni-MOF/multi-walled CNTs/GCE in the aspects of sensitivity, stability, and conductivity. Initially, synthesis of Ni-p-benzenedicarboxylic acid MOFs and multi-walled CNTs was performed through the solvothermal technique. These prepared nanocomposites were reformed through glassy carbon electrodes and applied for the electrochemical observation of hydrogen peroxide. At the time of evolution of composites, the nanosized MOF particles between 2 and 3 nm were uniformly distributed on the conducting CNTs. This phenomenon proved the proper connection of MOF nanoparticles with the conducting CNTs. The limit of detection of this technique was identified as 2.1 mM and linear bound to be 0.01–51.6 mM. More significantly, nickel-based MOFs were much greater than pure MOFs. Sensitivity of Ni-MOF/multi-walled CNTs/GCE was greater than other three combinations of nanocomposites [19].

In order to attain the selective identification of volatile organic compounds, a cross-reactive chemiresistive sensor has been constructed by the team through MOFs 1–3 in an order like the procedure of CNT composites preparation. When related to CNT composites, MOFs ensured the exact atomic level influence over the structure about controlling and eventually permitting an upgraded approach to sensor materials customized for appropriate operation. Furthermore, the MOFs mentioned above have the merit of a much more narrowly described proportion and structure, which is favorable on the subject of reproducibility and lifelong system stability. A group

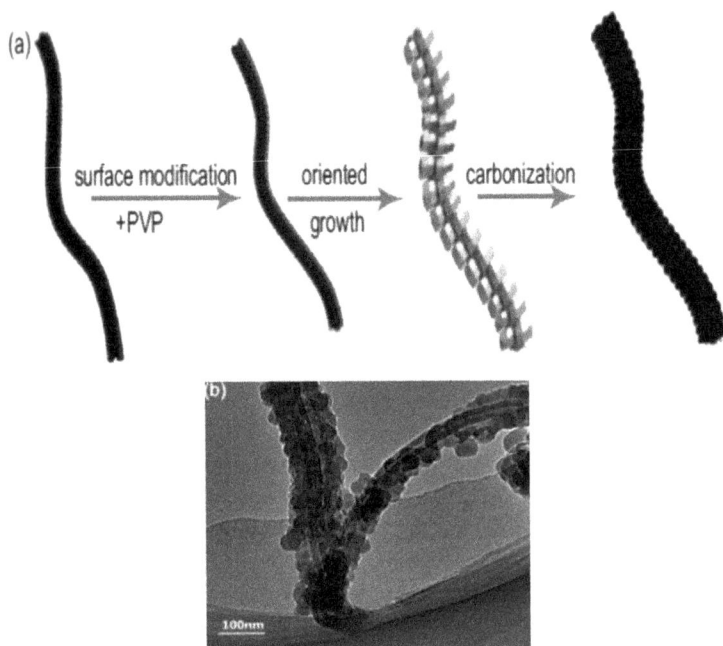

FIGURE 9.8 Schematic representation for (a) the synthesis of C at multi-walled CNTs. (b) As-resulted TEM image [50].

of chemically accessed resistors were developed from MOFs 1–3 was brought out to vapors of volatile organic compounds during the array sensing process, and their reactions were registered. In the following step, those results were analyzed statistically that permit for volatile organic compounds categorization by functional group. When considering the CNT-based sensor arrays discussed earlier, they attained a very similar volatile organic compounds distinction and were typically composed of various CNT composites that were planned to combine with definitive forms of functional categories [20–22].

Lin et al. produced the Co_3O_4 particles from the ZIF-9 and filled the same on a hybridized nanostructure of multi-walled CNTs/N-doped graphitic carbon layer. The electrocatalyst attained the optimal oxygen reduction rate by adjusting the proportion of multi-walled CNTs to ZIF-9 at the onset potential of 0.89 V to reversible hydrogen electrode. Moreover, this extremely severe catalyst agent also catalyzed the oxygen evolution reaction at the onset potential of 1.50 V to reversible hydrogen electrode. Such kinds of outcomes indicate the significant play of multi-walled CNTs equipped with active sites and having impact on the characteristics of electron-conductive hybridized materials [23].

9.2.2 Environmental Applications

Fabrication of the zeolitic imidazolate framework was done to study its adsorptive removal processes in which a hybridized nanocomposite was synthesized with the zeolitic imidazolate framework through facile method at closed room temperature. The required level of graphene oxide and CNTs taken and increment in dispersive forces were analyzed. The sufficient amount of these substrates determined the growth of nanoscale MOFs. The developed nanocomponents were employed as adsorbents for eliminating the cationic dye from wastewater. As a result, the rate of elimination of hybrid nanocomposites was better when compared to sole MOF. Malachite green is the color of cationic dye. At closed room temperature of about 20°C, the maximum adsorption capacity of zeolitic imidazolate framework (ZIF-8) was registered as 1667 mg/g. On the other hand, in this same environmental condition, zeolitic imidazolate framework with CNTs and zeolitic imidazolate framework with graphene oxide were recorded the results about of 2034 and 3300 mg/g respectively. The value of these results could be increased if the temperature conditions were maintained at higher level. The impact of various functional factors was analyzed well and optimized by batch adsorption study. Dosage of adsorbent, strength of solution at initial level and its pH, estimated quantity of loading, and condition of temperature are some of the factors that influenced the task. By employing the ethanol-washing technique, the regeneration of ZIF and its hybridized nanocomposites was done. The regeneration about of four cycles were shown by these hybridized nanocomposites. Eventually, these hybridized nanocomposites were ensured as greater components to adsorptive removal treatments through their attributes of simplified synthesis methods, effectivity, stability, and reformability [24].

Incorporation of multi-walled CNTs in the combining activity of zinc/MOF-74 and copper/BTC would involve molecular formation and attributes. Nevertheless MOF-74 is being selected for the current analysis due to its appreciable porous property and adsorption capability. In the task of catching carbon dioxide from flue gases, it has been carefully counted that the CNTs accompanied amine because of its chemic nature, specific physicochemical properties, and thermal behavior attributes [25–27].

In order to apply for the various environmental activities such as water remediation a combination of novel magnetic carbon nanostructures with MOFs was prepared and introduced. Magnetic nanoparticles of ferrous-ferric oxide (Fe_3O_4) and copper-benzene tricarboxylate (Cu-BTC) MOFs were developed through a green solvothermal technique. As earlier, by applying the same technique, nanoscale absorbents such as CNTs and graphene oxide were also developed for using as a platform to carry both of these nanostructured materials. Characterization of these hybridized nanocomposites was done by various techniques. X-ray diffraction (XRD) tests revealed that these hybridized nanocomposites exhibited a greater rate of crystalline structure. Morphological investigations ensured the productive development of Fe_3O_4 magnetic nanoparticles and nanoparticulate of Cu-BTC MOFs over the platforms of CNTs and graphene oxide. Bonding and compositeness of parent materials were inspected through various analyses. Nitrogen isotherms provide the cumulative pore volume for the CNT-based hybrid nanocomposite of Fe_3O_4/Cu-BTC about of 0.360 cm³/g whereas sole Cu-BTC MOF showed of 0.030 cm³/g. This result revealed the potential application for the purpose of MOF small molecule adsorption. This experimental work deduced that the application of CNTs and graphene oxide substrates reduces the aggregation and raises the dispersive forces inside the MOF; taking the MOFs to various sizes and structures; reaction in development of micropores in middle of the MOF and the platforms. Methylene blue is well known as an organic pollutant which was used to examine adsorption capacity. As a result, when comparing the parent materials, the prepared hybridized nanomaterials exhibited improved adsorption capacity, which was because of the synergistic integration of base materials and the exclusive attributes of nanoscale MOFs. Eventually, these modern materials were proved as superior candidates for environmental applications [28].

In water splitting, the oxygen evolution reaction is understood as the key rate-limiting step. In the preparation of electrocatalysts of oxygen evolution reactions, MOF-derived charged carriers of concentrated heteroatom CNTs are known for the hopeful OER electrocatalysts. In an experimental study, the MOF of ZIF-67 was initially chosen and annealed at 700°C in an inert condition (with the mixture of argon and hydrogen gases), and catalysts were developed. The as-prepared material showed tremendous electrocatalytic oxygen evolution reaction performance, significantly about lower Tafel slope when compared to the platinum-carbon electrode. In the same atmosphere, the endurance of N-doped CNT framework catalyst for the OER was greater when related to a commercial platinum-carbon electrode [29].

The compounds of cobalt phosphide at nitrogen and phosphorous-doped carbon/CNT hybrid were obtained by employing the phosphidation strategy via pyrolysis of cobalt phosphate MOFs and additional incorporation of CNTs by effective sonication treatment. The resulting catalysts exhibited a greater onset potential against Reversible hydrogen electrode (RHE) for OER and over potential value against RHE for Hydrogen evolution reaction (HER). When employed as an electrocatalyst in the water electrolyzer, this hybridized nanocomposite ($Co_2P@N$, P-doped C/CNT) kept the greater stability for 25 hours. In spite of lower stability than iridium-oxide/platinum couple, cobalt phosphide had registered its potential for water splitting. Also, note the significance of mutual interactivity of carbon structure and cobalt phosphide to uphold its activity as a bifunctional electrocatalyst [30].

To meet the demand for the economic and effective sensitive sensor for the detection of water vapor, Chappanda et al. developed a very sensitive humidity sensor. In which a nanocomposite HKUST-1 MOFs was synthesized with CNTs as very thin layered material. Such a thin layer was prepared by the spin-coating method where quartz-crystal micro-balance was covered. Herein relative humidity of adsorbed water vapor between 5 and 75% was found which was due to the relative shift in resonance frequency. As-prepared exhibited about 230% of increment in sensitivity was recorded when related the to the same with bare HKUST-1 film. And its performance on mean sensitivity was registered as ten times greater than that of quartz-crystal microbalance humidity sensors. The working methodology applied here was very simple and ensured a way for improving the sensitivity of MOF-grounded sensors [31].

Analysis work on MOF-5/multi-walled CNTs nanocomposite has been done to accumulate the H_2. As a result, 50% of further capability enhancement was found compared to bare MOF-5. The capability of H_2 at different conditions has been noticed and found betterment in adsorption outcomes when considered the bare MOF-5 [32]. Anbia et al. fabricated the multi-walled CNT/MIL-101 nanocomposite to enhance the capturing of carbon dioxide (CO_2). As-prepared material possesses the same morphology as bare MIL-101, at the same time 60% improvement was registered in CO_2 capturing. The enhancement of CO_2 capturing was due to the expanded nanopore volume of MIL-101 via multi-walled CNTs [33] (Figure 9.9).

Though MOFs have high potential to employ in various applications, their moisture adsorption suppresses their structure which leads to other limitations in properties. Enhancing the hydro-stability of MOF-5 by not changing any of its attributes could be the right idea to use this material. The solvothermal technique was applied to precipitate MOF-5 in quite-distributed carboxy-formed multi-walled CNTs. The same solvothermal process was used for precipitating MOF-5 in quite-distributed carboxy-functionalized multi-walled (FMW) CNTs for fabricating the MOF-5/FMWCNTs. Further investigations indicated that FMWCNTs supported to produce nanocomposites with good stability through MOF-5 nanoparticles and the subsequent hybridized MOF-5@FMWCNTs nanocomposite showed strong stable moisture. Though the stability extended at 28°C temperature to a long time with the humidity of about 55%, it indicated very minimal degradation. Hence, this experimental task established a simple, easy, synthetic way to produce stable MOF-5 through the aid of FMWCNTs without the sacrifice of surface area [34].

In order to gain the Fe_3C nanosized molecules implanted with N-doped CNTs, Guan and team prepared the dual MOFs contained pyrolysis scheme as a competent electrocatalyst for the oxygen reduction reaction. A prepared nanocomposite showed the significant electrocatalytic reaction for the oxygen reduction reaction in

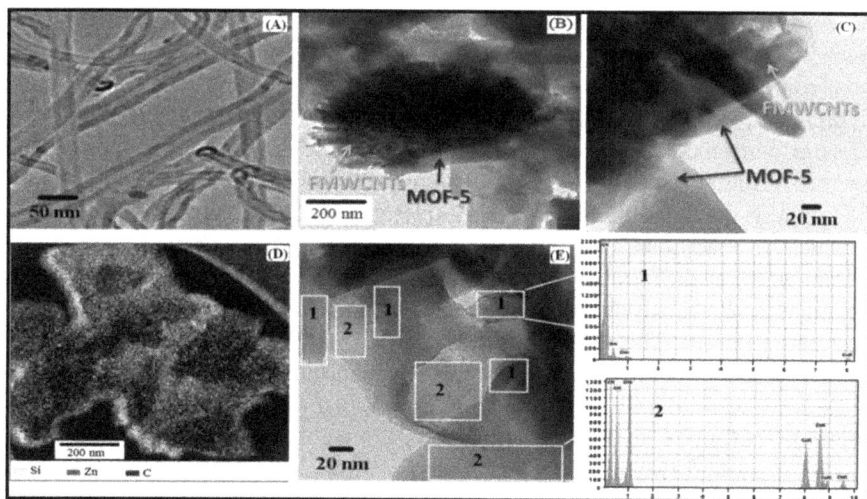

FIGURE 9.9 Synthesis of MOF-5 with CNTs: (a) TEM image of functionalized multi-walled CNTs, (b) and (c) TEM images of MOF-5@FMWCNTs, indicating functionalized multi-walled CNTs clustering over MOF-5, (d) energy-filtered TEM image of MOF-5@ functionalized multi-walled CNTs, and (e) high-resolution TEM image of functionalized multi-walled CNTs joined to MOF-5 particle with the analysis of energy dispersive X-ray spectroscopy [34].

the alkaline electrolyte owing to the short-sized iron carbide nanocrystallites and the strong formatted and poriferous N-doped ground substance. In particular, hybridized nanocomposite showed greater oxygen reduction reaction when related with platinum-carbon electrocatalysts [35].

9.2.3 BIOLOGICAL/MEDICAL APPLICATIONS

Biomolecules are well known for their plentiful convenient availability, accessibility in nature, economy, firmness, adjustability with various classifications of sites, providing distinct constructions, and agreeable combinations of biological MOFs. Calcium, magnesium, iron, and zinc are some of the harmless endogenous cations which are applied to synthesize the MOFs in which ligands comprise the derivatives or biomolecules [36]. In general bio-MOFs are employed in biomedical applications due to their bio-consistent and adaptability [36,37].

With the objective of making the biosensors for glucose detection, CNTs were synthesized with mesoporous MOFs in which terbium was used as the metal candidate to produce the advanced nanocomposite. This nanocomposite was characterized through various experimental analysis techniques and the results indicated that a thin film of Tb@meso MOFs unvaryingly sealed the outer faces of the CNTs. It acts as a supportive substance to carry both methylene green (MG) and glucose dehydrogenases (GDH). GDH/MG–Tb@MOF-CNTs/GCE exhibited a better operation for glucose detection. Here MG was the electron mediator; on the glassy carbon electrode (GCE), GDH acted as electrocatalysts to make the electrochemical glucose biosensor. Because of the superior electrical conductance of CNTs and the greater amount of Tb@meso-MOFs the nanopores united with the very huge surface area of the prepared nanoproduct [38].

MOFs of Cu-BTC and multi-walled CNTs were synthesized and the GCE reformed with the above to use in the pharmaceutical field. The traditional hydrothermal technique and ultrasonic irradiation technique are the two ways to perform the synthesis of Cu-BTCs. Both the techniques were applied to produce the Cu-BTC, in which the traditional one registered a mean crystallite size between the 20 and 40 mm, whereas the ultrasonic irradiation recorded 40 and 100 nm. Hence, the later technique had a relatively greater influence on the electrocatalytic reaction of the reformed electrode. This MOF/multi-walled CNT nanocomposite reformed by GCE was employed to compute the metformin in the pharmaceutical samples [39].

Rusling, who initially progressed and established the single-walled CNT forest electrode, permitted an effortless linkage of a huge quantity of captured antibodies. So, it showed two orders of magnitude of Low density lipoprotein (LDL) for immunoglobulin [40]. A coupled, single-walled CNT forest electrode and microfluidic sample deliverance magnetic tape records to understand multiplexed protein detection at high sensitivity as a series of concurrent events [41]. In order to realize the enterochromaffin-like (cells) visualization of single cells, a transparent and a greater conducting attributed CNT-based electrode interacted with microfluidic chips [42]. Many researchers reported that CNTs play a prominent part in the improvement of ECL intensity. Even though, it can also play a role of quencher of enterochromaffin-like cells signal. For instance, Tang et al. progressed this by quenching at first using multi-walled CNTs, followed by performing the sequence-specific DNA detection [43]. The multi-walled CNT-quenched luminescence were mainly assigned to O_2 molecules of the outer face of multi-walled CNTs united with intrinsic electron characteristics of the same. In particular, the author demonstrated a well-ordered and planned exploration on the impacts of CNTs for either improving or quenching ECL intensities. It yielded the procedure to apply the CNTs efficiently in ECL sensors for the intended action [44].

MOFs of $Ni_3(BTC)_2$ and the porous nanocomposite nickel were taken and synthesized with multi-walled CNTs and were applied to reform the indium tin oxide electrode. The as-prepared product was employed for detection of non-enzymatic urea. There are many interference species such as amino acids, creatinine, glucose, and urine analysis. The sensor showed a maximum selectivity towards urea in the existence of other interferences. When comparing with the urea signal, the amperometric reactions of other interferents were below 10% of the urea signal. The empirical experimental result of the sensor was inspected by finding the urea in spiked urine samples [45].

Synthetization of Cu(II)-based MOF nanocomposite was prepared and immobilized on multi-walled CNTs, followed by characterization of the product carried out through various analysis tests. Based on the results, the candidate was altered over the face of glassy carbon material and indicated two peaks in a phosphate buffer solution and assigned to redox process. This altered candidate material was then employed for producing a non-enzymatic hydrogen peroxide biosensor. As a result, a stable, repeatable, good respondent to electrocatalytic actions and better detectable biosensor was produced. Furthermore, this sensor has been employed to find out the accurate presence of hydrogen peroxide in H_2O sources in a wide range [46].

A nanosized metal-biomolecule-coordination polymer-ceramic was taken and a desired nanocomposite was synthesized in the combination of manganese-tyrosine-cathode nanotubes (Mn-tyr/CNTs) by a hydrothermic method. A GCE composite was employed for the oxidation and reduction of hydrogen peroxide. At the time of electrocatalytic reaction of hydrogen peroxide, the feasible reaction of oxidation and reduction progress is given below [47].

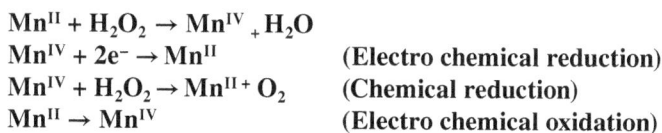

$$Mn^{II} + H_2O_2 \rightarrow Mn^{IV} + H_2O$$
$$Mn^{IV} + 2e^- \rightarrow Mn^{II} \qquad \text{(Electro chemical reduction)}$$
$$Mn^{IV} + H_2O_2 \rightarrow Mn^{II} + O_2 \qquad \text{(Chemical reduction)}$$
$$Mn^{II} \rightarrow Mn^{IV} \qquad \text{(Electro chemical oxidation)}$$

After the chemical oxidation of Mn^{II} due to hydrogen peroxide, a part of Mn^{IV} ions was involved in the rising of the reduction peak because of electrochemical reduction. The rest of the Mn^{IV} were reversed as Mn^{II} ions owing to hydrogen peroxide. Eventually, the electrochemical oxidation reaction led the Mn^{II} to Mn^{IV} to provide the peak of oxidation. Bidirectional electrocatalytic behavior of this material against oxidation (0.8 V) and reduction (0.4 V) of hydrogen peroxide must enable more choices to skip feasible intervening and interruption at the time of electrochemical determination [47].

In a bio-application to sense the glucose, an electrochemical-sensing candidate material was proposed, in which porphyrin-adsorbed Co-benzimidazole MOFs with multi-walled CNTs were applied in order to reform a glassy carbon electrode. When related to the individual counterparts, this hybridized nanocomposite showed finer oxidation behavior. Such a result was probably because of the free-conducting character of the multi-walled CNTs and the huge density-effective spots of MOFs. As a result, the line of detection of the reformed glassy carbon electrodes for electrochemical sensing of glucose was about 0.28 ppb and the liner range was between 1 and 400 [48].

In order to identify the hydrogen peroxide, Zhan and his team designed a biosensor by adopting a self-template technique. Due to its application, this sensor was specifically classified as a photo-electrochemical biosensor. A glass was prepared with Fluorine-doped Tin Oxide (FTO) coating and on its outer face, zinc oxide arrays were produced electrochemically in a systematic arrangement. As-prepared templates were submerged in a solution consisting of 2-methylimidazole $C_4H_6N_2$ as ligands in a solvent, dimethyl fumarate:water, followed by thermal combining processing, eventually received the ZnO-ZIF-8 nanotube arrangement as the core-shell structure. Zinc oxide supplied the holes and electrons subjected to light; the synthetization of zinc oxide and ZIF-8 permitted the limited proportion of reductive species within the micropores of MOFs, in respect of current generated in this hybridized arrangement. However, the limit of detection for the analysis substances was not defined, this team was capable of executing the identification of the concentration limit between 0 and 4 micrometers. This team assigned to the characterization of careful molecule-size selectability of MOFs, hence no other investigations were carried out about selectivity as well as reproducibility within batches [49].

9.2.4 ENERGY STORAGE AND OTHER APPLICATIONS

ZIF-8 developed carbon onto multi-walled carbon nanotubes (MWCNTs) to make a new constructional frame that resembles a necklace. The synthesis was done by mixing both the materials, preserved for 24 hours, then annealed at about 800°C for three hours followed by surface treatment with carbon. Hence, the free carbons were enclosed firmly on the crystalline face of MWCNTs. MOF-developed carbon/MWCNTs permitted a little more reachable outer area for ion transportation which did not form into one cluster, at the same time, when related other core-shell arrangements with this type, MWCNT-based structural materials can hardly see such properties. MOF-developed nanocomposites have a few enhanced behaviors, for instance 99.7% capacitance retention, high specific capacitance, and better rate capability [50].

Zou et al. had produced the ZnO particles from ZIF-8 and distributed those porous nanocomponents over a wide area on MWCNTs. When preparing the nanocomposite with the support of CNTs and applying the $C_4H_6N_2$ organic substances, the resulting product showed a greater capability rate from 100 to 1000 mA/g and very minimal reversible capacity at 200 mA/g. It held back the specific capacity of 326.8 mAh/g when mass normalized current attained 1000 mA/g [51]. Xie et al.

prepared the cobalt with nickel and carbon nanotubes implanted in carbon nano-cages for the major purpose of high efficiency counter electrode via pyrolysis of ZIF-67. When discussing the power conversions of this candidate material to be used in dye-sensitized solar cells, it attained a superior efficiency about of 9.04%, compared to 7.88% recorded by a platinum counter electrode [52].

The joined function of the nanoparticles into the MOF architectures produced novel composites with numerous properties. CNTs and MoFs were synthesized by adding CNTs into 1,3,5-benzenetricarboxylate. Because of the characteristics of sorption, stability, and metal sideways, this hybridized combination exhibited the productive carbon dioxide and methane uptakes that were improved by nearly three times and two times respectively when related to the sole MoF [53]. An inorganic-organic hybridized framework was fabricated in which zinc and single-walled CNTs were synthesized. Earlier, covalent binding of single-walled carbon nanotubes (SWCNTs) were confirmed with benzoic acid and interactive integrating was done with transition-metal ions in order to make the pattern of three-dimensional porous hybridized frameworks. Owing to the inherent conductive nature, structural poros-ity, and good distribution, SWCNT/zinc candidates showed greater electrocatalytic reduction of methyl parathion, and it acted as the separator and detector of organo-phosphate compounds [20].

The fastest advancements are taking place in the MOFs due to the micro/nano-porous materials. Many more developments are looking anticipated in the area of potential storage media for fossil gas and hydrogen. Nevertheless, MOFs are limited due to their instability in atmospheric moisture conditions too. The storage capa-bility of hydrogen is lower at 298 K itself. Hence, it's not used in a broad range of applications. To solve this issue, a hybridized nanocomposite was fabricated by syn-thesizing the MWCNT@MOFMC in which well-distributed MWCNTs in dimeth-ylformamide solution were immixed on same solution of $H_8N_2O_{10}Zn$ and $C_8H_6O_4$. Here, the MOFMC is known as MOF-5 and denoted as $Zn_4O(bdc)3$; bdc = 1,4-ben-zenedicarbocylate. The crystalline structure and appearance of MOFMCs is similar to the pure MOF-5, however it showed a larger Langmuir specific surface area. The storage capability of hydrogen was improved to 50% and the stability was enhanced at a better level under the existence of closed surrounding moisture conditions [54].

Oveisia et.al. developed these nanocomposites for the application of photocata-lytic dye degradation where porous nanostructured composites of MIL-125(Ti)) and CNT and synthesized the same through the hydrothermal technique. Various pro-portions of CNT were handled to prepare the composites and three different speci-mens were obtained for experimental needs, named as sole MIL, MIL/CNT(0.01) and MIL/CNT(0.03) and utilized for characterization purposes. In this experimental task, reactive black 5 was employed as the pollutant. As a result, because of the interactive integration of CNTs, MIL/CNT(0.01) indicated higher photocatalytic dye degradation. The zeta potential was registered as greater in MIL/CNT (0.01) about of 19.2 mV. Among the three nanomaterials, the rate of decolorization was higher in MIL/CNT(0.03) as 0.024 mg/L min; maximum correlation coefficient was about 0.9906 in MIL/CNT(0.01). Over two cycles, these photo-catalytic dye degradation catalysts can be reused [10] (Figure 9.10).

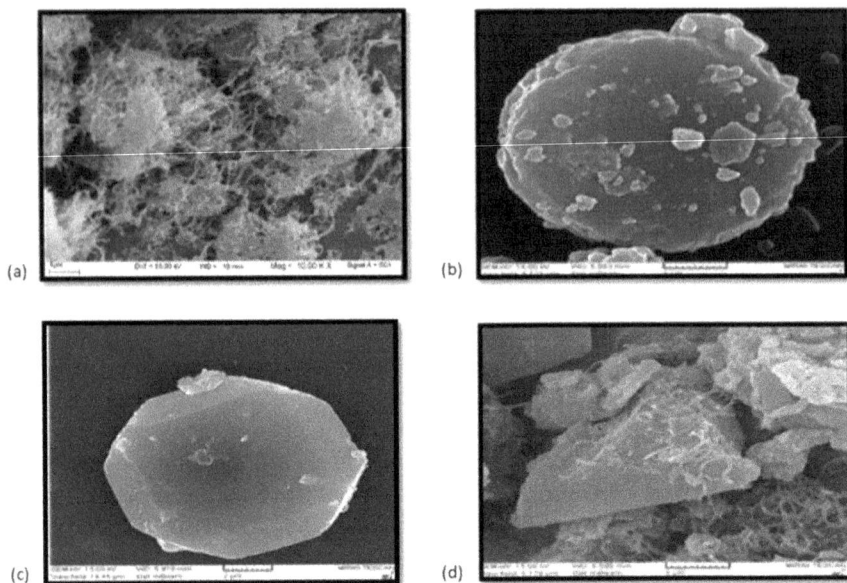

FIGURE 9.10 SEM images of synthesis of MIL/CNT (a) CNT, (b) MIL, (c) MIL/CNT(0.01), and (d) MIL/CNT(0.03) [10].

In order to design novel materials for the adsorption purposes, a simulation technique was applied. This technique supported this research analysis in a good way. Duren et al. estimated the storage capability of methane based on modeling of isoreticular MOFs, two zeolites, CNTs, and mesoporous material family of silicate and alumosilicate solids. As a result, as per unit volume, the quantity of adsorbed methane was enhanced by 23% when using 1,4-tetrabromobenzenedicarboxylate as an organic linker, and 36% for 9,10-anthracenedicarboxylate [55].

MOF nanocomposites are also used for gas storage. Some of the research tasks discuss storage of hydrogen through CNT/MOF composites [55,56]. Gas storage by adsorption is being widely analyzed with the support of MOF composites. Observations revealed that MOF-5 provides a wide capturing capability for hydrogen. Even though, this candidate is unstable when subjected to natural humidity. A MWCNT was synthesized with MOF-5 to enhance the stability in humidity and was mentioned as MOFMC [55]. This hybridized nanocomposite has a property of mutual penetrating construction and hierarchical nanovoids that increases the storable capacitance of the composite. This mesoporous MOFMC was well stabilized and expanded the accumulating capability to 2.02 wt.% at 77 K. Some other experimental analyses revealed that platinum placed MWCNTs/MOFs-5 with the improvement in hydrogen accumulation of about 1.20 wt.% in bare MOF to 1.89 wt.% in the as-prepared nanocomposite [57].

Huang et al. combined MWCNTs with cobalt (II, III) oxide through the annealing process as-prepared ZIF-67 with MWCNTs inserted at normal atmospheric conditions. In particular, the MWCNTs were first functionalized with COOH and the pretreated MWCNTs were distributed in methanol with ZIF-67 precursors where for the development of this MOF, MWCNTs worked as the nucleation center and polyvinylpyrrolidone as surfactant. Hence the MWCNTs were placed into the MOF to get the MWCNTs/ZIF-67 as nanocomposite. Eventually, the hybridized nano-structure was gained (MWCNTs/Co_3O_4) after the calcination process [58].

ZIF-67 deduced the Co_3O_4 nanotubes and developed the CNTs in several steps. Initially by applying the electrospinning technique, PAN-cobalt acetate nanofibers were produced followed by ZIF-67 crystals grown on the prepared nanofibers, hence producing the core-shell structure. Then the as-prepared core-shell structure was modified as ZIF-67 microtubes by keeping the PAN and Co $(Ac)_2$ in hot dimethyl-formamide solution. The microtubes obtained were furthermore altered as CNTs/cobalt-carbon through annealing in an inert environment at high temperature for couple of hours. The was followed by calcination for about 10 min and eventually CNT/Co_3O_4 microtubes were received. Due to the tubular construction, large volume changes could be efficiently relieved physically. As a result, the electrical conduc-tance and rate capability of the entire candidate materials were improved to enhance the quality of Li ion batteries [59].

For instance, a tubular-constructed arrangement from the higher order rank with the integration of cobalt (II, III) oxide hollow nanobits and CNTs have been gradu-ally prepared through a MOF-committed technique. Herein, polymer-Co acetate composite was initially produced; later on, reacting to the unsaturated acid $C_4H_6O_2$ ethanol solution, as-resulted core-shell nanofibers were developed, followed by application of dimethylformamide solution which eliminated the core of polymer-Co $(Ac)_2$ by dissolution. Accordingly, through the calcining technique the hollow MOF ZIF-67 (the shell nanofiber) was received in the inert condition (mixture of argon and hydrogen gases), produced the order in ranked hybridized CNTs-cobalt carbon candidates. Oxidizing the Co-specific molecules by the follow-up thermal method was conducted in atmosphere; hence the final CNT/Co_3O_4 microsized tubes were obtained. Desirably, the final product of hybridized nanocomposite for using in the Li ion batteries registered a highly reversible capacity with excellent rated capabili-ties and longer life of about 200 cycles as the anode candidate [60].

Iqbal et al. registered the actual developmental observations of successive levels of micropores in Cu-BTC synthesized with MOFs which provided the higher level of adsorption and improved the quality of bonding. Hence, these authors feel that realizing this CO_2 catching mechanism is leading to conflict and is highly important for the next level of progress to novel study. The synthesized MOFs show greater specific surface area and better storing capability, because of the existence of acid and amine-operated cathode nanotubes and a higher number of very limited sized nanopores (less than 1.0 nanometers) [61]. The latest analysis on nanocomposites of MOFs and various kinds of substrata graphene oxide, polymers, ceramics and, CNTs [62, 63], covers the issues connected to MOFs. Significantly, addition of CNTs, syn-thesis with MOFs must receive greater crystals for possessing a distinctive feature to a heightened degree on composite behavior, owing to the exceptional mechani-cal, thermo-electro conductance, and hydrophobic attributes of the CNTs [64,65].

Copper-BTC is generally applied for gas adsorption, however the extraordinary water sensitivity of copper-BTC leads to poor results subjected to atmospheric circumstances too; it also exhibits a sudden shortfall in surface area after subjecting it to humid air [66].

9.3 CONCLUSION

In this chapter the successful possible fields of MOF-derived nanostructures in chemical, electrochemical, biological, environmental, energy, and others have been discussed in the aspects of investigations on source materials, consequences of newly applied materials, design of procedure, processing techniques, characteristics derived, special features of nanocomposites on employed fields. The template approach has its radically distinctive merits in the synthesizing of nanostructures. Determining the stable, economical, and toxic templates is the present target in the analysis of template techniques. A lot of analyses are required for choosing the suitable template approaches based on various practical measures and conditions. Many developments have occurred in the MOF-derived nanostructure methods in recent years. At present, the major issues in this method can be solved by enhancing the conceptualization and treating of MOF precursors and comprehending the clear meaning of establishment of mechanisms of different structures of nanomaterials. Also, the stability of MOF-derived structures is comparatively lower than pure nanostructures, amongst various methods for the fabrication of such carbon materials, directly carbonizing through organic precursors. It is a technique applied very often to fabricate nanoporous carbons such as CNTs because of its adaptability and ease of procedure. Some of the present research tasks on nanocomposites with the support of CNT-based MOFs are biomedical micro-robots, air conditioning, carbon capture, gas sensors, eliminations of huge content of heavy metals from water, refrigeration, capture of nuclear waste, vaccines, nutrient sensors, etc. Even though there are some limitations, for instance, poor surface areas, unorganized structures, and irregular sizes which do not highly restrict their applications. Overall, the attitude of using the CNT-based MOF-derived nanostructures is good, whereas the next level of development is needed in the view of performance which enhances the chances of using metal-organic derived structures in the above-mentioned fields.

REFERENCES

1. Iijima S. Helical microtubules of graphitic carbon. *Nature* 1991;354:56–8.
2. Xia Z, Riester L, Curtin WA, Li H, Sheldon BW, Liang J, Chang B, Xu JM. Direct observation of toughening mechanisms in carbon nanotube ceramic matrix composites. *Acta Mater* 2004;52(4):931–44.
3. Saito R, Dresselhaus G, Dresselhaus MS. *Physical Properties of Carbon Nanotubes.* London, UK: Imperial College Press; 1998.
4. Dresselhaus MS, Dresselhaus G, Avouris P. *Carbon Nanotubes: Synthesis and Structure and Properties and Applications.* Berlin, Germany: Springer-Verlag; 2001.

5. Dean KA, Chalamala BR. Experimental studies of the cap structure of single-walled carbon nanotubes. *J Vacuum Sci Technol B: Microelectron Nanometer Struct Process Meas Phenom* 2003;21:868–71.

6. Xie Y, Kocaefe D, Chen C, Kocaefe Y. Review of research on template methods in preparation of nanomaterials. *J Nanomater* 2016:AID2302595. https://doi.org/10.1155/2016/2302595

7. Martens JA, Jammaer J, Bajpe S, Aerts A, Lorgouilloux Y, Kirschhock CEA. Simple synthesis recipes of porous materials. *Micropor Mesopor Mater* 2011;140(1–3):2–8.

8. Pal N, Bhaumik A. Soft templating strategies for the synthesis of mesoporous materials: Inorganic, organic–inorganic hybrid and purely organic solids. *Adv Colloid Interface Sci* 2013;189–190:21–41.

9. Chaikittisilp W, Ariga K, Yamauchi Y. A new family of carbon materials: Synthesis of MOF-derived nanoporous carbons and their promising applications. *J Mater Chem A* 2013;1:14–9.

10. Oveisi M, Asli MA, Mahmoodi NM. Carbon nanotube-based metal-organic framework nanocomposites: Synthesis and their photocatalytic activity for decolorization of colored wastewater. *Inorg Chim Acta* 2019;487:169–76.

11. Jiang Z, Lu W, Li Z, Ho KH, Li X, Jiao X, Chen D. Synthesis of amorphous cobalt sulfide polyhedral nanocages for high performance supercapacitors. *J Mater Chem A* 2014;2(23):8603–6.

12. Guan BY, Yu L, Wang X, Song S, Lou XW. Formation of onion-like $NiCo_2S_4$ particles via sequential Ion-exchange for hybrid supercapacitors. *Adv Mater* 2017;29(6):1–5.

13. Yu L, Yang JF, Lou XW. Formation of CoS_2 nanobubble hollow prisms for highly reversible lithium storage. *Angew Chem Int Ed* 2016;55(43):13422–6.

14. Chen T, Dai L. Carbon nanomaterials for high performance supercapacitors. *Mater Today* 2013;16(7–8):272–80.

15. Wang Y, Wu Y, Xie J, Ge H, Hu X. Multi-walled carbon nanotubes and metal-organic framework nanocomposites as novel hybrid electrode materials for the determination of nano-molar levels of lead in a lab-on-valve format. *Analyst* 2013;138:5113–20.

16. Zhou J, Li X, Yang L, Yan S, Wang M, Cheng D, Chen Q, Dong Y, Liu P, Cai W, Zhang C. The Cu-MOF-199/single-walled carbon nanotubes modified electrode for simultaneous determination of hydroquinone and catechol with extended linear ranges and lower detection limits. *Anal Chim Acta* 2015;899:57–65.

17. Ling P, Hao Q, Lei J, Ju H. Porphyrin functionalized porous carbon derived from metal-organic framework as a biomimetic catalyst for electrochemical biosensing. *J Mater Chem B* 2015;3:1335–41.

18. Aijaz A, Masa J, Rosler C, Xia W, Weide P, Botz AJR, Fischer RA, Schuhmann W, Muhler M. $Co@Co_3O_4$ Encapsulated in carbon nanotube grafted nitrogen doped carbon polyhedra as an advanced bifunctional oxygen electrode. *Angew Chem Int Ed* 2016;55(12):4087–91.

19. Wang MQ, Zhang Y, Bao SJ, Yu YN, Ye C. Ni(II)-based metal-organic framework anchored on carbon nanotubes for highly sensitive non-enzymatic hydrogen peroxide sensing. *Electrochim Acta* 2016;190:365–70.

20. Wang F, Zhao J, Gong J, Wen L, Zhou L, Li D. New multifunctional porous materials based on inorganic-organic hybrid single-walled carbon nanotubes: Gas storage and high sensitive detection of pesticides. *Chem Eur J* 2012;18(37):11804–10.

21. Mirica KA, Azzarelli JM, Weis JG, Schnorr JM, Swager TM. Rapid prototyping of carbon-based chemiresistive gas sensors on paper. *Proc Nat Acad Sci U S A* 2013;110(35):E3265–E3270.

22. Liu SF, Moh LCH, Swager TM. Single-walled carbon nanotube–metalloporphyrin chemiresistive gas sensor arrays for volatile organic compounds. *Chem Mater* 2015;27(10):3560–3.
23. Lin X, Li X, Li F, Fang Y, Tian M, An X, Fu Y, Jin J, Jiantai M. Precious-metal-free Co-Fe-Ox coupled nitrogen-enriched porous carbon nanosheets derived from Schiff-base porous polymers as superior electrocatalysts for the oxygen evolution reaction. *J Mater Chem A* 2016;3(17):6505–12.
24. Abdi J, Manouchehr V, Mahmoodi NM, Alemzadeh I. Synthesis of metal-organic framework hybrid nanocomposites based on GO and CNT with high adsorption capacity for dye removal. *Chem Eng J* 2017;326:1145–58.
25. Chen T, Deng S, Wang B, Huang J, Wang Y, Yu G. CO2 adsorption on crab shell derived activated carbons: Contribution of micropores and nitrogen-containing groups. *RSC Adv* 2015;5:48323–30.
26. Yu J, Yao L, Cheng B. Fabrication and CO_2 adsorption performance of bimodal porous silica hollow spheres with amine-modified surfaces. *RSC Adv* 2012;2(17):6784–91.
27. Zhou J, Wen L, Zhang Z, Xing W, Zhuo S. Carbon dioxide adsorption performance of N-doped zeolite Y templated carbons. *RSC Adv* 2012;2(1):161–8.
28. Jabbari V, Veleta JM., Chaleshtori MZ, Torresdey JG, Villagrán D. Green synthesis of magnetic MOF@GO and MOF@CNT hybrid nanocomposites with high adsorption capacity towards organic pollutants. *Chem Eng J* 2016;304:774–83.
29. Xia BY, Yan Y, Li N, Wu HB, Lou XW, Wang X. A metal-organic framework-derived bifunctional oxygen electrocatalyst. *Nat Energy* 2016;1(1):1–8.
30. Li X, Fang Y, Li F, Tian M, Long X, Jin J, Ma J. Ultrafine Co2P nanoparticles encapsulated in nitrogen and phosphorus dual-doped porous carbon nanosheet/carbon nanotube hybrids: High-performance bifunctional electrocatalysts for overall water splitting. *J Mater Chem A* 2016;4(40):15501–10.
31. Chappanda KN, Shekhah O, Yassine O, Patolea SP, Eddaoudi M, Salama KN. The quest for highly sensitive QCM humidity sensors: The coating of CNT/MOF composite sensing films as case study. *Sensor Actuat B Chem* 2018;257:609–19.
32. Landau O, Rothschild A, Zussman E. Processing-microstructure-properties correlation of ultrasensitive gas sensors produced by electrospinning. *Chem Mater* 2009;21(1):9–11.
33. Anbia M, Hoseini V. Development of MWCNT@MIL-101 hybrid composite with enhanced adsorption capacity for carbon dioxide. *Chem Eng J* 2012;191:326–30.
34. Rehman AU, Tirmizi SA, Badshah A , Ammad HM , Jawad M, Abbas SM , Rana UA, Khan SUD. Synthesis of highly stable MOF-5@MWCNTs nanocomposite with improved hydrophobic properties. *Arab J Chem* 2018;11:26–33.
35. Guan BY, Yu L, Lou XW. A dual-metal-organic-framework derived electrocatalyst for oxygen reduction. *Energy Environ Sci* 2016;9:3092–6.
36. Imaz I, Martinez MR, An J, Font IS, Rosi NL, Maspoch D. Metal-biomolecule frameworks (MBioFs). *Chem Commun* 2011;47:7287–302.
37. Briones D, Fernandez B, Calahorro AJ, Jimenez DF, Sanz R, Martínez F, Orcajo G, Sebastián ES, Seco JM, González CS, Llopis J, Dieguez AR. Highly active anti-diabetic metal-organic framework. *Cryst Growth Des* 2016;16(2):537–40.
38. Song Y, Shen Y, Gong C, Chen J, Xu M, Wang L, Wang PL. A novel glucose biosensor based on tb@mesoporous metal-organic frameworks/carbon nanotube nanocomposites. *Chem Electro Chem* 2017;4(6):1457–62.
39. Hadi M, Poorgholi, Mostaanzadeh H. Determination of metformin at metal-organic framework (Cu-BTC) Nanocrystals/multi-walled carbon nanotubes modified glassy carbon electrode. *S Afr J Chem* 2016;69:132–9.

40. Venkatanarayanan A, Crowley K, Lestini E, Keyes TE, Rusling JF, Forster RJ. High sensitivity carbon nanotube based electrochemiluminescence sensor array. *Biosens Bioelectron* 2012;31(1):233–9.
41. Kadimisetty K, Malla S, Sardesai NP, Joshi AA, Faria RC, Lee NH, Rusling JF. Automated multiplexed ECL immunoarrays for cancer biomarker proteins. *Anal Chem* 2015;87(8):4472–8.
42. Valenti G, Zangheri M, Sansaloni SE, Mirasoli M, Penicaud A, Roda A, Paolucci F. Transparent carbon nanotube network for efficient electrochemiluminescence devices. *Chem Eur J* 2015;21:12640–-5.
43. Tang X, Zhao D, He J, Li F, Peng J, Zhang M. Quenching of the electrochemiluminescence of Tris(2,2′-bipyridine)ruthenium(II)/Tri-n-propylamine by pristine carbon nanotube and its application to quantitative detection of DNA. *Anal Chem* 2013;85(3):1711–8.
44. Wusimanjiang Y, Meyer A, Lu L, Miao W. Effects of multi-walled carbon nanotubes on the electrogenerated chemiluminescence and fluorescence of CdTe quantum dots. *Anal Bioanal Chem* 2016;408(25):7049–57.
45. Tran TQN, Das G, Yoon HH. Nickel-metal organic framework/MWCNT composite electrode for non-enzymatic urea detection. *Sensor Actuat B: Chem* 2017;243:78–83.
46. Zhou E, Zang Y, Li Y, He X. Cu(II) based MOF immobilized on multiwalled carbon nanotubes: Synthesis and application for nonenzymatic detection of hydrogen peroxide with high sensitivity. *Electroanalysis* 2014;26(11):2526–33.
47. Yang J, Zhou B, Yao J, Jiang XQ. Nanorods of a new metal-biomolecule coordination polymer showing novel bidirectional electrocatalytic activity and excellent performance in electrochemical sensing. *Biosens Bioelectron* 2015;67:7.
48. Yiping HU, Duoliang S, Xiaoquan L. Nonenzyme sensor based on metal organic frameworks/porphyrin/multiwalled carbon nanotubes for detection of glucose. *Chem J Chin Univ* 2016;37:6.
49. Zhan WZ, Kuang Q, Zhou JZ, Kong XJ, Xie ZX, Zheng LS. Semiconductor@metal-organic framework core–shell heterostructures: A case of ZnO@ZIF-8 nanorods with selective photoelectrochemical response. *J Am Chem Soc* 2013;135(5):1926–33.
50. Wang Y, Chen B, Zhang Y, Fu L, Zhu Y, Zhang L, Wu Y. ZIF-8@MWCNT-derived carbon composite as electrode of high performance for supercapacitor. *Electrochim Acta* 2016;213:260–9.
51. Zou Y, Qi Z, Ma Z, Jiang W, Hu R, Duan J. MOF-derived porous ZnO/MWCNTs nanocomposite as anode materials for lithium-ion batteries. *J Electroanal Chem* 2017;788:184–91.
52. Xie Z, Cui X, Xu W, Wang Y. Metal-organic framework derived CoNi@CNTs embedded carbon nanocages for efficient dye-sensitized solar cells. *Electrochim Acta* 2017;229:361–70.
53. Xiang Z, Hu Z, Cao D, Yang W, Lu J, Han B, Wang W. Metal-organic frameworks with incorporated carbon nanotubes: Improving carbon dioxide and methane storage capacities by lithium doping. *Angew Chem Int Ed* 2011;50(2):491–4.
54. Yang SJ, Choi JY, Chae HK, Cho JH, Nahm KS, Park CR. Preparation and enhanced hydrostability and hydrogen storage capacity of CNT@MOF-5 hybrid composite. *Chem Mater* 2009;21(9):1893–7.
55. Duren T, Sarkisov L, Yaghi OM, Snurr RQ. Design of new materials for methane storage. *Langmuir* 2004;20(7):2683–9.
56. Jiang H, Feng Y, Chen M, Wang Y. Synthesis and hydrogen-storage performance of interpenetrated MOF-5/MWCNTs hybrid composite with high mesoporosity. *Int J Hydrogen Energy* 2013;38(25):10950–5.

57. Yang SJ, Cho JH, Nahm KS, Park CR. Enhanced hydrogen storage capacity of Pt-loaded CNT@MOF-5 hybrid composites. *Int J Hydrogen Energy* 2010;35(23):13062–7.
58. Huang G, Zhang F, Du X, Qin Y, Yin D, Wang L. Metal organic frameworks route to in situ insertion of multiwalled carbon nanotubes in Co$_3$O$_4$ polyhedra as anode materials for lithium-ion batteries. *ACS Nano* 2015;9(2):1592–9.
59. Chen YM, Yu XY, Li Z, Paik U, Lou XW. Hierarchical MoS2 tubular structures internally wired by carbon nanotubes as a highly stable anode material for lithium-ion batteries. *Sci Adv* 2016;2(7):1–8.
60. Chen YM, Yu L, Lou XW. Hierarchical tubular structures composed of Co$_3$O$_4$ hollow nanoparticles and carbon nanotubes for lithium storage. *Angew Chem Int Ed* 2016;128(20):6094–7.
61. Iqbal N, Wang X, Yu J, Jabeen N, Ullah H, Ding B. In situ synthesis of carbon nanotube doped metal-organic frameworks for CO$_2$ capture. *RSC Adv* 2016;6:4382–6.
62. Lei J, Qian R, Ling P, Cui L, Ju H. Design and sensing applications of metal-organic framework composites. *TrAC Trend Anal Chem* 2014;58:71–8.
63. Han JT, Kim SY, Woo JS, Lee GW. Transparent, conductive, and superhydrophobic films from stabilized carbon nanotube/silane sol mixture solution. *Adv Mater* 2008;20(19):3724–7.
64. Berson S, Bettignies RD, Bailly S, Guillerez S. Poly(3-hexylthiophene) fibers for photovoltaic applications. *Adv Funct Mater* 2007;17(8):1377–84.
65. Lin KYA, Yang H, Petitc C, Hsu FK. Removing oil droplets from water using a copper-based metal organic frameworks. *Chem Eng J* 2014;249:293–301.
66. Wang G, Chen L, Sun Y, Wu J, Fu M, Ye D. Carbon dioxide hydrogenation to methanol over Cu/ZrO$_2$/CNTs: Effect of carbon surface chemistry. *RSC Adv* 2015;5(56):45320–30.

10 Preparation and Characterization of Magnetic Metal-Organic Framework Nanocomposites

Afzal Ansari, Vasi Uddin Siddiqui, and Weqar Ahmad Siddiqi

CONTENTS

10.1 INTRODUCTION

Metal-organic frameworks (MOFs) are a complex type of porous inorganic-organic hybrid materials synthesized from metal ions (or cluster) with a multidentate organic ligand via coordination linkages [1]. Magnetic MOFs (MMOFs) nanocomposites or magnetic framework composites (MFCs) have emerged as new functional materials in recent years [2]. The MMOFs have attracted tremendous interest due to their diverse structures, flexible pore size, coordinated metal sites, high porosity, great mechanical strength, and adaptable surface properties [3,4]. The selection of material in the field of MOFs has become a major challenge as a large number of compounds are reported every year [5]. Over the past decades, more than twenty thousand different MOFs have been synthesized, which have presented a wide range of uses in the field of complex coordination chemistry and effective applications such as catalysis, gas storage, separation, drug delivery, electronic/optoelectronic devices, and sensors. The MMOFs nanocomposites are of great interest in different environmental applications, particularly in the field of water purification and green energy [6–8].

The magnetic property is one of the several features of the MOFs that can be used by incorporating magnetic moment carriers such as paramagnetic metals or organic ligands, or both [9]. This is not satisfactory to present a magnetic material because magnetism is a common phenomenon and hence requires an unusual mode of exchange between moment carriers. Therefore, connections between moment carriers are required at distances within the interaction range [10]. The interaction mechanisms between MOFs and substrate are also diverse, due to the structural features of MOFs, including electrostatic interactions, hydrogen bonding, acid-base interactions, the effect of framework metal, pore size-selective adsorption, π-π interactions, hydrophobic interactions, π-complexation, and vacuum expansion [11]. Furthermore, it allows the precise design and alteration of the chemical and physical properties of MOFs by selecting appropriate metal centers, ligands, and synthetic conditions to obtain specific pore sizes and spatial cavity patterns for diverse applications [12]. The resulting three-dimensional outline shows an extraordinary surface area of more than 2000–8000 square meters per gram of material [13].

The fabrication of magnetic-based MOF structures is an example of multifunctional network materials where the magnetic properties coexist with the porosity. These magnetic porous materials act as external chemical stimuli and offer to insert extra molecules into the pore that may serve to tune the magnetism of the structure. The coexistence of magnetism and porosity is a challenging aspect from a chemical design point of view, as these two properties are ineligible for each other, while magnetic exchange interactions require short distances between metal centers, which are typically spin inhibitors [14]. Porosity is generally preferred with the use of long linkers that often exist at temperatures above zero for too long for magnetic ordering. However, various synthetic pathways can be hypothesized to develop porous magnetic materials. In this chapter, the typical synthesis routes of MMOF materials are discussed, including their structure and characterization. Furthermore, this chapter aims to provide an overview of the MMOFs based on the recent advancement of nanocomposite development.

10.2 METHODS OF PREPARATION

There are different approaches to MMOF or MFC preparation depending on the design of the MOFs. Particles/fibers with magnetic properties are synthesized with MOFs by the process of growing or attaching. These MMOFs are examples of the mutual coexistence of porosity with magnetic properties. Typical magnetic nanoparticles (MNPs) are iron, cobalt, nickel, and their oxides. Usually, iron oxides such as Fe_3O_4 and $\gamma\text{-}Fe_2O_3$ are used as ferromagnetic and super ferromagnetic MNPs in MMOFs [15,16]. Gao et al. in a systematic review on MMOFs in adsorption and enrichment removal of food and environmental pollutants application, discussed that surface modification and sealing are two magnetic functionalized methods for loading MNPs to MMOFs [17]. In the surface modification method, organic molecules conjugated on the surface of magnetic particles and core-shell structures are grown via complexation and coordination of dispersed MNPs in a precursor of MOFs [18]. Besides getting the advantages of forming a core-shell structure with a large specific surface area and high adsorption performance the surface modification process has certain limitations or disadvantages such as getting an uneven thickness of the core-shell with poor magnetic properties and a tedious modification process. In the sealing methods, magnetic particles are embedded on MOF surfaces via continuous adsorption of magnetic particles during the growth of MOFs [19]. Since the magnetic particles are adsorbed continuously on the MOF surfaces, they provide a robust magnetism and rapid separation, but this synthesis process reduces the specific surface area and adsorption capacity as the magnetic particles enter the core. However, core-shell and embedded structures of MMOFs can be obtained by mixing the MNPs with MOFs via coprecipitation by the mixing method. So, the mixing process is easy to conduct however, in adsorption applications the adsorbed target molecules are difficult to elute. Bellusci et al. reported a new synthesis process (Figure 10.1) via ball milling induced mechanochemical reaction for targeted drug delivery and several other applications such as catalysis, sensing, and selective sequestration processes [20]. In this approach iron (III) 1,3,5-benzenetricarboxylate magnetic composite (MN@Fe-BTC) was studied in the loading and release of doxorubicin drug. Stable coordination bonds were formed between the drug-loaded in Fe-BTC and magnetic composite and it is delivered at the target by a mechanism controlled by diffusion and material degradation.

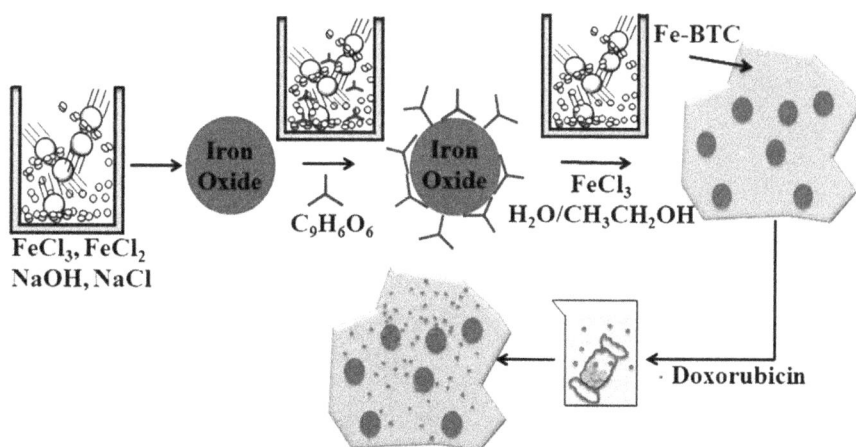

FIGURE 10.1   Synthesis scheme. (Reprinted with permission from ref. [20].)

Espallargas and Coronado also investigated the different approaches developed for preparing MOFs with magnetic and electronic properties [21]. Since MOFs are a porous material, the intrinsic molecule can stimulate or tune the magnetic behavior of the MOFs. Therefore, how the magnetism occurs in MOFs can be classified as four types of MOFs based on magnetic structures (Figure 10.2): (i) Magnetic MOFs, where magnetic cooperativity occurs via magnetic exchange through the ligand, (ii) the spin-crossover MOFs, where the nodes have the appropriate coordination environment for the existence of this phenomenon, (iii) MOFs with magnetic relaxation, where nodes are clusters with the anisotropic spin ground state, with single-molecule magnetic behavior, and (iv) MOFs with a magnetocaloric effect, where nodes are coupled to an isotropic spin ground state. Magnetic MOFs are designed by using short linkers with selected examples, viz. formats, cyanides, azolates, diazines, and lactates using the metallo-ligand precursor complexes, and the radical ligand to achieve coupling and porosity. Nadar et al. classified the MFCs on the basis of their structure into two forms (Figure 10.3): (i) core-shell MFCs and (ii) non-core-shell MFCs [4]. Here, the non-core-shell structure can be obtained by the in situ and embedded mechanism of synthesis in which the outer surface of the previously synthesized MOFs was attached to a MNP unit or synthesized within the MOFs via the solvothermal technique. Furthermore, in a core-shell type structure approach, the MNPs are considered as a core unit of the core-shell structure while the MOFs will be considered as the shell that was grown in a restrained condition by the organic precursor and ligands [22]. In this chapter, we discuss the four different main approaches usually used for designing MFCs as shown in Figure 10.4.

10.2.1 EMBEDDING

It is also a form of non-core shell-type approach for the synthesis of MFCs. In this method, functionalized magnetic nanoparticles are attached/embedded

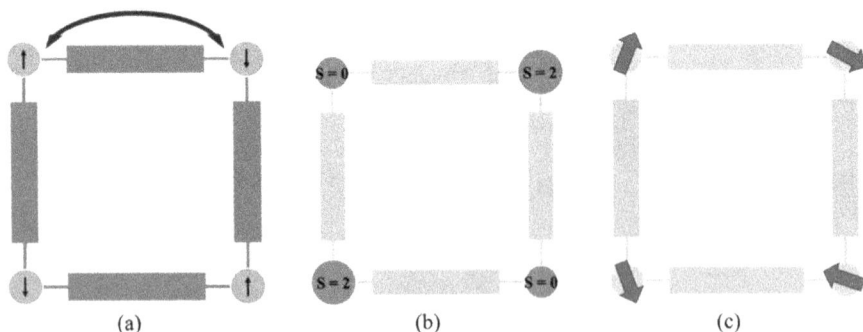

(a)   (b)   (c)

FIGURE 10.2 Schematic representation of the different strategies to incorporate magnetic functionalities into MOFs: (a) magnetic exchange via the ligands, (b) spin-crossover at the nodes, and (c) MOFs with magnetic clusters in the nodes (anisotropic spin for SMM, isotropic spin for magnetocaloric. (Reprinted with permission from ref. [21].)

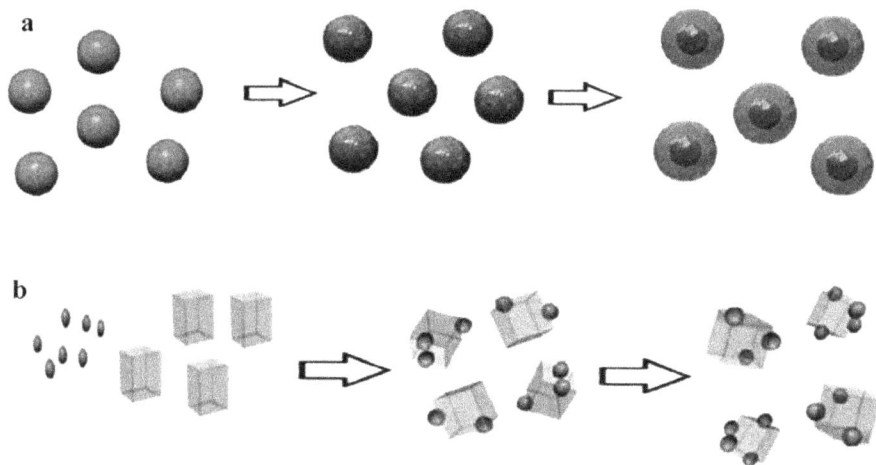

FIGURE 10.3　The design of magnetic framework composites. (a) Core-shell MFCs and (b) non-core-shell MFCs. (Reprinted with permission from ref. [4].)

on the surface of MOFs by the process of nucleation and growth of the MOFs. Most MOFs have the applicability of magnetization by this simple and one-step method. However, this growth of nanoparticles reduces the BET surface area and lowers the pore volume of MOFs by uniform dispersion of nanoparticles in a composite. As a result, the extensive use of MFCs is curbed [23,24]. The synthesized MFCs structure morphology is similar to the MOFs crystal that is used and defined by the route of the synthesis process. Mostly the synthesis procedure is from conventional heating to sonochemistry using elevated temperature. In self-template synthesis strategy, inorganic oxide template and MOFs precursor is mixed in a solvent or mixture of solvent in one pot. The surface of magnetic a nanoparticle is used for MOFs as stabilizer or bridge in the framework component which enables the functionalization of MNPs and particle size and its distribution carried out in the one-step process. The MOFs coating restricts the growth of MNPs within MOFs [25]. Wu et al. reported the one-step in situ pyrolysis route for the fabrication of MFCs. The MNP was produced by ferric tris(acetylacetonate) $[Fe(C_5H_7O_2)_3]$ precursor in the ZIF-8 system (Figure 10.5). The MFCs produced maintain the superparamagnetic characteristics with a porous structure, having a large surface area along with thermal stability [26]. A solvothermal and hydrothermal method is also practiced depending on the solvent used. The solvothermal method uses a high boiling point polar non-protic solvent such as N, N dimethylformamide (DMF), N, N-diethyl formamide (DEF), N, N-dimethylacetamide (DMA) or, less frequently, dimethyl sulfoxide (DMSO). Cosolvents such as methanol, ethanol or water can be also added. The hydrothermal process uses water or a mixture of solvents in which water is the main component. Generally, a sealed vessel such as an autoclave is used for the formation of MFCs at above room temperature.

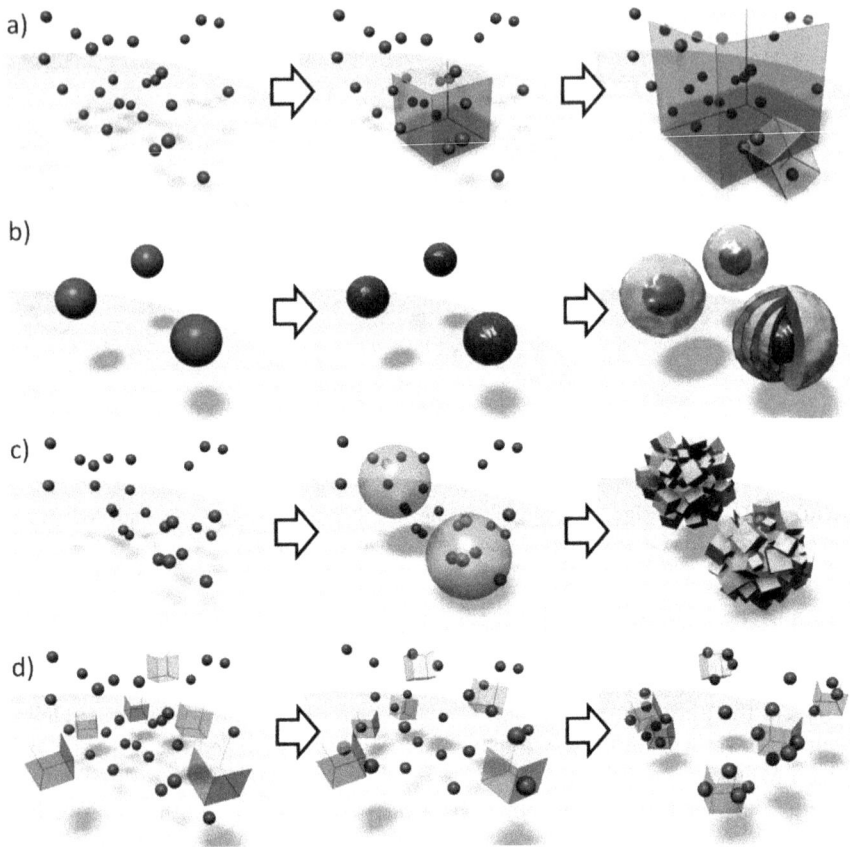

FIGURE 10.4 Different approaches to the design of magnetic framework composites: (a) embedding, (b) layer-by-layer, (c) encapsulation, and (d) mixing. (Reprinted with permission from ref. [22].)

10.2.2 LAYER-BY-LAYER (LBL)

In this method, an appropriate functional group is added on the surface of MNPs by stepwise liquid phase epitaxy (LPE) to form an MFCs material (Figure 10.4b). This modification ensures the controlled growth of the crystal and higher stability of MFCs with the formation of a core-shell structure. The thickness and shape of MOFs are controlled by the number of cycles and shapes of MNPs used respectively. In this case, by using the solvo/hydrothermal process, bi/trivalent metal ions as ligand and nitrate or acetate salts as inorganic precursors are combined with MNPs in a solvent or mixture of solvents. Nadar et al. reported the three ways of functionalization of MNPs—(i) simultaneous synthesis and functionalization of MNPs in one step, (ii) synthesis of bare MNPs and then coating them with functionalizing agents, and (iii) synthesis and consecutive linking with an encapsulating agent. Generally, carboxylate, phosphonate, thiol, and hydroxyl functional groups can bind to the surface

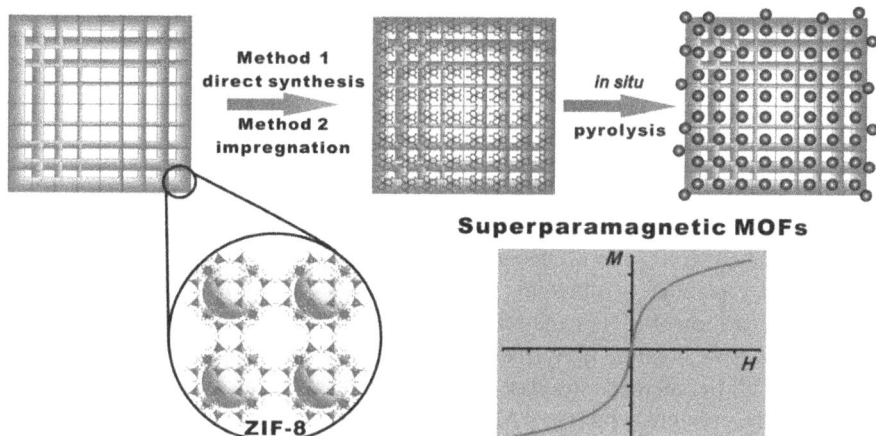

FIGURE 10.5 Schematic illustration of the preparation of superparamagnetic metal-organic frameworks of γ-Fe_2O_3 @ZIF-8. (Reprinted with permission from ref. [26].)

of MNPs. Polyacrylate, polydopamine, dipyridyl amine, and (3-Aminopropyl) triethoxysilane (APTES) are used for $-NH_2$, Benzene-1,3,5-tricarboxylate, mercaptoacetic acid and trisodium citrate are used for $-COOH$ while 4-(thiazolyl azo) resorcinol used for $-SH$ surface functionalization [4]. Besides having the advantages of precise control of thickness and an easy preparation method at room temperature in the LbL process, there are some limitations or disadvantages of this process such as obtaining a specific thickness to need to achieve a specific surface area and a large growth cycle which is limited to the MOFs based on divalent metals. HKUST-1 is the most common example prepared by the LbL method [27]. Ke et al. used to synthesize the novel MOFs (Fe_3O_4@[$Cu_3(BTC)_2$]) based on this method and also studied the relation between the thickness and temperature. The MOFs shell of the microspheres prepared at 70°C is approximately 575 nm after 50 assembly cycles, which is nearly two times as large as that of the MOFs microspheres prepared at 25°C with the same assembly cycle [28]. Not only a simple hierarchical level but multi-level hierarchical core-shell MOFs have been synthesized by this simple LbL method [29].

10.2.3 ENCAPSULATION

In this method, MFCs can be obtained by a buffer interface between the MOFs and MNPs that encourage the growth of MOFs girdled at the magnetic nanostructure (particles or fibers) and the carrier. These magnetic functionalized nanostructures provide better compatibility to trap the material, consequently these nanocomposites exercised for nucleation and growth of MOFs (Figure 10.4c). This method has a simple processing step and by changing the MNPs getting the advantage of controlling the MFCs core-shell structure. Furthermore, this method limits the control over the compatibility of the MOFs coating and requires pre-embedded MNPs in the polymer for a better agreement of the MFCs [3]. An innovative and fast synthesis of MOFs and MFCs was recently achieved by Kim and his coworker [30]. A microfluidic device was used for continuous microdroplet production. The microdroplets were used as microreactors in which MOFs and MFCs were produced in a few minutes. In particular, iron oxide nanoparticles have been successfully encapsulated into ZIF-8 using a polystyrene sulfonate (PSS) buffer interface.

10.2.4 MIXING

In this method, MNPs and MOFs are synthesized separately and mixed uniformly under high temperature or ultrasonic for preparing MFCs (Figure 10.4d). Aggregation and polymerization of MNPs and MOFs create a strong electrostatic interaction that leads to a durable bonding performance. This approach is one of the simplest methods among all the methods discussed for preparing MFCs as it could be applied to almost all the previously prepared MOFs however, in between MOFs and MNPs the MNPs are easily detachable. Additionally, a large amount of MOFs are required for this method of mixing which makes it an expensive method. The mixing method for the preparation of MFCs has been reported by Huo and Yang, 2012 [31]. The unreacted MNPs are washed by the centrifuge technique several times and a pure MOF crystal is separated from the synthesized MFCs effectively.

10.3 CHARACTERIZATION OF MAGNETIC MOFs

The resulting MOFs that are synthesized by different methods should further describe their properties with the help of various physicochemical techniques to determine their properties. From the design perspective of MOFs, it is also important to know its structural and textural properties, studying its homogeneity and stability. The useful techniques for the characterization of MMOFs are explained below.

10.3.1 X-RAY DIFFRACTION (XRD)

XRD is generally used to define structural parameters and the crystallinity of magnetic MOFs. The structural description of the synthesized MOFs can be performed by comparison to diffraction with the previous one reported in the literature, or a single-crystal X-ray pattern and included in a database or using computational modeling [32]. It is possible with this technique to identify the crystallographic parameters (unit cell size, lattice, and crystallite size), distinctive polymorphic forms, differentiate between amorphous and crystalline materials and to estimate the percentage of its crystallinity [33]. The crystalline size is usually calculated using the Debay-Scherrer equation with the most intense no-overlapped peak after the diffraction peaks are identified.

$$D = \frac{K \cdot \lambda}{\beta \cdot \cos\theta}$$

Where D is the crystalline size, K is the size factor (value 0.84–0.98, which is dependent on equipment), β the full width at half maximum height of the peak (FWHM) and λ the Bragg angle pattern for the peak identified in diffraction.

This characterization technique is used extensively because it allows an under-standing of the crystallization of MOFs as well as verifying the formation of new MOFs or composite. Besides, there are other techniques based on the absorption of X-rays such as EXAFS or XAFS (extended X-ray absorption fine structure) which are less used in the characterization of MOFs, mainly because they require the use of complex synchrotron radiation installations. However, X-ray absorption near-edge structure (XANES) gives information about the energy bandwidth, valence state, and bond angles [34,35]. The XRD patterns of the MIL-100(Fe) and M@MIL-100(Fe) (M = Au, Pd, and Pt) are exhibited in Figure 10.6 as prepared by Liang and his coworker. It is evident that the M@MIL-100(Fe) nanocomposites with different metal loadings exhibit similar XRD patterns, suggesting that the integrity of the MIL-100(Fe) framework is maintained [36].

10.3.2 SCANNING ELECTRON MICROSCOPY (SEM)

The SEM technique produces high-resolution two-dimensional (2D) images that are routinely used for MOFs characterization. It displays the shape of the material and its spatial variations, revealing information about external morphology, dispersion, and mixing of phases. The porous structure exhibited by the MOFs gives rise to particles of curious shapes such as cubes, bars, rhombohedral, and yields a diverse morphology (Figure 10.7a). SEM characterization usually requires gold or platinum for the coating of the surface with a conductive material based on the insulating nature of MOFs. The equipment works with a field emission gun that provides a highly focused electron beam, which increases the spatial resolution working at low potentials. This charac-teristic prevents damage by reducing the charging effect on the insulating material that the electron beam may induce in some sensitive MOFs [37,38].

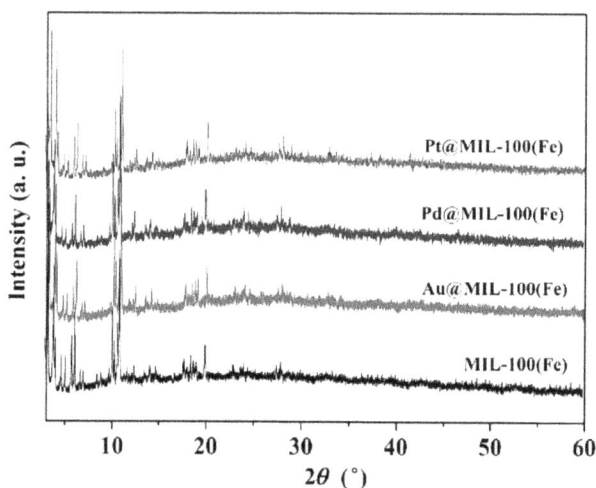

FIGURE 10.6   XRD patterns of MIL-100(Fe) and M@MIL-100(Fe) (M = Au, Pd, and Pt) (Reprinted with permission from ref..[36]).

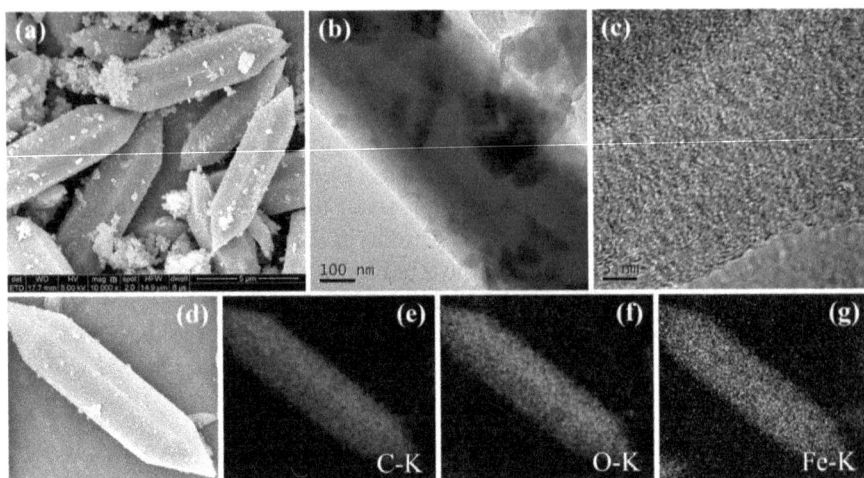

FIGURE 10.7 (a) Field Emission Scanning Electron Microscopy (FE-SEM) image of Fe-MOF, (b) Transmission Electron Microscopy (TEM) image of Fe-MOF, (c) High-Resolution TEM (HRTEM) image of Fe-MOF, and, (e–g) Energy Dispersive X-ray Analysis (EDX) element mapping of C, O, and Fe obtained from (d). (Reprinted with permission from ref. [38].)

10.3.3 Transmission Electron Microscopy (TEM)

TEM has been generally used to determine different particle sizes and to create a histogram with crystallographic data, such as face indices and dislocations. This technique is highly beneficial for the characterization of modified MOFs by the composition of nanoparticles, as the images obtained present data about the size and dispersion of those nanoparticles. These techniques can relate to energy dispersive spectroscopy (EDS) or energy dispersive X-ray analysis (EDX or EDAX), which allows determining the elemental analysis and qualitative composition of MOFs. Gao et al. synthesized a Fe-MOF which was a uniform axis-like shape and a similar microstructure and lattice structure characterized by a field emission SEM (FE-SEM) and high-resolution TEM (HRTEM) which is shown in Figure 10.7b. The chemical structure is studied by the EDX yield mapping results of each element which allows knowing the symmetry of the synthesized MOFs [38].

10.3.4 Thermogravimetric Analysis (TGA)/ Differential Thermal Analysis (DTA)

This technique is useful in estimating the thermal stability of MOFs and determining their solvent-accessible pore volume and their mass loss of materials as a function of temperature in a controlled environment. The MOFs disintegration varies depending on the carrier gas (N_2, air, or O_2) that is used for TGA analysis. Taghizadeh et al. synthesized the MMOFs nanocomposite and evaluated thermal stability by

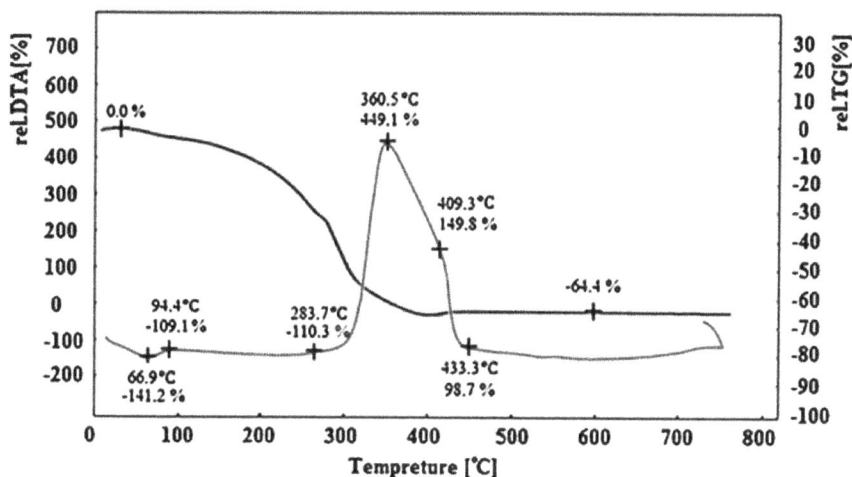

FIGURE 10.8 The TGA/DTA plots of magnetic MOF. (Reprinted with permission from ref. [39].)

thermogravimetric analysis as shown in Figure 10.8. In DTA plots of MMOFs, exothermic peaks were observed at 283.7, 360.5, 409.3, and 433.3°C. As shown in the TGA plots, the weight loss for MOFs was about 72%, and the remaining 28% is related to the presence of Cu^{2+} in the MOFs composition. However, the weight loss for the magnetic nanocomposite was about 64.4%. In the remaining 35.6%, about 28% is related to Cu^{2+}, and about 7.6% is due to the presence of Fe^{2+} and Fe^{3+} ions in the structure of the magnetic nanocomposite [39].

The TGA curve of MOFs indicates the weight loss (%) versus temperature in the study. This plot normally describes the first loss of mass in the 40–110°C range that corresponds to the removal of solvent molecules and a second loss between 300–400°C that typically results in structural modification of the MOFs framework. It should be kept in mind that this is not due to a structural change to reduce weight, so it is necessary to supplement this study with analysis by XRD at variable temperatures to examine the change of structure [40].

10.4 CONCLUSION

The magnetic-based MOFs or MFCs are a novel type of materials that can be obtained with the different particles, beads, fibers, or nanorods, porous crystal with magnetic and functional properties. In this chapter, systematic studies have been mentioned to understand the recent advancements based on the preparation and characterization of MMOFs nanocomposites. Among all the above-mentioned methods, the mixing method has been found as the most suitable technique for the fabrication of MMOFs based on cost-effectiveness and high yield. Furthermore, a summary of different characterization techniques is also described to identify their structural and textural properties.

ACKNOWLEDGMENT

The author, Afzal Ansari gratefully acknowledges the financial assistance in terms of "Non-NET fellowship" by University Grant Commission (UGC), New Delhi. Further, the authors are also grateful to the Jamia Millia Islamia, New Delhi, for providing the facility.

REFERENCES

1. Furukawa H, Cordova KE, O'Keeffe M, Yaghi OM. The chemistry and applications of metal-organic frameworks. *Science (80-)* 2013;341:1230444. doi:10.1126/science.1230444
2. Wu M-X, Gao J, Wang F, Yang J, Song N, Jin X, et al. Multistimuli responsive core-shell nanoplatform constructed from Fe3O4 @MOF equipped with pillar[6]arene nano-valves. *Small* 2018;14:1704440. doi:10.1002/smll.201704440
3. Ma YJ, Jiang XX, Lv YK. Recent advances in preparation and applications of magnetic framework composites. *Chem Asian J* 2019;14:3515–30. doi:10.1002/asia.201901139
4. Nadar SS, Varadan N, Suresh S, Rao P, Ahirrao DJ, Adsare S. Recent progress in nanostructured magnetic framework composites (MFCs): Synthesis and applications. *J Taiwan Inst Chem Eng* 2018;91:653–77. doi:10.1016/j.jtice.2018.06.029
5. Watanabe T, Sholl DS. Accelerating applications of metal-organic frameworks for gas adsorption and separation by computational screening of materials. *Langmuir* 2012;28:14114–28. doi:10.1021/la301915s
6. Zhao G, Qin N, Pan A, Wu X, Peng C, Ke F, et al. Magnetic nanoparticles@metal-organic framework composites as sustainable environment adsorbents. *J Nanomater* 2019;2019:1–11. doi:10.1155/2019/1454358
7. Ansari A, Siddiqui VU, Khan I, Akram MK, Siddiqi WA, Khan A, et al. Metal-Organic Frameworks (MOFs) for industrial wastewater treatment. In Anish Khan, Baha M Abu-Zaid, Mahmoud A Hussein, Abdullah M Asiri, Mohammad Azam, *Met. Framew. Compos. Vol. I, vol. 53*, Materials Research Forum LLC, USA; 2019, pp. 1–28. doi:10.21741/9781644900291-1
8. Siddiqui VU, Ansari A, Ahmad I, Khan I, Akram MK, Siddiqi WA, et al. Designing metal-organic frameworks for clean energy applications. In Anish Khan, Baha M Abu-Zaid, Mahmoud A Hussein, Abdullah M Asiri, Mohammad Azam, *Met. Framew. Compos. Vol. I, vol. 53*, Materials Research Forum LLC, USA; 2019, pp. 140–69. doi:10.21741/9781644900291-7
9. Kurmoo M. Magnetic metal-organic frameworks. *Chem Soc Rev* 2009;38:1353. doi:10.1039/b804757j
10. Blundell S, Thouless D. Magnetism in condensed matter. *Am J Phys* 2003;71:94–5. doi:10.1119/1.1522704
11. Wang B, Xie L-H, Wang X, Liu X-M, Li J, Li J-R. Applications of metal-organic frameworks for green energy and environment: New advances in adsorptive gas separation, storage and removal. *Green Energy Environ* 2018;3:191–228. doi:10.1016/j.gee.2018.03.001
12. Yaghi OM, Li H, Davis C, Richardson D, Groy TL. Synthetic strategies, structure patterns, and emerging properties in the chemistry of modular porous solids †. *Acc Chem Res* 1998;31:474–84. doi:10.1021/ar970151f
13. Stoeck U, Krause S, Bon V, Senkovska I, Kaskel S. A highly porous metal-organic framework, constructed from a cuboctahedral super-molecular building block, with exceptionally high methane uptake. *Chem Commun* 2012;48:10841. doi:10.1039/c2cc34840c

14. Coronado E, Mínguez Espallargas G. Dynamic magnetic MOFs. *Chem Soc Rev* 2013;42:1525–39. doi:10.1039/C2CS35278H
15. Ghosh S, Badruddoza AZM, Hidajat K, Uddin MS. Adsorptive removal of emerging contaminants from water using superparamagnetic Fe_3O_4 nanoparticles bearing aminated β-cyclodextrin. *J Environ Chem Eng* 2013;1:122–30. doi:10.1016/j.jece.2013.04.00
16. Afkhami A, Moosavi R. Adsorptive removal of Congo red, a carcinogenic textile dye, from aqueous solutions by maghemite nanoparticles. *J Hazard Mater* 2010;174:398–403. doi:10.1016/j.jhazmat.2009.09.066
17. Gao Y, Liu G, Gao M, Huang X, Xu D. Recent Advances and applications of magnetic metal-organic frameworks in adsorption and enrichment removal of food and environmental pollutants. *Crit Rev Anal Chem* 2019:1–13. doi:10.1080/10408347.2019.1653166
18. Yin Z, Wan S, Yang J, Kurmoo M, Zeng M-H. Recent advances in post-synthetic modification of metal-organic frameworks: New types and tandem reactions. *Coord Chem Rev* 2019;378:500–12. doi:10.1016/j.ccr.2017.11.015
19. Bagheri A, Taghizadeh M, Behbahani M, Akbar Asgharinezhad A, Salarian M, Dehghani A, et al. Synthesis and characterization of magnetic metal-organic framework (MOF) as a novel sorbent, and its optimization by experimental design methodology for determination of palladium in environmental samples. *Talanta* 2012;99:132–9. doi:10.1016/j.talanta.2012.05.030
20. Bellusci M, Guglielmi P, Masi A, Padella F, Singh G, Yaacoub N, et al. Magnetic metal-organic framework composite by fast and facile mechanochemical process. *Inorg Chem* 2018;57:1806–14. doi:10.1021/acs.inorgchem.7b02697
21. Mínguez Espallargas G, Coronado E. Magnetic functionalities in MOFs: From the framework to the pore. *Chem Soc Rev* 2018;47:533–57. doi:10.1039/C7CS00653E
22. Ricco R, Malfatti L, Takahashi M, Hill AJ, Falcaro P. Applications of magnetic metal-organic framework composites. *J Mater Chem A* 2013;1:13033. doi:10.1039/c3ta13140h
23. Falcaro P, Lapierre F, Marmiroli B, Styles M, Zhu Y, Takahashi M, et al. Positioning an individual metal-organic framework particle using a magnetic field. *J Mater Chem C* 2013;1:42–5. doi:10.1039/C2TC00241H
24. Falcaro P, Normandin F, Takahashi M, Scopece P, Amenitsch H, Costacurta S, et al. Dynamic control of MOF-5 crystal positioning using a magnetic field. *Adv Mater* 2011;23:3901–6. doi:10.1002/adma.201101233
25. Stock N, Biswas S. Synthesis of Metal-Organic Frameworks (MOFs): Routes to various MOF topologies, morphologies, and composites. *Chem Rev* 2012;112:933–69. doi:10.1021/cr200304e
26. Wu Y, Zhou M, Li S, Li Z, Li J, Wu B, et al. Magnetic Metal-Organic Frameworks: γ-Fe2O3@MOFs via confined in situ pyrolysis method for drug delivery. *Small* 2014;10:2927–36. doi:10.1002/smll.201400362
27. Shekhah O, Wang H, Kowarik S, Schreiber F, Paulus M, Tolan M, et al. Step-by-step route for the synthesis of Metal-Organic Frameworks. *J Am Chem Soc* 2007;129:15118–9. doi:10.1021/ja076210u
28. Ke F, Qiu L-G, Yuan Y-P, Jiang X, Zhu J-F. Fe3O4@MOF core-shell magnetic microspheres with a designable metal–organic framework shell. *J Mater Chem* 2012;22:9497. doi:10.1039/c2jm31167d
29. Ma R, Yang P, Ma Y, Bian F. Facile synthesis of magnetic hierarchical core-shell structured Fe3O4@PDA-Pd@MOF nanocomposites: Highly integrated multifunctional catalysts. *ChemCatChem* 2018;10:1446–54. doi:10.1002/cctc.201701693
30. Faustini M, Kim J, Jeong G-Y, Kim JY, Moon HR, Ahn W-S, et al. Microfluidic spproach toward continuous and ultrafast synthesis of metal-organic framework crystals and hetero structures in confined microdroplets. *J Am Chem Soc* 2013;135:14619–26. doi:10.1021/ja4039642

31. Huo S-H, Yan X-P. Facile magnetization of metal-organic framework MIL-101 for magnetic solid-phase extraction of polycyclic aromatic hydrocarbons in environmental water samples. *Analyst* 2012;137:3445. doi:10.1039/c2an35429b

32. Qin J, Wang S, Wang X. Visible-light reduction CO 2 with dodecahedral zeolitic imidazolate framework ZIF-67 as an efficient co-catalyst. *Appl Catal B Environ* 2017;209:476–82. doi:10.1016/j.apcatb.2017.03.018

33. Wilmer CE, Leaf M, Lee CY, Farha OK, Hauser BG, Hupp JT, et al. Large-scale screening of hypothetical metal-organic frameworks. *Nat Chem* 2012;4:83–9. doi:10.1038/nchem.1192

34. Burtch NC, Jasuja H, Walton KS. Water stability and adsorption in metal–organic frameworks. *Chem Rev* 2014;114:10575–612. doi:10.1021/cr5002589

35. Lee PA, Citrin PH, Eisenberger P, Kincaid BM. Extended x-ray absorption fine structure—its strengths and limitations as a structural tool. *Rev Mod Phys* 1981;53:769–806. doi:10.1103/RevModPhys.53.769

36. Liang R, Jing F, Shen L, Qin N, Wu L. M@MIL-100(Fe) (M = Au, Pd, Pt) nanocomposites fabricated by a facile photodeposition process: Efficient visible-light photocatalysts for redox reactions in water. *Nano Res* 2015;8:3237–49. doi:10.1007/s12274-015-0824-9

37. Bedia J, Muelas-Ramos V, Peñas-Garzón M, Gómez-Avilés A, Rodríguez JJ, Belver C. A review on the synthesis and characterization of metal organic frameworks for photocatalytic water purification. *Catalysts* 2019;9:52. doi:10.3390/catal9010052

38. Gao G, Nie L, Yang S, Jin P, Chen R, Ding D, et al. Well-defined strategy for development of adsorbent using metal organic frameworks (MOF) template for high performance removal of hexavalent chromium. *Appl Surf Sci* 2018;457:1208–17. doi:10.1016/j.apsusc.2018.06.278

39. Taghizadeh M, Asgharinezhad AA, Pooladi M, Barzin M, Abbaszadeh A, Tadjarodi A. A novel magnetic metal organic framework nanocomposite for extraction and preconcentration of heavy metal ions, and its optimization via experimental design methodology. *Microchim Acta* 2013;180:1073–84. doi:10.1007/s00604-013-1010-y

40. Brunauer S, Emmett PH, Teller E. Adsorption of gases in multimolecular layers. *J Am Chem Soc* 1938;60:309–19. doi:10.1021/ja01269a023

11 Metal-Organic Framework with Immobilized Nanoparticles
Synthesis and Applications in Hydrogen Production

*Neslihan Karaman, Kemal Cellat, Hilal Acıdereli,
Anish Khan, and Fatih Şen*

CONTENTS

11.1 INTRODUCTION

Hydrogen is a clean, renewable, environmentally friendly alternative energy source and 21st-century energy carrier. In order to apply hydrogen energy, technical difficulties need to be overcome and improved in terms of the use of fuel cells in the production and storage of hydrogen. The use of nanomaterials for the production and storage of hydrogen to increase the efficiency of fuel cells is essential to the use of nanotechnology to improve the hydrogen economy [1]. Hydrogen gas is an alternative source of energy that can be obtained from internal sources and can be used as fuel in zero-emission energy generators. However, there are concerns about the requirement for a robust hydrogen storage medium to resist high pressure. Some methods have been developed to solve this situation. One of these methods is storage in porous structures. Metal-organic frameworks (MOFs) provide a highly porous medium for storage. MOFs store the hydrogen through the interaction of weak van der Waals bonds. MOFs have become a promising material for hydrogen storage applications due to the following advantages: improving activity on delivering and facilitating the emission of hydrogen by using a simple technique, significantly increased hydrogen storage capacity, and 4% by weight storage at room temperature. The most important advantage of MOFs is their ability to form the desired molecular cavities and excellent crystal structures [2–4].

The use of MOF-based materials for nanostructures, energy storage, and conversion are important by effectively immobilizing active functional materials and improving material properties. There are remarkable improvements in energy applications based on MOFs, composites, and derivatives [5]. Unlike known storage systems, hydrogen is known to be physically adsorbed in storage environments either chemically or by weak van der Waals interactions. Hydrogen can be stored in carbon nanotubes by physical adsorption and in an MOF with a high surface area. Functionalization of MOFs with various nanoparticles (NPs) is an effective way to strengthen. A typical hydrogen energy system consists of following parts: production of hydrogen from renewable sources, storage of high volumes in low volumes, and the realization of hydrogen energy conversion with high efficiency [6–8]. The optimal calcination process with the rational combination and production of MOF precursors works synergistically to improve MOF performance. Considering environmental damage caused by fossil fuels and low-efficiency energy transformations, it is inevitable to search for systems that use renewable energy sources that perform energy conversion with higher efficiency in the future [9]. MOFs are a kind of three-dimensional porous material. MOFs are well suited for hydrogen storage due to their advantages, including structural diversity, functionality, ease of loading, and reverse laying applications [10] (Figure 11.1).

Focusing on the synthesis, structure, and properties of MOFs is important for this new class of porous materials due to large pore size, high surface area, and selectivity of small molecules. When synthesizing a new MOF, various factors should be considered, as well as the geometric principles considered during the design. The main purpose of MOF synthesis is to obtain high-quality single crystals for structural analysis.

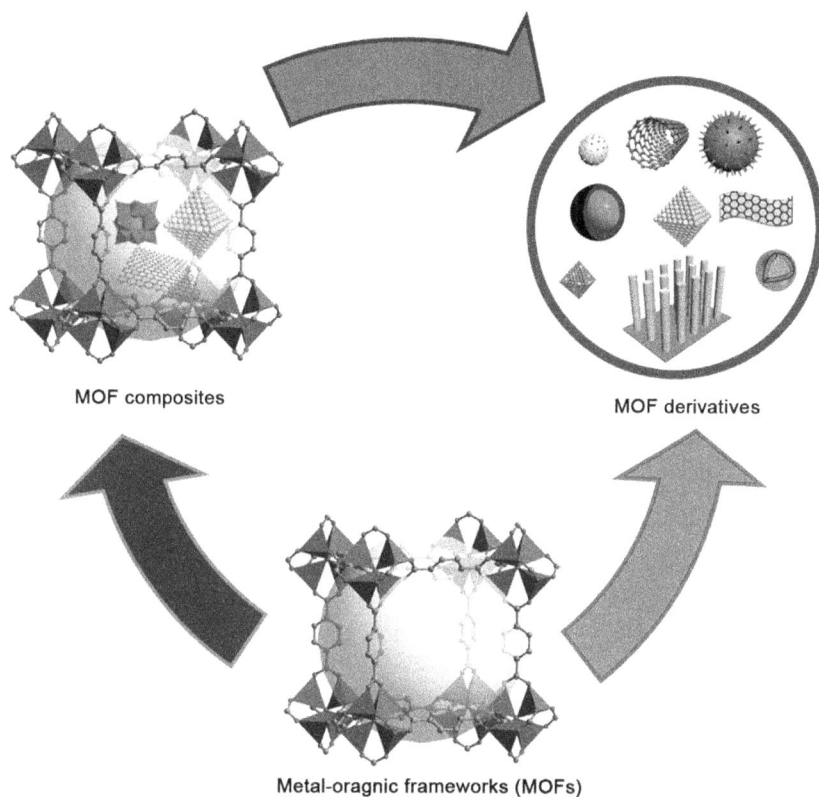

FIGURE 11.1 Representation and transformations of MOFs, MOF composite, and MOF derivatives [4].

For the storage of hydrogen, the resulting MOFs, crystal, porous, high surface area, and low cost with high efficiency are preferred [11]. The functionalized MOFs of the NPs in the MOFs are called NP@MOFs. Composite materials containing MOFs and NP@MOFs, which contain inert NPs, are preferred for hydrogen production and storage. The surface area of the integrated MOFs with high porosity, interchangeability, and immobilized NPs offers unique functionality and versatility. Functionalization of MOFs with various NPs is an effective way to strengthen the function of MOFs. The production of metal NPs in MOFs has attracted great interest in recent years due to its wide application field. Progress in the field of NP@MOF materials and different preparation methods, as well as appropriate characterization techniques, are discussed [12].

11.1.1 SYNTHESIS OF METAL-ORGANIC FRAMEWORK STRUCTURES

Metal-organic framework structures are nanoporous structures formed by the coordination of organic binders with metal ions, which can be one, two, or three dimensional. Interest in MOF structures has increased as a result of their efficient use in areas such as H_2 transport, CO_2 adsorption, and drug release. MOFs are generally synthesized by dissolving metal salts and binders in a solvent at a certain pH, temperature, and pressure. In addition to this method, new methods have been developed to obtain more stable and porous structures for MOFs in less time. The methods of MOF synthesis can be gathered under the following topics:

1. Solvothermal Method MOF
2. Microwave-Assisted MOF
3. MOF Synthesis with Sound Waves
4. MOF Synthesis by Electrochemical Method
5. Mechanical Agitator (Mechanochemical) MOF Synthesis
6. Ionic Fluid Assisted MOF Synthesis
7. MOF Synthesis by Microfluidic Method
8. MOF Synthesis by Dry Gel Method

11.1.1.1 Solvothermal MOF Synthesis

Hydrothermal or solvothermal synthesis is the most common and effective method used to obtain MOFs. It is carried out in a closed system by dissolving metal salts with organic binders in the appropriate temperature range (usually 80–260°C) using water or other organic solvents. After the formation of MOF structures, the solvent is removed from the medium by centrifugation and washing [13].

11.1.1.2 Microwave-Assisted MOF

Some MOF syntheses are carried out at a high temperature and over a long period by the solvothermal method. This means a waste of time and a high cost for energy in industrial-scale applications. The biggest advantage of the microwave method is the synthesis of MOF structures completed in a very short time. The microwave method is important for the performance of high-speed syntheses [13].

11.1.1.3 MOF Synthesis with Ultrasonication

The sound waves method achieves homogenous and accelerated nucleation with small particle sizes and shortened crystallization time. The main advantages of this method are small particle size and short reaction time. Due to the high energy emitted by ultrasonic waves, the mixing in the reaction medium is rapid. The ultrasonic wave frequency is preferred between 20 kHz and 10 MHz, and the reaction is occurred rapidly [13].

11.1.1.4 MOF Synthesis by Electrochemical Method

In the electrochemical synthesis of MOF, the metal ions are obtained from the anodic solution as the metal source instead of the dissolved binder molecules and the metal salts that react with the conductive salts in the reaction medium. The metal residue in the cathode is removed by solvents, and as a result of this process, H_2 is obtained. Electrochemical synthesis has many advantages, such as low cost, low temperature, and outdoor pressure requirement. The production speed can be adjusted to the desired levels [13].

11.1.1.5   Mechanochemical Synthesis of MOF

The physical and chemical processes take place simultaneously by the utilization of this method. Mechanochemical synthesis is a method of powder metallurgy in which controlled and fine composite powders can be produced as a result of repeated cold welding, rupture, and re-welding in a high energy ball mill. The first step of this process is mechanical alloying with the help of a mill or grinder to convert the raw material into the desired microstructure grain size and content as a result of a mechanical effect in the presence of grinding media and balls. No solvent is required in this method. Hence, this method is preferred to avoid the adverse effects of the solvent. Moreover, the reaction occurs at room temperature; no heating is required. In some cases, metal oxides may be preferred instead of metal salts. In such cases, pure water is obtained as a byproduct. Adding a small amount to the medium accelerates the reaction due to the movement of molecules which increases in the liquid medium. It has been proved that keeping the reaction method short increases the pore diameter of the pores of the frame structure [13].

11.1.1.6   Ionic Fluid-Assisted MOF Synthesis

It is similar to the solvothermal method. Unlike the solvothermal method, the solvent medium is prepared using the ionic liquid. The ionic liquid can be used instead of other solvents due to its superior properties such as low vapor pressure, high solvent property, and high thermal resistance, etc. The most important advantages of this method are the low solvent cost and the fact that the solvent is not affected by air [13].

11.1.1.7   MOF Synthesis by Microfluidic Method

In this method, two different solutions are prepared, and these two solutions are sent to a reservoir via a pump system. The droplets of the two solutions are mixed in a chamber, and this mixture forms a lattice structure. These droplets then pass through Teflon immersed in a hot water bath. Following the Teflon column, the droplets are transferred to a bath with ethanol. In this bath, methanol is removed, and lattice structures are obtained. Although there is no industrial production of MOF structures, this method is the most suitable method to be adapted to the industrial scale [13].

11.1.1.8   MOF Synthesis by Dry Gel Method

This method is generally used in the synthesis of zeolite NPs and zeolite membranes. The dry amorphous silica gel is placed on a platform. Under the platform, there is water or a volatile amine compound. This solvent evaporates over time and drops onto the silica gel on the platform, and MOF crystals are formed. The main advantage of this method is that the volume of MOF synthesized is very low, and the system requires little space [13].

11.2   CHARACTERIZATION METHODS

A variety of devices have been developed to examine the surface areas and pore structures of MOFs as shown below:

1. Scanning Electron Microscope (SEM)
2. X-Ray Diffraction (XRD)
3. Thermal Analysis
4. Fourier Transform-Infrared (FTIR)

11.2.1   Scanning Electron Microscope (SEM)

Scanning electron microscopy or Scanning Electron Microscope (SEM) works with the principle of scanning the surface with high energy electrons that focus on a very small area. The focusing of high-voltage accelerated electrons on the sample is achieved by collecting the effects of various interferences between the electron and the sample atoms during scanning of the electron beam on the sample surface and transferring them to the screen of a cathode ray tube after passing through the signal amplifiers. In modern systems, the signals from these sensors are translated and given to the computer monitor. Thus, the surface characterization of the material is also carried out by SEM [13].

11.2.2   X-Ray Diffraction (XRD)Method

It is possible to visualize the atomic structure of a material using a variety of high-resolution microscopes. However, it is necessary to use diffraction techniques to specify unknown structures or to determine structural parameters. X-ray diffraction is the most widely used diffraction technique to study the crystal structure of solids. With the X-ray working principle, when X-rays are dropped onto the crystal structure, the rays are fully reflected from the solid surface at small angles of incidence and are scattered by the parallel planes of the atoms in the crystal. This crystalline scattering is called diffraction, and diffraction consists of scattering containing many atoms [13].

11.2.3   Thermal Analysis

Thermal analysis methods can be grouped under two main headings. These are based on mass loss measurement, thermogravimetry (TG), temperature difference measurement, differential thermal analysis (DTA), and differential scanning calorimetry (DSC). In the TG method, the decreases in the mass of the substance to be analyzed as a result of the programmatically increased temperature are examined as a function of temperature or time. The resulting temperature mass curves are called thermogram or thermal decomposition curves. Mass losses resulting from a temperature rise are generally the separation of volatile compounds such as water from the structure or the decomposition of the substance. The most obvious changes examined in the thermogravimetric analysis method are limited to physical processes such as degradation and oxidation reactions, evaporation sublimation, and desorption. In the DTA method, the same temperature program is applied to the sample and the thermally inert reference material. DTA is performed to record the change of very small thermal effects. DTA provides the change in temperature by establishing the relationship between the main part and other standard parts [13].

11.2.4 FOURIER TRANSFORM-INFRARED SPECTROSCOPY (FTIR)

As a working principle, infrared light is absorbed by the measured material, and the wave number against the infrared intensity of light is measured by the mathematical Fourier transform method. The infrared region of the electromagnetic light array is between 14,000 cm^{-1} and 10 cm^{-1}, which consists of the following three main regions: near wavelength infrared (NIR; 4000 ~ 14,000 cm^{-1}, medium wavelength infrared (MIR; 400 ~ 4000 cm^{-1}) and far wavelength infrared (FIR; 4 ~ 400 cm^{-1}) [13].

11.3 IMMOBILIZATION

A material having functional groups capable of carrying out the desired physico-chemical process is attached to a solid support material which is superior to it in some physical properties by a suitable method [14].

11.3.1 IMMOBILIZATION BY PHYSICAL ADSORPTION

The desired compound binds to the surface of the solid support by weak van der Waals interactions. Due to poor interactions, the solubility of the compound can be reduced [14].

11.3.2 IMMOBILIZATION BY CHEMICAL REACTIONS

A chemical reaction is carried out between the functional groups of the compounds to be immobilized and the functional groups in the solid support material, resulting in a chemical bond between the solid support material and the material. Because of the strong interactions, the solubility of the material can be significantly reduced [14].

11.3.3 THE SYNTHESIS OF MOF CONTAINS IMMOBILIZED NANOPARTICLES

MOFs are materials that emerge as a three-dimensional porous species. High porosity, low density, and large surface area act as promoters or containers to contain metal NPs. In the latter case, active species can be designed as nodes, and one- or two-dimensional MOFs are synthesized due to their superior properties such as stability and reuse capacity, to increase catalytic activity and increase their insolubility [15]. In 2018, Zhang-Hui Lu successfully released hydrogen emissions of 2.8 nm, with NPs of rhodium nickel alloy immobilized with the MOF. Bimetallic composites with rhodium nickel alloy NPs immobilized on MOFs produce a porous structure with a high surface area. The $Rh_{0.8}$ $Ni_{0.2}$/MIL-101 catalyst, 100% conversion, and hydrogen selectivity, the catalytic performance of $N_2H_4BH_3$ and $N_2H_4 \bullet H_2O$ dehydrogenation in aqueous solution are enormous [16].

11.3.4 SYNTHESIS OF MIL-101

MOFs are unique materials with high surface area, high metal content, and crystal structures suitable to form molecular cavities. These materials are highly promising for gas storage, separation processes, and catalysis applications. MIL-101 is used in hydrogen storage areas due to its high surface area and porosity. The MIL-101, a member of the MOF family, was the first synthesized in 2005 by Material Institute Lavoisier [17]. In the literature, studies on the synthesis of MIL-101 are limited and discussed the improvement of the present method. The MIL-101 material containing about 50% chromium in its structure is formulated as chromium terephthalate, Cr_3F $(H_2O)_2O$ $[(O_2C)$ $-C_6H_4-$ $(CO_2)]3.nH_2O$ ($n \sim 25$). MIL-101 is synthesized according to the following procedure: the aqueous solution containing 1.661 g p-phthalic acid, 4.002 g Cr $(NO_3)_3$, $9H_2O$, 0.6 ml aqueous hydrogen fluoride and 70 ml water is placed in a 100 ml Teflon vessel and heated at 220°C for eight hours. The resulting green powder has the formula $KCr_3F(H_2O)_2[(O_2C)C_6H_4(CO_2)]3H_2O$ and is filtered using a porous glass filter and secondary cooling. The resulting green powder is washed with hot water-ethanol to remove unreacted excess terephthalic acid and dried in an oven at 150°C for 24 hours [18].

11.3.5 SYNTHESIS OF RHNI/MIL-101 CATALYSTS

To synthesize RhNi/MIL-101 catalysts, 50 mg of MIL-101 is dispersed in 5 ml of H_2O and sonicated for ten minutes, stirred for eight hours at 20°C. A reduction process with a total molar ratio reduction ratio of 0.1 mmol of $RhCl_3 \bullet 3H_2O$ and $NiCl_2 \bullet 6H_2O$ yields Rh_xNi_{1-x}/MIL-101 catalysts with different content of bimetallic compounds were given in detail. Besides, the synthesis of $Rh_{0.8}Ni_{0.2}NPsRhCl_3 \bullet 3H_2O$ (0.08 mmol) and $NiCl_2 \bullet 6H_2O$ (0.02 mmol) in 5 ml of deionized water containing 30 mg of cetyl trimethyl ammonium bromide was performed by using ultrasonication for 2–3 minutes. 30 mg of $NaBH_4$ was added to the prepared solution and stirred for 30 minutes to form $Rh_{0.8}Ni_{0.2}NPs$. The catalytic performance of the catalysts synthesized for the production of hydrogen from $N_2H_4BH_3$ or N_2H_4 solution is tested in the water-filled burette system, and the first stage of the H_2 production reaction is completed. $Rh_{0.8}$ can be converted entirely to H_2 and N_2 in 2.8 minutes on $Ni_{0.2}$/MIL-101. The synthesized RhNi/MIL-101 is a good catalyst for H_2 production from both $N_2H_4BH_3$ and N_2H_4. The metal/MOF component regions obtained by immobilization of rhodium nickel alloy NPs on the MOF have rhodium nickel NP dimensions (2.8 nm), high surfactants, and porous structure transfer. A heterogeneous catalyst with high yield and excellent power is obtained for the production of NH_3-free hydrogen from $Rh_{0.8}Ni_{0.2}$/ MIL-101, $N_2H_4BH_3$, and N_2H_4 solution. Compared to pure $Rh_{0.8}Ni_{0.2}$ NPs, the resulting $Rh_{0.8}Ni_{0.2}$/MIL-101 catalyst exhibits high catalytic activity for H_2 production from dehydrogenation of $N_2H_4BH_3$ solution with 100% hydrogen selectivity. RhNi/MIL-101 has high activity, durability, and reusability performance. MOF rhodium-nickel alloy NPs immobilized to provide H_2 production from hydrazine borane and aqueous hydrazine are successfully synthesized [18] (Figure 11.2).

Metal NPs have very small sizes and can increase activity with a high surface area [19]. There are also parts where the hydrogen energy system needs to be improved. One of the priorities is the storage of high volumes of hydrogen in low volumes. Hydrogen can be stored in many different forms. Hydrogen gas can be stored in pressurized tubes, liquid tanks, solid storage media, carbon nanotubes, zeolites, and MOF compounds. MOFs are highly porous organic coordination compounds containing metal ions. The most important advantage of MOFs is their ability to form molecular cavities of the desired properties and excellent crystal structures [20]. Yaghi et al. reported that MOF-5, one of the MOFs, stores 5% by mass under 77 K and 90 bar pressure, and IRMOF-20 (iso-reticular; networked) stores 6.7% by mass under the same conditions. It is synthesized according to the room temperature synthesis method, characterized by different analysis techniques and hydrogen storage performance. Among chemical hydrogen storage materials, hydrazine monohydrate ($N_2H_4 \bullet H_2O$) has a hydrogen content (8.0% by weight) developed as a hydrogen carrier and easy charging [21]. AuNi alloy NPs consist of hydrogen and ammonia boron from catalytic hydrolysis of high activity of AuNi alloy NP mesoporous MIL-101 with controlled double phase method (DPM) fixed by size and position control to MIL-101 [7] (Figure 11.3).

Ammonia borane (NH_3BH_3, AB) is a promising material for chemical hydrogen storage, and hydrogen is released by hydrolysis. AB hydrolysis is used to evaluate catalytic activities. The introduction of the aqueous AB solution into the reaction flask is achieved by hydrolysis of H_2 AB produced by vigorously shaking at room temperature containing synthesized M@MIL-101 catalysts. AuNi@NPs are used as storage materials for hydrogen production and clean energy applications from ammonia boron, which can serve as a high-performance catalyst. AuNi alloy NPs

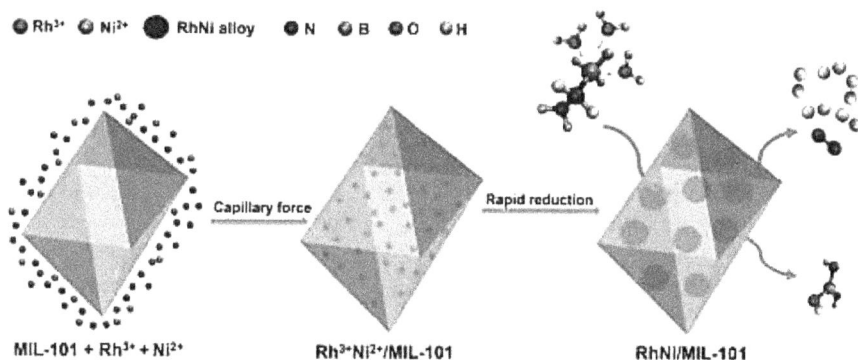

FIGURE 11.2 Demonstration of catalytic dehydrogenation of $N_2H_4BH_3$ for H_2 production by immobilizing MIL-101 with RhNi NPs [18].

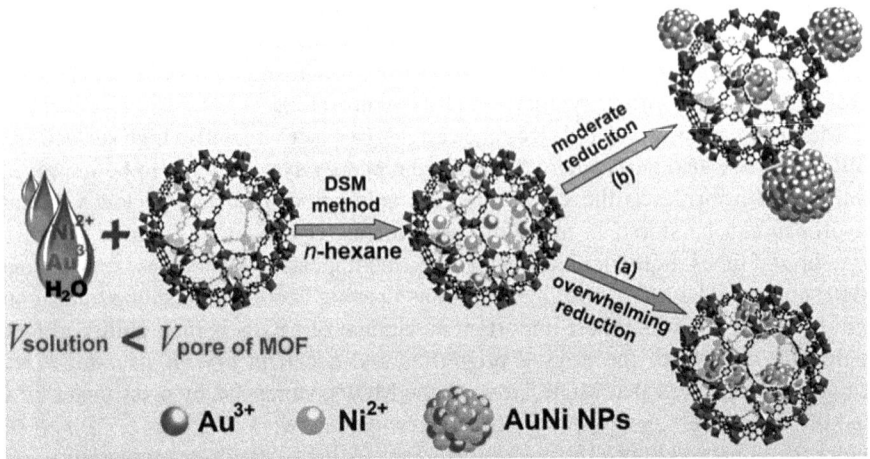

FIGURE 11.3 Schematic representation of the immobilization of AuNi [7].

have been successfully immobilized to MIL-101 by the dual-solvent method (DSM) combined with a liquid phase concentration-controlled reduction strategy. Uniform three-dimensional distribution of ultra-fine AuNi NPs encapsulated in MIL-101 pores. Ultra-fine non-noble metal-based NPs in the internal pores of MOFs shedding light on new opportunities, ultra-fine AuNi alloy NPs in mesoporous MIL-101 ammonia showed extremely high activity for hydrogen production from catalytic hydrolysis of borane [7] (Table 11.1).

MOFs containing immobilized NPs for chemical hydrogen storage are promising materials. Hydrazine monohydrate ($N_2H_4 \bullet H_2O$), as a safe and efficient storage material, has been developed as a promising hydrogen carrier [32–36]. Hydrogen content (8.0% by weight) and easy charging are very important for these types of materials. MOF-supported metal NP catalysts increase the hydrogen release kinetics and a high amount of hydrogen storage material [37–40]. As with any system, the hydrogen energy system needs to be improved. It is essential to store high volumes of hydrogen in low volumes of storage material. Hydrogen can be stored in gas-filled pressurized tubes, liquid-form tanks, solid storage media, carbon nanotubes, and MOF compounds by chemical bonding to surface-interacting metals. Values are increasing from 10 to 100 bar pressure at 77 Kelvin to form adsorption isotherms. The data were processed using the software of the analyzer and recorded as National Institute of Standards and Technology (NIST) isotherms. MOFs are enormous crystal structures for hydrogen storage. Catalysts for heterogeneous catalysis that would provide NPs with microstructures in functional MOF pore walls in functional MOFs, size, and shape of pores with broad surface adjustable values. The structures of MOFs are supported as NPs of metals, NP/MOF, and MOF. Composite materials containing MOFs and inert NPs, NP@MOFs, are preferred for hydrogen production and storage. The surface area of the integrated MOFs with high porosity, interchangeability, and immobilized NPs offers unmatched functionality and versatility [41–43].

TABLE 11.1
Hydrogen Adsorption Capacities of Different Metal-Organic Frameworks with Surface Areas

References	Metal-Organic Framework Type	SSA (m²/g)	Hydrogen (Capacity % by Weight)	Conditions (Temperature/ Pressure)
Furukawa et al. [22]	Metal-organic framework -177		7.5	77 Kelvin/70 bar
Han et al. [23]	Mg-Metal-organic framework -C30		8.08	77 Kelvin/20 bar
	Zn-Metal-organic framework -C30		6.47	77 Kelvin/20 bar
	Be-Metal-organic framework -C30		7.61	77 Kelvin/20 bar
Hirscher and Panella [24]	Metal organic framework -5	1014	1.6	77 Kelvin/10–20 bar
			0.2	RT 65 bar
Kraweic et al. [25]	$Cu_3(BTC)_2$	1.239 (BET)	2.18	77 Kelvin/1 bar
	Iso-reticular metal-organic framework-8	890 (BET)	1.45	77 Kelvin/1 bar
Li and Yang [26]	Metal-organic framework -5 (pure)		1.28	77 Kelvin/1 bar
	Iso-reticular metal-organic framework -8 (pure)		1.48	77 Kelvin/1 bar
	Metal-organic framework -5 (pure)		0.4	298 Kelvin/100 bar
	Metal-organic framework -5 (physically mixed with 10 with % Pt/AC)		1.32	298 Kelvin/100 bar
	Iso-reticular metal-organic framework -8 (pure)		0.5	298 Kelvin/100 bar
	Iso-reticular metal-organic framework -8 (physically mixed with 10% Pt/ C)		1.8	298 Kelvin/100 bar
Li and Yang [27]	Iso-reticular metal-organic framework -8 (pure)		0.5	298 Kelvin /100 bar
	Iso-reticular metal-organic framework -8 (Modified by Pt/AC)		4.0	298 Kelvin/100 bar
Panella et al [28]	Metal-organic framework -5	3840 (2296 BET)	5.1	77 Kelvin saturation pressure
	$Cu_3(BTC)_2$	1958 (1154 BET)	3.6	77 Kelvin- saturation pressure
Rossi et al. [16]	Metal-organic framework -5		4.5	78 Kelvin/0.75 bar
			1.0	298 Kelvin/20 bar
Rowsell et al [29]	Iso-reticular metal-organic framework -1	3362	1.32	77 Kelvin
	Iso-reticular metal-organic framework -8	1466	1.5	
	Iso-reticular metal-organic framework -11	1911	1.6	
	Iso-reticular metal-organic framework -18	1501	0.89	
	Metal-organic framework -177	4526	1.25	

(Continued)

TABLE 11.1 (CONTINUED)
Hydrogen Adsorption Capacities of Different Metal-Organic Frameworks with Surface Areas

References	Metal-Organic Framework Type	SSA (m²/g)	Hydrogen (Capacity % by Weight)	Conditions (Temperature/ Pressure)
Surble et al. [30]	MIL-102		1.0	77 Kelvin/35 bar
Wong-Foy et al [31]	Iso-reticular metal-organic framework -1	4170	5.2	77 Kelvin/50 bar
	Iso-reticular metal-organic framework -6	3,300	4.8	77 Kelvin/50 bar
	Iso-reticular metaloorganic framework -11	2340	3.5	77 Kelvin/34 bar
	Iso-reticular metal-organic framework -20	4590	6.7	77 Kelvin/70–80 bar
	Metal organic framework -177	5640	7.5	77 Kelvin/70–80 bar
	Metal organic framework -74	1070	2.3	77 Kelvin/26 bar
	HKUST-1	2280	3.2	77 Kelvin/75 bar

11.4 CONCLUSIONS

The usage of hydrogen as a fuel cell energy source is limited by the lack of a viable hydrogen storage system. MOFs belong to a new class of microporous materials that have recently been introduced for hydrogen storage. However, no significant hydrogen storage capacity was obtained at ambient temperature in the MOFs. Hydrogen storage capacities of modified MOFs have been significantly increased using a simple technique that provides the emission of hydrogen. Adsorption is reversible, and rates are fast. These properties made MOFs promising material for hydrogen storage. The hydrogen energy system, also called hydrogen economy, is important to produce hydrogen from renewable sources, to store high volumes in low volumes, and to realize hydrogen-energy conversion with high efficiency. As with any system, the hydrogen energy system needs to be improved. Composite materials containing MOFs and inert NPs, NP@MOFs, are preferred for hydrogen production and storage. The surface area of the integrated MOFs with high porosity, interchangeability, and immobilized NPs offers unmatched functionality and versatility. Hydrogen energy systems, production of hydrogen from renewable sources, storage of high volumes in low volumes, and the realization of hydrogen-energy conversion with high efficiency are very important for future applications. As in all systems, the hydrogen energy system needs to be improved. NP@MOFs advances in hydrogen production and storage. The role of nanotechnology in improving fuel cell efficiency and reducing costs is important.

REFERENCES

1. Sahaym U, Norton MG. Advances in the application of nanotechnology in enabling a 'hydrogen economy.' *J Mater Sci* 2008;43:5395–429. doi:10.1007/s10853-008-2749-0

2. Rowsell JLC, Yaghi OM. Metal–organic frameworks: A new class of porous materials. *Microporous Mesoporous Mater* 2004;73:3–14. doi:10.1016/J.MICROMESO.2004.03.034

3. Zhu Q-L, Xu Q. Metal-organic framework composites. *Chem Soc Rev* 2014;43:5468–512. doi:10.1039/C3CS60472A

4. Wang H, Zhu Q-L, Zou R, Xu Q. Metal-organic frameworks for energy applications. *Chem* 2017;2:52–80. doi:10.1016/J.CHEMPR.2016.12.002

5. Railey P, Song Y, Liu T, Li Y. Metal-organic frameworks with immobilized nanoparticles: Synthesis and applications in photocatalytic hydrogen generation and energy storage. *Mater Res Bull* 2017;96:385–94. doi:10.1016/J.MATERRESBULL.2017.04.020

6. Zhu B, Zou R, Xu Q. Metal-organic framework-based catalysts for hydrogen evolution. *Adv Energy Mater* 2018;8:1801193. doi:10.1002/aenm.201801193

7. Zhu Q-L, Li J, Xu Q. Immobilizing metal nanoparticles to metal-organic frameworks with size and location control for optimizing catalytic performance. *J Am Chem Soc* 2013;135:10210–3. doi:10.1021/ja403330m

8. Aijaz A, Akita T, Tsumori N, Xu Q. Metal-organic framework-immobilized polyhedral metal nanocrystals: Reduction at solid–gas interface, metal segregation, core-shell structure, and high catalytic activity. *J Am Chem Soc* 2013;135:16356–9. doi:10.1021/ja4093055

9. Li S, Huo F. Metal–organic framework composites: From fundamentals to applications. *Nanoscale* 2015;7:7482–501. doi:10.1039/C5NR00518C

10. Zhu Q-L, Xia W, Akita T, Zou R, Xu Q. Metal-organic framework-derived honeycomb-like open porous nanostructures as precious-metal-free catalysts for highly efficient oxygen electroreduction. *Adv Mater* 2016;28:6391–8. doi:10.1002/adma.201600979

11. Moon HR, Lim D-W, Suh MP. Fabrication of metal nanoparticles in metal-organic frameworks. *Chem Soc Rev* 2013;42:1807–24. doi:10.1039/C2CS35320B

12. Zheng F, Zhang C, Gao X, Du C, Zhuang Z, Chen W. Immobilizing Pd nanoclusters into electronically conductive metal-organic frameworks as bi-functional electrocatalysts for hydrogen evolution and oxygen reduction reactions. *Electrochim Acta* 2019;306:627–34. doi:10.1016/J.ELECTACTA.2019.03.175

13. Lee Yu-Ri, Kim Jun, Ahn Wha-Seung. Synthesis of metal-organic frameworks: A mini review. Department of Chemistry and Chemical Engineering, Inha University, Incheon 402-751, *Korea Korean J. Chem. Eng.* 2013;30(9):1667–80. doi:10.1007/s11814-013-0140-6

14. Jiang H-L, Liu B, Akita T, Haruta M, Sakurai H, Xu Q. Au@ZIF-8: CO oxidation over gold nanoparticles deposited to metal–organic framework. *J Am Chem Soc* 2009;131:11302–3. doi:10.1021/ja9047653

15. Wang LJ, Deng H, Furukawa H, Gándara F, Cordova KE, Peri D, et al. Synthesis and characterization of metal–organic framework-74 containing 2, 4, 6, 8, and 10 different metals. *Inorg Chem* 2014;53:5881–3. doi:10.1021/ic500434a

16. Rosi NL. Hydrogen storage in microporous metal-organic frameworks. *Science (80-)* 2003;300:1127–9. doi:10.1126/science.1083440

17. Li B, Wen H-M, Cui Y, Zhou W, Qian G, Chen B. Emerging multifunctional metal-organic framework materials. *Adv Mater* 2016;28:8819–60. doi:10.1002/adma.201601133

18. Zhang Z, Zhang S, Yao Q, Feng G, Zhu M, Lu Z-H. Metal-organic framework immobilized RhNi alloy nanoparticles for complete H_2 evolution from hydrazine borane and hydrous hydrazine. *Inorg Chem Front* 2018;5:370–7. doi:10.1039/C7QI00555E

19. Lee J, Farha OK, Roberts J, Scheidt KA, Nguyen ST, Hupp JT. Metal-organic framework materials as catalysts. *Chem Soc Rev* 2009;38:1450. doi:10.1039/b807080f
20. Farrusseng D, Aguado S, Pinel C. Metal-organic frameworks: Opportunities for catalysis. *Angew Chem Int Ed* 2009;48:7502–13. doi:10.1002/anie.200806063
21. Zhu Q-L, Xu Q. Liquid organic and inorganic chemical hydrides for high-capacity hydrogen storage. *Energy Environ Sci* 2015;8:478–512. doi:10.1039/C4EE03690E
22. Furukawa H, Miller MA, Yaghi OM. Independent verification of the saturation hydrogen uptake in MOF-177 and establishment of a benchmark for hydrogen adsorption in metal-organic frameworks. *J Mater Chem* 2007;17:3197. doi:10.1039/b703608f
23. Han SS, Deng W-Q, Goddard WA. Improved designs of metal-organic frameworks for hydrogen storage. *Angew Chem Int Ed* 2007;46:6289–92. doi:10.1002/anie.200700303
24. Hirscher M, Panella B. Nanostructures with high surface area for hydrogen storage. *J Alloys Compd* 2005;404–406:399–401. doi:10.1016/j.jallcom.2004.11.109
25. Krawiec P, Kramer M, Sabo M, Kunschke R, Fröde H, Kaskel S. Improved hydrogen storage in the metal-organic framework Cu3(BTC)2. *Adv Eng Mater* 2006;8:293–6. doi:10.1002/adem.200500223
26. Li Y, Yang RT. Significantly enhanced hydrogen storage in metal-organic frameworks via spillover. *J Am Chem Soc* 2006;128:726–7. doi:10.1021/ja056831s
27. Li Y, Yang RT. Hydrogen storage in metal-organic frameworks by bridged hydrogen spillover. *J Am Chem Soc* 2006;128:8136–7. doi:10.1021/ja061681m
28. Panella B, Hirscher M, Pütter H, Müller U. Hydrogen adsorption in metal-organic frameworks: Cu-MOFs and Zn-MOFs compared. *Adv Funct Mater* 2006;16:520–4. doi:10.1002/adfm.200500561
29. Rowsell JL, Millward AR, Park KS, Yaghi OM. Hydrogen sorption in functionalized metal–organic frameworks. *J Am Chem Soc.* 2004;126(18):5666–7. doi:10.1021/JA049408C
30. Surblé S, Millange F, Serre C, Düren T, Latroche M, Bourrelly S, et al. Synthesis of MIL-102, a chromium carboxylate metal–organic framework, with gas sorption analysis. *J Am Chem Soc* 2006;128:14889–96. doi:10.1021/ja064343u
31. Wong-Foy AG, Matzger AJ, Yaghi OM. Exceptional H2 saturation uptake in microporous metal-organic frameworks. *J Am Chem Soc* 2006. doi:10.1021/ja058213h
32. Sen B, Demirkan B, Şavk A, Karahan Gülbay S, Sen F. Trimetallic PdRuNi nanocomposites decorated on graphene oxide: A superior catalyst for the hydrogen evolution reaction. *Int J Hydrogen Energy* 2018;43:17984–92. doi:10.1016/j.ijhydene.2018.07.122
33. Şen B, Demirkan B, Savk A, Kartop R, Nas MS, Alma MH, et al. High-performance graphite-supported ruthenium nanocatalyst for hydrogen evolution reaction. *J Mol Liq* 2018;268:807–12. doi:10.1016/j.molliq.2018.07.117
34. Şen B, Aygün A, Şavk A, Akocak S, Şen F. Bimetallic palladium-iridium alloy nanoparticles as highly efficient and stable catalyst for the hydrogen evolution reaction. *Int J Hydrogen Energy* 2018;43:20183–91. doi:10.1016/j.ijhydene.2018.07.081
35. Sen B, Şavk A, Kuyuldar E, Karahan Gülbay S, Sen F. Hydrogen liberation from the hydrolytic dehydrogenation of hydrazine borane in acidic media. *Int J Hydrogen Energy* 2018;43:17978–83. doi:10.1016/j.ijhydene.2018.03.225
36. Sen B, Kuzu S, Demir E, Onal Okyay T, Sen F. Hydrogen liberation from the dehydrocoupling of dimethylamine–borane at room temperature by using novel and highly monodispersed RuPtNi nanocatalysts decorated with graphene oxide. *Int J Hydrogen Energy* 2017;42:23299–306. doi:10.1016/j.ijhydene.2017.04.213
37. Dhakshinamoorthy A, Garcia H. Catalysis by metal nanoparticles embedded on metal-organic frameworks. *Chem Soc Rev* 2012;41:5262. doi:10.1039/c2cs35047e

38. Li Z, Zhu G, Lu G, Qiu S, Yao X. Ammonia borane confined by a metal-organic framework for chemical hydrogen storage: Enhancing kinetics and eliminating ammonia. *J Am Chem Soc* 2010;132:1490–1. doi:10.1021/ja9103217

39. Zhao D, Yuan D, Sun D, Zhou H-C. Stabilization of metal-organic frameworks with high surface areas by the incorporation of mesocavities with microwindows. *J Am Chem Soc* 2009;131:9186–8. doi:10.1021/ja901109t.

40. Li S-L, Xu Q. Metal-organic frameworks as platforms for clean energy. *Energy Environ Sci* 2013;6:1656. doi:10.1039/c3ee40507a

41. Zhu Q-L, Xu Q. Immobilization of ultrafine metal nanoparticles to high-surface-area materials and their catalytic applications. *Chem* 2016;1:220–45. doi:10.1016/J.CHEMPR.2016.07.005

42. Rossin A, Tuci G, Luconi L, Giambastiani G. Metal-organic frameworks as heterogeneous catalysts in hydrogen production from lightweight ınorganic hydrides. *ACS Catal* 2017;7:5035–45. doi:10.1021/acscatal.7b01495

43. Şen B, Aygün A, Şavk A, Yenikaya C, Cevik S, Şen F. Metal-organic frameworks based on monodisperse palladium cobalt nanohybrids as highly active and reusable nanocatalysts for hydrogen generation. *Int J Hydrogen Energy* 2019;44:2988–96. doi:10.1016/j.ijhydene.2018.12.051

12 Metal-Organic Frameworks with Immobilized Nanoparticles for Hydrogen Generation

Kabelo E. Ramohlola, Tshaamano C. Morudu, Thabiso C. Maponya, Gobeng R. Monama, Edwin Makhado, Phuti S. Ramaripa, Mpitloane J. Hato, Emmanuel I. Iwouha, Anish Khan, and Kwena D. Modibane

CONTENTS

12.1 INTRODUCTION

A rapid and sudden increase in the global population has caused a severe stress on the global economy as well as the energy sector [1,2]. This is because of the continuous reliance on conventional fossil fuels to facilitate the means of production (during industrialization processes) in order to meet human needs [3,4]. Dependency on a fossil fuel-based energy supply faces threats in the future, i.e. depletion of fossil reservoirs [5,6]. Moreover, carbon dioxide (CO_2), methane (CH_4), sulfur hexafluoride, hydrofluorocarbons (HFCs), and perfluorocarbons (PFCs) gases accumulate in the atmosphere and result in a rise in global temperatures and acidic rains [3]. More efforts have been devoted to finding a sustainable and clean energy supply. In this regard, hydrogen gas has been recognized as an alternative energy carrier owing to zero carbon emission, highly gravimetric energy, and its renewable nature [7–11]. However, the key aspect in commercialization of hydrogen gas as an energy carrier is to find a suitable hydrogen generation/production method.

Currently, hydrogen gas is produced mainly from CO_2 emission technologies such as coal gasification and steam reforming of natural gases [8,12]. Ecofriendly electrochemical water splitting (EWS) offers a favorable solution to generate hydrogen of high purity [11,13,14]. During EWS, there are two reactions (half-cell) taking place, cathodic hydrogen evolution (HER) and anodic oxygen evolution reaction (OER) [15–17]. HER is the most sluggish reaction which requires a large driving overpotential (η) to take place [9]. Nowadays, precious/platinum group metals (PGMs) have been used to facilitate HER reactions, thus reducing the required overpotential and kinetically accelerate the process [10,18,19]. However, the market prices and lack of earth abundance for PGMs hampers their high-scale utilizations [20,21]. Therefore, it is significant and imperative to find an alternative cheap, earth-abundant, highly stable, and environmentally friendly electrocatalyst for HER.

Metal-free electrocatalysts such as intrinsic conducting polymers [22], carbon allotropes [23,24], and heterocyclic aromatic compounds [25–28] have been explored for HER. These metal-free materials have advantages such as superior synthetic flexibility, low-cost in manufacturing and great chemical stability. However, the catalytic efficiency does not match that of PGM-based materials which still need the doping of metals to enhance their catalytic activities [29,30]. Other materials that gained enormous attention and became hot research for HER are the PGM-free transitional metal-based compounds (TMCs) due to their excellent stability, conductivity, and electronic structures placing them center stage to replace PGMs [21,31–34]. Among various TMCs, molybdenum disulphide (MoS_2) has been regarded as the best candidate as its binding energy of hydrogen atoms is closer to that of Pt [23,35–37]. However, limited active sites and poor intrinsic conductivity hinder its application for HER [38–41]. Various technical strategies to improve MoS_2's active sites and conductivity are reported in our recent review on MoS_2 [42]. Taking into account those strategies, it can be seen that the HER properties are affected by factors such as active sites which can be linked with the surface area of the material. In this context, it is important to evaluate HER on materials possessing high surface area and porosity as an electron transport channel such as carbonaceous compounds which include single- and multi-walled carbon nanotubes (CNTs) and carbon black as well as inorganic porous materials zeolites and metal-organic frameworks (MOFs).

MOFs are coordinative porous compounds built from interaction between metal node and di-/polydendate organic linkers which have found great uses in gas (CO_2, CH_4, H_2) storage, purification, separation, and sensing (during energy production technologies), drug delivery (for biological and medicinal purposes), and catalysis [43–49]. This is because MOFs exhibit excellent characteristics such as high surface area (Brunauer-Emmett-Teller (BET) surface area of approximate 10,000 m^2/g), permanent porosity, and size/shape selectivity as well as tunable functionality [50–52]. Most importantly, MOFs exhibit types of catalytic sites which make them attract interest in heterogeneous catalysis and electrochemical sensors [53–55]. The metal node and organic linker can play multiple roles which includes Lewis acid or base and redox mediators [56]. In addition, the MOF pores offer a route for migration of electrons [56]. However, MOFs have limited applications in electrocatalysis, and this is due to their low chemical stability and poor electrical conductivities [49,57–61]. Interestingly, MOFs' high porosity and open framework structures allow full exposure of active sites and fine tuning of their framework to boost their intrinsic properties [62,63]. This allows incorporation of secondary components, for example carbonaceous materials, metal NPs, conducting polymers, and metal oxides/sulfides/selenides and derivation of MOFs into porous carbon material which are of high catalytic activity [62]. Hence, this chapter deliberates on the contemporary status of improving the electrochemical applications of MOFs. This is achieved by looking first at the fundamental mechanisms and factors affecting HER. Critical parameters which are used to measure HER properties of electrocatalysts are discussed, then structures and synthetic approaches of MOFs are introduced. Lastly, the challenges and approaches to address the limitations of MOFs for electrocatalysis are outlined.

12.2 ELECTROCHEMICAL WATER SPLITTING

12.2.1 FUNDAMENTALS FOR ELECTROCHEMICAL WATER SPLITTING

During EWS, there are two critical and important half-reactions taking place [64–66]. Oxygen gas (O_2) is produced at the anode half-cell and it is expressed as follows [15–17,67–74]:

$$2H_2O \rightarrow O_2 + 4H^+ + 4e^-$$ (12.1)

The actual potential needed to drive the OER, E^o_{OER} is given by:

$$E_{OER} = E^O_{OER} + iR + \eta_{OER}$$ (12.2)

where $E^O{}_{OER} = 1.23\ V$, iR defines the ohmic potential drop and η is the overpotential needed. The other half-cell reaction at the cathode is where hydrogen gas (H_2) is produced and can be shown by:

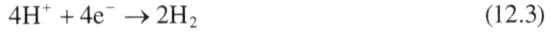

$$4H^+ + 4e^- \rightarrow 2H_2 \tag{12.3}$$

and its actual potential is given by:

$$E_{HER} = E^O{}_{HER} + iR + \eta_{HER} \tag{12.4}$$

and $E^O{}_{HER} = 0.0\ V$.

Combining the two half-cell reactions, anodic OER and cathodic HER, the complete EWS reaction is obtained and given as:

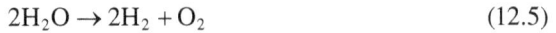

$$2H_2O \rightarrow 2H_2 + O_2 \tag{12.5}$$

Consequently, the standard potential of the EWS ($E_{cell} = E_{HER} - E_{OER}$) is −1.23 V at 25°C at pH 0 ([H$^+$] = 1.0 M).

12.2.2 MECHANISM OF ELECTROCHEMICAL HER

HER takes place on the surface of a cathode electrode commonly referred as an electrocatalyst. During the HER process, two electrons are transferred between the electrode surface and the electrolyte and combine with hydrogen protons to produce molecular hydrogen, H_2, as described in Equations 12.6 and 12.7 depending on the electrolyte type (acidic or alkaline) [75,76]:

$$2H^+ + 2e^- \rightarrow H_2\ (acidic) \tag{12.6}$$

$$2H_2O + 2e^- \rightarrow 2OH^- + H_2\ (alkaline) \tag{12.7}$$

In an acidic medium, readily available hydrogen protons react either catalytically (Tafel) or electrochemically (Heyrovsky) to yield H_2 whereas in alkaline electrolyte, a water molecule is dissociated first to generate protons needed to facilitate HER [8,13]. In principle, HER takes place in a number of steps and the mechanism is different in acidic and alkaline media. The two multistep mechanisms can be either Volmer-Tafel or Volmer-Heyrovsky (Figure 12.1) [77,78].

In a Volmer-Tafel mechanism [75,77]:

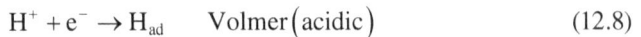

$$H^+ + e^- \rightarrow H_{ad} \quad Volmer\,(acidic) \tag{12.8}$$

$$H_2O + e^- \rightarrow OH^- + H_{ad} \quad Volmer\,(alkaline) \tag{12.9}$$

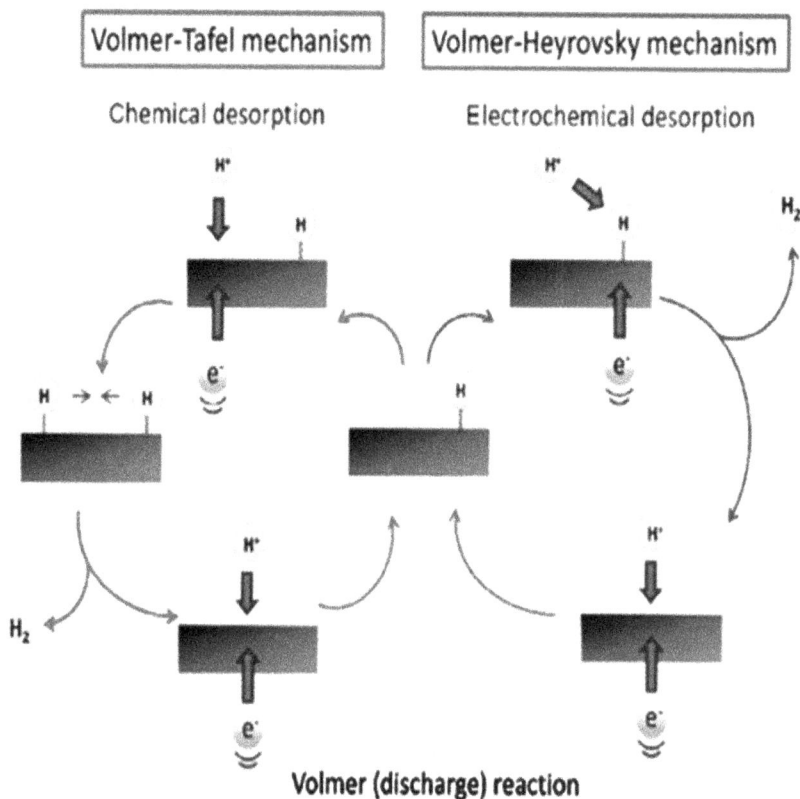

FIGURE 12.1   HER mechanism [13].

$$H_{ad} + H_{ad} \rightarrow H_2 \quad \text{Tafel} \left(\text{Both alkaline and acidic} \right) \tag{12.10}$$

The reaction starts by adsorption of hydrogen protons onto the electrocatalyst surface (acidic) or dissociation of water molecule to generate hydrogen protons (alkaline) necessary to initiate the HER process [79]. The latter process is slow and requires high overpotential (to break the covalent H-OH bond) as compared to the former (the dative covalent bond in H^+-H_2O) and HER kinetics decrease with increase in pH [80]. Moreover, the OH^- anions may poison the catalyst by adsorbing onto the surface of the electrocatalyst suppressing the reaction to take place, hence, HER is more favored in acidic medium than alkaline medium [75]. In addition, for the overall HER process to take place smoothly, the electrode must adsorb the hydrogen proton which is neither weak nor strong as shown in Figure 12.2 [81]. Weak adsorption of hydrogen protons will result in desorption of the proton before the process starts whereas strong ones will suppress the catalytic combination of neighboring hydrogen proton adsorbed onto the surface of the electrocatalyst to produce molecular H_2 as demonstrated in Equation 12.10. By the way, a cathode electrode must

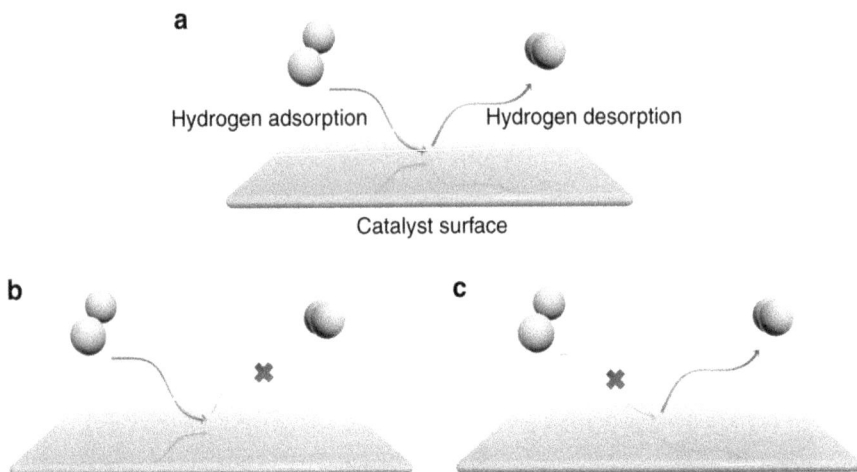

FIGURE 12.2 Schematic description of (a) balanced and (b) too strong and too weak hydrogen adsorption and desorption on a catalyst surface [81].

possess a great balance between hydrogen proton adsorption and molecular hydrogen desorption process [42,57].

For the second mechanism, Volmer-Heyrovsky, the first phase is similar to the one in the Volmer mechanism illustrated by Equations 12.8 and 12.9 for acidic and alkaline media, respectively. The succession step is the electrochemical desorption in which the adsorbed proton reacts with the proton in electrolyte (acidic) or dissociated proton from a water molecule (alkaline) to yield H_2.

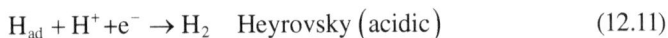

$$H_{ad} + H^+ + e^- \rightarrow H_2 \quad \text{Heyrovsky (acidic)} \tag{12.11}$$

$$H_{ad} + H_2O + e^- \rightarrow OH^- + H_2 \quad \text{Heyrovsky (alkaline)} \tag{12.12}$$

In both electrolytic media, if a hydrogen proton adsorbs slowly, the rate-determining step will be a Volmer step with a Tafel slope of 120 mV dec^{-1} [8,82,83]. And for faster molecular hydrogen desorption, either a Heyrovsky step with a Tafel slope of 40 mV dec^{-1} or 30 mV dec^{-1} will be a rate-determining step [8,75]. Hence, a faster hydrogen desorption process is preferred during HER.

12.2.3 ASSESSMENT OF THE HER ACTIVITY

It was seen in electrochemistry that the HER process is one of the most studied electrochemical methods for hydrogen production [75,76]. Numerous aspects such as overpotential, Tafel plot (to deduce Tafel slopes and exchange current density), faradaic efficiency, stability, and impedance have been used to evaluate HER activity. Profound understanding of these parameters is essential to design an active, efficient, and robust catalyst.

12.2.3.1 Overpotential

In electrochemistry, overpotential (η) is the potential difference (voltage). Its main purpose is to solve the problem related to the energy barrier that originates from the mass diffusion in the solution, and electron transfer as well as connections on the surface of the electrode in practical electrolysis [84,85]. There are three probable sources for the rise of overpotential which are activation overpotential, concentration overpotential, and resistance overpotential [85,86].

I. Activation overpotential, also known as onset potential, is an intrinsic prop- erty of an electrocatalyst and it is vital for initiation of the HER process. This activation overpotential varies with components of the electrode and it can be altered from one material to another depending on the modification of the electrocatalyst. The value of the onset potential must ideally be closer to zero (0.00 V) like the ones of a Pt-based electrocatalyst.

II. Concentration overpotential occurs immediately the electrode reaction starts as a consequence of the rapid drop in concentration proximate the electrode/electrolyte interfaces. This is caused by difference in concen- tration resulting from the slow ion diffusion. This can be minimized by homogenizing the electrolyte through stirring the solution.

III. Resistance overpotentials, sometimes referred as junction overpotentials, occur at electrode surfaces and interfaces like electrolyte membranes. The junction overpotential can be minimized by conducting ohmic drop com- pensation as shown by Equation 12.13

$$E_{corrected} = E_{measured} - iR_{\Omega} \tag{12.13}$$

where $E_{corrected}$ defines the measured potential versus reversible hydrogen electrode (RHE), i denotes the measured current and R_{Ω} represents the uncompensated resis- tance which can be obtained from the electrochemical impedance spectroscopy (EIS).

During HER studies, the ideal electrocatalyst must give the current density of 10 mA/cm² at low overpotentials derived mostly from linear sweep voltammetry (LSV). For example, Yu et al. [87] used an LSV polarization curve to determine the electrocatalytic efficiency of electrocatalysts and results are shown in Figure 12.3a. They obtained the Co@NC/CNT-700 composite possessing a smaller overpotential of 137 mV which is better than Co@NC-700 with the overpotential of 178 mV at a current density of 10 mA cm⁻².

12.2.3.2 Exchange Current Density and Tafel Plot

Generally, the catalytic efficiency of the electrode materials can be investigated by the Tafel parameters, i.e. the exchange current density, i_0, slope, b, and transfer coef- ficient, α, collected from the LSV polarization and Tafel plot [88]. The exchange current density, i_o is basically the most significant parameter which is used to probe the HER performance of the electrocatalyst. The i_o value is mostly related with the surface state/compositions of the electrode as well as the surface area of the elec- trocatalyst [89]. The large i_o obtained from the LSV polarization curves indicates

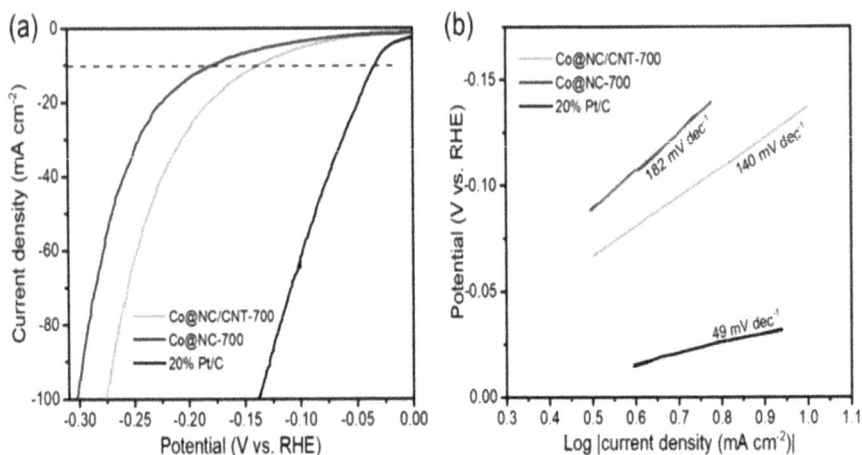

FIGURE 12.3   (a) LSV polarization curve and corresponding (b) Tafel plots of MOF-derived electrocatalysts [87].

the fast electron transfer process and large catalytic surface area [90,91]. From LSV curves, the i_o can be determined from the two points, thus the Tafel plot at a higher overpotential region and the linear portion of low overpotential [86]. For HER electrocatalyst, i_o is deduced from the low overpotential region as surface of the catalyst changes with potential [92, 93]. The i_o is obtained from the simplified Tafel Equation 12.14:

$$\eta = a + b \log i \tag{12.14}$$

where i is the current density and b is the Tafel slope.

Another parameter, Tafel slope is normally used to determine the rate-limiting step and the reaction kinetics. For example, for Volmer, Heyrovsky, and Tafel steps (see Mechanisms of HER section) to be a rate-determining step, the value of the Tafel slope must be approximately 120, 40, and 30 mV/dec, respectively [8,75]. For faster HER kinetics, the Tafel slope must be low, thus the Tafel step (30 mV/dec) is the fastest step during HER [70]. Figure 12.3b shows the Tafel slopes derived from Co-based electrocatalysts in 0.50 M H_2SO_4. In summary, a good HER electrode material must possess high exchange current density at low overpotential and a low Tafel slope [94].

12.2.3.3   Impedance

EIS is a technique used to study the electrode/electrolyte reactions occurring during HER, consequently revealing the kinetics of the reaction [95]. In order to deduce the information about energy consumed and the size of electrical resistance, the charge transfer resistance, R_{ct} is obtained and associated to the electrochemical reactions taking place on the electrode material surface. The value of R_{ct} determined corresponds with the kinetics of the HER process and low R_{ct} value corresponds to faster HER processes [96].

12.2.4 Factors Affecting HER Performance

12.2.4.1 Intrinsic Property of the Electrode Material

The intrinsic property of the electrocatalyst plays important role during the HER applications in which H protons can be absorbed on the surface of the electrocatalyst and converted into hydrogen gas [42]. Consequently, the active sites of the materials determine the electrocatalytic HER. It was shown that the intrinsic properties of the electrocatalysts are associated with their nature, structural, and electronic properties, which mostly display a discrete chemisorbed bond between electrode material and H intermediate [88,92]. Electrocatalysts which are electron deficient (cationic species) normally experience strong hydrogen adsorption, whereas some anionic species (electron-rich) are likely to trap H protons weakly, allowing faster hydrogen desorption. Furthermore, it was demonstrated that the intrinsic property of the electrode materials can be influenced by factors such as ligand/strain effects, crystal defects, and/or metal-anion. Accordingly, it is important to consider the chemical composition and crystal nature of a material in order to obtain a material of superior efficiency during electrocatalytic studies/applications.

12.2.4.2 The Number of Active Sites

One of the important factors in searching for a suitable electrocatalyst for HER, it is the number of active sites of the materials. Ramohlola et al. [42] reported that an electrochemical active surface area (ECSA) can be related with the number of active sites possessed by the electrocatalyst. The electrocatalyst must have high ECSA as it will tell the number of hydrogen protons to be adsorbed and then related to the turnover frequency (TOF). TOF is the quantity of hydrogen formed in a unit of time and is reliant on the active sites available. Hence, materials with high specific surface areas are a well-known method of increasing the number of active sites.

12.2.4.3 Electron Transfer Ability

It is well documented that the electrocatalytic reaction takes place on the electrode material surface and two electrons are transferred, as depicted during the HER mechanism [42,57]. Even though the transport of electrons is a discrete process from the charge-transfer, it affects the overall HER process and it can be revealed in Tafel slopes. Since there is a direct relation between electron transfer ability and the electrical conductivity of the catalytic structure of the material, developing material with high conductivity is of paramount importance to enhance the activity of the materials. In this regard, the proposed competent approach is to prepare a material with the outstanding characteristic of electrical conductivity. Alternatively, the electrocatalysts can be loaded on special matrix materials that possess high electrical conductivity to enhance their conductivities, which will subsequently improve the electron transport ability of the electrocatalyst.

12.3 OVERVIEW OF METAL-ORGANIC FRAMEWORKS

12.3.1 CONTEXTUAL BACKGROUND OF METAL-ORGANIC FRAMEWORKS

MOFs are coordination polymers with a crystalline structure made up of a transition metal node coordinated to multidentate organic linkers [97–99]. The organic linkers usually attach to the metal ion clusters through oxygen atoms to yield an inorganic-organic hybrid framework with interesting morphology [100,101]. This interaction combines the physical and chemical functionalities of the individual components to produce a MOF-type polymer with exceptional properties for a wide range of possible applications [102]. MOFs were first reported in 1995 by Yaghi and coworker wherein, they synthesized copper-4,4'-bypyridyl complex [103]. When compared to other conventional porous materials such as zeolites and CNTs, MOFs have distinct characteristics such as tunable pore sizes, remarkable porosities, and a variety of reactive sites [104,105]. Furthermore, MOF polymers have been identified to have a high BET surface area of up to 5000 m^2/g [106,107]. Depending on the metal node and organic linkers used, MOFs can form various structural geometries such as linear, tetrahedral, square-planar, and octahedral [108–110]. The metal nodes or clusters which are mostly used during the synthesis of MOFs are transition metals belonging in the first row of the periodic table [111,112]. These metals exhibit a variety of oxidation states which give rise to a variety of coordination numbers, hence different geometries. Furthermore, this oxidation state can form other coordination geometries due to the organic ligands/linkers that are coordinated to it. MOFs which are prepared from lanthanides are rarely studied as compared to transition metal-based MOFs [113]. Lanthanide ions are usually employed in producing MOFs with unusual structural morphologies and flexible coordination geometries [114,115]. The behavior of these metals results from high coordination numbers (usually above seven) and their ability to form porous solids which have Lewis centers and unsaturated coordination sites [113,114]. For the choice of the organic linker, the important property which is preferred is the rigidity of the ligand, which is responsible for stability in the MOF structure. Furthermore, the rigidness of the ligand backbone allows for easy prediction of MOF coordination geometry and also assists in maintaining the open-pore structure after removing the incorporated solvent. These organic ligands which serve as pillars in forming 3D networks, have different electrical charges such as cationic, anionic, and neutral [113]. The commonly used linkers are pyrazine and 4,4'bipyridine which are neutral and carboxylates which are anionic [111]. Cationic organic linkers are rarely used because they have low affinity towards cationic metal ions [115].

MOFs can be synthesized using various methods which lead to different morphologies and properties [116–118]. The main features of MOF syntheses are to institute synthetic routes that may give to well-defined inorganic building blocks without disintegration of the organic moiety. Concurrently, crystallization time (kinetics) must be suitable to permit the process of nucleation and growth to take place [118]. Conventionally, MOFs are normally synthesized through hydro-/solvothermal preparation methods by heating a Teflon autoclave consisting of metal ion and organic linker solution in an oven for a certain reaction time [119,120]. Different MOF fabrication routes which include microemulsion, sono-/mechano-chemical, electrochemical, and microwave-assisted were investigated in order to find synthetic methods with shortened reaction time and fabricate ultrathin, uniform MOF crystals [119,121–127]. Furthermore, MOFs have been extensively studied for various applications which include drug delivery, heterogeneous catalysis, gas sensing, separation, and storage [54,128]. This is due to ultra-high surface area, porosity, and tunable functionality [128,129]. However, there are few studies on MOFs for electrochemical application which involve reduction-oxidation (redox) reactions which include HER, OER, oxygen reduction reaction (ORR), and carbon dioxide reduction reaction (CO_2RR) [54,85,129].

12.3.2 MOF AS HER ELECTROCATALYST

The HER process involves two electron transfer processes and MOFs possess low charge transfer abilities [130]. The low charge transfer property of MOFs results from the use of organic linkers with high electronegative atoms such as oxygen and nitrogen in the carboxylate and imidazolate linkers, respectively. Furthermore, the above-mentioned reactions require an electrocatalyst with high conductivity which MOFs lack [49]. The low conductivity of MOFs arises from the poor overlap between the π-orbital (organic linker) and d-orbital (metal node) [57,131]. Taking into account their tunable functionality, great effort has been made to address the challenges faced by MOFs in redox reaction-based applications. The reported approaches comprise of (i) pyrolysis of MOF to form carbonized-MOFs, (ii) compositing with secondary compounds such as conducting polymers, carbon-based materials and porphyrins to generate MOF hybrid compounds, and (iii) immobilizing metal NPs on the surface of MOFs [52,132]. This chapter looks at those strategies and provide a literature summary on (a) pure MOFs, (b) carbonization of MOFs, and (c) MOF immobilized metal NPs.

12.3.2.1 Pure MOFs as HER Electrocatalysts

The first report on the synthesis of MOFs for electrocatalytic HER was pioneered by Nohra et al. [133] using polyoxomelate-based MOF (POMOF). The POMOF was fabricated via the hydrothermal method and their structures were found that they are insoluble which is a useful property required in electrolysis. This is inconsistent with other POMOF-based studies [134]. The electrocatalytic studies revealed that the metal centers of POMOF are electroactive. This is credited to the sequential electron processes of the chemically reversible reduction of Mo^{VI} centers. Furthermore, the stability test showed that as-synthesized POMOF can be cycled several times without significant loss of activity. Comprehensive HER studies of POMOF were reported by Qin et al. [135]. They studied the HER activities of three different POMOFs, $[TBA]_4[\epsilon-PMo^V_8Mo^{VI}_4O_{37}(OH)_3Zn_4]Cl_4$ (NENU-499), $[TBA]_3[\epsilon-PMo^V_8Mo^{VI}_4O_{36}(OH)_4Zn_4][BTB]_{4/3} \cdot x$Guest (NENU-500), and $[TBA]_3[\epsilon-PMo^V_8Mo^{VI}_4O_{37}(OH)_3Zn_4][BPT]$ (NENU-501), where TBA, BTB, and BPT are tetrabutylammonium, benzene tribenzoate and [1,1'-biphenyl]-3,4',5-tricarboxylate, respectively. HER activities of the synthesized materials and other types of MOFs, NENU-5, and HKUST-1 were studied in 0.50 M H_2SO_4 at a scan rate of 5 mV/s using LSV and the results obtained are shown in Figure 12.4. NENU-500 showed a superior HER performance as compared to other catalysts with an onset potential of 180 mV as depicted in Figure 12.4a, overpotential (η) of 237 mV to attain current density (i) of 10 mA.cm^{-2}, Tafel slope (Figure 12.4b) of 96 mV dec^{-1} and charge transfer resistance (R_{CT}) of 28 Ω (Figure 12.4c–d). The excellent HER performance of NENU-500 was attributed to its excellent stability, porosity, and exposed active sites. Following the successful studies of POMOFs, more effort was placed by different researchers to study the HER properties of MOF. Wang et al. [136] fabricated two new isostructural 3D MOF materials, Co-MOF and Ni-MOF ([M(ddbp)$_{0.5}$(4,4'-biby)$_{0.5}$(H$_2$O)$_2$] (M = Co/Ni)), based on mixed-ligands, 1, 3-di(3',5'-dicarboxyl phenyl)benzene and 4,4'-bipyridine. The two as-prepared compounds exhibit four-fold polyrotaxane-like (Figure 12.5a) and (3,4)-connected **dmd** networks (Figure 12.5b). With unique isostructural networks, two MOFs showed remarkable HER electrocatalytic activity in 0.50 M H_2SO_4. Co-MOF exhibited best electrocatalytic ability as compared to Ni-MOF with onset potential of 249 mV vs. RHE and overpotential of 357 mV at 10 mA cm^{-2} (Figure 12.5c), Tafel slope of 107 mV dec^{-1} (Figure 12.5d), and smaller charge transfer resistance (Figure 12.5e). The excellent HER performance of Co-MOF as compared to Ni-MOF can be attributed to the intermediate binding affinity of Co metal as compared to weak binding of Ni. Durability studies (Figure 12.5f) demonstrated

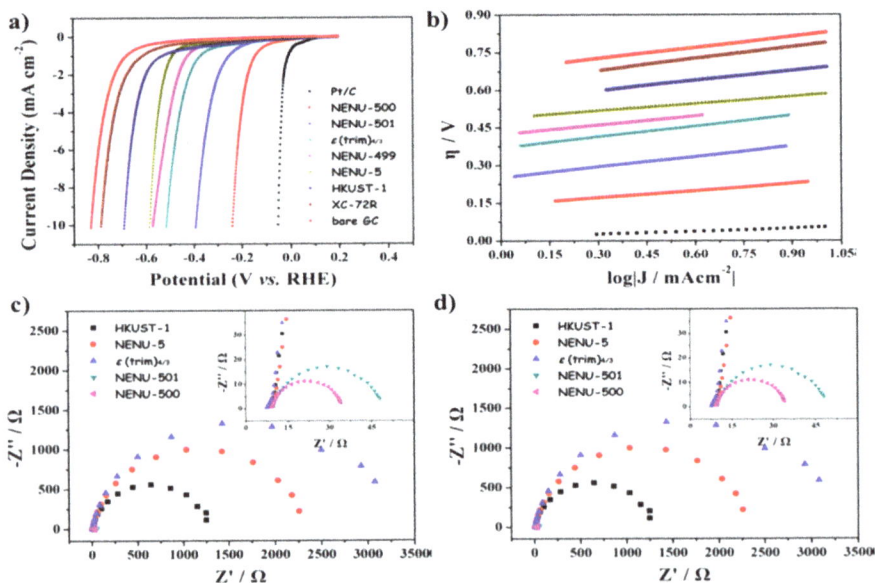

FIGURE 12.4 (a) iR-compensation LSV and (b) corresponding Tafel slopes of the prepared catalysts in 0.50 M H_2SO_4. EIS Nyquist plot of (c) different catalysts examined at −0.55 V (vs. Ag/AgCl) and of NENU-500 recorded at different potentials [135].

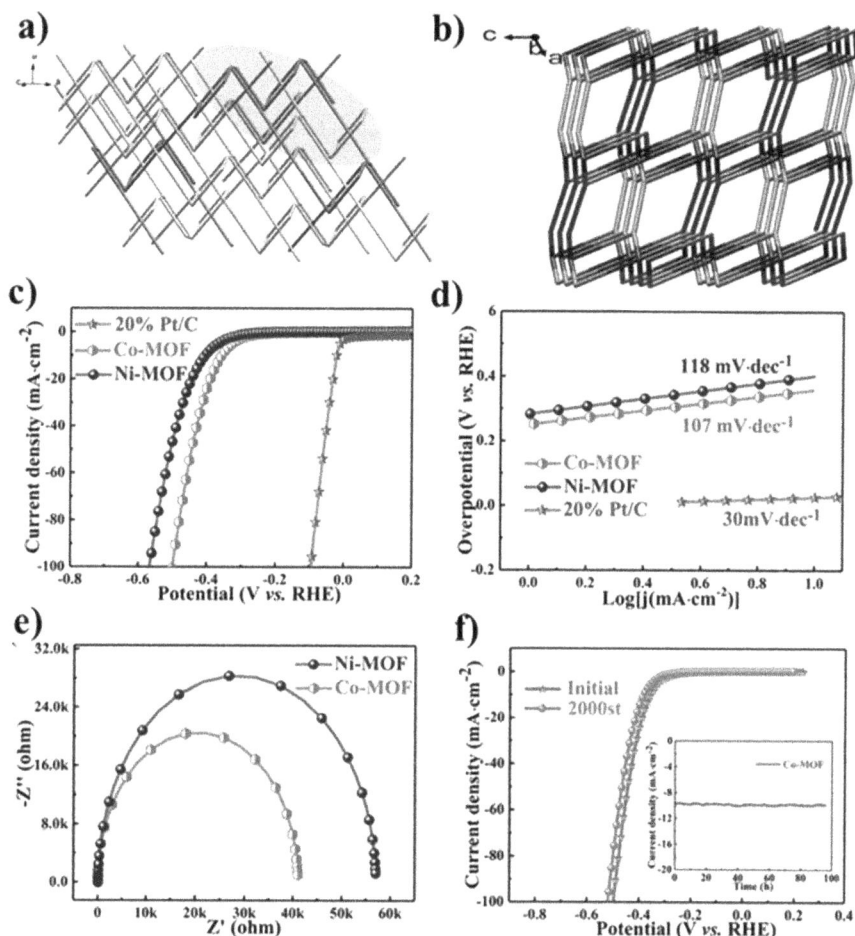

FIGURE 12.5 (a) Four-fold polyrotaxane network view and (b) illustration of 2-nodal (3,4)-connected *dmd* net. (c) LSV, (d) Tafel plots, (e) EIS plot, and (f) durability studies of Ni-/Co-MOF with comparison to Pt/C in acidic (0.50 M H_2SO_4) medium [136].

that the catalytic activity can be maintained for more than 96 h and 72 h for Co-MOF and Ni-MOF, respectively.

Duan and coworkers have investigated two dimensional (2D) MOFs for HER electrocatalysts [137]. The 2D MOFs were prepared using a one-step facile chemical bath deposition approach using 2.6-naphthalenecarboxylic acid dipotassium as organic linker, and nickel acetate and iron nitrate as metal precursors (Figure 12.6a). The obtained ultrathin NiFe-MOF array was found to have macropores (Figure 12.6b) with a distinct crystalline structure showing the clear lattice fringes (Figure 12.6c) with d-species of ~1.4 nm. The electrocatalytic studies of NiFe-MOF was performed in alkaline medium (1.0 M KOH). The LSV polarization curves (Figure 12.6d) revealed that NiFe-MOF exhibited overpotential of 134 mV at the current density

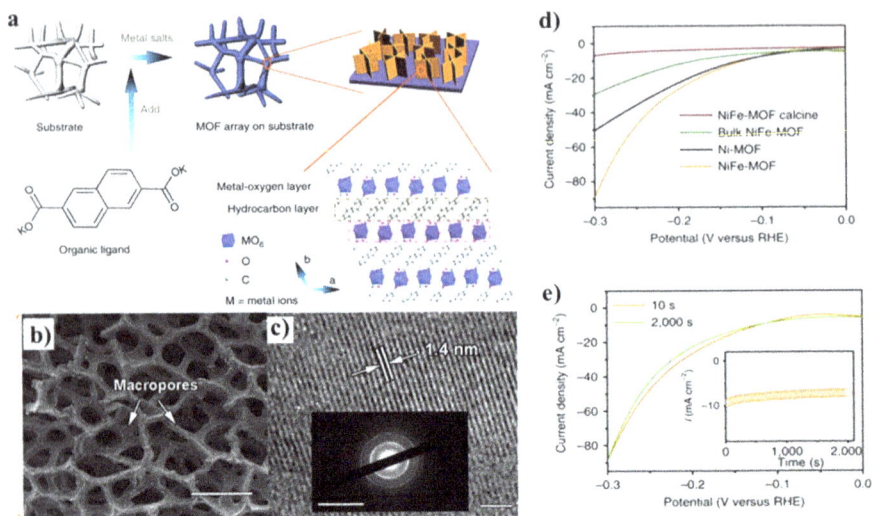

FIGURE 12.6 (a) Schematic diagram for the preparation of bimetallic NiFe-MOF and its (b) SEM and (c) TEM images (inset: selected area electron diffraction (SAED) photograph). (d) LSV curves and (e) cycle durability test in 0.50 M H$_2$SO$_4$ [143].

of 10 mA.cm^{-2} which is smaller compared to 177, 196, and 255 mV of Ni-MOF, bulk NiFe-MOF, and calcined NiFe-MOF respectively. In addition, the turnover frequency (TOF) at the NiFe-MOF was obtained to be 2.8 s^{-1} which is four times higher than other materials. Furthermore, NiFe-MOFs indicated excellent durability (Figure 12.6e).

In another study, Lin and coworkers prepared a bimetallic NiFe-MOF using solvothermal synthesis (Figure 12.7a) [138] other than one-step facile chemical bath deposition used by Duan et al. [137] (Figure 12.6a). It deduced that fabrication of a bimetallic MOF electrocatalyst via in situ hydrothermal growth on the surfaces of the conductive material offers steady close contact between the bimetallic MOFs and conducting Ni foam and this enhances both the charge transport across the electrocatalyst/substrate interface and the mechanical integrity of the electrode to improve the stability. In addition, with the correct optimization of the mole ratio of metal precursors, inter- and intra-molecular synergistic effects can be realized, and this will enhance the electrocatalytic property. From Figure 12.7b, it can be deduced that both bimetallic MOF/NF electrodes (5:1 and 1:1) exhibited lower overpotentials than the other as-prepared materials (Fe-MOF and Ni-MOF). However, the FeNi(BDC) (DMF,F)/NF electrocatalyst of molar ratio Fe:Ni (5:1), showed excellent performance, hence it was said, with correct molar ratio, a bimetallic electrocatalyst with high activity can be achieved. Other parameters that can access the electrocatalytic ability of the material are Tafel (Figure 12.7c) and Nyquist plots (Figure 12.7d) which give information about reaction mechanisms and charge transfer resistances respectively of the electrocatalyst toward the HER. Unsurprisingly, the FeNi(BDC)(DMF,F)

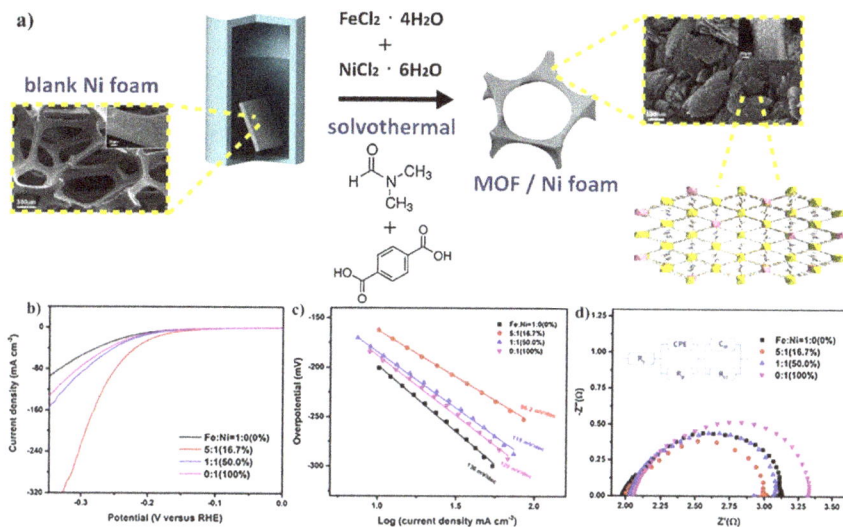

FIGURE 12.7 (a) Schematic illustration of the hydrothermal synthesis of bimetallic FeNi-MOF. (b) LSV polarization curves, (c) Tafel plots, and (d) EIS spectrum of the as-prepared materials in 1.0 M KOH [138].

electrode of Fe:Ni of 5:1 demonstrated the lowest Tafel slope of 96.2 mV.dec^{-1} and Rct value of 0.799 Ω among all as-synthesized four MOF/NF electrodes.

Roy and group designed a cobaloxime-based MOF, UU-100 (Co). UU-100 (Co) consisting of tetranucleating cabaloxime linker (Figure 12.8a) with carboxylates as metal (Co) anchoring groups [139]. The structural mode of UU-100 (Co) MOF is shown in Figure 12.8b. It can clearly be seen the six-coordinated centers of Co (III) metal center with cobaloxime linker. The mechanistic illustration of HER on UU-100 (Co) MOF is illustrated in Figure 12.8c in acidic medium and glassy carbon substrate. Scan-rate dependent studies (Figure 12.8d) shows that the as-prepared material is electroactive and diffusion-controlled. LSV polarization curve (Figure 12.8e) showed that the UU-100 (Co) requires onset potential of −150 mV vs RHE in an acetate butter at pH 4. The Tafel slope obtained was 250 mV.dec^{-1} (Figure 12.8f).

Very recently, Li et al. [140] prepared 2D polyhalogenated Co (II) MOFs for HER. Halogens have the electron-withdrawing characteristic, and in this case they have the ability to enhance the affinity for the reactants to intermediate. Halogen-substituted linker consisted of a phthalic acid (X$_4$-H$_2$pta = tetrahalogenphthalic acid and X is F, Cl, Cl, and Br). The three synthesized MOFs, Co-Br$_4$-MOF, Co-Cl$_4$-MOF, and Co-F$_4$-MOF with general representation [CoCX$_4$-pta) (bpy) (H$_2$O)$_2$]$_n$ (bpy = 4.4-dipyridyl) were prepared via hydrothermal methods. The HER activities were investigated in 0.50 M H$_2$SO$_4$ and from the three Co-X$_4$-MOFs, Co-Cl4-MOF displayed excellent HER electrocatalytic activity with overpotential of 24 mV at 10 mA.cm^{-2} and Tafel slope of 125 mV.dec^{-1}. The high active Co-Cl$_4$-MOF was then subjected to acetylene black (AB) and the corresponding HER results are shown in Figure 12.9a–d.

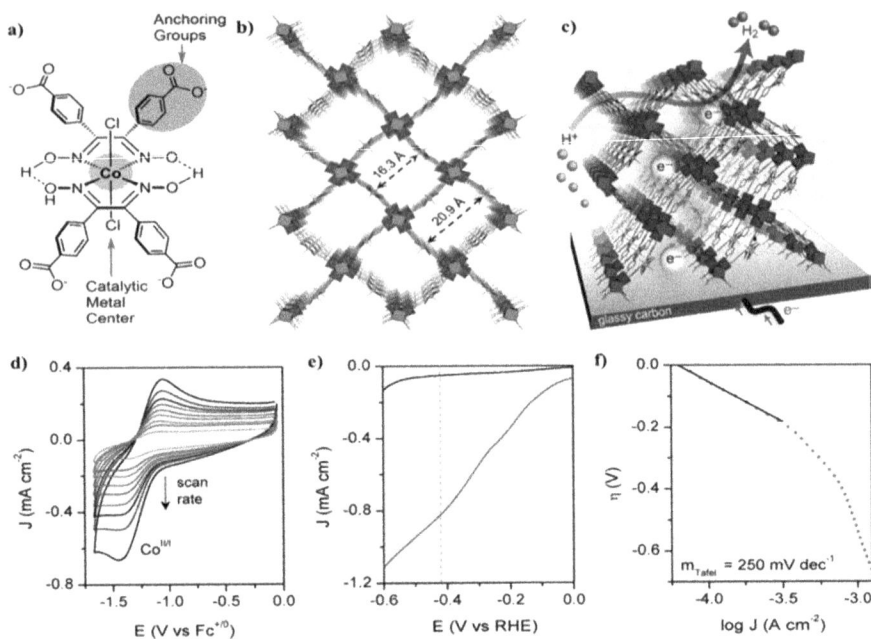

FIGURE 12.8 (a) Cobaloxime-linker in UU-100(Co) structure and (b) framework porous structure of UU-100(C). (c) Illustration of HER mechanism, (d) scan-rate dependent studies, (e) LSV curve, and (f) corresponding Tafel slope of prepared UU-100(Co) in acetate butter [139].

12.3.2.2 MOF Composite for HER

The electrocatalytic performance of MOF is hampered by its poor conductivity, and stability in different media [49,61]. In order to improve MOF electrocatalytic activity, most studies were conducted to address the poor conductivity and enhance its stability. These approaches include encapsulating metal NPs which are deemed good electrical conductors in the porous frameworks of MOFs. These metal NPs bring new properties such as new active sites, stabilization of MOF frameworks, and new channels for electron transport mechanism. Another strategy is to derive the MOF framework into a porous carbon material, using the MOF as a precursor to achieve meso/macroporous carbons. The carbon porosity derived from the MOF framework is then used as a support to immobilize the metal NPs to further enhance its catalytic activity. These metal NPs deposited on meso/macroporous carbons derived from MOFs were found to be highly active as these immobilized metal NPs expose more active sites and enhance the mass transfer rate. Finally, MOFs can form a composite with other conducting materials in order to enhance its activity. Herein, MOF composites will be comprehensively discussed.

12.3.2.2.1 Metal Nanoparticles@MOF Composite

MOFs structure has a permanent porosity which offers inherent condition to trap metal NPs. Porous MOFs will stabilize the formation of metal NPs and thus bring

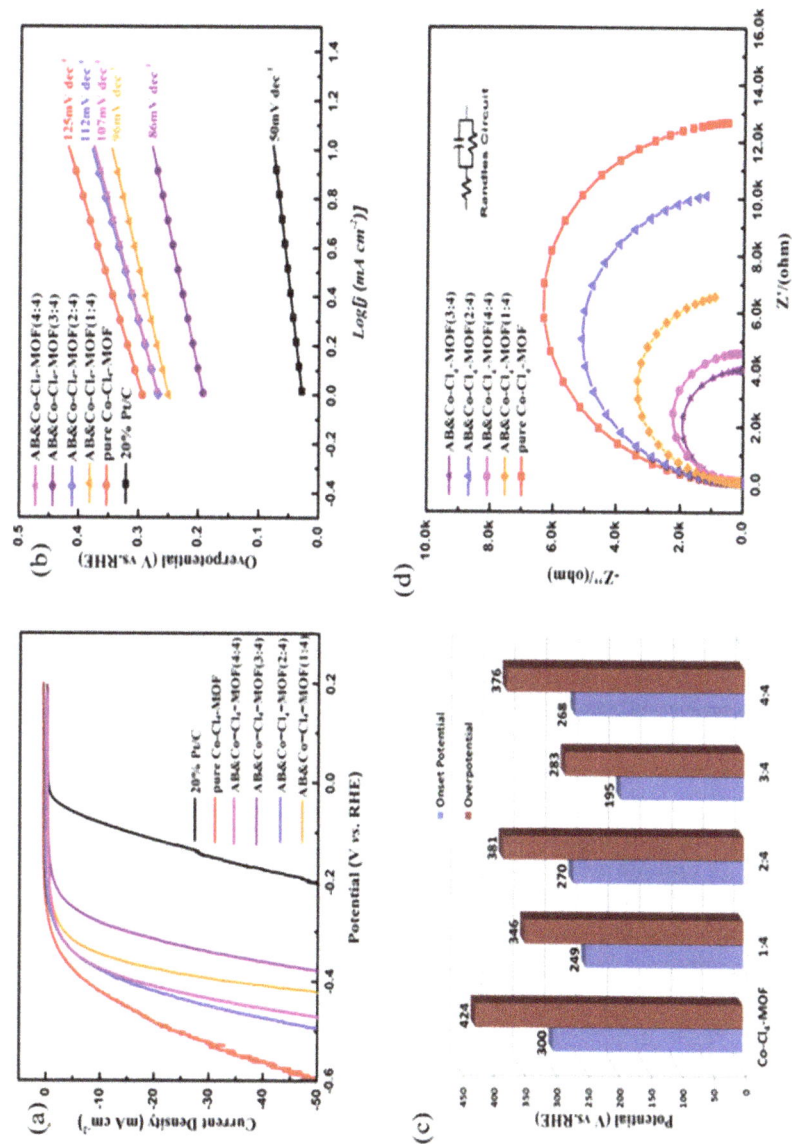

FIGURE 12.9 (a) iR-compensation LSV polarization curves, (b) corresponding Tafel slopes, (c) onset potential and overpotential comparison, and (e) Nyquist plot of polyhalogenated catalysts in 0.50 M H_2SO_4 [140].

generation of specific adsorption/catalytic sites [129]. Metal NPs have an advantage over their bulk counterparts as they give a large surface area at low metal loading, making them cheaper to use [141]. Furthermore, metal NPs in MOFs provide size and shape selectivity during catalysis [142]. Noble (Pt, Ru, Au), non- noble (Ni, Co) and bimetallic NPs have been immobilized in the pores of MOFs to improve their electrocatalytic HER activity [143]. This is because NPs can easily transfer electrons to MOFs [144]. There are various approaches of loading metal NPs in the pores of MOFs which include solid grinding, solution infiltration, and layer-by-layer assembly. In addition to approaches that are conventionally used to immobilize metal NPs into the pores of MOFs, there are two well-known strategies that are mostly used, i.e. (i) deposition of metal ions into MOF pores and reducing them into metal clusters or NPs, and (ii) immobilizing the metal NPs into the pores of MOFs using the "ship in a bottle" or "bottle around a ship". Rui and coworkers [145] presented direct hybridization of Pt NPs into the pores of 2D-Ni-MOF using a surfactant-free approach. Remarkably, by integrating the Pt NPs into the porous framework of Ni-MOF, the as-prepared Ni-MOF@Pt heterostructures exhibited excellent electrocatalytic activity in both acidic and alkaline media. In acidic media (0.50 M H_2SO_4), the Ni-MOF@Pt electrocatalyst displayed a low overpotential of 43 mV to achieve current density of 10 mA.cm^{-2} (Figure 12.10a), Tafel slope of 30 mV.dec^{-1} (Figure 12.10b) which are comparable to commercial Pt/C catalyst (overpotential of 43 mV at 10 mA.cm^{-2} and Tafel slope of 31.0 mV.dec^{-1}). In alkaline medium (1.0 M KOH), Ni-MOF@Pt demonstrated superior HER activity as compared to Pt/C with overpotential of 102 mV (Pt/C = 172 mV) (Figure 12.10c) at 10 mA.cm^{-2} and Tafel slope of 86 mV.dec^{-1} (Pt/C = 146 mV.dec^{-1}) (Figure 12.10d). Excellent performance in an alkaline medium suggests that the water dissociation step was enhanced as this step is the one that limits usage of various electrocatalyst for HER application [42]. As compared to pure Pt NPs, the pure NPs out-performed the as-prepared Ni-MOF@Pt in both media (Figure 12.10e). The mass activity studies shown at Figure 10f show that Ni-MOF@Pt out-shined Pt NPs (146 mA/mg at −100 mV in 1.0 M KOH and 54 mA/mg at −35 mV in 0.50 M H_2SO_4) with current density of 241.0 MA/mg at −100 mV in 1.0 M KOH and 126 mA/mg at −35 mV in 0.50 M H_2SO_4. From the data obtained, it can be deduced that Ni-MOF@Pt possess enhanced electrical conductivity and more active sites that resulted in improvement in HER. There are several studies on noble metals, particularly Pd deposited on MOF for HER since the binding affinity of Pd is more favorable for hydrogen than other noble metals [57,146–148].

In another work, Ye et al. [149] immobilized Pt quantum dots (QDs) (QDs have large surface area and are more conductive than NPs) on a Fe-MOF nanosheet array supported with conductive core-shell Ni foam (NF) using a one-step facile hydrothermal technique (Figure 12.11a). The electrochemical HER performance of the as-prepared NF, Fe-MOF/NF and Pt DQs @Fe-MOF/NF were studied in alkaline medium, 1.0 M KOH and displayed in Figure 12b–e. From the iR-compensation LSV polarization curve, Figure 12.11b, it can be seen that the MOF encapsulated with Pt QDs supported on a conductive NF electrode (Pt DQs @Fe-MOF/NF) exhibited the lowest overpotential (33 mV of Pt DQs @Fe-MOF/NF) at a current

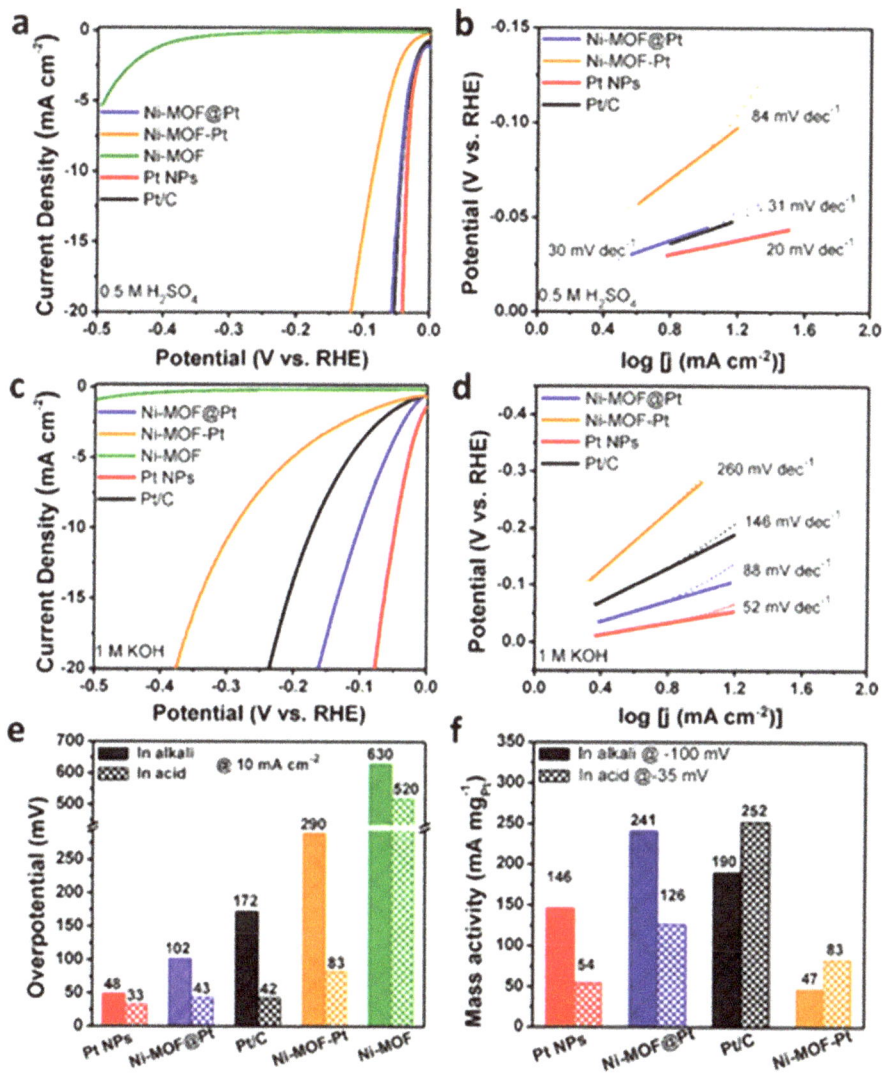

FIGURE 12.10 (a) LSV polarization curves and (b) Tafel plots of prepared catalysts in 0.50 M H$_2$SO$_4$. (c, d) LSV curves and Tafel plots of prepared catalysts in 1.0 M KOH. (e) Comparison of overpotential at 10 mA cm^{-2} as well as (f) mass activity of fabricated electrocatalysts in both acidic and alkaline media [145].

density of 10 mA cm^{-2} which is a measure for HER activity as compared to 219 and 176 mV for NF and Fe-MOF/NF. The significant enhancement of HER on the Pt DQs @Fe-MOF/NF was attributed to Pt QDs bringing (1) new active centers and (2) creation of a porous ion buffer reservoir which helps MOFs to be in intimate contact with the electrolyte, which results in a faster electron transfer process and improved conductivity. The enhancement of the HER activity was supported by

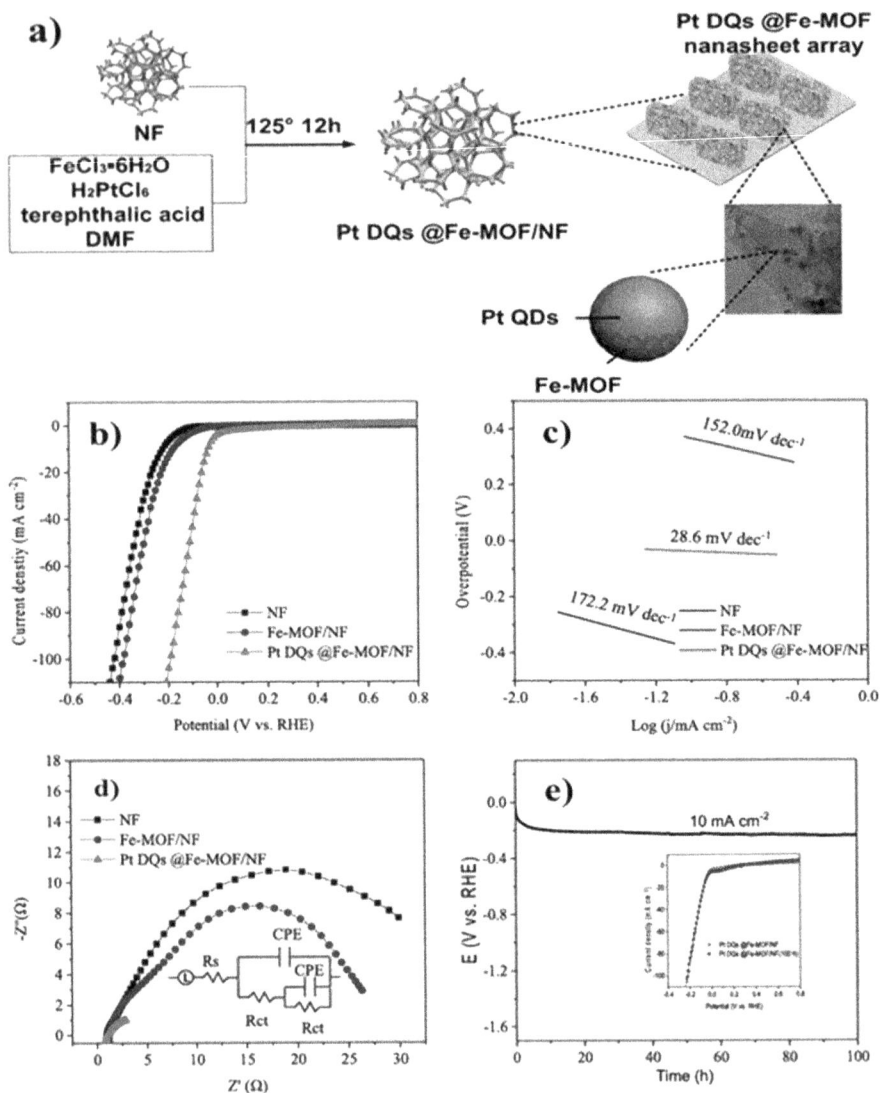

FIGURE 12.11 Schematic diagram for the one-step facile hydrothermal synthesis of Pt QDs@Fe-MOF/NF. (b) LSV polarization curves, (c) corresponding Tafel slopes, (d) EIS plots, and (e) stability investigation for the NF, Fe-NOF/NF and Pt QDs@Fe-MOF/NF in 1.0 M solution [149].

Tafel slopes (Figure 12.11c). Pt DQs @Fe-MOF/NF showed the lowest Tafel slope of 28.6 mV.dec^{-1} as compared to other catalysts, NF (172.2 mV.dec^{-1}) and Fe-MOF/NF (152.0 mV.dec^{-1}). The Nyquist EIS plot (Figure 12.11d) which is essential to probe the kinetics of electrocatalyst (lower R_{ct} value shows faster kinetics), showed that Pt QDs@Fe-MOF/NF exhibited lower R_{ct} value of 7.678 Ω as compared to 12.322 and

11.847 Ω of NF and Fe-MOF/NF, respectively. This suggests that the prepared Pt QDs@Fe-MOF/NF displays high conductivity and faster HER reaction kinetics. The chronopotentiometry studies (Figure 12.11e) at a constant current density of 10 mA cm^{-2} and scan rate of 5 mV s^{-1} for 100 h in 1.00 M KOH displayed a slight decay for Pt DQs @Fe-MOF/NF. The observations showed that deposition of metal NPs/ quantum dots on the MOF porous framework offers fascinating electrocatalyst with improved kinetics, stability, and conductivity.

Although monometallic NPs have been encapsulated in MOF porous frameworks and displayed superior catalytic activity, the use of bimetallic NPs has gained great attention. This is because conversion of monometallic NPs to bimetallic NPs results in enhancement of catalytic activity due to synergism between the two metals [150,151]. The synergistic effect between MOF and bimetallic NPs on the resultant composite results in improved catalytic activity [143]. This is because the MOF porous structures and their large surface area will offer a dynamic space for dispersion of bimetallic NPs which will suppress the agglomeration and leaching of the bimetallic NPs [143]. Furthermore, the porous structure allows the substrate to be more fully contacted to the catalyst [143]. Ding and coworkers [152] encapsulated bimetallic RhRu alloy (bimetallic) NPs into the framework of monometallic MOFs instead of bimetallic MOFs. The as-prepared RhRu@UiO-66-NH$_2$ was investigated for HER studies and it was obtained that it has superior ability in all pH (Figures 12.12–12.14). The HER performance of the prepared catalysts was first investigated in 0.50 M H$_2$SO$_4$. The LSV electrolytic studies (Figure 12.12a.) revealed that UiO-66-NH$_2$ showed inefficient HER studies because of their poor conductivity. It can be noted that upon encapsulating the pores of MOF with Ru and Rh, the HER activity significantly improved. Rh$_{50}$Ru$_{50}$@UiO-66-NH$_2$ exhibited an overpotential of 77 mV @ 10 mA.cm^{-2} which is low as compared to 134, 124, 98, 185, and 210 mV of Rh$_{100}$@UiO-66-NH$_2$, Rh$_{67}$Ru$_{33}$@UiO-66-NH$_2$, Rh$_{33}$Ru$_{67}$@UiO-66-NH$_2$, Ru$_{100}$@ UiO-66-NH$_2$, and Ru/C, respectively. This is an indicative that Rh$_{50}$Ru$_{50}$@UiO-66-NH$_2$ demonstrated the excellent HER properties in 0.50 M H$_2$SO$_4$ but did not match the activity of commercial Pt/C catalyst (31.0 mV). The corresponding Tafel slope (Figure 12.12b) of highly active Rh$_{50}$Ru$_{50}$@UiO-66-NH$_2$ was 79.0 mV.dec^{-1}. Comparatively, the prepared RhRu alloy immobilized MOF, Rh$_{100}$@UiO-66-NH$_2$, Rh$_{67}$Ru$_{33}$@UiO-66-NH$_2$, Rh$_{33}$Ru$_{67}$@UiO-66-NH$_2$, Ru$_{100}$@UiO-66-NH$_2$, Ru/C and Pt/C showed Tafel slope of 103.0, 98.3, 92.3, 109.5, 128.1, and 60.3 mV.dec^{-1}, respectively. The value of Tafel slope of Rh$_{50}$Ru$_{50}$@UiO-66-NH$_2$ is closer to one of commercial Pt/C and this tells that the HER kinetics were enhanced. To further reveal the intrinsic properties, TOF was investigated and it was obtained that Rh$_{50}$Ru$_{50}$@ UiO-66-NH$_2$ possesses the highest TOF as compared to other Ru and Rh based catalysts (Figure 12.12c). The LSV polarization curves (Figure 12.12)d–f) of Rh$_{50}$Ru$_{50}$@ UiO-66-NH$_2$ (Figure 12.12)d), Ru/C (Figure 12.12e), and Pt/C (Figure 12.12f) before and after 3000 CV cycles at the current density of 10 mA.cm^{-2}, that the overpotential shifted negatively at 56, 149, and 107 mV, respectively and this suggests that Rh$_{50}$Ru$_{50}$@UiO-66-NH$_2$ outclassed an Ru/C and Pt/C for long-term operation.

Afterwards, 1.0 M PBS was used as the electrolyte (pH = 7.0, neutral) to test the as-prepared catalysts. As in acidic solution, UiO-66-NH$_2$ showed poor electrocatalytic

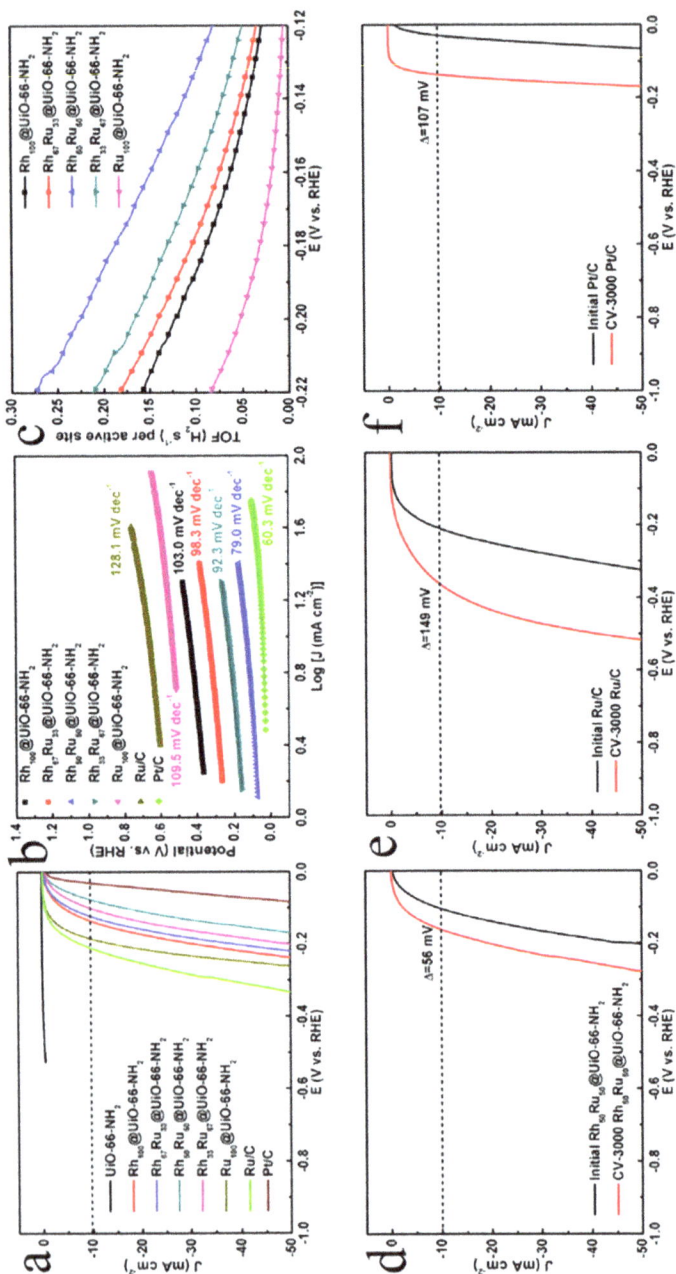

FIGURE 12.12 (a, b, c) LSV curves, Tafel plots and turnover frequency (TOF) per surface metal site for RhRu alloyed encapsulated MOF in 0.50 M H_2SO_4 and (d, e, f) LSV curves of $Rh_{50}Ru_{50}$@UiO-66-NH_2, Ru/C and Pt/C before and after cycling for 3000 cycles [152].

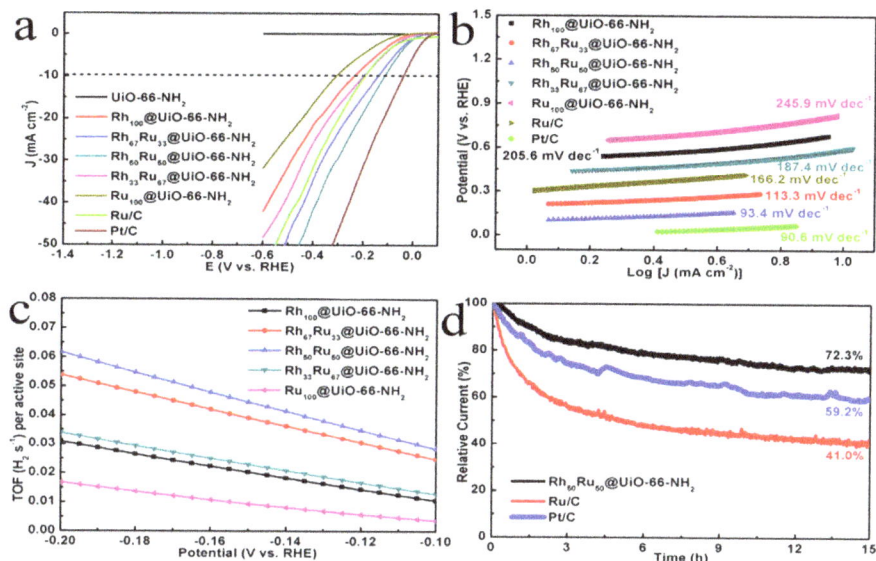

FIGURE 12.13 (a) LSV polarization curves, (b) corresponding Tafel slopes and TOF per surface metal site for the synthesized catalysts in 1.0 M PBS. (d) Stability test for $Rh_{50}Ru_{50}@$UiO-66-NH$_2$, Ru/C and Pt/C in 1.0 M PBS [152].

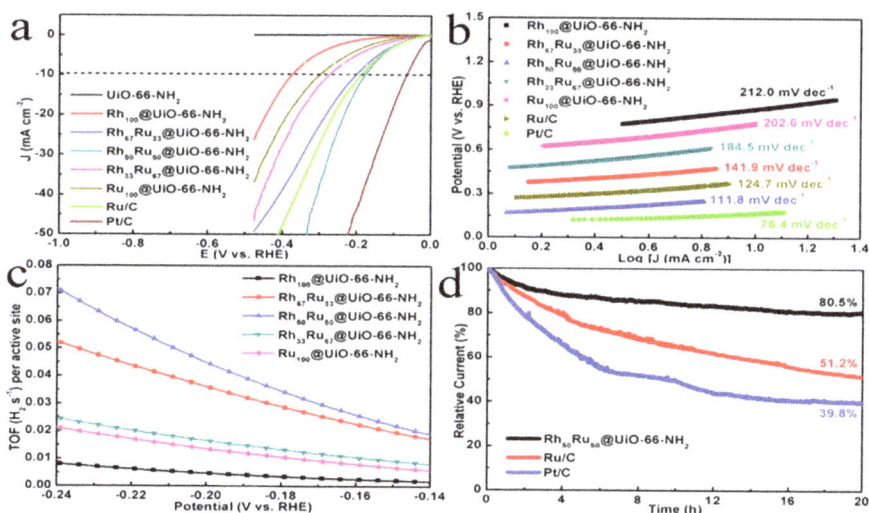

FIGURE 12.14 (a, b, c) LSV polarization curves, Tafel plots, and TOF per surface metal site for fabricated catalysts in 1.0 M KOH. (d) The long-term stability test for $Rh_{50}Ru_{50}@$UiO-66-NH$_2$, Ru/C and Pt/C in 1.0 M KOH [152].

HER activities. Rh_{100}@UiO-66-NH$_2$, $Rh_{67}Ru_{33}$@UiO-66-NH$_2$, $Rh_{50}Ru_{50}$@UiO-66-NH$_2$, $Rh_{33}Ru_{67}$@UiO-66-NH$_2$, Ru_{100}@UiO-66-NH$_2$, Ru/C, and Pt/C exhibited the overpotential of 232, 137, 111, 197, 303, 195, and 40 mV to attain the current density of 10 mA.cm^{-2} (Figure 12.13a). The corresponding Tafel slopes (Figure 12.13b) were 205.6, 113.3, 93.4, 187.4, 245.9, 166.2, and 90.6 mV for Rh_{100}@UiO-66-NH$_2$, $Rh_{67}Ru_{33}$@UiO-66-NH$_2$, $Rh_{50}Ru_{50}$@UiO-66-NH$_2$, $Rh_{33}Ru_{67}$@UiO-66-NH$_2$, Ru_{100}@UiO-66-NH$_2$, Ru/C, and Pt/C, respectively. TOF (Figure 12.13c) values obtained revealed that the intrinsic properties of the UiO-66-NH$_2$ were enhanced and stability test (Figure 12.13d) showed that $Rh_{50}Ru_{50}$@UiO-66-NH$_2$ possessed an enormously enhanced stability than that of Ru/C and Pt/C for HER in 1.0 M PBS solution. Consequently, the HER activities were performed in alkaline solution of 1.0 M KOH. The similar trends were observed as the one of acidic and neutral solutions (Figure 12.14a–d). From the three studies, it can be concluded that the as-prepared RhRu alloyed MOF can be used as an alternative to Pt/C electrocatalysts.

12.3.2.2.2 Metal Nanoparticles Deposited on Porous Carbon Derived from MOF as Electrocatalyst

One of the limitations of using MOF for electrochemical/catalytic applications is their poor electrical conductivity [153] and this hampers their usage as electrocatalysts. In that case, it is essential to find alternative routes to enhance the conductivity of MOFs. Recently, MOFs have been used as a precursor for carbon-based materials [154,155]. Carbonization or pyrolysis of MOF results in degradation/decomposition of the metal-organic ligand bond and from porous carbon-based materials [153,156]. The hierarchical porous carbon obtained exhibited excellent stability and conductivity [153]. The metal centers of MOFs are then transformed to metal/metal oxide NPs [156]. In short, the pyrolysis or carbonization of MOFs results in graphitization, growth, agglomeration, oxidation, and carbon reduction of metal oxide into metal NPs [154,157]. Graphitization of MOF to porous carbon matrix results in meso/macropores which are essential for diffusion/electron transport mechanism [153]. One other thing, the metal NPs are distributed/dispersed uniformly on the carbon matrix [154]. This leads to addition of active sites which are significant during HER [154]. In 2016, Fan et al. [158] fabricated nickel-carbon (Ni-C)-based catalyst from a direct carbonization of Ni-MOF and activated their structural properties by leaching with HCl (HCl-Ni@C) and further activation to obtain A-Ni-C (Figure 12.15a). The HER studies of HCl-Ni@C, activated Ni-C (A-Ni-C) together with commercial Pt/C were investigated in 0.50 M H$_2$SO$_4$ and the LSV polarization curves are displayed in Figure 12.15b. HCl-Ni@C and A-Ni-C exhibited the overpotentials of −440 mV and −34 mV to achieve the current density of 10 mA/cm^2. Consequently, the A-Ni-C showed a small Tafel slope (Figure 12.15c) of 41 mV/dec as compared to 194 mV/dec of HCl-Ni@C displaying the faster HER process occurring via a Volmer-Heyrovsky mechanism. The faster HER kinetics were further supported by the EIS Nyquist plot (Figure 12.15d) which shows that the A-Ni-C exhibited the smallest charge transfer resistance similar to that of Pt/C. Most significantly, the A-Ni-C displayed greatest durability of 25 hours with no significant current loss (Figure 12.15e).

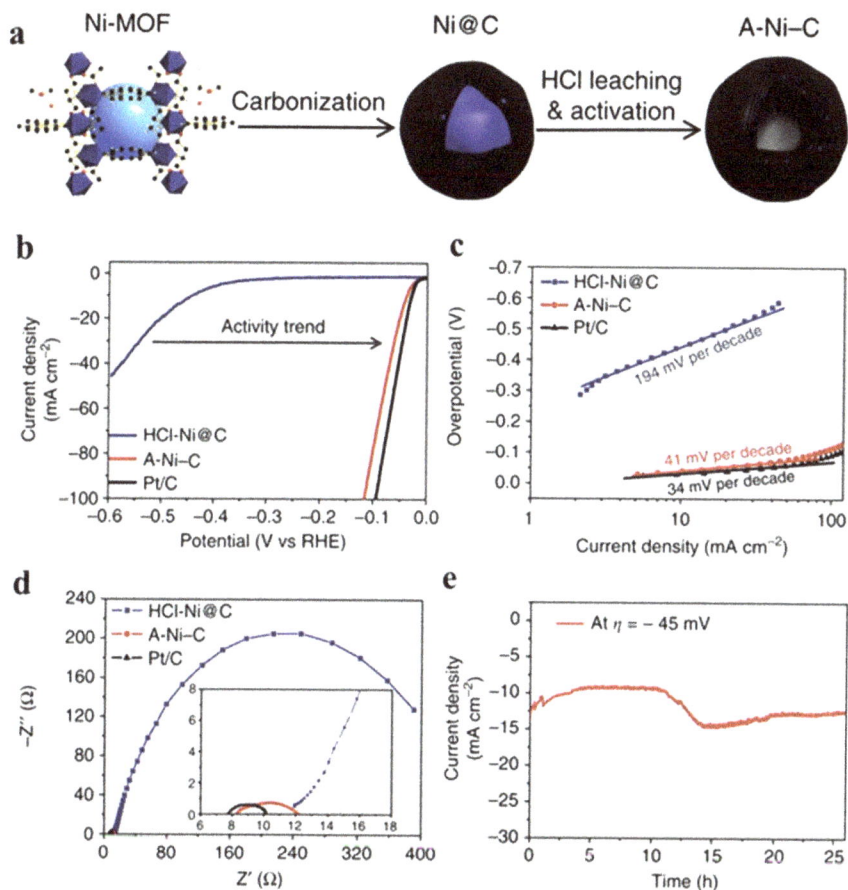

FIGURE 12.15 (a) Schematic diagram for preparation of pyrolysis, HCl leaching and activation of Ni-MOF. (b) LSV, (c) corresponding Tafel plots, (d) EIS Nyquist plot and (e) durability test of prepared materials [158].

Recently, Qiu et al. [159] reported a bimetallic MOF to fabricate the highly exposed active site Ru-based electrocatalysts using facile pyrolysis. MOFs will alter advantages such as (i) high surface area and porosity which are essential for mass transfer and diffusion of charges during the electrochemical process, (ii) metal centers in MOFs can be fine-tuned into ultrathin and NPs, and (iii) carbons in MOF organic linkers can promote electron transfer. In addition, utilization of CuRu-MOF as a precursor for preparation of Ru-based electrocatalyst plays a significant role as (i) the Ru and Cu sites are uniformly distributed and will suppress aggregation during pyrolysis, (ii) large surface area and rich porosity will generate meso/macropores during removal of Cu, and (iii) carbon-based linker acts as an electron-transport chamber. The pyrolysis process CuRu-MOF is shown in Figure 12.16a. For comparison, monometallic MOF, Cu-MOF and Ru-MOF were synthesized and subjected to

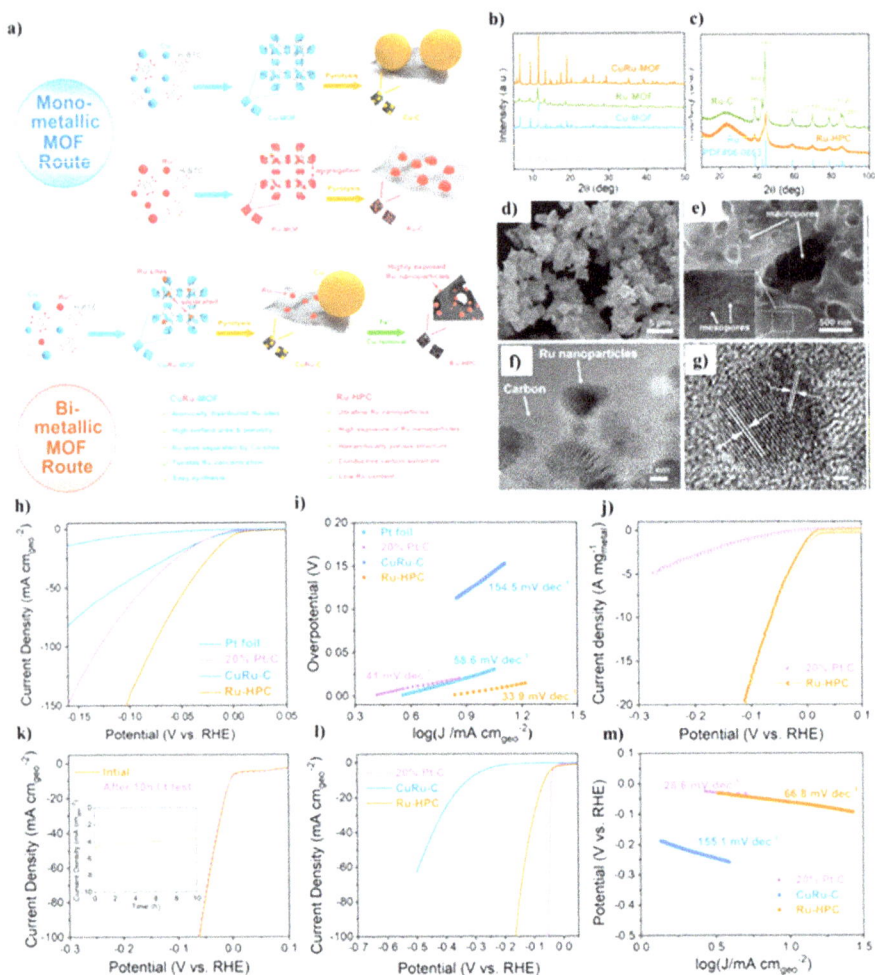

FIGURE 12.16 Illustration of a fabrication procedure of monometallic, Ru-MOF and Cu-MOF, and bimetallic, CuRu-MOF as well as their respective simple pyrolysis methods. XRD of (a) as-synthesized MOFs and (b) pyrolyzed MOF products. (d and e) SEM images and (f and g) high-resolution (HR) TEM for the MOF-derived Ru-HPC. (h) LSV curves, (i) Tafel plots, (j) mass current density, and (k) stability and durability test for prepared materials in 0.50 M H_2SO_4. (l) LSV and (m) corresponding Tafel slopes in 1.0 M KOH for the prepared catalysts [159].

pyrolysis. The XRD patterns of the CuRu-MOFs indicated the similar as monometallic MOFs (Figure 12.16b). This indicated that CuRu-MOF was obtained without altering the framework structure. After pyrolysis of CuRu-MOF and monometallic MOFs, the XRD pattern revealed loss of phases of pristine MOF with appearance on metal (Cu and Ru) phases. The XRD of pyrolyzed CuRu-MOF denoted by Ru-HPC (Figure 12.16c) showed the disappearance of Cu phases after treatment

$FeCl_3$. The SEM micrographs of Ru-HPC indicated the inherent porosity resulted from the use of MOF as precursors (Figure 12.16d) and large meso/micropores appearing (Figure 12.16e). TEM images illustrated the uniform distribution of Ru NPs on the porous carbon network (Figure 12.16f). Furthermore, the high-resolution TEM (Figure 12.16g) showed d-spacing of 0.234 and 0.214 nm which correspond to the Ru [100] and [002], respectively. When investigating the HER carbonized materials as HER electrocatalysts in 1.0 M KOH (Figure 12.16h), Ru-HPC exhibited low overpotentials (22.7 and 44.6 at current densities of 25 and 50 mA cm^{-2}) as compared to commercial Pt/C electrocatalyst (45.4 and 84.4 mV at 25 and 50 mA cm^{-2}). The Tafel slopes (Figure 12.16i) of Ru-HPC (33.9 mV/dec) was also lower that of Pt/C (41.0 mV dec^{-1}). The small Tafel slope tells that the pyrolysis of MOF results in fast HER rates and is desirable for practical applications. Furthermore, dispersion of Ru NPs increased the active sites on the Ru-HPC electrocatalyst. This is evident from the mass current density studies shown in Figure 12.16j). The mass current density of Ru-HPC (7.8 A/mg) was 16 times higher than the one of Pt/C (0.46 A/mg) at an overpotential of 50 mV. Durability studies (Figure 12.16k) showed that the Ru-HPC can be used for 10 h without losing its activity. In acidic medium, 0.50 M H_2SO_4 (Figure 12.16l–m), Ru-HPC showed superior activities as compared to CuRu-C. The excellent HER activities of Ru-HPC was due to uniform distribution of Ru NPs on carbon support.

12.3.2.3   Other MOF Composites as HER Electrocatalysts

Formation of MOF hybrid composite can be another strategy to address the limitation of MOF for HER applications [132]. Highly conductive materials such as intrinsic conductive polymers and carbon-based compounds have been studied and were found to enhance the MOF activity.

12.3.2.3.1   Conducting Polymers@MOF Composites

Intrinsic conducting polymers (ICPs) such as polypyrrole (Ppy), polyaniline (PANI), and polythiopene (PTh) have been investigated extensively for electrochemical energy storage applications such as supercapacitors and lithium-ion batteries [42,83]. This is due to their metal-like electrical properties that makes them fascinating for different applications. Moreover, the resonance-stabilized structures of ICPs allow incorporation of secondary components. This is achieved by ability of ICPs to act as a support/template of secondary components and impart their properties and this improves the functionality of the secondary component [30]. Furthermore, the electronic and physical properties of the material can be influenced by the type of the interactions between the ICP and secondary component [30]. In this regard, one of the main things to consider when integrating MOFs and ICPs is the way in which they interact which result from the approach used for fabrication of the MOF-polymer. Several methods for preparing MOF-polymer composite include polymer grafted through MOF, mixed matrix membranes, polymer-templating MOF growth, and using polymer ligands as organic linkers to yield polyMOFs. Our group prepared MOF-PANI/ PANI derivative, poly(3-aminobenzoic acid) (PABA)-based electrocatalysts over the years using two different approaches, viz. (i) in situ polymerization of aniline in the presence of MOF/zeolitic imidazolate frameworks (ZIFs) and (ii) facile hydrother-mal fabrication of MOF in presence of PANI/poly-3-amino benzoic acid (PABA) [22,82,83,160,161]. The enhancement of electrocatalytic HER in the as-fabricated MOF/ZIF-PANI/PABA composite was attributed to polymer acting as a channel for electron transport to allow the diffusion process to take place. In addition, the decrease in band gap obtained showed that the MOF-polymer composite exhibited high conductivity which is essential for HER. In another study of enhancing the activity of MOF-polymer composite, Khalid and coworkers [162] prepared PANI/ZIF-67 using different weight percentage of PANI via in situ preparation (Figure 12.17a). The as-prepared PANI/ZIF-67 was further subjected to a simple pyrolysis method. From the as-prepared PANI/ZIF-67, 50 wt.% PANI/ZIF-67 displayed superior HER activity in 1.0 M KOH but never matched the activity of Pt/C (Figure 12.17b). The Tafel slope of 50 wt.% PANI/ZIF-67 (83 mV.dec^{-1}) is also less than the other PANI/ZIF-67 composite at a different weight percentage (Figure 12.17c). This illustrates that the kinetics of 50 wt.% PANI/ZIF-67 is higher. The stability test (Figure 12.17d) indicated a gradual decay of HER activity after cycling but the loss is not significant. This tells that the stability was enhanced, and formation of MOF-polymer composite can be used to address the limitation of MOF for HER applications.

12.3.2.3.2   Carbon-Based MOF Composite

MOF-carbon-based composites present a new fascinating hybrid material as it combines the state-of-the-art nature of MOFs and the classical nature of carbonaceous materials [163,164]. Integration of MOFs with carbon-based materials brings new functionalities which include improved stabilities, enhanced electrical conductivities, and template effects [127]. Various methods like in situ (one-pot and stepwise

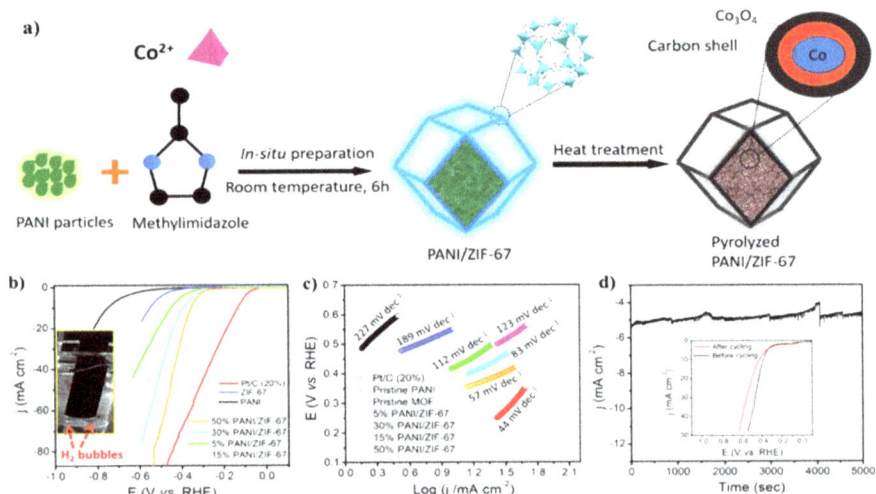

FIGURE 12.17   (a) Schematic diagram for the synthesis of PANI/ZIF-67 and corresponding pyrolysis route to obtain pyrolyzed PANI/ZIF-67. (b, c) LSV polarization curves and Tafel plots of the prepared catalysts in 1.0 M KOH. (d) Stability test of the 50%PANI/ZIF-67 [162].

synthesis) and ex situ approaches have been used for preparation of fascinating MOF-carbon-based composites [165]. Carbon-based materials such as graphene oxides (GO), carbon nanotubes (CNTs), and acetylene black (AB) have been used for fabrication of MOF-carbon-based composite for electrocatalytic HER studies [166]. Johan et al. [164] prepared Cu-centered MOF-GO composite via a facile hydrothermal method with different GO weight percentage (2.4 and 8 wt.%). Experimental results showed that increasing the amount of GO resulted in increasing the electron transfer rate. The HER studies in 0.50 M H_2SO_4 indicated that (GO 8 wt.%) Cu-MOF exhibited high electrocatalytic activity due to decrease in activation energy attributed to presence of GO. In another study, Micheroni et al. [167] investigated MOF-CNT hybrid composite. The as-prepared Hf_{12}-CoDBP/CNT composite significantly enhanced electrical conductivity which resulted in improved HER activity (Figure 12.18) as revealed by the low onset potential (Figure 12.18a), Tafel slope of 178 mV.dec^{-1} (Figure 12.18b), turnover number of 3.2×10^4 after 30 minutes (Figure 12.18c) as well as time-dependent current densities (Figure 12.18d). To be more specific, Wang et al. [168] used two facile routes to prepare AB&Cu-MOF with enhanced HER activities. The hydrothermally synthesized 1.7 wt.% AB&Cu-MOF showed remarkable HER activity in 0.50 M H_2SO_4. The fabricated 1.7 wt.% AB&Cu-MOF exhibited the onset potential of 148 mV (Figure 12.19a) and Tafel slope of 80 mV.dec^{-1} (Figure 12.19b). The superior HER activity was further supported by the EIS data which displayed a faster kinetics, low R_{ct} (Figure 12.19c) and durability studies displayed negligible decay (Figure 12.19d) and withstand harsh conditions

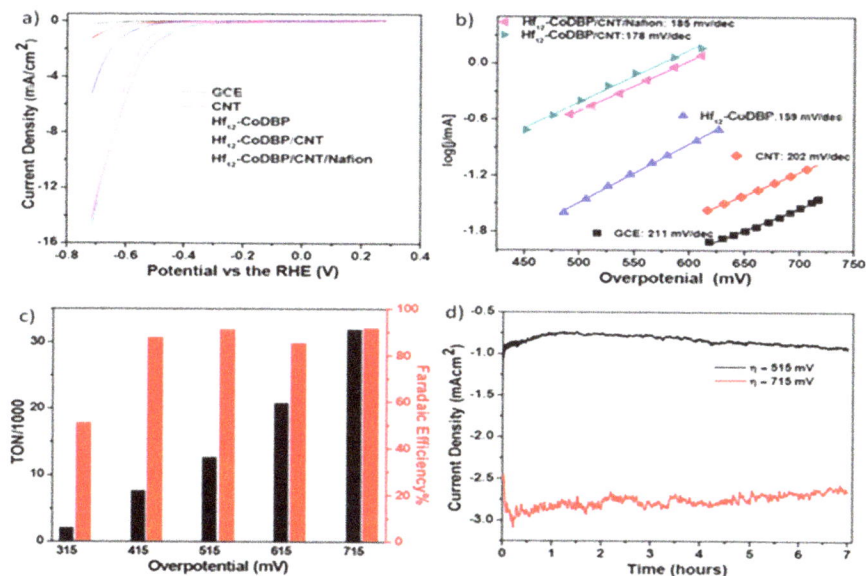

FIGURE 12.18 (a) LSV curves and (b) corresponding Tafel plots of prepared MOF/CNTs catalysts compared to CNT and GCE in 0.10 M perchloric acid. (c, d) Turnover number (TON) and current-density-time studies of hf$_{12}$-CoDBP/CNT/Nafion [167].

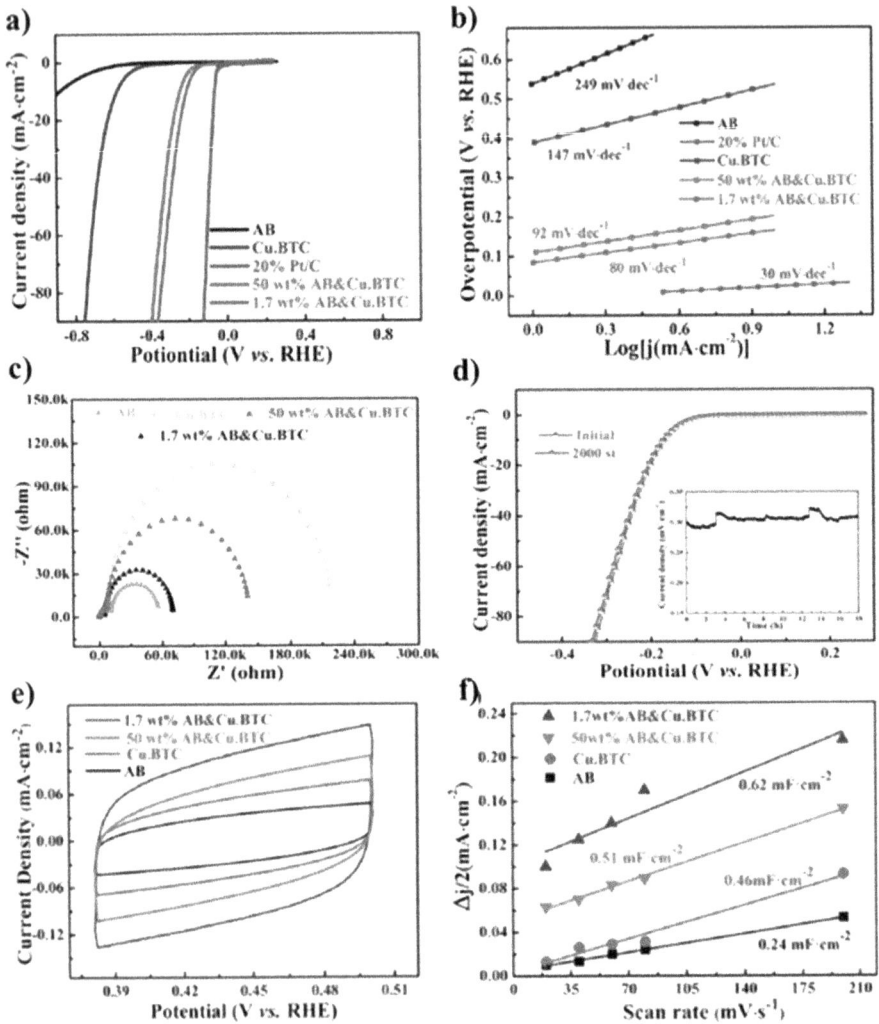

FIGURE 12.19 (a) LSV polarization, (b) corresponding Tafel plots, (c) EIS Nyquist plot, (d) durability test, (e) cyclic voltammograms at 100 mV s⁻¹, and (f) capacitance double-layer plots for the synthesized catalysts in 0.50 M H_2SO_4 [168].

(stability) for 18 h (inset). Furthermore, the 1.7 wt.% AB&Cu-MOF exhibited good capacitive behavior and double capacitance layer (Figure 12.19f) of 0.62 mF.cm⁻² which shows that new active sites were added. These studies revealed that conductive carbons are suitable materials for preparing MOF composite for HER applications.

12.4   CONCLUSIONS AND FUTURE PERSPECTIVES

Hydrogen evolution reaction plays an important role in driving hydrogen production economy towards commercialization. It was seen that manufacture of cost-effective, stable, durable, and highly catalytic active electrocatalysts is a crucial step in realizing the fruit of hydrogen energy. In this chapter, we clearly looked at the insight reaction mechanisms during electrolysis splitting of water to produce hydrogen gas in order to understand better the vital aspects which need to be taken into consideration during selection of electrocatalysts. It was observed that HER process requires an electrocatalyst with balanced energies in order for it to proceed. It was also noted that the HER performance of the material can be monitored by studying electrochemical parameters such as onset potential, over-potential, Tafel analysis, and charge resistance transfer. Hence in this chapter, the theoretical background of the above-mentioned electrochemical parameters were illustrated. Furthermore, we reviewed the HER mechanisms in both alkaline and acidic media where it was demonstrated that the HER process may take place through Volmer-Heyrovsky or Volmer-Tafel mechanisms as the rate-determining step. The HER mechanisms were mostly associated with the number of active sites of the electrocatalyst to adsorb hydrogen intermediately. This was shown to correspond with the Tafel slopes of 30, 40, and 120 mV dec^{-1} for Tafel, Heyrovsky, and Volmer, respectively. It was seen that the HER mechanism depends on the electrical conductivity of the catalytic structure of the material, therefore, developing materials with high electrical conductivity will facilitate the electron transfer process. In this regard, the contemporary approach was based on fine tuning MOFs and their composites for electrocatalysis owing to their large surface area. However, the HER performance of MOF is hampered by its poor conductivity caused by organic linkers and stability in different media. The encapsulation of metal NPs was reported in order to improve MOF electrocatalytic activity by adding new active sites, stabilization of MOF and new channels for the electron transport mechanism. In addition, the carbon porosity derived from the MOF is then used as a support to immobilize the metal NPs to further enhance its catalytic activity. The other approach is to introduce conducting polymers, where it was shown to improve the electrical conductivity of MOFs resulting in enhanced HER properties. In future, more efforts are needed in order to attain MOF nanostructures with superior HER activity.

12.5 ACKNOWLEDGMENT

MJH and KDM would like to thank the financial support from the National Research Foundation (NRF) under the Thuthuka programme (UID Nos. 117727 and 118113), Sasol Foundation for purchasing both STA and UV-vis instruments and University of Limpopo (Research Development Grants R202 and R232), South Africa.

LIST OF ABBREVIATIONS AND ACRONYMS

2-, 3-D	two-, three dimensional
4,4′-biby	4,4′-bipyridine
AB	acetylene black
BET	Brunauer-Emmett-Teller
BTB	benzene tribenzoate

BPT	[1,1′-biphenyl]-3,4′,5-tricarboxylate
bpy	4.4-dipyridyl
CNTs	carbon nanotubes
CO$_2$RR	carbon dioxide reduction reaction
CV	cyclic voltammetry
ddbp	1,3-di(3′,5′-dicarboxyl phenyl) benzene
EIS	electrochemical impedance spectroscopy
EWS	electrochemical water splitting
GO	graphene oxide
GCE	glassy carbon electrode
H$_{ad}$	adsorbed hydrogen atom
HER	hydrogen evolution reaction
HFCs	hydrofluorocarbons
HKUST-1	Hong Kong University of Science and Technology-1
HRTEM	high resolution transmission electron microscope
ICPs	intrinsic conducting polymers
LSV	linear sweep voltammetry
M	metal
MOFs	metal-organic frameworks
NENU-499	[TBA]$_4$[ε-PMo$^V_8$Mo$^{VI}_4$O$_{37}$(OH)$_3$Zn$_4$]Cl$_4$
NENU-500	[TBA]$_3$[ε-PMo$^V_8$Mo$^{VI}_4$O$_{36}$(OH)$_4$Zn$_4$][BTB]$_{4/3}$. xGuest
NENU-501	[TBA]$_3$[ε-PMo$^V_8$Mo$^{VI}_4$O$_{37}$(OH)$_3$Zn$_4$][BPT] (NENU-501)
NF	Ni foam
NPs	nanoparticles
OER	oxygen evolution reaction
ORR	oxygen reduction reaction
PABA	poly(3-aminobenzoic acid)
PANI	polyaniline
PFCs	perfluorocarbons
PGMs	platinum/precious group metals
POMOF	polyoxomelate based MOF
PPy	polypyrrole
PTh	polythiopene
QD	quantum dots
RHE	reversible hydrogen electrode
SAED	selected area electron diffraction
TBA	tetrabutylammonium
TEM	transmission electron microscope
TMCs	transitional metal-based compounds
TON	turnover number
TOF	turnover frequencies
X	halogen
X$_4$-H$_2$pta	tetrahalogenphthalic acid
XRD	powder X-ray diffraction

REFERENCES

1. He Y, Chen F, Li B, Qian G, Zhou W, Chen B. Porous metal-organic frameworks for fuel storage. *Coord Chem Rev* 2018; 373: 167–8.
2. Luo H, Zeng Z, Zeng G, Zhang C, Xiao R, Huang D, et al. Tian S. Recent progress on metal-organic framework-based and derived photocatalysts for water splitting. *Chem Eng J* 2020; 383: 123196.
3. Murthy AP, Madhavan J, Murugan K. Recent advances in hydrogen evolution reaction catalysts on carbon/carbon-based supports in acid media. *J Power Sources* 2018; 398: 9–26.
4. Song Y, Xin X, Guo S, Zhang Y, Yang L, Wang B, et al. One-step MOFs-assisted synthesis of intimate contact MoP-Cu$_3$P hybrids for photocatalytic water splitting. *Chem Eng J* 2020; 384: 123337.
5. Jamesh MI, Sun X. Recent progress on earth abundant electrocatalysts for hydrogen evolution reaction (HER) in alkaline medium to achieve efficient water splitting –A review. *J Energy Chem* 2019; 34: 111–60.
6. Lei L, Huang D, Zhou C, Chen S, Yan X, Li Z, et al. Demystifying the active roles of NiFe-based oxides/(oxy)hydroxides for electrochemical water splitting under alkaline conditions. *Coord Chem Rev* 2020; 408: 213177.
7. Lin K, Adhikari AK, Ku C, Chiang C, Kuo H. Synthesis and characterization of porous HKUST-1 metal organic frameworks for hydrogen storage. *Int J Hydrogen Energy* 2012; 37: 13865–71.
8. Wang T, Xie H, Chen M, Aloi AD, Cho J, Wu G, et al. Precious metal-free approach to hydrogen electrocatalysis for energy conversion: From mechanism understanding to catalyst design. *Nano Energy* 2017; 42: 69–89.
9. Zhu Y, Chen G, Zhong Y, Zhou W, Liu M, Shao Z. An extremely active and durable Mo$_2$C/graphene-like carbon based electrocatalyst for hydrogen evolution reaction. *Mater Today Energy* 2017; 6: 230–7.
10. Modibane KD, Lototskyy M, Davids MW, Williams M, Hato MJ, Molapo KM. Influence of co-milling with palladium black on hydrogen sorption performance and poisoning tolerance of surface modified AB$_5$-type hydrogen storage alloy. *J Alloys Compd* 2018; 750: 523–9.
11. Pukazhselvan D, Kumar V, Sighn SK. High capacity hydrogen storage: Basic aspects new developments and milestones. *Nano Energy* 2012; 1: 566–89.
12. Cardoso DSP, Amaral L, Santos DMF, Šljukić B, Sequeira CAC, Macciò D, et al. Enhancement of hydrogen evolution in alkaline water electrolysis by using nickel-rare earth alloys. *Int J Hydrogen Energy* 2015; 40: 4295–302.
13. Su J, Zhou J, Wang L, Liu C, Chen Y. Synthesis and application of transition metal phosphides as electrocatalyst for water splitting. *Sci Bull* 2017; 62: 633–44.
14. Smiljanic M, Rakocevic Z, Strbac S. Electrocatalysis of hydrogen evolution reaction on tri-metallic Rh@Pd/Pt(poly) electrode. *Int J Hydrogen Energy* 2018; 43: 2763–71.
15. Abdolmaleki A, Mohamadi Z, Ensafi AA, Atashbar NZ, Rezaei B. Efficient and stable HER electrocatalyst using Pt nanoparticles@ poly(34eethylene dioxythiophene) modified sulfonated graphene nanocomposite. *Int J Hydrogen Energy* 2018; 43: 8323–32.
16. Wu Z, Zou Z, Huang J, Gao F. Fe-doped NiO mesoporous nanosheets array for highly efficient overall water splitting. *J Catal* 2018; 358: 243–52.
17. Li Y, Wang H, Wang R, He B, Gong Y. 3D self-supported Fe-O-P film on nickel foam as a highly active bifunctional electrocatalyst for urea-assisted overall water splitting. *Mater Res Bull* 2018; 100: 72–5.

18. Mahale NK, Ingle S. Electrocatalytic hydrogen evolution reaction on nano-nickel decorated graphene electrode. *Energy* 2017; 119: 872–8.

19. Jamesh MI, Sun X. Recent progress on earth abundant electrocatalysts for oxygen evolution reaction (OER) in alkaline medium to achieve efficient water splitting—A review. *J Power Sources* 2018; 400: 31–68.

20. Ma B, Chen TT, Li QY, Qin HY, Dong XY, Zang SQ. Bimetallic-organic-framework derived nanohybrid $Cu_{0.9}Co_{2.1}S_4$@MoS_2 for highly performance hydrogen evolution reaction. *Materials* 2019; 2: 1134–48.

21. Zhang P, Xu B, Chen G, Gao C, Gao M. Large-scale synthesis of nitrogen doped MoS_2 quantum dots for efficient hydrogen evolution reaction. *Electrochim Acta* 2018; 270: 256–63.

22. Mashao G, Ramohlola KE, Mdluli SB, Monama GR, Hato MJ, Makgopa K, et al. Zinc-based zeolitic benzimidazolate framework/polyaniline nanocomposite for electrochemical sensing of hydrogen gas. *Mater Chem Phys* 2019; 230: 287–98.

23. Zhang X, Yang Y, Ding S, Que W, Zheng Z, Du Y. Construction of high-quality SnO_2@MoS_2 nanohybrids for promising photoelectrocatalytic applications. *Inorg Chem* 2017; 56: 3386–93.

24. Theerthagiri J, Sudha R, Premnath K, Arunachalam P, Madhavan J, Al-Mayouf AM. Growth of iron diselenide nanorods on graphene oxide nanosheets as advanced electrocatalyst for hydrogen evolution reaction. *Int J Hydrogen Energy* 2017; 42: 13020–30.

25. Dolganov AV, Tanaseichuk BS, Ivantsova PM, Tsebulaeva YV, Kostrukov SG, Moiseeva DN, et al. Metal-free electrocatalyst for hydrogen production from water. *Int J Electrochem Sci* 2016; 11: 9559–65.

26. Dolganov AV, Tanaseichuk BS, Moiseeva DN, Yurova VY, Sakanyan JR, Shmelkova NS, et al. Acridinium salts as metal-free electrocatalyst for hydrogen evolution reaction. *Electrochem Commun* 2016; 68: 59–61.

27. Dolganov AV, Muryumin EE, Chernyaeva OY, Chugunova EA, Mishkin VP, Nishcev KN. Fabrication of new metal-free materials for the hydrogen evolution reaction on base of the acridine derivatives immobilized on carbon materials. *Mater Chem Phys* 2019; 224: 148–55.

28. Dolganov AV, Tanaseichuk BS, Yurova VY, Chernyaeva OY, Okina EV, Balandina AV, et al. Moving from acridinium to pyridinium: From complex to simple. 246-Triphenylpyridine and 246-triphenylpyrilium perchlorate as "metal-free" electrocatalysts of hydrogen evolution reaction (HER): The influence of the nature of the heteroatom and acid on the pathway HER. *Int J Hydrogen Energy* 2019; 44: 21495–505.

29. Dalla-Corte DA, Torres C, Correa PS, Rieder ES, Malfatti CF. The hydrogen evolution reaction on nickel-polyaniline composite electrodes. *Int J Hydrogen Energy* 2012; 37: 3025–32.

30. Hatchett DW, Josowicz M. Composites of intrinsically conducting polymers as sensing nanomaterials. *Chem Rev* 2018; 108: 746–69.

31. Yang F, Kang N, Yan J, Wang X, He J, Huo S, et al. Hydrogen evolution reaction property of molybdenum disulphide/nickel phosphide hybrids in alkaline solution. *Metals* 2018; 8: 359–76.

32. Wang D, Pan Z, Wu Z, Wang Z, Liu Z. Hydrothermal synthesis of MoS_2 nanoflowers as highly efficient hydrogen evolution reaction catalysts *J Power Sources* 2014; 264: 229–34.

33. Li H, Tsai C, Koh AL, Contryman AW, Fragapane AH, Zhao J, et al. Activating and optimizing MoS_2 basal planes for hydrogen evolution through formation of strained sulphur vacancies. *Nat Mater* 2016; 15: 48–53.

34. Wang D, Xie Y, Wu Z. Amorphous phosphorus-doped MoS_2 catalyst for efficient hydrogen evolution reaction. *Nanotechnology* 2019; 30: 205401–407.

35. Li X, Zhu H. Two-dimensional MoS_2: Properties preparation and applications. *J Materiomics* 2015; 1: 33–44.

36. Xiang ZC, Zhang Z, Xu XJ, Zhang Q, Yuan C. MoS_2 nanosheets array on carbon cloth as a 3D electrode for highly efficient electrochemical hydrogen evolution. *Carbon* 2016; 98: 84–9.

37. Kong Q, Wang X, Tang A, Duan W, Liu B. Three-dimensional hierarchical MoS_2 nanosheet arrays/carbon cloth as flexible electrodes for high-performance hydrogen evolution reaction. *Mater Lett* 2016; 177: 139–42.

38. Dai X, Du K, Li Z, Sun H, Yang Y, Zhang X, et al. Wang H. Highly efficient hydrogen evolution catalyst by MoS_2-MoN/carbonitride composites derived from tetrathiomolybdate/polymer hybrids. *Chem Eng Sci* 2015; 134: 572–80.

39. Su C, Xiang J, Wen F, Song L, Mu C, Xu D, et al. Microwave synthesized three-dimensional hierarchical nanostructure CoS_2/MoS_2 growth on carbon fiber cloth: A bifunctional electrode for hydrogen evolution and supercapacitor. *Electrochim Acta* 2016; 212: 941–9.

40. Karikalan N, Sundaresan P, Chen SM, Karthik R, Karuppiah C. Cobalt molybdenum sulfide decorated with highly conductive sulfur-doped carbon as an electrocatalyst for the enhanced activity of hydrogen evolution reaction. *Int J Hydrogen Energy* 2019;.44: 9164–73.

41. Shi Y, Zhou Y, Yang DR, Xu WX, Wang C, Wang FB, et al. Energy level engineering of MoS_2 by transition-metal doping for accelerating hydrogen evolution reaction. *J Am Chem Soc* 2017; 139: 15479–85.

42. Ramohlola KE, Hato MJ, Monama GR, Makhado E, Iwuoha EI, Modibane KD. State-of-the-art advances and perspectives for electrocatalysis. In Inamuddin, Boddula R Asiri A, editors. *Methods for Electrocatalysis*. Cham, Switzerland: Springer. 2020. pp. 311–52.

43. Pal TK, De D, Bharadwaj PK. Metal–organic frameworks for the chemical fixation of CO_2 into cyclic carbonates. *Coord Chem Rev* 2020; 408: 213173.

44. Wen X, Zhang Q, Guan J. Applications of metal-organic framework-derived materials in fuel cells and metal-air batteries. *Coord Chem Rev* 2020; 409: 213214.

45. Chen Y, Qiao Z, Wu H, Lv D, Shi R, Xia Q, et al. An ethane-trapping MOF PCN-250 for highly selective adsorption of ethane over ethylene. *Chem Eng Sci* 2018; 175: 110–7.

46. Xue D, Wang Q, Bai J. Amide-functionalized metal-organic frameworks: Syntheses structures and improved gas storage and separation properties. *Coord Chem Rev* 2019; 378: 2–16.

47. Zhao J, Liu X, Wua Y, Li D, Zhang Q. Surfactants as promising media in the field of metal-organic frameworks. *Coord Chem Rev* 2019; 391: 30–43.

48. Xu C, Fang R, Luque R, Chen L, Li Y. Functional metal–organic frameworks for catalytic applications. *Coord Chem Rev* 2019; 388: 268–92.

49. Wang B, Xie L, Wang X, Liu X, Li J, Li J. Applications of metal-organic frameworks for green energy and environment: New advances in adsorptive gas separation storage and removal. *Green Energy Environ* 2018; 3: 191–228.

50. Bazer-Bachi D, Assié L, Lecocqa V, Harbuzarua B, Falk V. Towards industrial use of metal-organic framework: Impact of shaping on the MOF properties. *Powder Technol* 2014; 255: 52–9.

51. Petit C. Present and future of MOF research in the field of adsorption and molecular separation. *Curr Opin Chem Eng* 2018; 20: 132–42.

52. Liao P, Shen J, Zhang J. Metal–organic frameworks for electrocatalysis. *Coord Chem Rev* 2018; 373: 22–48.

53. Amini A, Kazemi S, Safarifard V, Metal-organic framework-based nanocomposites for sensing applications—A review. *Polyhedron* 2020; 177: 114260.

54. Zhu C, Tang H, Yang K, Wu X, Luo Y, Wang J, et al. A urea-containing metal-organic framework as a multifunctional heterogeneous hydrogen bond-donating catalyst. *Catal Commun* 2020; 135: 105837.

55. Mir SH, Nagahara LA, Thundat T, Mokarian-Tabari P, Furukawa H, Khosla A. Review- Organic-inorganic hybrid functional materials: An integrated platform for applied technologies. *J Electrochem Soc* 2018; 165: B3137–56.

56. Jin Z, Yang H. Exploration of Zr-metal-organic framework as efficient photocatalyst for hydrogen production. *Nanoscale Res Lett* 2017; 12: 539–49.

57. Monama GR, Mdluli SB, Mashao G, Makhafola MD, Ramohlola KE, Molapo KM, et al. Palladium deposition on copper (II) phthalocyanine/metal-organic framework composite and electrocatalytic activity of the modified electrode towards the hydrogen evolution reaction. *Renew Energy* 2018; 119: 62–72.

58. Monama GR, Hato MJ, Ramohlola KE, Maponya TC, Mdluli SB, Molapo KM, et al. Hierarchiral 4-tetranitro copper (II) phthalocyanine based metal-organic framework hybrid composite with improved electrocatalytic efficiency towards hydrogen evolution reaction. *Results Phys* 2019; 15: 102564.

59. Rahmanifar MS, Hesari H, Noori A, Masoomi MY, Morsali A, Mousavi MF. A dual Ni/Co-MOF-reduced graphene oxide nanocomposite as a high-performance superca- pacitor electrode material. *Electrochim Acta* 2018; 275: 76–86.

60. Sundriyal S, Kaur H, Bhardwaj SK, Mishra S, Kim K, Deep A. Metal-organic frame- works and their composites as efficient electrodes for supercapacitor applications. *Coord Chem Rev* 2018; 369: 15–38.

61. Wen X, Guan J. Recent progress on MOF-derived electrocatalysts for hydrogen evolu- tion reaction. *Appl Mater Today* 2019; 16: 146–68.

62. He X, Yin F, Wang H, Chen B, Li G. Metal-organic frameworks for highly efficient oxygen electrocatalysis Chinese J Catal 2018; 39: 207–27.

63. Mao X, Ling C, Tang C, Yan C, Zhu Z, Du A. Predicting a new class of metal-organic frameworks as efficient catalyst for bi-functional oxygen evolution/reduction reactions *J Catal* 2018; 367: 206–11.

64. Darband GB, Aliofkhazraei M, Shanmugam S. Recent advances in methods and tech- nologies for enhancing bubble detachment during electrochemical water splitting. *Renew Sustain Energy Rev* 2019; 114: 109300.

65. Balogun MS, Huang Y, Qiu W, Yang H, Ji H, Tong Y. Updates on the development of nanostructured transition metal nitrides for electrochemical energy storage and water splitting. *Mater Today* 2017; 20: 425–51.

66. Munonde TS, Zheng H, Matseke MS, Nomngongo PN, Wang Y, Tsiakaras P. A green approach for enhancing the electrocatalytic activity and stability of $NiFe_2O_4$/CB nano- spheres towards hydrogen production. *Renew Energy* 2020; 154: 704–14.

67. Zhang W, Cui L, Jingquan L. Recent advances in cobalt-based electrocatalysts for hydrogen and oxygen evolution reactions. *J Alloys Compd* 2020; 821: 153542.

68. Liu W, Zhang H, Li C, Wang X, Liu J, Zhang X. Non-noble metal single-atom catalysts prepared by wet chemical method and their applications in electrochemical water split- ting. *J Energy Chem* 2020; 47: 333–45. doi: 10.1016/j.jechem.2020.02.020

69. Wang J, Yue X, Yang Y, Sirisomboonchai S, Wang P, Ma X, et al. Earth-abundant transition-metal-based bifunctional catalysts for overall electrochemical water split- ting: A review. *J Alloys Compd* 2020; 819: 153346.

70. Cao LM, Lu D, Zhong DC, Lu TB. Prussian blue analogues and their derived nanoma- terials for electrocatalytic water splitting. *Coord Chem Rev* 2020; 407: 213156.

71. Wen L, Sun Y, Zhang T, Bai Y, Li X, Lyu X, et al. $MnMoO_4$ nanosheet array: An effi- cient electrocatalyst for hydrogen evolution reaction with enhanced activity over a wide pH range. *Nanotechnology* 2018; 29: 335403. doi:10.1088/1361-6528/aac851

72. Cheng Y, Jiang SP. Advances in electrocatalysts for oxygen evolution reaction of water electrolysis-from metal oxides to carbon nanotubes. *Prog Nat Sci Mat Int* 2015; 25: 545–53.

73. Tong SS, Wang XJ, Li QC, Han XJ. Progress on electrocatalysts of hydrogen evolution reaction based on carbon fiber materials. *Chin J Anal Chem* 2016; 44: 1447–57.

74. Xu W, Wang H. Earth-abundant amorphous catalysts for electrolysis of water. *Chin J Catal* 2017; 38: 991–1005.

75. Mosconi D, Till P, Calvillo L, Kosmala T, Garoli D, Debellis D, et al. Effect of Ni Doping on the MoS_2 structure and its hydrogen evolution activity in acid and alkaline electrolytes. *Surfaces* 2019; 2: 531–45.

76. Lasai A. Mechanism and kinetics of the hydrogen evolution reaction. *Int J Hydrogen Energy* 2019; 44: 19484–518.

77. Yang Y, Luo M, Zhang W, Sun Y, Chen X, Guo S. Metal surface and interface energy electrocatalysis: Fundamentals, performance, engineering, and opportunities. *Chem* 2018; 4: 2054–83.

78. Habrioux A, Morais C, Napporn TW, Kokoh B. Recent trends in hydrogen and oxygen electrocatalysis for anion exchange membrane technologies. *Curr Opin Electrochem* 2020; 21: 146–59. doi:10.1016/j.coelec.2020.01.018

79. Monama GR, Modibane KD, Ramohlola KE, Molapo KM, Hato MJ, Makhafola MD, et al. Copper (II) phthalocyanine/metal organic framework electrocatalyst for hydrogen evolution reaction application. *Int J Hydrogen Energy* 2019; 44: 18891–902.

80. Wang X, Xu C, Jaroniec M, Zheng Y, Qiao SZ. Anomalous hydrogen evolution behavior in high-pH environment induced by locally generated hydronium ions. *Nat Commun* 2019; 10: 4876. doi:10.1038/s41467-019-12773-7

81. Li F, Han G, Noh H, Jeon J, Ahmad I, Chen S, et al. Balancing hydrogen adsorption/desorption by orbital modulation for efficient hydrogen evolution catalysis. *Nat Commun* 2019; 10: 4060

82. Ramohlola KE, Masikini M, Mdluli SB, Monama GR, Hato MJ, Molapo KM, et al. Electrocatalytic hydrogen evolution reaction of metal-organic frameworks decorated with poly (3-aminobenzoic acid). *Electrochim Acta* 2017; 246: 1174–82.

83. Ramohlola KE, Masikini M, Mdluli SB, Monama GR, Hato MJ, Molapo KM, et al. Electrocatalytic hydrogen production properties of poly (3-aminobenzoic acid) doped with metal organic frameworks. *Int J Electrochem Sci* 2017; 12: 4392–405.

84. Peng Y, Fengmei W, Tofik AS, Xueying Z, Xiaoding L, Fan X, et al. Earth abundant materials beyond transition metal dichalcogenides: A focus on electrocatalyzing hydrogen evolution reaction. *Nano Energy* 2019; 58: 244–76.

85. Ziliang C, Huilin Q, Kun Z, Dalin S, Renbing W. Metal-organic framework-derived nanocomposites for electrocatalytic hydrogen evolution reaction. *Prog Mater Sci* 2020; 108: 100618.

86. Theerthagiri J, Lee SJ, Murthy AP, Madhavan J, Choi MY. Fundamental aspects and recent advances in transition metal nitrides as electrocatalysts for hydrogen evolution reaction: A review. *Curr Opin Solid State Mater Sci* 2020; 24: 100805 (1–22). doi:10.1016/j.cossms.2020.100805

87. Yu D, Ilango PR, Han S, Ye M, Hu Y, Li L, Peng S. Metal-organic framework derived Co@NC/CNT hybrid as a multifunctional electrocatalyst for hydrogen and oxygen evolution reaction and oxygen reduction reaction. *Int J Hydrogen Energy* 2019; 44: 32054–65.

88. Miroslava K, Vitalii L, Viktor K, Hoydoo Y, Serhii V, Alexandra K, et al. Evaluation of hydrogen evolution reaction activity of molybdenum nitride thin films on their nitrogen content. *Electrochim Acta* 2019; 315: 9–16.

89. Kumar SS, Himabindu V. Hydrogen production by PEM water electrolysis—A review. *Mater Sci Energy Technol* 2019; 2: 442–54.

90. Jakob K, Thomas FJ. Molybdenum Phosphosulfide: An active acid-stable earth-abundant catalyst for the hydrogen evolution reaction. *Angew Chem Int Ed* 2014; 53: 1–6.

91. Qianfeng L, Erdong W, Gongquan S. Layered transition-metal hydroxides for alkaline hydrogen evolution reaction. *Chin J Catal* 2020; 41: 574–91.

92. Andrzej L. Mechanism and kinetics of the hydrogen evolution reaction. *Int J Hydrogen Energy* 2019; 44: 19484–518.

93. Vesborg PCK, Seger B, Chorkendorff IB. Recent development in hydrogen evolution reaction catalysts and their practical implementation. *J Phys Chem* 2015; 6: 951–7.

94. Safizadeh F, Ghali E, Houlachi G. Electrocatalysis developments for hydrogen evolution reaction in alkaline solutions-A Review. *Int J Hydrogen Energy* 2015; 40: 256–74.

95. Delgado D, Minakshi M, Kim DJ. Electrochemical impedance spectroscopy studies on hydrogen evolution from porous raney cobalt in alkaline solution. *Int J Electrochem Sci.* 2015; 10: 9379–94.

96. Wu L, Xu X, Zhao Y, Zhang K, Sun Y, Wang T, et al. Mn doped MoS_2/reduced graphene oxide hybrid for enhanced hydrogen evolution. *Appl Surf Sci* 2017; 425: 470–77.

97. Kuppler RJ, Timmons DJ, Fang QR, Li JR, Makal TA, Young MD, et al. Potential applications of metal-organic frameworks. *Coord Chem Rev* 2009; 253: 3042–66.

98. Srimuk P, Luanwuthu S, Krittayavathanon A, Sawangphruk M. Solid-type supercapacitor of reduced graphene oxide-metal organic framework composite coated on carbon fibre paper. *Electrochim Acta* 2015; 157: 69–77.

99. Xamena FXL, Corma A, Garcia H. Applications of metal-organic frameworks (MOFs) as quantum dot semiconductors. *J Phys Chem C* 2007; 111: 80–5.

100. Bagheri H, Javanmardi H, Abbasi A, Banihashemi S. A metal organic framework-polyaniline nanocomposite as a fiber coating for solid phase microextraction. *J Chromatogr A* 2016; 1431: 27–35.

101. Guo SN, Zhu Y, Yan YY, Min YL, Fan JC, Xu QJ, et al. (Metal-organic framework)-polyaniline sandwich structure composites as novel hybrid electrode materials for high-performance supercapacitor. *J Power Sources* 2016; 316: 176–82.

102. Zhou HC, Long JR, Yaghi OM. Introduction to metal-organic frameworks. *Chem Rev* 2012; 112: 673–74.

103. Rowsell JLC, Yaghi OM. Metal-organic frameworks: A new class of porous materials. *Microporous Mesoporous Mater* 2004; 73: 3–14.

104. Gangu KK, Maddila S, Mukkamala SB, Jonnalagadda SB. A review on contemporary metal-organic framework materials. *Inorg Chim Acta* 2016; 446: 61–74.

105. Cui Y, Li B, He H, Zhou W, Chen B, Qian G. Metal-organic frameworks as platforms for functional materials. *Acc Chem Res* 2016; 49: 483–93.

106. Salunke-Gawali S Kathawate L Puranik VG. MOF with hydroxynaphthoquinone as organic linker: Molecular structure of [Zn $(Chlorolawsone)_2(H_2O)_2$] and thermogravimetric studies. *J Mol Struct* 2012; 1022: 189–96.

107. Kreno LE, Leong K, Farha OK, Allendorf M, Van Duyne PR, Hupp JT. Metal-organic framework materials as chemical sensors. *Chem Rev* 2012; 112: 1105–25.

108. Butova VV, Soldatov MA, Guda AA, Lomachenko KA, Lamberti C. Metal-organic frameworks: Structure properties methods of synthesis and characterisation. *Russ Chem Rev* 2016; 85: 280–307.

109. Schoedel A, Li M, Li D, O'Keeffe M, Yaghi OM. Structures of metal-organic frameworks with rod secondary building units. *Chem Rev* 2016; 116: 12466–535.

110. Qiu S, Zhu G. Molecular engineering for synthesising novel structures of metal-organic frameworks with multifunctional properties. *Coord Chem Rev* 2009; 235: 2891–911.

111. Alhamami M, Doan H, Cheng CH. A review on breathing behaviour of metal-organic frameworks (MOFs) for gas adsorption. *Materials* 2014; 7: 3198–250.
112. Furukawa H, Cordova KE, O'Keeffe M, Yaghi OM. The chemistry and applications of metal-organic frameworks. *Science* 2013; 341: 1–12.
113. Pagis C, Ferbinteanu M, Rothenberg G, Tenase S. Lanthanide-based metal organic frameworks: Synthetic strategies and catalytic applications. *ACS Catal* 2016; 6: 6063–72.
114. Cui Y, Chen B, Qian G. Lanthanide metal-organic frameworks for luminiscent sensing and light-emitting applications. *Coord Chem Rev* 2014; 273–274: 76–86.
115. Abdelbaky MS, Amghouz Z, Garcia-Granda S, Garcia JR. Synthesis structures and luminescence properties of metal-organic frameworks-based lithium-lanthanide and terephthalate. *Polymer* 2016; 8: 86–100.
116. Raja DS, Luo JH, Wu CY, Cheng YJ, Yeh CT, Chen YT, et al. Solvothermal synthesis structural diversity and properties of alkali metal-organic frameworks based on V-shaped ligands. *Crys Growth Des* 2013; 13: 3785–93.
117. Ahmed I, Jeon J, Khan NA, Jhung SH. Synthesis of a metal-organic framework iron-benzenetricarboxylate from dry gels in the absence of acid and salt. *Crys Growth Des* 2012; 12: 5878–81.
118. Stock N, Biswas S. Synthesis of metal-organic frameworks (MOFs): Route to various MOF topologies morphologies and composites. *Chem Rev* 2012; 112: 933–69.
119. Lee YR, Kim J, Ahn WS. Synthesis of metal-organic frameworks: A mini review. *Korean J Chem Eng* 2013; 30: 1667–80.
120. Song YS, Yan B, Chen ZX. Hydrothermal synthesis crystal structure and luminescence of four novel metal-organic frameworks. *J Solid State Chem* 2006; 179: 4037–46.
121. Flugel EA, Ranft A, Haase F, Lotsch BV. Synthetic routes toward MOF nanomorphologies. *J Mater Chem* 2012; 22: 10119–33.
122. Sun Y, Zhou HC. Recent progress in the synthesis of metal-organic frameworks. *Sci Technol Adv Mater* 2015; 16: 054202 (1–11). doi:10.1088/1468-6996/16/5/054202
123. Seo YK, Hundal G, Jang IT, Hwany YK, Jun CN, Chang J-S. Microwave synthesis of hybrid inorganic-organic materials including porous $Cu_3(BTC)_2$ from Cu(II)-trimesate mixture. *Microporous Mesoporous Mater* 2009; 119: 331–7.
124. Lee JS, Halligudi SB, Jang NH, Hwang DW, Chang JS, Hwang YK. Microwave synthesis of a porous metal-organic framework Nickel(II) dihydroxyterephthalate and its catalytic properties in oxidation of cyclohexene. *Bull Korean Chem Soc* 2010; 31: 1489–95.
125. Lin ZJ, Yang Z, Liu TF, Huang YB, Cao R, Microwave-assisted synthesis of a series of lanthanide metal-organic frameworks and gas sorption properties. *Inorg Chem* 2012; 51: 1813–20.
126. Choi TY, Kim J, Jhung SH, Kim HK, Cheng JS, Chae HK. Microwave synthesis of a porous metal-organic framework zinc terephthalate MOF-5. *Bull Korean Chem Soc* 2006; 27: 1523–24.
127. Liu X, Sun T, Hu J, Wang S. Composites of metal-organic frameworks and carbon-based materials: Preparations functionalities and applications. *J Mater Chem A* 2016; 4: 3584–616.
128. Banerjee D, Simon CM, Elsaidi SK, Haranczyk M, Thallapally PK. Xenon gas separation and storage using metal-organic frameworks. *Chem* 2018; 4: 466–94.
129. Chen L, Xu Q. Metal-organic framework composites for catalysis. *Matter* 2019; 57–89.
130. Xia Z, Fang J, Zhang X, Fan L, Barlow AJ, Lin T, Wang S, et al. Pt nanoparticles embedded metal-organic framework nanosheets: A synergistic strategy towards bifunctional oxygen electrocatalysis. *Appl Catal B: Environ* 2019; 245: 389–98.

131. Talin AA, Centrone A, Ford AC, Foster ME, Stavila V, Haney P, et al. Tunable electrical conductivity in metal-organic framework thin film devices. *Science* 2014; 343: 66–9.

132. Zhou W, Wu Y, Wang X, Tian J, Huang D, Zhao J, et al. Improved conductivity of a new Co(II)-MOF by assembled acetylene black for efficient hydrogen evolution reaction. *R Soc Chem* 2018; 30: 4804–9. doi:10.1039/C8CE00921J

133. Nohra B, Moll HE, Albelo LMR, Mialane P, Marrot J, Mellot-Draznieks C, et al. Polyoxometalate-based metal organic frameworks (POMOFs): Structural trends energetics and high electrocatalytic efficiency for hydrogen evolution reaction. *J Am Chem Soc* 2011; 133: 13363–74.

134. Gong Y, Wu T, Jiang PG, Lin JH, Yang YX. Octamolybdate-based metal-organic framework with unsaturated coordinated metal center as electrocatalyst for generating hydrogen from water. *Inorg Chem* 2012; 52: 777–84.

135. Qin J, Du D, Guan W, Bo X, Li Y, Guo L, et al. Ultrastable polymolybdate-based metal-organic frameworks as highly active electrocatalysts for hydrogen generation from water. *J Am Chem Soc* 2015; 137: 7169–77.

136. Wang X, Luo J, Tian J, Huang D, Wu Y, Li S, et al. Two new 3D isostructural Co/Ni-MOFs showing four-fold polyrotaxane-like networks: Synthesis crystal structures and hydrogen evolution reaction. *Inorg Chem Commun* 2018; 98: 141–4.

137. Duan J, Chen S, Zhao C. Ultrathin metal-organic framework array for efficient electrocatalytic water splitting. *Nat Commun* 2017; 8: 15341.

138. Lin H, Raja DS, Chuah X, Hsieh C, Chen Y, Lu S. Bi-metallic MOFs possessing hierarchical synergistic effects as high performance electrocatalysts for overall water splitting at high current densities. *Appl Catal B Environ* 2019; 258: 118023.

139. Roy S, Huang Z, Bhunia A, Castner A, Gupta AK, Zou X, et al. Electrocatalytic hydrogen evolution from a cobaloxime-based metal-organic framework thin film. *Am Chem Soc* 2019; 141: 15942–50.

140. Li Y, Yi J, Wei J, Wu Y, Li B, Liu S, et al. Three 2D polyhalogenated Co(II)-based MOFs: Synthesis crystal structure and electrocatalytic hydrogen evolution reaction. *J Solid State Chem* 2020; 281: 121052.

141. Fenoy GE, Maza E, Zelay E, Marmisollé WA, Azzaroni O. Layer-by-layer assemblies of highly connected polyelectrolyte capped-Pt nanoparticles for electrocatalysis of hydrogen evolution reaction. *Appl Surf Sci* 2017; 416: 24–32.

142. Koo W, Kim S, Jang J, Kim D, Kim I. Catalytic metal nanoparticles embedded in conductive metal-organic frameworks for chemiresistors: Highly active and conductive porous materials. *Adv Sci* 2019; 6: 1900250.

143. Duan M, Jiang L, Zeng G, Wang D, Tang W, Liang J, et al. Bimetallic nanoparticles/metal-organic frameworks: Synthesis applications and challenges. *Appl Mater Today* 2020; 19: 100564.

144. Shi Y, Yang A, Cao C, Zhao B. Applications of MOFs: Recent advances in photocatalytic hydrogen production from water. *Coord Chem Rev* 2019; 390: 50–75.

145. Rui K, Zhao G, Lao M, Cui P, Zheng X, Zheng X, et al. Direct hybridization of noble metal nanostructures on 2D metal organic framework nanosheets to catalyse hydrogen evolution. *Nano Lett* 2019; 9: 8447–53.

146. Ghiamaty Z, Ghaffarinejad A, Faryadras M, Abdolmaleki A, Kazemi H. Synthesis of palladium-carbon nanotube-metal organic framework composite and its application as electrocatalyst for hydrogen production. *J Nanostruct Chem* 2016; 6: 299–308.

147. Le HV, Nguyen QTT, Co. TT, Nguyen PKT, Nguyen HT. A composite based on Pd nanoparticles incorporated into a Zirconium-based metal-organic frameworks Zr-AzoBDC and its electrocatalytic activity for hydrogen evolution reaction. *J Electron Mater* 2018; 47: 6918–22.

148. Le HV, Doan THL, Tran BQ, Nguyen HHT, Co TT, Nguyen HT, et al. Selective incorporation of Pd nanoparticles into the pores of an alkyne-containing metal-organic framework VNU1 for enhanced electrocatalytic hydrogen evolution reaction at near neutral pH. *Mater Chem Phys* 2019; 233: 16–20.

149. Ye B, Jiang R, Yu Z, Hou Y, Huang J, Zhang B, et al. Pt (111) quantum dot engineered Fe-MOF nanosheet arrays with porous core-shell as an electrocatalyst for efficient overall water splitting. *J Catal* 2019; 380: 307–17.

150. Chen Y, Gu B, Uchida T, Liu J, Liu X, Ye B, et al. Location determination of metal nanoparticles relative to a metal-organic framework. *Nature Commun* 2019; 10: 3462.

151. Karuppasamy K, Jothi VR, Vikraman D, Prasanna K, Maiyalagan T, Sang B, et al. Metal-organic framework derived NiMo polyhedron as an efficient hydrogen evolution reaction electrocatalyst. *Appl Surf Sci* 2019; 478: 916–23.

152. Ding Z, Wang K, Mai Z, He G, Lui Z, Tang Z. RhRu alloyed nanoparticles confined within metal organic frameworks for electrochemical hydrogen evolution at all pH values. *Int J Hydrogen Energy* 2019; 44: 24680–89.

153. Maina JW, Pozo-Gonzalo C, Schutz JA, Wang J, Dumee LF. Tuning CO_2 conversion product selectivity of metal-organic frameworks derived hybrid carbon photoelectrocatalytic reactors. *Carbon* 2019; 148: 80–90.

154. Nadeem M, Yasin G, Bhatti MH, Mehmood M, Arif M, Dai L. Pt-M bimetallic nanoparticles (M = Ni Cu Er) supported on metal-organic frameworks-derived N-doped nanostructured carbon for hydrogen evolution and oxygen evolution reaction. *J Power Sources* 2018; 402: 34–42.

155. Zhao S, Yin H, Du L, He L, Zhao K, Chang L, et al. Carbonized nanoscale metal-organic frameworks as high performance electrocatalyst for oxygen reduction reaction. *ACS Nano* 2014; 8 12660–8.

156. Barylak M, Cendrowski K, Mijowska E. Application of carbonized metal-organic framework as efficient adsorbent of cationic dye. *Ind Eng Chem Res* 2018; 57: 4867–79.

157. Trukawka M, Cendrowski K, Peruzynska M, Augustyniak A, Nawrotek P, Drozdzik M, et al. Carbonized metal-organic frameworks with trapped cobalt nanoparticles as biocompatible and efficient azo-dye adsorbent. *Environ Sci Eur* 2019; 31: 56.

158. Fan L, Liu PF, Yan X, Gu L, Yang ZZ, , Yang HG, et al. Atomically isolated nickel species anchored on graphitized carbon for efficient hydrogen evolution electrocatalysis. *Nat Commun* 2016; 7: 10667. doi: 10.1038/ncomms10667

159. Qui T, Liang Z, Guo W, Gao S, Qu C, Tabassum H, et al. Highly exposed ruthenium-based electrocatalysts from bimetallic metal-organic frameworks for overall water splitting. *Nano Energy* 2019; 58: 1–10.

160. Ramohlola KE, Monana GR, Hato MJ, Modibane KD, Molapo KM, Masikini M, et al. Polyaniline-metal organic framework nanocomposite as an efficient electrocatalyst for hydrogen evolution reaction. *Compos B Eng* 2018; 137: 129–39.

161. Ramohlola KE, Masikini M, Mdluli SB, Monama GR, Hato MJ, Molapo KM, et al. Electrocatalytic hydrogen production properties of polyaniline doped with metal-organic frameworks. In Kaneko S et al. (eds). *Carbon-related Materials in Recognition of Nobel Lectures by Prof. Akira Suzuki in ICCE*. Cham, Switzerland: Springer. 2017, pp. 373–89.

162. Khalid M, Honorato AMB, Varela H, Dai L. Multifunctional electrocatalysis derived from conducting polymer and metal-organic framework complexes. *Nano Energy* 2018; 45: 127–35.

163. Ghen Z, Qing H, Zhou K, Sun D, Wu R. Metal-organic framework-derived nanocomposites for electrocatalytic hydrogen evolution reaction. *Prog Mater Sci* 2020; 108: 100618.

164. Jahan M, Liu Z, Loh KP. A graphene oxide and copper-centered metal-organic framework composite as a tri-functional catalyst for HER OER and ORR. *Adv Funct Mater* 2013; 23: 5363–72. doi:10.1002/adfm.201300510

165. Ngo TV, Moussa M, Tung TT, Coghlan C, Losic D. Hybridization of MOFs and gra-
     phene: A new strategy for the synthesis of porous 3D carbon composites for high per-
     forming supercapacitors. *Electrochim Acta* 2020; 329: 135104.
166. Wu Y, Zhou W, Zhao J, Dong W, Lan Y, Li D, et al. Surfactant-assisted phase-selective
     synthesis of new cobalt MOFs and their efficient electrocatalytic hydrogen evolution
     reaction. *Angew Chem Int Ed* 2017; 56: 13001–5.
167. Micheroni D, Lan G, Lin W. Efficient electrocatalytic proton reduction with carbon
     nanotube supported metal-organic frameworks. *J Am Chem Soc* 2018; 140: 15591–5.
168. Wang X, Zhou W, Wu Y, Wu J, Tian J, Wang X, et al. Two facile routes to an AB&Cu-
     MOF composite with improved hydrogen evolution reaction. *J Alloys Compd* 2018; 753:
     228–33.

Index

Taylor & Francis Group
an **informa** business

Taylor & Francis eBooks

www.taylorfrancis.com

A single destination for eBooks from Taylor & Francis
with increased functionality and an improved user
experience to meet the needs of our customers.

90,000+ eBooks of award-winning academic content in
Humanities, Social Science, Science, Technology, Engineering,
and Medical written by a global network of editors and authors.

TAYLOR & FRANCIS EBOOKS OFFERS:

A streamlined
experience for
our library
customers

A single point
of discovery
for all of our
eBook content

Improved
search and
discovery of
content at both
book and
chapter level

REQUEST A FREE TRIAL
support@taylorfrancis.com

Routledge
Taylor & Francis Group

CRC Press
Taylor & Francis Group

For Product Safety Concerns and Information please contact our EU
representative GPSR@taylorandfrancis.com
Taylor & Francis Verlag GmbH, Kaufingerstraße 24, 80331 München, Germany

www.ingramcontent.com/pod-product-compliance
Lightning Source LLC
Chambersburg PA
CBHW060803220326
41598CB00022B/2529

* 9 7 8 0 3 6 7 6 2 7 9 5 9 *